Essential Statistics for Economics, Business and Management

Essential Statistics for Economics, Business and Management

Teresa Bradley

BICENTENNIAL
1807
WILEY
2007
BICENTENNIAL

John Wiley & Sons, Ltd

Other Wiley Editorial Offices

John Wiley & Sons, Inc., 111 River Street, Hoboken, NJ 07030, USA

Jossey-Bass, 989 Market Street, San Francisco, CA 94103-1741, USA

Wiley-VCH Verlag GmbH, Boschstr. 12, D-69469 Weinheim, Germany

John Wiley & Sons Australia Ltd, 42 McDougall Street, Milton, Queensland 4064, Australia

John Wiley & Sons (Asia) Pte Ltd, 2 Clementi Loop #02-01, Jin Xing Distripark, Singapore 129809

John Wiley & Sons Canada Ltd, 6045 Freemont Blvd, Mississauga, ONT, L5R 4J3

Wiley also publishes its books in a variety of electronic formats. Some content that appears in print may not be available in electronic books.

Anniversary logo design: Richard J. Pacifico

Library of Congress Cataloging-in-Publication Data

Bradley, Teresa.
    Essential statistics for economics, business and management / Teresa Bradley.
        p.   cm.
    Includes bibliographical references and index.
    ISBN-13: 978-0-470-85079-4 (pbk : alk. paper)
    ISBN-10: 0-470-85079-5 (pbk. : alk. paper)
    1. Economics – Statistical methods.   2. Commercial statistics.   I. Title.

HB137.B73   2007
519.5 – dc22   2006036194

A Catalogue record for this book is available from the British Library

ISBN: 978-0-470-85079-4

Typeset in 10/12pt Times and Helvetica by TechBooks, New Delhi, India

To Sonas

CONTENTS

CONTENTS

INTRODUCTION

What is statistics?

The word 'statistics' has its origin in the Latin *statisticus*, meaning 'affairs of state'. As far back as five thousand years ago the Sumerians collected population statistics for taxation purposes.

The definition given in Collins Concise Dictionary is as follows:

Statistics

1. a science concerned with the collection, classification and interpretation of quantitative data and with the application of probability theory to the analysis and estimation of population parameters.
2. the quantitative data themselves

Statistic: a datum capable of exact numerical representation such as a correlation coefficient of two series or the standard deviation of a sample.

Throughout the text, these definitions *will be explored and explained starting with the collection* and classification in Chapter 1, the calculation of statistics to represent data numerically in Chapters 2 and 3, probability theory in Chapters 4, 5 and 6. The application of probability theory to the estimation of population parameters forms the basis for statistical inference in the remaining chapters.

Who uses statistics?

Just about everyone! Pick up any newspaper, magazine, journal or browse internet sites – you will see statistics quoted everywhere: from opinion polls, tam ratings, consumer price indices, trends in inflation, unemployment rates, social statistics, health statistics, research, the relationship between internet usage and age, etc.

Who needs statistics?

Researchers, decision makers, policy makers, organisations and companies, manufacturers, etc. The following are just some examples. Individuals or research groups may require information about society or the economy, both past and present. Governments need statistics to view the state of the nation in order to make informed decisions and introduce effective policies. Companies use statistics in market research, auditing, and in manufacture and production. Statistical process control (SPC) was introduced by W. Edward Deming and other statisticians[1] in the 1950s to increase profitability by reducing variation and continually improving goods and/or services at every stage in its life cycle from the design through production/implementation, distribution, marketing, maintenance and the recycling. Medical, scientific and other research requires statistical methods to design experiments, collect data, analyse the results and give conclusions with associated levels of confidence.

The potential impact and power of statistics was realised over 100 years ago (when there were no computers or electronic calculators) by the famous English author and one of the first science fiction writers, **H. G. Wells**[2] **(1866–1946)** who proclaimed that

'Statistical thinking will one day be as necessary for efficient citizenship as the ability to read and write.'

Should we believe statistics?

'There are three kinds of lies: lies, damned lies and statistics.'

(Evidence is now available to conclude that the quote originally appeared in 1895 in an article by Leonard H. Courtney and not Disraeli as originally thought.)

Non–statisticians' love to bandy this famous quote around. However, it is unfortunate that there are occasions when this quote seems justified and 'quoted' statistics are misleading. For example, the median should be used as the representative value for skewed data but almost invariably the average value is quoted. In correlation analysis a strong relationship between two variables is often assumed to imply causation when in fact the correlation is spurious. In research, erroneous conclusions may be reached if the data is biased or inappropriate statistical models are used in the analysis of the data. In most cases the misuse of statistics is unintentional and due to the use of statistics by non-statisticians with 'a little knowledge'.

Studying statistics

Start at the beginning!

To study any subject, it is imperative to get the basics right and understand the associated terminology. The basic descriptive statistics may not appear to be very interesting – sampling, data collection, calculating means, medians, standard deviations, regression lines, probability, etc. However, the basic statistics are the statistics that are usually quoted in the media, in reports (business, financial etc.) and by politicians. To motivate the reader through the earlier chapters in the text, real data from various sources is used when introducing topics but particularly in the exercises. A comprehensive understanding of descriptive statistics and probability equips the reader with the foundation that is essential to progress easily through the various topics on statistical inference covered in the later chapters.

---

[1] Logothetis (1992) *From Deming to Taguchi and SPC*. Prentice Hall. ISBN 013553512 3.
[2] Herbert George Wells, born in Bromley, Kent in 1866, first studied science, often credited with being the father of science fiction. He was a prolific writer. As early as 1895, the publication of 'The Time Machine' revealed his fascination with technological change and the potential for a reordered human society. Even the most vivid imaginations of 1895 could not have foreseen the technological changes that were to take place in the 20th century or the associated changes in society.

Worked Examples (WE)

Throughout the text, each topic is explained through a simple example using a small data set so that the calculations are simple and do not detract from the method. Following the introduction of each new topic, a Worked Example is used, not only to demonstrate methods and calculations but also how to read the question, to decipher the information given in the question that is essential to the solution rather than background information. Finally, a verbal description of the answer is given.

Skill Development Exercises (SKD)

Solving problems in statistics requires a certain amount of drill and practise in order to become 'skilled' at recognising the problem, the method of solution and working systematically through the method. To help the reader acquire these skills a 'Skill Development Exercise (SKD)' is given at certain stages in the text, usually following a Worked Example. The Skill Development Exercise is usually similar to the previous Worked Example or it may pull together several strands of a topic covered in earlier Worked Examples. Answers to intermediate stages of SKD's are given as a guide to the reader. Comprehensive answers are given on the web site that accompanies the text.

Progress Exercises (PE)

These exercises are given within chapters following the coverage of one or more topics. Many of the questions are based on real data from various sources. The data for most of the questions is also given in Excel files that may be accessed on the web site to accompany this text. It is vitally important to attempt a reasonable number of the questions in each exercise in order to become proficient at reading questions, applying the methods to real data and stating the results in the context of the question. For larger data sets, it is expected that the student will use software such as Excel and/or Minitab.

Solutions for each Progress Exercise are given for question one and even numbered questions at the back of the book. Solutions for the odd numbered questions are available on the web site (www.wileyeurope.com/college/bradley).

Calculator

Most scientific calculators have inbuilt statistical functions for mean, standard deviation, regression and correlation. Initially, these descriptive statistics are calculated manually to enhance understanding, but thereafter students are encouraged to use the calculator.

Excel[†]

To enhance understanding, students are again encouraged to work through the various topics in the text and attempt most of the exercises manually (for small data sets) before using Excel. Excel's inbuilt statistical functions are then explained and used to confirm results already obtained manually and to attempt similar problems for larger sets of data. Excel is also used for time-consuming tasks such as plotting graphs and charts. Screen dumps are given to guide the reader through the steps in using Excel.

[†]Microsoft product screen shots reprinted with permission from Microsoft Corporation.

Statistical analysis, such as hypothesis tests and confidence intervals, ANOVA, regression analysis, etc. are available in Excel's add-in 'Data analysis'. Data analysis (which also gives printouts of descriptive statistics) gives the calculated statistics such as test statistic for hypothesis testing, the ANOVA table in analysis of variance, and so on. The reader is expected to explain the calculated statistics in the printout and use them to make inference about the population before stating the conclusion verbally. In this text, it is not intended to give a comprehensive course in Excel as it would detract from its main purpose – statistics. However, further examples using additional features and functions in Excel are given on the web site that accompanies this text.

Minitab and other statistical software

Minitab is one of the many statistical software packages available commercially. While Minitab is used in this text, familiarity with one software package makes it easy to adapt to another. Again, Minitab is used in a supportive role for the topics covered in the text. Like Excel, Minitab printouts give the descriptive statistics and statistical analysis which the reader is expected to explain and use to reach a conclusion. Minitab has a wide range of graphing capabilities that allows visual checks on model assumptions such as ANOVA, regressions, etc. In Excel, Minitab or any software the user is expected to have sufficient knowledge to choose the appropriate test for the analysis and draw conclusions.

An introduction to SPSS is given on the website that accompanies this text. Selected Worked Examples from the text are given in SPSS.

Statistical tables

From Chapter 5 onwards, it will be necessary to use various Statistical tables. Extract from selected tables are given at the back of the text but the reader should also be familiar with books of published statistical tables. Two sets of tables that are recommended are Statistical tables by J. Murdock and J.A. Barnes, ISBN 0-333-55859-6 (Macmillan Press Ltd) and New Cambridge Statistical Tables by C. V. Lindley and W. F. Scott, ISBN 0-521-48485-5 (Cambridge University Press).

Mathematics

Some mathematics is necessary at certain stages in the text, but explanations are given and mathematical symbols are explained.

Finally. . .

It is important to develop a common sense understanding of statistics and its use in decision making.

Acknowledgements

A special thanks to Michael Hayes who not only reviewed the text but also assisted with the solutions to many of the exercises. I would also like to thank Dr Martin Knott, Helen Paul, Dr Oliver Hyde, Professor Diarmuid Bradley, Mary O'Sullivan, Siobhan Bradley, Chris Naughton, Elizabeth Fahy, Margaret Meehan and Brendan Kelleher who reviewed various chapters in the text and made valuable suggestions and contributions. The Wiley team Steve Hardman, Deborah Egleton, Mark Styles, Emma Cooper, Debbie Scott and Flick Williams.

Data Collection and Its Graphical Presentation

1

Chapter Objectives

At the end of this chapter you should be able to do the following

- Appreciate issues that arise in data collection and how bias arises at different stages in statistical studies
- Describe the stages in survey design
- Describe and compare the methods of sample collection
- Access data from a range of reliable sources
- Sort and classify data into stem and leaf plots and frequency tables
- Plot bar charts, histograms, Ogives, Lorenz curves, scatter plots and line graphs, both manually and in Excel
- Verbally describe the information depicted in charts and plots
- Choose the most appropriate graph to convey the pertinent information in a set of data
- Read and interpret charts and graphs accurately

1.1 Introduction to statistics

An extract from the dictionary definition given in the introduction defined '**statistics**' as '*a science concerned with the collection, classification and interpretation of quantitative data and with the application of probability theory to the analysis and estimation of population parameters.*

The collection of 'good' data is the essential basis for any statistical study. Good data should be representative of the population from which it is selected. For example, there is little point in interviewing 50 city dwellers in a study on lifestyle of coastal communities. Chart 1.1 illustrates how data (either census or sample) is the basis for all statistical analysis. Data are summarised to describe the characteristics of the population or sample. The chart also illustrates how sample data may be further analysed to make inference about the population.

It is also important to build up a thorough understanding of the vocabulary and terminology associated with the subject throughout the text. For example, in the preceding paragraph the terms census, population, data and sample were used. These and some other terminologies are explained below and as they arise within the text.

Statistics (singular) is the science of collecting, summarising, analysing and interpreting data with the application of probability theory to the analysis and estimation of population parameters.

Statistics are one or more sets of numerical data relating to populations, such as labour force statistics, birth rates, and mortality rates, taxes; finance, agricultural output, industrial output and so on.

A Statistic is a sample characteristic, such as a sample average.

A population is the entire set of people, animals or objects that is under study.

Some examples of populations are

(a) The individuals who live in a given country on a given date and time. (This is the everyday meaning of the word 'population').
(b) The individuals with high cholesterol.
(c) The members of a football supporters club.
(d) The cars manufactured by a given company during a specified year.
(e) The wild salmon in a river.

A parameter is a population characteristic, such as a population average.

A sample is any subset of a population. Here are two examples; a sample of 100 students selected from all the students in a college or a sample of 20 components selected each hour from a production line.

Each individual in a population or a sample is known as **a member** or **an element** of that population or sample.

Data are values, measurements, facts, or observations. Data may come in many forms depending on the type of statistical study being undertaken.

A datum is the singular of data, such as a single measurement, fact or observation.

A measurement is a numerical characteristic such as a height, a weight, volume, time

A measurement is said to be **accurate** if it is equal or very close to the true value.

A sample is referred to as accurate if the sample characteristics are the same as or very close to those of the population.

The purpose of a statistical study is to obtain information and/or investigate characteristics of a population.

Statistical studies may broadly be classified as **observational studies** or **designed experiments**.

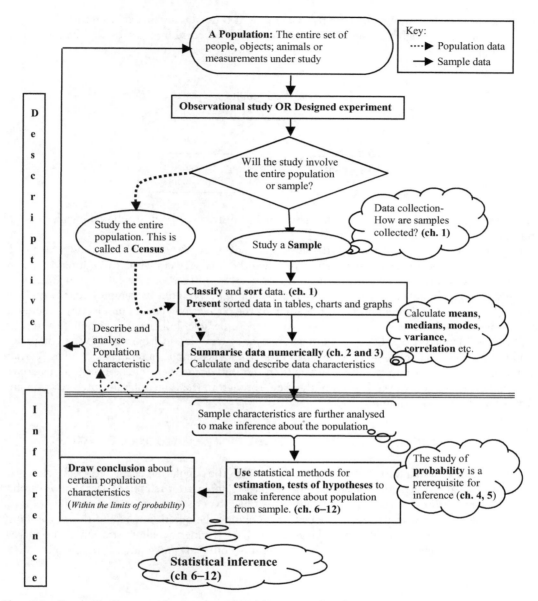

Chart 1.1   Data collection provides the raw material for statistical studies.

Observational studies: data characteristics are measured and/or recorded. No attempt is made to influence or change the population characteristics being studied. Examples of observational studies include the national census, opinion polls, government surveys such as household budget surveys, labour force surveys, etc.

To carry out a census a prerequisite is a list of the entire population. A census can take a long time to complete and costs can be high. For example, the national census of the population conducted at five or even ten year intervals requires huge resources to plan, organise, collect the data and finally sort, summarise, analyse and publish the results. Hence, it is usually impractical in terms of time and cost to collect data through a census: instead data is usually collected from a subset of the population called a sample.

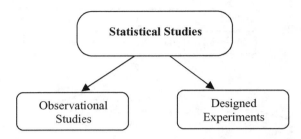

The gathering of sample data through questionnaires distributed by post, phone, email, personal interview, etc. is called **a survey.**

The person who designs and conducts the statistical study is called the **researcher:** those who collect survey data by means of personal interviews are called the **interviewers** while those interviewed are called **interviewees;** those who reply to the questionnaires by other means are called **respondents.**

Ideally a sample should be a 'miniature' of the population (a representative sample): so that studying the sample will yield the same information as studying the population. The art of selecting a representative sample is a subject in its own right and requires considerable skill, experience and ingenuity. A sample whose characteristics differ substantially in some systematic way from those of the population is said to be a **biased sample** (the everyday use of the word '**bias**' implies a tendency to favour certain views). Bias can arise at any stage in a survey from the initial statement of the research objectives, the design of the survey, the collection of the data, through to the final analysis, interpretation and inference made from the sample.

1.2   Data collection: Samples, surveys and experiments

The objective in sampling is to collect a sample that reflects the population characteristics as closely or accurately as possible. In practice selecting a representative sample is not a simple matter. The researcher must consider all aspects of the survey including the objectives of the study, the availability of a list of the members/elements of the target population, designing the questionnaire and the analysability of the data. Then the practicalities must be considered: organising the administration of the survey; collecting the data taking account of the time and cost of collecting and analysing the sample data.

1.2.1   Planning a survey

The following points briefly describe the main steps involved in planning a survey and the precautions which should be taken to reduce bias.

1. Start by defining the **objectives of the study**.
   What are you trying to find out? How will this survey help? Who or what is the target population? Have similar studies been carried out before and how successful were they?
2. **Define the target population**. The target population is the population of people or things that is the subject of the study. It is important to ensure that each group relevant to the study is included in the target population: otherwise the sample will be biased. Check the availability of lists of individuals, groups or elements of the target population (such a list is called a **sampling frame**).

Lists could be electoral registers; company employees; cars (or products) produced, imported or sold by a company; herds of cattle; college registers of students, investors, bank accounts, etc. Omission of any key group from these lists can give rise to biased samples (this is called a **sampling error**).

Note: lists may not be available to the researcher for various reasons such as confidentiality (student registers; patient records, etc.). Sometimes it may not be possible to enumerate the population in the first place. For example, how do you enumerate all those with high cholesterol or the population of wild deer in a given region?

3. **Sampling methods**: these are the methods used to select a sample from the target population. The method of selection may be random or non-random. Random samples are called probability samples and are particularly important as randomisation minimises bias and sampling error may be estimated (see Chapter 6 and later). Hence, random samples are the basis for the inferential statistics in Chapters 6 to 12. In a random (or probability) sample each element of the population has a known, non-zero chance of being selected. The disadvantage of random samples from a practical point of view is that they can be expensive and take considerable time to collect (see next section). Therefore, if a researcher requires a quick descriptive view of a population it is faster and cheaper to select a non-random sample. Quota sampling, where the researcher specifies the numbers (quotas) of individuals of a given age, profession, gender or other characteristic, that should be selected according to the interviewers own judgement, is a well known type of a non-random sample. Sections 1.2.2 and 1.2.3 give further detail on sampling methods.

4. **Conduct of the survey**. Decide how the respondents are to be contacted. The usual methods include (a) postal survey (b) interview (c) email (d) telephone and others. Consider the accessibility of the respondents and expected response rate by methods (a), (b), (c) or (d) since non-response is a major source of bias.

   A brief overview of the methods of contact is given in section 1.2.4.

5. **Prepare the questionnaire** or fact sheet. Keep the purpose of the study in mind. What questions do you need to ask to get the information you require? Questions should be short, clear, and relevant. Avoid general terms such as 'single parent' and offer suitable alternatives such as 'unmarried; widowed; separated; other'

   For analysis of the results, make sure that the questions produce answers which are quantifiable. Avoid too many open-ended questions, though a small number can provide very useful insights and occasionally inspire further research.

6. **Administer the survey**. Training of those who will administer the survey by any method is particularly important to reduce bias. Face-to face personal interview is sometimes used by market research companies and in major government surveys. Other surveys use telephone interviews. Well trained interviewers will not lead or influence the respondent in answering questions, but will know how to employ questions unambiguously.

   Response bias (error) arises when questions or terminology are misunderstood or answers are inaccurate.

7. **Carry out a pilot survey**: Pilot surveys give very important information on problems that arise at all stages in the conduct of the survey. For example, the pilot study will highlight problems in contacting respondents, whether non-response is likely to be high, whether the interviewers administer the questionnaire properly, whether the questionnaire meets the survey objectives, whether the survey data is analysable, etc.

8. **Carry out a post enumeration survey** (usually for large surveys such as the household expenditure survey): a sub sample of respondents is re-interviewed to confirm that the main survey is accurate and reproducible. For example, it is well known that understatement of alcohol consumption is as high as 50 % (when checked against customs and excise figures). Responses to questions on age tend

to be rounded to the nearest 5 years: certain questions such as alcohol consumption or practice of religion tend to be inaccurate.

At each stage, 1 to 8, pay particular attention to ways in which bias can be reduced or minimised.

1.2.2 Random samples

Randomisation is fundamental to minimising bias. Random samples are defined as follows:

(i) **A simple random sample** is a sample in which every member of the population has an equal, non-zero and known probability of being selected.
(ii) **A random sample** is a sample in which every member of the population has a non-zero, known, but not necessarily equal probability of being selected.

A simple random sample

Selecting a random sample requires a population list, so that each element may be assigned a number from which the sample is randomly selected by use of random numbers. The following Worked Example illustrates the procedure.

Worked Example 1.1: How to select a simple random sample

Describe how you would select a simple random sample of 10 students from a group of 850 first year business students, using (a) a drum (b) random number tables (c) random numbers on the calculator.

Solution

Assign every member of the population a number. For example, in the population of 850 first year students, each student is given a number from 1 to 850.
    The sample of 10 students may be selected as follows:

(a) The drum: put all the names or numbers corresponding to the names in a drum and select 10 names, one on each roll of the drum.
(b) Use **Random number tables**. Random number tables are given in books of statistical tables such as Table 37, Statistical Tables Murdock and Barnes; Table 27, Cambridge Elementary Statistical Tables. Random numbers are available on most calculators (see part (c)) and in Microsoft Excel. A short table of random numbers is given at the end of this text.
    To use the random number tables: State the population size. In this example the population size is 850, so three digits or three columns of numbers from the tables are required when selecting the sample. To select the sample, start from any randomly selected point on the random number tables, taking any three consecutive columns from the six digit column. Write down successive numbers that fall between 001 and 850, discarding numbers greater than 850. For example, an extract from the random number tables is given in Table 1.1, where the randomly selected starting point was 634 278. Taking the first three digits in successive numbers (column by column) we **obtain 634; 249; 059; 533; 074; 207; 691; 831; 691; 180.** The individuals whose names correspond to these numbers are selected for the sample.

Table 1.1   Extract from random number tables

634 278	594 502	819 590
249 144	209 611	804 028
059 294	026 391	400 671
533 333	920 195	446 213
074 330	755 009	156 302
207 386	771 462	739 813
951 188	410 173	525 254
831 967	937 780	148 159
954 636	750 072	664 082
691 992	378 425	721 898
269 114	946 931	018 712
180 293	460 024	110 901

(c) **From the Calculator**: on most calculators there is a random number key: this is Ran# on the Casio model. Pressing this key produces 3 digit random numbers; for example, the numbers below were obtained in this instance. The 10 students whose numbers are given below (within 1 to 850) are selected for the sample 290, 432, 421, 153, 955, 021, 354, 376, 190, 005 and 785.

(d) **Use Excel**: place the cursor in any cell: type '=RAN()' to generate a ten digit random number. Auto-fill columns or rows adjoining this cell to generate a table of random numbers. Use the HELP function in Excel for further information on variations on this function.

Problems with simple random samples

1. A population list may not be available. For example, how do you get a list of the all those with high cholesterol or a list of customers in a shopping centre or a list of commuters or the wild salmon in a river? It may not be possible to locate individual members of a population that are required for the sample. For example, how do you locate individual salmon, even if all the wild salmon have been tagged electronically?

2. Collecting a random sample may be time consuming and therefore costly. Those selected may not be available or accessible or they may be dispersed over a large geographical area. Some individuals may refuse to participate in the survey.

3. Non-response may be high. If this happens the characteristics of non-respondents should always be investigated. In some surveys there are explanations for non-response from certain groups. For example, busy practicing accountants may not have time to fill in a postal questionnaire. Other methods of contact such as a telephone interview may be more successful. The nature of non-respondents must be noted in the survey report.

4. There is a chance that a random sample will be biased! This is because the selection is random; hence certain subgroups that are numerically small but important for the current study may not be included in the sample.

The last point is extremely important if there are distinct subgroups within the population. For example, suppose the management of a company is considering the introduction of new work practices and decides to take the views of employees into consideration by conducting a random survey. Because the selection is random, certain key groups within the organisation that have distinctly different views on the new proposals may not be selected. This problem is addressed in stratified random sampling.

Stratified random sampling

A stratified random sample is used to ensure that all the important key sub-groups in the population are included in the sample. The population is subdivided into the key subgroups or **strata**. Then the number of subjects required from each stratum is calculated so that each stratum is proportionately represented in the sample. The required numbers from each stratum are then selected randomly. Overall, a stratified sample is likely to reflect the population characteristics more accurately than a simple random sample.

Worked Example 1.2: Stratified random sampling

The employees of a company are classified (in Table 1.2), as follows:

Table 1.2   Classification of employees

Category	Male	Female	Total
Management	10	20	30
Professional	50	40	90
Administration	40	60	100
Services	60	20	80
Total	160	140	300

A random sample of 30 individuals is required. Calculate how many should be selected from each stratum. How should the required number from each stratum be selected?

Solution

Step 1: Calculate the proportion of the population which is required for the sample.

The proportion of the population to be sampled is $\frac{\text{Total sample size}}{\text{Total population size}} = \frac{30}{300} = \frac{1}{10}$.

Then $\frac{1}{10}$ of each group is selected for the sample.

Step 2: The number from each stratum or group to be included in the sample is calculated by multiplying the group size by $\frac{1}{10}$. Remember that sample sizes are calculated to the nearest whole number. The results are given in brackets in Table 1.3.

Step 3: The required number from each group must be selected **'randomly'**, as described in Worked Example 2.1.

Table 1.3   Numbers to be selected randomly from each group are given in brackets

Category	Male	Female	Total
Management	10 [1]	20 [2]	30
Professional	50 [5]	40 [4]	90
Administration	40 [4]	60 [6]	100
Services	60 [6]	20 [2]	80
Total	160 [16]	140 [14]	300 [30]

Remember, because of stratification, stratified random samples are more accurate than simple random samples. Hence smaller sample sizes are required.

Problems with stratified random samples

Since the sample is random, the stratified sample suffers from the same problems as simple random samples which are listed following Worked Example 1.1.

Cluster sampling

One of the problems with simple random sampling and stratified sampling particularly for surveys that involve face to face interviews, is the time and money required to reach and interview those selected. A solution to this problem is the use of clusters. **Clusters** may consist of geographical areas or units such as schools, households, factories, hospitals, etc. When the clusters have been defined (for example, schools could be clusters) the required number of clusters is selected *randomly*. Then, depending on the research involved, all or some of the individuals within the cluster are surveyed. This is called **one-stage cluster sampling**.

Two-stage cluster sampling: Each cluster is divided into smaller clusters from which further *random* samples are chosen.

Multi-stage cluster sampling: This method involves several stages of cluster sampling. The household budget survey is a survey of this type. The sampling methodologies for any government survey are described on their web site: simply click on 'Methodologies'.

Problems with cluster sampling

If a cluster sample is to be representative of the population, each cluster should be representative of the population. This is sometimes difficult to achieve. For example, in a survey on extra curricular activities of schoolchildren where schools are clusters, there may be substantial differences between schools. One way to overcome this problem is to stratify the population of schools and then select clusters randomly from each stratum.

Sequential or systematic sampling

This type of random sampling uses existing lists, such as the electoral registers, lists of employees, telephone lists, lists of customers, etc. A starting pointing in the list is randomly selected, and then every *i*th member of the list is selected to obtain the required sample.

The problems that can arise with systematic sampling are (i) every possible sample does not have a chance of being selected and (ii) the sample may be unrepresentative because of patterns in the list. For example, in a housing estate every tenth house may be a corner house.

However, when such lists are available sequential sampling is convenient and saves time and money.

1.2.3   Non random sampling

If a population list is not available and random sampling is not possible, the researcher and/or interviewer must use their own 'judgement' to devise methods for the selection of the sample. Hence non random samples may be called **'judgement' samples:** The most common type of non-random or judgement sample is the quota sample. Other non-random samples include focus groups, opportunity samples, convenience samples, 'self selected' samples.

Quota samples

Stratified quota samples. This type of sampling may be used even when a sampling frame is available. The calculation of the numbers required from each stratum is the same as for stratified random sampling described in the previous section but the selection of the individuals from each stratum is **not random;**

the interviewer selects the required number using their own judgement. **When a sampling frame is not available,** the researcher estimates the numbers required from each stratum and issues these to the interviewers as 'quota controls'. For example, the quota controls for a survey on 'commuters' might be given numbers of males and females who commute by bus, by train and by car. More accurate samples could be achieved by further refinement of the quota controls. For example instead of giving quotas of males and females, give quotas of males and females within specified age brackets. The accuracy of the sample is highly dependent on the accuracy of the quota controls and the experience of the interviewer.

Note: quota sampling is also used when a sample is required quickly, to reduce cost or when sample will not be used further statistical analysis.

Problems with quota sampling

1. The data collected in this way cannot be used for further statistical inference: there in no objective way of determining the variability of the data.
2. The risk of collecting biased information is high.
3. Non-response bias may be high. Some researchers record non-response.

Other non-random samples

A self-selected sample is often useful as an initial indicator of views on certain topics. The simplest type of sample is one where TV and radio stations invite the public to respond to a small number of short questions or give their views on topical issues. This type of sample will almost certainly be biased: only those who are watching or listening will reply and then it depends on whether they have the time, feel strongly about the topic etc.

Focus groups consist of individuals selected because of their experience and knowledge (or lack of these) in relation to the subject of the research. Focus groups are useful if the views of individuals belonging to a profession, a company, an organisation, or other group are required. The results can suffer from bias as the individuals selected may not be representative. Focus groups may serve as a basis for further research.

A snowball or opportunity sample is used when there is no sampling frame available and it is difficult to locate members of the target population. For example, in research on active drug users, the researcher may be able to locate one or two individuals who are willing to be interviewed and who may suggest others also willing to participate and so on. Such samples are likely to be highly biased, but there is no way of checking this.

Mixed sampling methods

In practice, combinations of the above sampling methods are used. To view some of methods used to collect samples for the various national surveys, search web sites such as

(a) www.cso.ie Search for *Surveys and methodology* to view some of the methods for designing and collecting data for national surveys such as the labour force survey, the national household survey, etc. Many of the sample designs are stratified multistage cluster samples.
(b) www.statistics.gov.uk Search for *sample design.* For example, the sample design for the 'Family and working lives survey' was a two stage stratified nationally representative random sample, of adults aged 16 to 69 in Great Britain.
(c) www.europa.eu.int/comm/eurostat View the annexes for some of the qualitative studies.

1.2.4   Methods of contact

This section gives a brief overview of the methods of contact.

Face-to-face interview is the most common method of contact. With a good interviewer the responses should be accurate. Non-response can be recorded and investigated. But disadvantages are that this method is expensive and time consuming: questions that require the respondent to look up technical information or details on past dates, etc. are not suitable.

Telephone surveys are less expensive and have become increasingly popular with the widespread ownership of mobile phones. Bias can easily arise as not everyone is available by phone, response rate may be low; responses may be inaccurate because of the anonymity. On the other hand, because of anonymity it may be easier to elicit responses to questions of a personal nature.

Postal surveys are not expensive. The survey may not reach those selected and the response rate is generally low. However, the postal survey does give the respondents time to look up answers to historical or technical type questions and time to reflect.

Surveys by email are convenient and inexpensive but are limited to those who have access to email.

Using Excel to select random samples from a population list

Statistics in Excel is introduced at the end of Chapter 2, section 2.7.

Excel will select random or sequential samples from a population list. With the population list in an Excel worksheet, click (from the menu bar) **Tools** → **Data Analysis**. Select 'Sampling' from the 'Data Analysis' dialogue box. See Figure 2.12a The 'Sampling' dialogue box requests the location of the population data, the type of sample required and the location for the output (the sample lists).

Skill Development Exercise SK_1.1: Census and surveys

Define a (i) census (ii) a survey.

What is the difference between random and non-random samples? A multinational company intends to conduct a survey on the views of staff on the activities organised by the 'social club'.

Explain how the company should conduct a (i) random survey (ii) a quota survey.

Organisers of the social club suggest that a questionnaire should be sent to all employees by email. Discuss the advantages and disadvantages of email and the other three survey methods given in section 1.2.4 which could be used when carrying out this survey.

1.2.5   Designed experiments

The purpose of an experiment is to test the effect of a treatment(s) on a group(s) of objects or individuals. A treatment is anything that can have an affect on people, animals or objects, such as medicines, etc. Experiments usually involve **'treatment' groups** and **a control group.** As the name implies the individuals in treatment groups are subjected to some treatment while the individuals in the control group are given no treatment at all or else a standard treatment.

An example of a designed experiment

An experiment is set up to test a new yogurt that is claimed to reduce blood cholesterol. Three groups of participants are involved:

Group 1: blood cholesterol over 8.0
Group 2: blood cholesterol between 6.0 and 8.0 inclusive
Group 3: blood cholesterol less than 6.0

Each individual is asked to take four pots of yogurt per day for six months. Blood cholesterol levels are measured at the end of each week and participants are asked to fill in a questionnaire on their feeling of wellbeing. These recorded results/responses may be analysed statistically by methods described later in this text, to determine whether the yogurt has an effect on blood cholesterol and whether there is a difference between the three groups.

Some issues on the planning and conduct of experiments

The planning of experiments is a subject in its own right. It is concerned with investigating and measuring the response to given treatments and the interaction between treatments. Great care must be taken to design the experiment so that the response under investigation can be attributed to the applied treatment(s) and is not caused by or confused with other factors. The following examples should give some brief insights into problems that can arise when planning and conducting experiments.

In experiments involving human participants (such as the cholesterol experiment above) there may be a tendency to report the 'expected' responses. For example if the cholesterol reducing yogurt is hailed as being a new breakthrough there is a real risk that the participants will tend to report the expected effect. Even those conducting the experiment may encourage (sometimes unintentionally) the expected responses. To deal with this type of 'bias' in a statistical study researchers may organise **'blind'** or **'double-blind'** experiments.

An experiment is said to be 'blind' when the participants do not know whether they are receiving a treatment or a placebo (a placebo is a medicine that contains no active ingredient).

An experiment is 'double-blind' when neither the participants nor those who administer and record the results know whether they are receiving or giving the treatment or the placebo.

Confounding variables: A confounding variable in a study or experiment is a variable that can affect the measured/recorded response or outcome but which was not taken into consideration by the researcher: For example, in the cholesterol experiment, other factors such as diet, medication, age, weight, blood pressure, etc. may also affect the level of cholesterol and feeling of wellbeing.

Ethics: ethics are an important consideration in conducting experiments. In the cholesterol experiment above it would be unethical to put patients at risk by stopping or denying a conventional treatment for high cholesterol while testing the effectiveness of the 'yogurt'.

1.3   Some sources of statistical data

Statistical data may be collected by researchers employed by governments through national statistics offices, by institutions such as universities, colleges, financial institutions, private companies, etc.

- Market research companies conduct research for private individuals, companies, newspapers. Companies such as Gallup and Mori are well known for research on voting intentions and opinion polls as well as research on issues ranging from politics to consumer products, and television programmes, etc. Some of the data is available on their respective websites: www.gallup.com, www.mori.com.
- Governments conduct a census of the population at 5 to 10 year intervals. Governments and state agencies also carry out surveys at regular intervals. For example, the household budget survey or household expenditure survey, labour force survey, prices, etc. Data is collected from industry, businesses, banking, government departments such as education, health, etc. on industrial and agricultural output, investment at home and abroad, energy consumption and output, tourism, climate, education, health, etc. Statistics and data is available on state statistics agency sites. Consult UK statistics at www.statistics.gov.uk, Ireland at www.cso.ie, Europe www.europa.eu, Singapore: www.singstat.gov.sg Hong Kong: www.censtatd.gov.hk, Australia (Australian Bureau of

Statistics): www.abs.gov.au and (Statistics Queensland) www.oesr.qld.gov.au and others: addresses are easily available since all web sites give links to similar sites.

Most of the sites provide very helpful introductory information for students, and are well worth viewing.

- Agencies, some of which are listed below, collect data and collate data supplied by governments and others worldwide on major social trends, economics, health, education, gender issues, etc.

    The United Nations Organisation has several sites, including UNECE United Nations Economic Commission for Europe: www.unece.org.

    The United Nations Economic and Social statistics for Asia and the Pacific (UNESCAP): www.unescap.org.

    World Health Organisation: www.who.int.

    Consult also World Bank www.worldbank.org European Environmental Agency www.eea.europa.eu Organisation for Economic and Co-operative Development (OECD) www.oecd.org, and www.sourceoecd.org.

- Statistical Bulletins, Statistical Highlights and various statistical reports summarise very interesting key economic and social statistics over a given period (year, quarter, etc.) and use historical statistics to highlight changing trends and patterns.

Industry and commerce

Some limited information is available from companies and manufacturers. Data is collected during production, sales, services etc. and used in statistical procedures to monitor quality and continual improvement of products and services starting at the research and design stage through to production, maintenance and disposal/recycle in order to produce quality goods and services, minimise losses and ultimately increase profits. There is a burgeoning industry which deals with the provision of courses, from introductory to advanced, on the methods and practices based on statistics. An internet search on SPC, quality control, experimental design will immediately produce a list of such courses, course content, etc.

Progress Exercises 1.1: Data collection and sources

1. Explain each of the following terms with an appropriate example. (a) Census (b) random sample (c) quota sample.
   Do an Internet search for the terms (a), (b) and (c). Do the results support your answers?
2. What is 'bias' in the statistical context? Explain how bias could arise in the selection of a sampling frame.
3. What is response bias? Explain how bias could arise at any of the eight stages in survey design given in the text.
4. (a) Explain the following (i) a simple random sample (ii) a stratified random sample (iii) a cluster sample (iv) a multistage sample.
   (b) Which type of sample would you use for the following surveys?
      (i) Views on statistics held by the current first year students attending the statistics course.
      (ii) Support for a football team by the members of the football supporters club.
      (iii) The views of primary school children on homework.
5. Explain the methods used to select the sample of your choice in 4(b) above
6. How would you contact those selected for your sample in 4(b) above. How would you deal with non-response?

7. What is a quota sample?

What are quota controls and why are they necessary?

Explain, stage by stage, how you would organise and conduct a survey on the views of those who use public transport.

8. Give the main reasons for using a quota sample. Explain how you would select a sample from the population of people who have high cholesterol.

9. Look up the web site www.mori.com.

What are the main research areas on this site?

Find information on elections.

On this site, look up Election Polls then 'What does Mori do and why?'

Summarise the sampling method used, the sample size and the method of contact in your own words. Is the sample random or non-random?

10. Go to the web site at www.cso.com. Look up the Quarterly National Household survey (QNHS). What is the purpose of this survey? Note down the sampling method (also called sample design), sample size, method of contact and how frequently data is collected.

11. Look up www.statistics.gov.uk In the search box enter 'household surveys'. Click on 'Integrated household surveys'. Describe the purpose of this survey. Describe the labour force survey: its coverage, frequency and method of contact.

12. On the eurostat site at www.europa.eu select 'Services' and find the section on statistics and opinion polls (Eurobarometer). Describe any opinion poll of your choice from this site, with methodology, sample size, definition of terms and any other interesting information.

13. **World Bank** www.worldbank.org. This is a global monitoring information systems.

Select 'Data' and the country about which you require information. Make sure you get the outline country statistics but also an indicator (out of 100) on the statistical services in that country. List the national surveys conducted by the government and the frequency of the census.

14. UNESCAP: United Nations Economic and Social Commission for Asia and the Pacific at www.unescap.org or www.unescap.org/Stat/.

Describe the coverage contained in this site. Select the country of your choice. Give statistics to describe this country.

15. www.unece.org is the web address for UNECE (United Nations Economic Commission for Europe).

On this site, select 'About us'. Describe the issues covered on this site. When was this agency set up?

16. Describe and give an example of the following sampling methods:

(a) self-selected samples (b) focus groups (c) opportunity samples. Explain how each of these sampling methods could result in biased samples?

17. Do a web search for SPC (Statistical Process Control). List four companies that offer courses in statistics. Can you find courses whose contents you will have covered by the time you finish your present course?

1.4   Sorting and classifying data

Data may come in many forms depending on the type of statistical study being undertaken. In this section the different types of data are described together with two techniques for sorting data into groups or tables.

1.4.1 Qualitative and quantitative data

There are two types of data: Qualitative or categorical data and Quantitative or numeric data.

Qualitative or categorical data are defined by some quality characteristic such as gender, eye colour, etc. Qualitative data is further classified as either nominal or ordinal. **Nominal data** defines some group characteristic such as gender, nationality, profession, membership of clubs or societies, etc. **Ordinal data** is the result of ranking products, goods, services; opinions, etc. in order of preference. For example, viewers may be asked to rank television programmes in order of preference: 1, 2, 3, etc. But the difference between ranks 1 and 2 may not be the same as between ranks 2 and 3.

Quantitative or numeric data are described numerically by counts or measurements. Numeric data may be discrete or continuous. **Discrete data** can only assume certain distinct values; for example the numbers that turns when a die is thrown can only be 1, 2, 3, 4, 5 or 6. **Continuous data** on the other hand can assume any value from a continuous set of values: for example the time between arrivals at a check-in desk could be a value such as 3.25 minutes. Quantitative data may be further classified as ratio or interval. An interval scale is based on measurements from an arbitrary zero – a classic example being temperature scale in Fahrenheit with zero at 32°F where, unlike ranks, a difference of say, 10°F between any two temperatures is the same whether it is between 40°F and 50°F or between 85°F and 95°F. **A ratio scale** is an interval scale with a natural zero so that measurements can be compared. Ratio scales include distances, heights, weights volumes, time. For example, a queuing time of 40 minutes is twice as long as a queuing time of 20 minutes. The levels of data measurement are summarised at the end of the chapter.

Accuracy of measurements

The number of decimal places given in a measured value depends on the type of measurement, the measuring instrument and the accuracy required. For example, time to the nearest minute may be sufficiently accurate to record times between arrivals at a check-in desk. But for various events in athletics, times are recorded to the nearest 0.001 second. To record times to the nearest 0.001 second, the time must be 'rounded' to three decimal places.

Rounding numbers

When processing continuous data, such as time, numbers are normally 'rounded' correct to a given number of decimal places. For example, to round a number correct to three decimal places note the digit in the fourth decimal place – if the value of this digit is five or greater, then the digit in the fourth decimal place is increased by 1; if it is less than 5, leave it as it is. So 0.3676 is rounded to 0.368, but 0.3674 is rounded to 0.367.

1.4.2 Sorting and classifying the data

Raw data refers to data that has been collected or recorded by the researcher and it is usually in no particular numeric order. For example, the numeric data in Table 1.4 is a record of the numbers of MP3 players sold in a music store from January 1st to 31st (store closes on Sundays). At a glance, the numbers per day range from 10-something to 50-something but it is difficult to see any pattern or read 'information' from the data. For example, the store manager might want to know 'the most usual number sold daily' or the average number sold per day or s/he may want to display a chart summarising the information contained in the raw data.

Table 1.4   Raw data: number of MP3 players sold daily   File: MP3 players

30	11	29	34	54	36
49	31	42	45	25	25
15	18	13	25	13	
55	55	38	31	43	
38	22	37	20	36	

To obtain more information and to plot charts it is necessary to start by sorting or classifying the data in some way. In the following sections, two methods for sorting data will be explained:

1. Sorting data into a Stem-and-leaf plot (or diagram).
2. Sorting data into a frequency distribution table.

Note: Categorical data is easily sorted according to the pertinent data characteristics such as gender, hair colour, Y/N, etc.

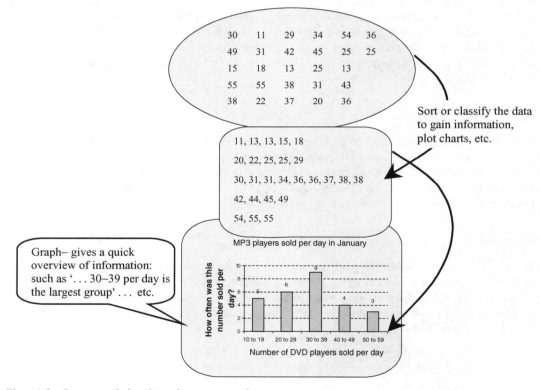

Chart 1.2   Sorting and classifying data to gain information.

1.4.3   Stem-and-leaf plot

This is a simple chart consisting of stems as the main groups while the leaves are elements within each main group/stem.

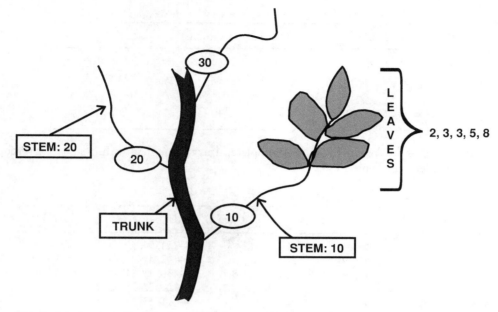

Figure 1.1   Reason for the name 'stem-and-leaf' plot.

For example, the picture in Figure 1.1 depicts a stem-and-leaf plot with 3 stems (or main groups): 10; 20; 30. The stem, 10, has leaves 2; 3; 3; 5; 8 representing the numbers 12; 13; 13; 15; 18.

It is not necessary to 'draw' a stem-and-leaf picture such as that in Figure 1.1. The sorted data may be presented in a table as illustrated in the following Worked Example 1.3.

Worked Example 1.3: Stem-and-leaf plot

Sort the data in Table 1.4 (MP3 players) in a stem-and-leaf plot.

Solution

Normally between 5 and 20 stems or categories are desirable-too few give very little information while too many stems give scrappy and disjointed information that is difficult to read. A good choice of stems for the data in Table 1.4 would be: 10, 20, 30, 40 and 50. Therefore, start with row 1 of Table 1.4 and distribute each number into the appropriate group (or stem), as follows:

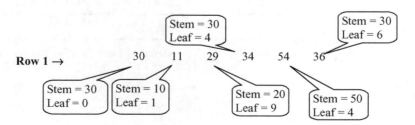

Table 1.5   Data from first row of Table 1.4 sorted into the stem-and leaf table

Stem (tens)	L	e	a	v	e	s	(units)	
5	4							← 54
4								←
3	0	4	6					← 30, 34, 36
2	9							← 29
1	1							← 11

When the sorting of the data in Table 1.4 is complete, array the numbers within each stem (group) in ascending order as in Table 1.6:

Table 1.6   The stem leaf plot for the data in Table 1.4

Stem (tens)	L	e	a	v	e	s		(units)	Read out as
5	4	5	5						← 54, 55, 55
4	2	4	5	9					← 42, 44, 45, 49
3	0	1	1	4	6	6	7	8 8	← 30, 31, 31, 34, 36, 36, 37, 38, 38
2	0	2	5	5	5	9			← 20, 22, 25, 25, 29
1	1	3	3	5	8				← 11, 13, 13, 15, 18

Information can be read from a stem and leaf plot

From this simple plot one can see at a glance:

- the range of values from the lowest to the highest.
- the main groups; the biggest single group in this data is the 30s. Between 30 and 39 MP3 players were sold on nine of the 27 days.

From the stem-and-leaf plot the entire data set may be written in ascending or descending order – a prerequisite for the calculation of medians and quartiles, as we shall see in the following sections.

1.4.4   Frequency distribution table

The data in this particular stem-and-leaf plot may readily be summarised in a table called a frequency distribution table as shown in Table 1.7a:

Table 1.7a   Frequency distribution table for the numbers of MP3 players sold in January

Intervals	Frequency
MP3 players sold per day ranged from....	The number of days when this number was sold
10 to 19	5
20 to 29	6
30 to 39	9
40 to 49	4
50 to 59	3

Table 1.7b   Frequency distribution table with a tally column and relative frequencies

Intervals	Tally	Frequency
10 to 19	‖‖	5
20 to 29	‖‖ |	6
30 to 39	‖‖ ‖‖	9
40 to 49	‖‖	4
50 to 59	‖‖	3
Total		27

In the absence of a ready-made stem-and-leaf plot, raw data may be sorted into a frequency distribution table as follows:

1. Find the highest and lowest value in the data; choose between 5 and 20 intervals to cover this range using equal intervals, if possible. The data in Table 1.4 ranged from 11 to 55 hence five intervals starting at 10, ending at 60 would cover this range. Since there are only 27 data, five intervals are sufficient.
2. Intervals may be defined in various ways but they must be defined clearly, and if possible use equal widths. There must be have no gaps or overlaps. In other words the intervals should be mutually exclusive and exhaustive over the data range.
3. Having defined the intervals, sort each datum in Table 1.4 into intervals by entering a 'tick' in the appropriate interval in what is called a 'tally column' see Figure 1.7b. When all the data is sorted, the ticks for each interval are totalled. The number of data in a given interval is called the **'frequency'** of that interval.

Unequal and open-ended intervals: Occasionally, when sorting data into a frequency table it will be necessary to use intervals of different widths, for example when the use of equal intervals would result in a large number of empty intervals. In other instances it may be necessary to use open-ended intervals at the beginning and/or end of a table. For example, in a company where the salary for most employees will be within a well defined range, but a small number of workers are be on a very low salary the first interval would be defined as 'less than X': at the other extreme, a small number of consultants may be on very large salary so the upper interval in the table would be defined as 'Y and over'. While unequal and open-ended intervals give a more compact table, information on extreme salaries is lost. Frequency distribution tables are a prerequisite for plotting bar charts and histograms.

1.5   Charts: Bar charts, pie charts and plotting in Excel

Charts and graphs are common features on the business pages in most daily newspapers, journals, internet sites, etc. Charts may be drawn for numeric and categorical data either manually or by using packages such as Microsoft Excel. To give the reader an appreciation of the factors that determine their accuracy, readability and clarity (scale, units, titles, labels, etc.) the manual plotting of basic bar charts is described first. Following the manual plotting a brief introduction is given to the very versatile plotting capabilities of Microsoft Excel.

1.5.1   Simple bar charts

To plot bar charts the data should be sorted into stem-and-leaf or frequency tables. Worked Example 1.4 outlines the steps involved in plotting a simple bar chart.

Worked Example 1.4: Plotting a Bar chart (manually)

Plot a bar chart from the data given in the frequency distribution Table 1.7a for daily sales of MP3 players. Give a verbal description of the information displayed in the chart.

Method

1. **Scale the axis**
   Horizontal axis: the intervals 'Numbers sold per day "10 to 19": "20 to 29", etc' are plotted on the horizontal axis. Mark out 5 intervals of equal width: label each interval to correspond with the frequency table.
   Vertical axis: frequency will be plotted on the vertical scale: the highest frequency is 9. Therefore scale the vertical axis from 0 up to about 12. You need to leave room at the top for titles, etc.
2. **Labels/Titles:** Write the label and units on the horizontal axis and vertical axis to correspond with those in the frequency distribution table. Give the graph an overall title and, if necessary for clarity, give subtitles. See Figure 1.2.
3. **Plot the chart.** Over each interval on the horizontal axis draw a bar, the height of which corresponds to the frequency or number of data within that interval.

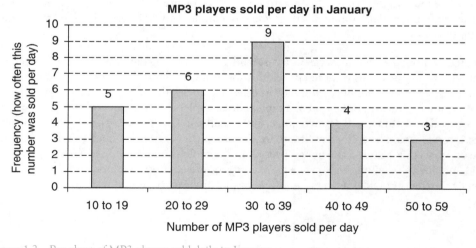

Figure 1.2   Bar chart of MP3 players sold daily in January.

4. **Give a verbal description of the information displayed in the chart.** The information here is the same as given by the stem-and-leaf plot. The numbers sold per day ranged from 10 to 59 and is distributed reasonably symmetrically about the central group. The largest single group is 30 to 39 since 30 to 39 MP3 players were sold on 9 out of the 27 days. Between 50 and 59 MP3 players were sold on only three days.

1.5.2   Variation on the simple bar charts and pie charts

The manual plotting of relative frequency and percentage bar charts is described below. Again, these and numerous other variations are readily available in Excel.

The relative frequency expresses the number of items in an interval as a proportion or fraction of the total number of items in the data set.

$$\text{Relative frequency for interval } i = \frac{\text{Number of items in interval } i}{\text{Total number of items}}$$

> Also described as the proportion of items in interval i.

$$= \frac{\text{Frequency for interval } i}{\text{Sum of all frequencies}}$$

See Table 1.8 where relative frequencies for each interval are given in column 3.

Percentage frequency bar chart

The percentage frequency for data in a given interval is simply its relative frequencies multiplied by 100.

> Percentages are proportions × 100

$$\text{Percentage frequency (interval } i) = \frac{\text{Frequency for interval } i}{\text{Sum of all frequencies}} \times 100$$

Calculations for percentage frequencies are in column 4 in Table 1.8:

Table 1.8   Frequency distribution table with relative frequencies expressed as proportions of 100 (i.e. %) and proportions of 360°

Intervals	Frequency	Relative Frequency	Percent frequency Relative frequency × 100	Angle of Pie slice Relative frequency × 360
MP3 players sold per day ranged from....	Number of days when this number was sold	$\dfrac{frequency}{Total\ freq.}$	$\dfrac{frequency}{Total\ freq.} \times 100$	$\dfrac{frequency}{Total\ freq.} \times 360$
10 to 19	5	$\dfrac{5}{27}$	$\dfrac{5}{27} \times 100 = 18.5\,\%$	$\dfrac{5}{27} \times 360 = 66.7°$
20 to 29	6	$\dfrac{6}{27}$	$\dfrac{6}{27} \times 100 = 22.2\,\%$	$\dfrac{6}{27} \times 360 = 80°$
30 to 39	9	$\dfrac{9}{27}$	$\dfrac{9}{27} \times 100 = 33.3\,\%$	$\dfrac{9}{27} \times 360 = 120°$
40 to 49	4	$\dfrac{4}{27}$	$\dfrac{4}{27} \times 100 = 14.8\,\%$	$\dfrac{4}{27} \times 360 = 53.3°$
50 to 59	3	$\dfrac{3}{27}$	$\dfrac{3}{27} \times 100 = 11.1\,\%$	$\dfrac{3}{27} \times 360 = 40°$
Total	27	1	100\,%	360°

A careful comparison of Figures 1.2 and 1.3 will show that the percentage frequency bar chart is a rescaled version of the simple bar chart.

Pie charts. In Column 5 of Table 1.8 the frequencies are expressed as proportions or fractions of 360° by multiplying the relative frequency by 360: these values are the angles of the slices of the pie

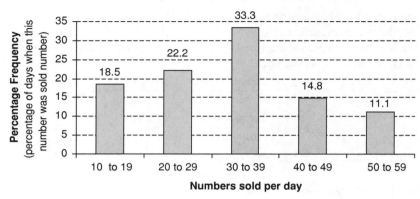

Figure 1.3   Percentage bar chart for MP3 players were sold January.

chart given in Figure 1.10(a). The drawing of the pie chart by hand involves the use of protractors, rulers, etc. – so this we leave to packages such as Excel in section 1.5.4.

Pie charts or Bar charts? Pie charts are used to display the relative size of subgroups within a set of data. However, visually it is difficult to see differences between slices of a pie that are close: a 90° slice will look the same as a 92° slice. Furthermore, accurate comparisons between two or more data sets are even more difficult for the same reason: especially as the data is plotted in separate 'pies'. Such small differences are immediately evident in bar charts. But, pie charts are popular; they look impressive and variations are easily plotted in colour, in 3 dimensions, etc. in Excel and other statistical packages.

Progress Exercises 1.2: Sorting and classifying data and bar charts

In questions 1 to 3

(a) Sort the data into (i) a stem-and-leaf plot and (ii) a frequency distribution table using the intervals specified in the question.
(b) Plot (i) a simple bar chart and (ii) percentage bar chart.
(c) Give a verbal description of the information in parts (a) and (b).

1. A survey of passengers who did not 'show' for 40 flights to certain destinations revealed that the majority booked flights more than six weeks in advance when fares were particularly low and the prospect of a break seemed like a good idea.

Table PE1.2 Q1   Numbers of 'no shows' per flight | File: No show |

31	12	8	6	25	8	18	15
29	34	15	13	9	9	21	10
5	7	16	20	13	21	14	12
18	21	16	14	9	12	17	30
8	22	14	15	15	37	27	25

Use intervals 0–4; 5–9; 10–14 etc.

2. The examination marks for a groups of students is given in Table PE1.2:

Table PE1.2 Q2   Examination marks for students on an SPC course | File: Exams

69	6	49	49	31	20	57	50	74	61	49	52	60	38	49
94	53	72	52	65	18	30	55	68	55	54	40	82	45	71
65	58	69	49	36	12	42	48	48	50	65	64	45	47	65
81	58	61	66	43	26	50	29	39	80	56	67	36	74	60

Use intervals 0 < 10; 10 < 20; 20 < 30 etc.

3. A mobile phone bill, summarised in Table PE1.3, gives the duration of calls that exceeded one minute.

File: Mobile calls

Table PE1.2 Q3   Duration of mobile calls that exceeded 1 minute

13.58	14.49	1.06	1.07	10.57	1.19	22.10	2.16	1.23	1.45	1.47	1.47	1.50
1.55	2.05	2.05	1.21	2.17	2.17	2.36	3.04	6.12	58.14	2.52	2.52	1.20
17.31	3.57	17.18	4.13	4.28	4.32	3.33	2.28	2.18	7.23	8.06	8.28	9.06
10.26	1.08	13.39	1.01	1.03	17.07	4.05	6.02	5.07	28.40	33.27	33.31	2.47

Source: My phone bill.

Use intervals 0 < 1.5; 1.5 < 3.0; 3.0 < 4.5 etc. Would an upper open-ended interval be sensible for this data?

4. The breakdown of road user fatalities for 2005 in Ireland is given in Table PE1.2 Q4

Table PE1.2 Q4   Breakdown of road user fatalities

Pedestrians	71
Drivers	170
Passengers	91
Motorcyclists	51
Pillion passengers	4
Pedal cyclists	11
Other	1
Total	399

Source: CSO – Department of the environment, heritage and local government.

Plot (i) a simple bar chart (ii) a percentage bar chart for the data.
Give a verbal description of the information displayed in the chart.

5. Write a brief report on the information displayed in Figure PE1.2 Q5. What other information would you require to give a more comprehensive report?

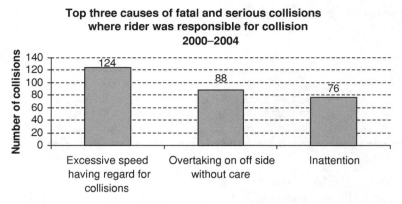

Figure PE1.2 Q5   Top three causes of fatal and serious collisions where motor cycle rider was responsible.
Source: PSNI.

6. (a) What information can you deduce from the following chart? What further detail should be given on the chart in order to read meaningful information from it?
   (b) The data and other information for the chart Figure PE1.2 Q6 is given in the file: Phone cost eurostat. Use data from this file to describe further information on the cost of phone calls in the various countries.

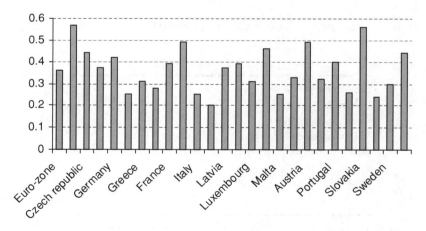

Figure PE1.2 Q6   Data source: Eurostat.

7. Write a brief account of the information displayed in Figure PE1.2 Q7. Suggest changes to this plot and/or explanatory notes that would provide the reader with more in-depth information.

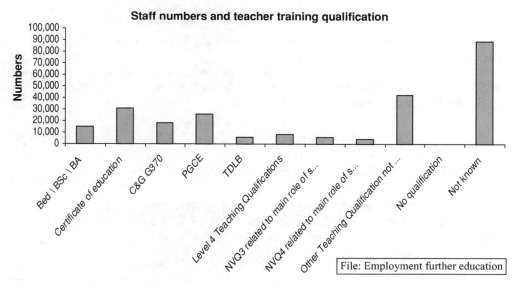

Figure PE1.2 Q7   Further Education colleges in England, 2004/05 (includes Specialist Designated Institutions). *Source:* Office for National Statistics. Crown copyright material is reproduced with the permission of the Controller of HMSO.

8. The total value of selected agricultural commodities (in $,000) produced and exported in 1997–98 in Australia is given in Table PE2.1 Q8

File: Agri products

Table PE1.2 Q8   Value of total production and exports of selected commodities

1997–98 Agricultural commodities	($,000) Production	($,000) Exports
Plant nurseries	506 866	5 624
Cut flower and flower seed growing	150 830	24 325
Vegetable growing	1803 250	244 149
Grape growing	917 041	82 438
Apple and pear growing	448 910	62 267
Stone fruit growing	154 176	24 479
Kiwifruit growing	5 135	2 892

Source: ABS Australia.

(a) Plot a bar chart for the commodities to display (i) production and exports (ii) production (iii) exports.
(b) Calculate the percentage of each commodity (in monetary terms) that is exported. Plot a bar chart to display this information. Comment on the information displayed in (a) and (b).

1.5.3 Plotting bar charts with Excel

To plot charts in Excel type the data into the Excel worksheet. This is illustrated in Figure 1.4 for the data for the MP3 players.

Next use the cursor to highlight the columns for ages and frequency, including the column headings as in Figure 1.5. Then click on the graph icon (See Figure 1.4) in the menu bar and follow the step-by-step instructions given in the drop down menus.

1. Click on the **graph icon** on the menu bar.
2. From the 'drop-down' menu (Figure 1.5) **Step 1 of 4** select bar chart.
3. Click **NEXT** from the menu bar at the bottom of the drop down menu

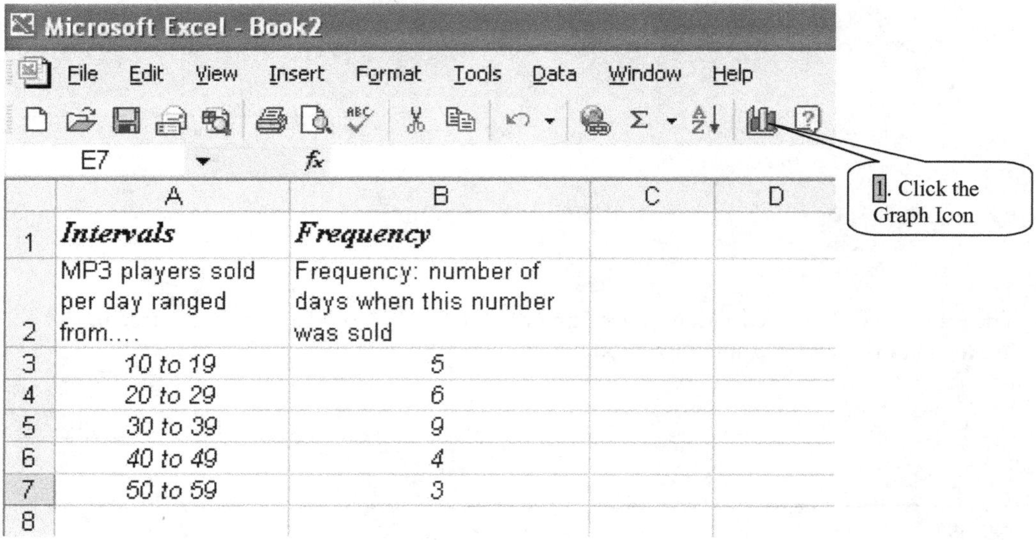

Figure 1.4   Enter the data into an Excel worksheet.

4. **Step 2 of 4.** Check the preview and confirm that the correct data is used in the chart. Click **NEXT** for the next window.
5. **Step 3 of 4.** Select titles, then type in the chart title, title for horizontal and vertical axis. Click **NEXT.**
6. **Step 4 of 4** – include the chart in the work sheet

When finished, you can adjust the size of the chart by dragging the borders. You can modify the appearance of the chart by pointing the mouse close to the chart area and right clicking the mouse: the dropdown menu gives several options to format plot area, change chart type, source data, etc. If 'Format Plot Area' is selected (See Figure 1.8) the background and borders of the inner plot area may be changed.

Point the mouse in the outer area of the chart; next right-click the mouse and then select 'chart options' to change titles, labels, and gridlines and so on.

DATA COLLECTION AND ITS GRAPHICAL PRESENTATION

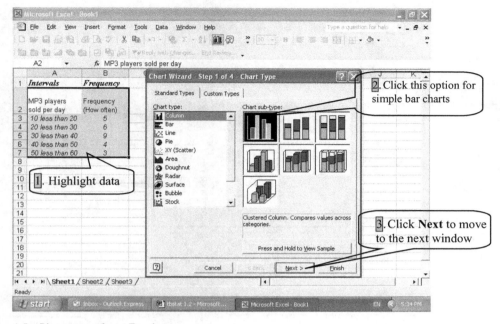

Figure 1.5   Plotting graphs in Excel.

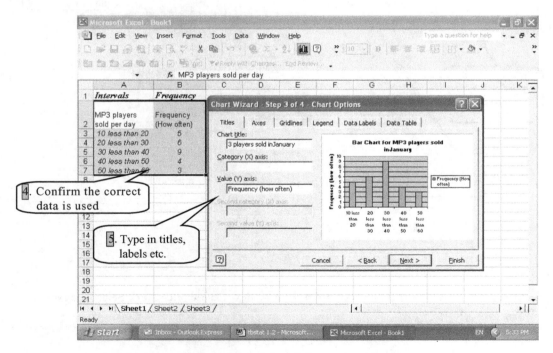

Figure 1.6   Plotting graphs in Excel: Step 3 of 4 enter titles and labels.

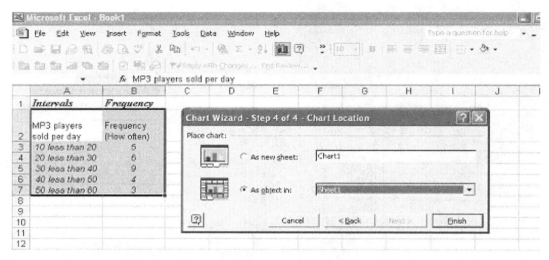

Figure 1.7   Plotting graphs in Excel: Step 4 of 4. Locate chart in separate graph window or in present worksheet.

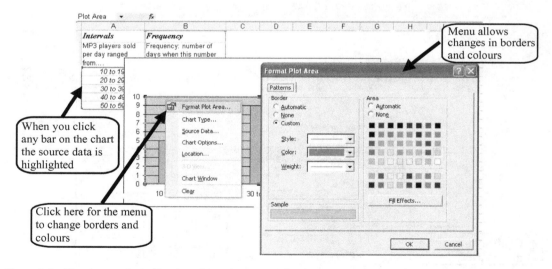

Figure 1.8   Plotting graphs in Excel: making changes to plots.

1.5.4   Plotting pie charts in Excel

Select the graph icon from the menu bar: from the drop down menu select 'pie'. Select the type of pie chart required: this is **Step of 1 of 4**.

Click next and work through the remaining 3 steps, filling in titles, etc. in the same way as for the bar chart.

Two examples of pie charts for the data in Table 1.5 are given in Figures 1.10 and 1.11.

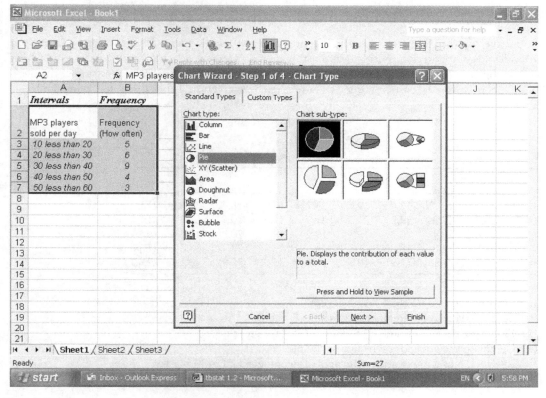

Figure 1.9   Plotting pie charts in Excel: Select pie chart.

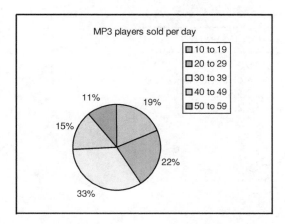

Figure 1.10   A pie chart for data in Table 1.8.

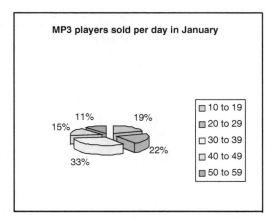

Figure 1.11   A 3D, blown-apart pie chart for data in Table 1.8.

Skill Development Exercise SK1_2: Sort data and plot bar charts

In a drive to attract new members to a sports club the management advertise the social, recreational and competitive activities of its existing members. Age is an important consideration when organising certain activities, particularly teams for interclub competitions. For example, the ages of members willing to participate in swimming teams is given in Table 1.9 (File: Swimmers). At a glance, it can be seen that practically all ages are in the higher teens to low twenties but it is difficult to see which group predominates unless the data is sorted.

Table 1.9   Ages of swimmers available for competitions

File: Swimmers

				Ages of swimmers					
18.8	18.7	20.2	17.4	29.6	23.7	19.6	18.4	20.6	25.5
17.9	24.2	20.9	19.2	23.1	21.5	21.9	20.7	22.3	21.7
20.5	21.6	20.4	20.2	19.5	19.9	19.9	20.8	17.2	19.1
18.1	19.2	18.8	18.8	21.2	24.3	22.3	22	18.8	21.7
18.1	20.9	21.1	17.9	18.6	18.8	20.1	20.9	20.1	22.4

(a) Sort the data into a stem-and-leaf plot.
(b) Sort the data into a frequency distribution table.
(c) Plot a bar chart for number of swimmers by age group.
(d) Calculate the angle of each slice of the pie.
(e) In Excel, plot a pie chart for the ages of swimmers data in Table 1.9.
(f) What information can you read from the table and graph above?

Answers:
(a)

Table SK1_2.1   Stem-and-leaf for ages

Stem (units of 1's)	Leaves (Decimals or units of 0.1)
17	2,4,9,9
18	1,1,4,6,7,8,8,8,8,8
19	1,2,2,5,6,9,9
20	1,1,2,2,4,5,6,7,8,9,9,9
21	1,2,5,6,7,7,9
22	0,3,3,4
23	1,7
24	2,3
25	5
26	
27	
28	
29	6

(b) and (c)

Table SK1_2.2   Frequency tables for ages of swimmers

Intervals	Tally column	Frequency
17 < 19	₭ ₭ IIII	14
19 < 21	₭ ₭ ₭ IIII	19
21 < 23	₭ ₭ I	11
23 < 25	IIII	4
25 < 27	I	1
27 < 29	I	1
Total		50

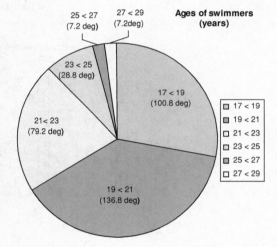

Number of swimmers in various age groups

Figure SK1_2.1   Bar chart for number of swimmers classified by age group.

(d) and (e)

Table SK1_2.3   Ages of swimmers in team with angles for slices of pie chart

Ages	Frequency	Angle in degrees
17 < 19	14	100.8
19 < 21	19	136.8
21 < 23	11	79.2
23 < 25	4	28.8
25 < 27	1	7.2
27 < 29	1	7.2
Total	50	360

Note: the intervals are defined so that '17 < 19' includes all data between 17.0 and up to but not including 19.0.

Figure SK1_2.2   Pie chart for ages of swimmers (plotting in Excel will be covered in the next section).

(f) Information from frequency distribution tables and bar charts

The first thing that strikes the reader when looking at the stem-and-leaf, table or bar or pie chart is the main groupings. In this example 19 out of the 50 swimmers are in the '19 < 21' age group. In fact, almost half of all swimmers are aged less than 21. The spread of ages within the team is clear; most (44 out of the 50) fall within 17 to 23 years, while only 6 swimmers have ages ranging from 23 to 29 years.

In general, when pie charts display the same information as bar charts, small differences between frequencies are not so easy to detect.

Table 1.10   Ages of swimmers and walkers

Ages	Frequency Swimmers	Frequency Walkers
17 < 19	14	8
19 < 21	19	12
21 < 23	11	18
23 < 25	4	37
25 < 27	1	15
27 < 29	1	7
29 < 31	0	3
Totals	50	100

1.5.5   Compound bar charts: making comparisons

Bar charts are particularly useful for comparing two or more sets of data on the same chart. For example, a comparison of ages for members of the walking team and the swimming team is given in Table 1.10 and Figure 1.12a, plotted in Excel.

In the latter example there are twice as many walkers as swimmers. An overall comparison of the *age distributions* would be clearer if we calculate and plot the percentage of swimmers in each age group and the percentage of walkers in each age group. See Table 1.11, Figure 1.12b.

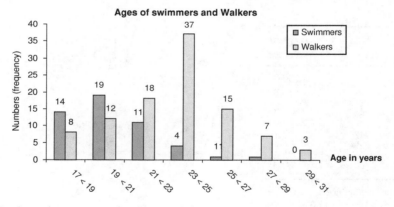

Figure 1.12a   Numbers of swimmers and walkers in each age group.

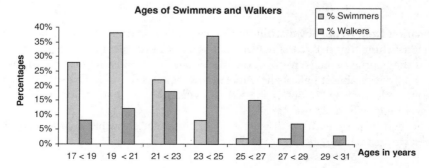

Figure 1.12b   Percentages of swimmers and walkers in each age group.

Calculation of percentages:

Table 1.11   Frequency distribution table giving percentage of swimmers in each age group, similarly for walkers

Reminder: $\dfrac{Number}{Total} \times 100$ expresses the *Number* as a percentage of the *Total*.

Ages	Swimmers		Walkers	
	Swimmers	% Swimmers	Walkers	% Walkers
17 < 19	14	$\dfrac{14}{50} \times 100 = 28\%$	8	8 %
19 < 21	19	$\dfrac{19}{50} \times 100 = 38\%$	12	12 %
21 < 23	11	$\dfrac{11}{50} \times 100 = 22\%$	18	18 %
23 < 25	4	$\dfrac{4}{50} \times 100 = 8\%$	37	37 %
25 < 27	1	$\dfrac{1}{50} \times 100 = 2\%$	15	15 %
27 < 29	1	$\dfrac{1}{50} \times 100 = 2\%$	7	7 %
29 < 31	0	$\dfrac{0}{50} \times 100 = 0\%$	3	3 %
Totals	50		100	

We can now see that walkers tend to be older than swimmers, with ages ranging from 17 to 31–37 % of walkers fall into the single largest age group – '23 < 25' while 36 % of swimmers fall into the '19 < 21' age group. The ages of walkers are fairly evenly distributed above and below the '23 < 25' group while most swimmers are concentrated in the younger age groups with very few in the higher age groups.

Another variation on compound bar charts. Figure 1.12c is a plot of the numbers of swimmers and walkers for each age group in the same bar. This would be useful if we want to look at the ratio of

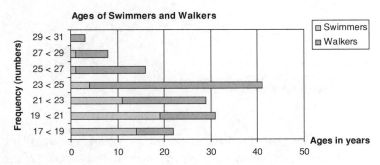

Figure 1.12c   Compound bar chart for ages of swimmers and walkers.

swimmers and walkers within the various age groups as well as between age groups. For example the ratio of swimmers to walkers is very small for age 23 < 25.

Progress Exercises 1.3: Sorting data, charts and graphs, Excel or otherwise

1. The data in Table PE1.3 Q1 gives the price in Euro of a 10 minute call at 11 am on a weekday (including VAT) for a local call (3 km). The prices refer to August each year. Normal tariffs without special rates are used.

   Plot a bar chart for this data to compare prices in the three countries from 1997 to 2004. Write a brief description of the plot.

Table PE1.3 Q1   The price, in Euro of a 10 minute call at 11 am on a weekday (including VAT) for a local call (3 km)

File: Euro phones

	1997	1998	1999	2000	2001	2002	2003	2004
Spain	0.2	0.32	0.32	0.28	0.28	0.28	0.28	0.28
France	0.45	0.42	0.42	0.42	0.39	0.39	0.39	0.39
Ireland	0.58	0.58	0.49	0.51	0.51	0.51	0.51	0.49

Source: EuroStat.

2. Plot the data in Table PE1.3 Q2 (File Absentees 1) to compare the absentee rate for men and women by age group. Give a brief verbal description of the plot.

Table PE1.3 Q2   Percentage absent from work for at least one day in the week preceding the survey

File: Absentees 1

Sickness absence (percentages) by age and sex

United Kingdom, Winter 2005			
Age	All	Men	Women
16–24	2.9	2.6	3.3
25–34	2.8	2.2	3.5
35–49	2.7	2.5	2.9
50–59/64	2.7	2.4	3.0
60+/65+	2.2	2.4	2.1
All	2.7	2.4	3.1

Source: National Statistics website: www.statistics.gov.uk. Crown copyright material is reproduced with permission of the Controller of HMSO.

3. The data in Table PE1.3 Q3 gives the absentee rate for various occupations. Plot a suitable chart to display this information graphically for (i) all absentees (ii) to compare of the absentee rate for men and women. In each case, explain the graphs verbally.

Table PE1.3 Q3   Percentage absent from work for at least one day in the week preceding the survey

United Kingdom, Winter 2005 Sickness absence rates: by occupation	Percentages		
	All	Men	Women
Managers and Senior Officials	2.1	1.8	2.4
Professional occupations	3.0	2.5	3.5
Associate Professional and Technical	3.2	2.9	3.4
Administrative and Secretarial	2.7	2.6	2.7
Skilled Trades Occupations	2.4	2.5	1.5
Personal Service Occupations	3.0	2.4	3.1
Sales and Customer Service Occupations	2.8	1.8	3.3
Process Plant and Machine Operatives	2.5	2.3	4.2
Elementary Occupations	3.0	2.8	3.3
Total	2.7	2.4	3.1

Source: National Statistics website: www.statistics.gov.uk. Crown copyright material is reproduced with the permission of the Controller of HMSO.

4. The Consumer Price Index for four categories of goods and services is given in Table PE1.3 Q4:

Table PE1.3 Q4   CPI series for 1996 to 2001 and 2001 to 2005

Consumer Price Index (Base Mid-November 1996 = 100)

Item	1996	1997	1998	1999	2000	2001
Food	100.1	101.8	106.1	109.6	113.7	121.2
Alcoholic Drink	98.6	101.1	104.8	108.7	114	119.5
Tobacco	99.6	103.4	108	112.7	130.5	133.9
Clothing and Footwear	99.7	92.5	87.2	81.1	76.5	73.8

Consumer Price Index (Base Mid-December 2001 = 100)

Item	2002	2003	2004	2005
Food and Non Alcoholic Beverages	101.7	103.1	102.8	102.1
Alcoholic Beverages and Tobacco	103.8	114	118	118.7
Clothing and Footwear	94.2	90.4	87.2	84.8

Source: CSO.

(a) Plot a suitable chart to compare the CPI series from 1996 to 2001 for (i) food and alcoholic drink (ii) tobacco and clothing and footwear.
(b) Plot a suitable chart for the 2002 to 2005 CPI series to compare the three given categories.
(c) Is it possible to plot a chart for the CPI from 1996 to 2004? State carefully the reasons for your answer.

5. The statistics relating to the England-Ecuador match in defensive phase for the 2006 World Cup is given in Table PE1.3 Q5.
   Plot a suitable to compare the performance of the teams. Comment verbally.

File: World Cup 2006

Table PE1.3 Q5   Match statistics England Ecuador World Cup 2006

England	Defensive phase	Ecuador
6	Off sides provoked	3
36	Headers	26
32	Balls won	32
2.46	Balls won / Fouls committed	1.28
22	Successful tackles	14
59 %	% Successful tackles	38 %
25	Interceptions	24
1	Saves	3

6. The economic activities of the Irish population aged 16 and older are classified according to gender in 2005. See Table PE1.3Q6 (extract from File: labour force).
   (a) Plot a pie chart to display the percentages in each economic activity for the female population.
   (b) Plot a suitable chart to compare the economic activities of the male and female populations.
   (c) Describe the information displayed (a) and (b) verbally.

File: Labour force 1

Table PE1.3 Q6   Quarterly national household survey 2005 Ireland

	Males	Females
At work	1080.60	776.8
Unemployed	77.1	32.5
Student	176.8	193.6
On home duties	5.1	539.4
Retired	208.7	72.2
Other	72.5	42.1
Total	1620.90	1656.60

Source: CSO Quarterly national household survey 2005.

7. The United Nations Economic Commission for Europe (UNECE) compiles statistics on global economic and social trends with particular reference to Europe and North America. An extract is given in Table PE1.3 Q7 for the main export commodities as a percentage of total exports from the Russian Federation.

File: UNECE Ex_Trade

Table PE1.3 Q7   Main external trade commodities as a % of total exports

Russian Federation Main external trade commodities	Year 2003 (% of total) exports of goods
Petroleum, petrol. product	38.1
Spec.transact.not classd	13.8
Gas, natural, manufactured	13.1
Iron and steel	6.1
Non-ferrous metals	5.6
Remainder	23.3

Source: United Nations Economic Commission for Europe.

(a) Plot a (i) bar (ii) pie chart for the data.

(b) Which of the charts is most effective in displaying the information graphically?

8. The estimated annual waste by sector in 2000 is given in Table PE1.3 Q8:

Table PE1.3 Q8   Estimated annual waste by sector

File: Waste 1

	Million tonnes	Percentage
Agriculture	87	20.1 %
Mining and quarrying	118	27.3 %
Sewage sludge	1	0.2 %
Dredged spoils	33	7.6 %
Household	30	6.9 %
Commercial	28	6.5 %
Demolition and construction	80	18.5 %
Industrial	55	12.7 %
Total	432	100.0 %

Source: Defra, Environment Agency, Water UK.

Plot (a) a pie chart (b) a bar chart for the data.

Visually, which chart gives the reader the more accurate view of the relative amounts of waste from the various sources?

9. Singapore 2006 Statistical highlights proclaim 'Strong growth in Foreign Direct Investment (FDI) and Direct Investment Abroad (DIA)' in 2004.

The data is given in percentages in Table PE1.3 Q9:

Table PE1.3 Q9   Singapore's FDI and DIA in 2004

File: Foreign Investment

Region	Percentage FDI	Percentage DIA
Asia	22.8	48
North America	16.5	5
Caribbean/Latin America	15.1	25
Europe	42.4	10
Others	3.1	12

Source: © Department of Statistics, Ministry of Trade and Industry, Republic of Singapore.

(a) Plot two pie charts to compare the FDI and DIA for the five regions.

(b) Plot a bar chart to compare the FDI and DIA for the five regions.

(c) Explain which chart best conveys the information given in Table PE1.3 Q9.

10. Personnel costs are defined as the total remuneration payable by an employer to an employee in return for work done by the latter. It includes taxes and employees' social security contributions. The costs in France, Poland, Slovenia and the UK for the construction sector and the hotel and restaurant sector are given in Table PE1.3 Q.10:

Table PE1.3 Q10   Personnel costs for construction and hotels and restaurants.   File: Personnel costs

	Construction			Hotels and restaurants		
	2001	2002	2003	2001	2002	2003
France	41 132.3	42 987.3	45 188.5	1 511.4	1 645.6	1 840.9
Poland	3 632.9	5 758.7	2 707.4	594.9	1 036.8	527.2
Slovenia	665.6	697.3	697.5	234.1	249	266.3
United Kingdom	38 123.9	41 680.7	39 664.8	20 965.1	21 553	20 482.5

Source: eurostat.

(a) Plot a bar chart to compare costs in the construction sector and the hotels and restaurants sector for the four countries. Describe briefly the information displayed in the chart. Suggest at least two reasons for the higher costs in France and the UK. What other information would give useful comparisons of personnel cost between countries?

(b) Would it be useful to plot separate charts for the high costs countries and the low costs countries? Give reasons.

Plot a bar chart to compare costs in the construction sector and the hotels and restaurants sector (i) for UK and France (ii) for Poland and Slovenia Describe briefly the information displayed in the charts.

1.6   Graphs: Histograms and Ogives. Graphs in Excel

Graphs are used to represent numeric data accurately. The graphs covered in this section are histograms, Ogives (a cumulative frequency graph) and line graphs.

1.6.1   Histograms: for data sorted into equal intervals

Histograms are graphical representations of frequency distributions for numeric data. Histograms are plotted from data sorted into frequency tables. If the frequency table consists of equal-width intervals the method for plotting histograms is similar to that described above for bar charts but there are 'no gaps' between the bars. To leave the reader in no doubt that there are 'no gaps', the intervals in Table 1.7a could be relabelled as '10 less than 20' which means the interval contains numbers ranging from 10 and up to, but not including 20 (since this is count data 10 less than 20 is the same as 10–19). The histogram plotted from the relabelled intervals from Table 1.7a is given in Figure 1.13. Note, on the graph the boundaries between intervals should be 9.5, 19.5, 29.5, etc. See Appendix A for further explanations on conventions for defining class boundaries, class intervals, etc.

Notes
- Histograms display the numbers (frequencies) of data that fall within a set of mutually exclusive intervals: the intervals span the entire range of data from the smallest to the largest.
- If the data is symmetrically distributed about a central value we say it is 'normally distributed'. Normality is a very important characteristic and will be discussed in later chapters.
- A different choice of intervals may alter the shape of the histogram.

Table 1.7a   MP3 players sold per day in January

Intervals	Frequency
Number of MP3 players sold per day	Frequency: Number of days on which stated quantity was sold
10 less than 20	5
20 less than 30	6
30 less than 40	9
40 less than 50	4
50 less than 60	3

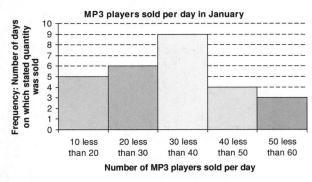

Figure 1.13   Frequency distribution table and histogram for data in Table 1.7a.

Plotting histograms for equal width intervals in Excel

In Excel, histograms for intervals of equal width are plotted under the heading 'Bar chart'. The gaps between the bars may then be reduced to gap size 0 as illustrated in Figure 1.14.

   Plot the bar chart for the data in Table 1.4 as above

   →Place the cursor on any one of the 'bars' in the bar chart

   →Right-click the mouse → Select format data series → Select Options

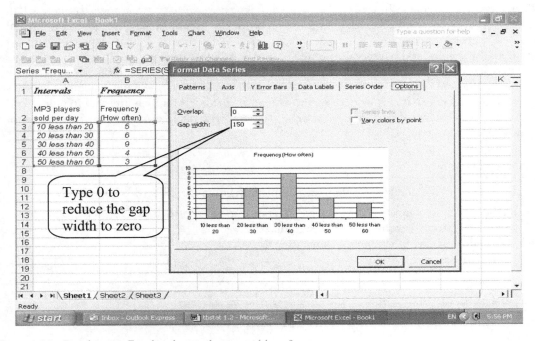

Figure 1.14   Bar charts in Excel: reducing the gap width to 0.

   In Options reduce the 'gap width' to 0. The bar chart is now a histogram.
   Note: while in Options you can also select 'vary colours by point'.

Skill Development Exercise SK1_3: Plot a histogram

The ages of members of the swimming team are given in Table 1.9. Summarise the data in a frequency distribution table. Plot a histogram for the ages of the swimmers.

Solution

Table SK1_2.2   Ages of swimmers in team

Intervals (Years)	Tally Column	Frequency
17 < 19	HHH HHH II	14
19 < 21	HHH HHH HHH IIII	19
21 < 23	HHH HHH	11
23 < 25	IIII	4
25 < 27	II	1
27 < 29	III	1
Total		50

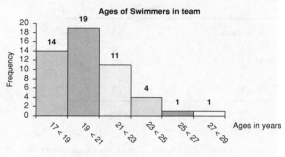

Figure SK1_3.1   Histogram of ages of swimmers in team.

1.6.2   Plot histograms for unequal intervals and open-ended intervals

When plotting histograms, the *area of a bar or rectangle is proportional to the frequency of the interval*. If the intervals are equal in width, then the height is proportional to the frequency, as stated above. Recall, that in section 1.4.4, it was stated that it is not always sensible to use equal intervals when sorting data into frequency tables. If the intervals (plotted on the horizontal) are not equal in width, the heights of rectangles must be calculated so that the area is proportional to frequency. The calculation of the heights for unequal intervals is illustrated in Worked Example 1.5. Some frequency tables will have open-ended intervals so it may be necessary to assume an end point for such intervals in order to plot graphs. For convenience the width of an open-ended interval is often assumed to be the same as that of the adjoining interval.

Worked Example 1.5: Histogram with open-ended and unequal intervals

Table 1.12 gives the numbers (in thousands) emigrating from the UK in 1995 and 2002.

(a) Plot a bar chart comparing the numbers emigrating in 1995 and 2002. Give a verbal description of the migration pattern by age in 1995 and 2002.

(b) Plot a histogram of the numbers emigrating in 2002.

Table 1.12   Total international emigration: by age and gender

File: Migration

	Outflow in thousands				
	Under 15	15–24	25–44	45–64	65 and over
1995	32.6	69.1	106.5	21.0	7.3
2002	25.0	91.9	186.4	46.2	9.9

Source: International Migration 2002- Series MN No. 29 Table 2.2, Migration Statistics, Office for National Statistics.

Solution

(a) This plot is straightforward – for each age group plot bars whose heights are proportional to the frequencies. The graph below was plotted in Excel. The graph reveals that the numbers emigrating in 2002 are higher than in 1995, except for the under 15 group. There is a large increase in the 25–44 and 45–64 age groups in 2002.

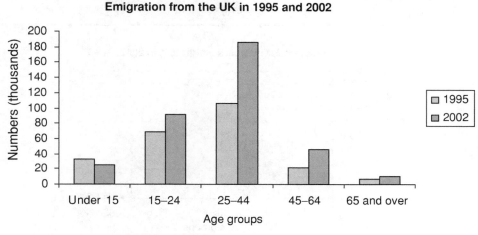

Figure 1.15   Number of persons (in thousands) emigrating from the UK in 1995 and 2002.

(b) Age group 15–24 contains all ages that round from 15 to 24, i.e. 14.5 to 24.49999 (written as 24.5⁻): the interval width is 10.

The interval 25–44 contains ages 24.5 to 44.5⁻, hence interval width is 20
The interval 45–64 contains ages 44.5 to 64.5⁻, hence interval width is 20
Assuming the interval '65 and over' is 64–84, this interval is also of width 20. The interval 'under 15' can only have a lower limit of zero (negative ages are nonsensical) and so the interval width is 14.5.

The calculation of height of the rectangles for this histogram with unequal widths intervals is explained as follows. The statement 'area is proportional to frequency' is written symbolically as Area $\propto$ frequency, where the symbol $\propto$ means 'is proportional to'

Height $\times$ width $\propto$ frequency ... where area of rectangle = height $\times$ width
$h \times w \propto f$ ..... where h = height, w = width, f = frequency
$h \times w = Kf$ ..... where K is a constant
$h = \frac{Kf}{w}$ .... solving for h

In Table 1.12 all intervals, except the first, are of width $w = 20$. Since K may be any value, for convenience select $K = 20$ (the width of majority of the intervals).

$h = \dfrac{20f}{20} \rightarrow h = f$, the frequency, for intervals of width 20

For the first interval of width 14.5, the height is, $h = \dfrac{20f}{14.5} = \dfrac{20(25)}{14.5} = 34.5$

The calculated heights are given in Table 1.13 below, the histogram in Figure 1.16.

Table 1.13 Calculation of heights of rectangles when intervals are not all equal

Outflow in thousands Age intervals	2002				
	Under 15	15–24	25–44	45–64	65 and over
Closed age intervals	0–14	15–24	26–44	46–64	65–84
Interval width	14.5	20	20	20	20
Frequency (thousands)	25.0	91.9	186.4	46.2	9.9
Height of rectangle	34.5	91.9	186.4	46.2	9.9

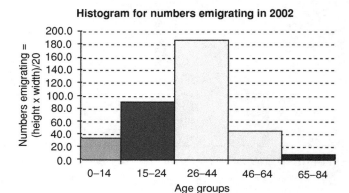

Figure 1.16 Histogram of emigration by age in 2002.

1.6.3 Cumulative frequency graph (Ogive)

Another graph that is useful in estimating quartiles, deciles, etc. (these will be explained in the next section) is the cumulative frequency graph or **Ogive**. This is a graph depicting the total number of data that have values less than the upper class boundary of each interval as given in the frequency distribution table. Accordingly, the intervals in the frequency distribution table are re-labelled as 'less than 10', 'less than 20', etc. as illustrated in Worked Example 1.6. The cumulative frequencies are plotted against this upper class boundary of each interval.

Worked Example 1.6: Ogives (cumulative frequency graph)

(a) Plot the Ogive for the numbers of MP3 players given in Table 1.7a
(b) (i) From the frequency table, calculate the percentage of days when fewer than 30 MP3 players were sold.
    (ii) From the Ogive, read off the daily sales on at least 50 % of days.

Solution

From the frequency distribution Table 1.7a, calculate the cumulative frequencies: that is, the sum of all frequencies that are less than the upper class boundary of the interval.

Table 1.14   Cumulative frequency distribution table

Daily sales of MP3 players	Less than...	Frequency	Cumulative frequency
0 to 9	10	0	0
10 to 19	20	5	5
20 to 29	30	6	11
30 to 39	40	9	20
40 to 49	50	4	24
50 to 59	60	3	27

> State the upper limits for cumulative frequencies: e.g. 'less than 20' means all daily sales from 0 but less than 20 etc.

> Total of all daily sales 'less than..'

To plot the graph, mark off intervals on the horizontal axis at the 'less than' limits, 10, 20, 30, 40, 50 and 60. Scale the vertical axis from 0 to 30, since the highest cumulative frequency is 27. Plot cumulative frequencies against the 'less than.' limits of 10, 20, 30, 40, 50 and 60. An additional vertical scale may be added on the right of the graph, treating the total cumulative frequency as 100 %. In this example the percentage cumulative frequency scale is divided into four equal parts at the 25 %, 50 %, 75 % and 100 % points called quartiles. In many examples the 100 % scale is divided into intervals at 10 % points, called deciles and even 20 % points called quintiles.

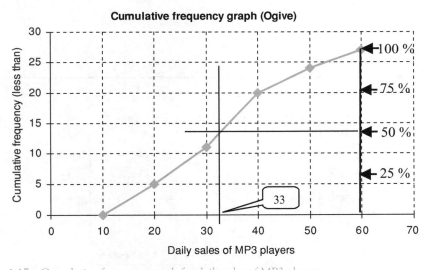

Figure 1.17   Cumulative frequency graph for daily sales of MP3 players.

(b) (i) From Table 1.14, fewer then 30 MP3 players were sold on 11 of 27 days, that is, on $\frac{11}{27} \times 100 = 40.7\%$ of days

(b) (ii) On Figure 1.17, draw a horizontal line at the 50 % mark on the percentage vertical scale. Through the point of intersection with the graph, drop a vertical line until it cuts the horizontal axis. This line cuts the horizontal axis at 33. Therefore, on 50 % of days fewer than 33 MP3 players were sold.

Note: cumulative histograms (not shown here) are useful to display the same information as Ogives and applications such as cumulative growth over time.

In Excel: The basic cumulative graph may be plotted in Excel by selecting 'XY Scatter'. XY (Scatter) graphs are covered in the following section.

Skill development Exercise SK1_4: Ogives

Plot the Ogive for the ages of the members of the swimming team given in Table SK1_2.3.

(a) If (i) 20 % are aged less than X (ii) 60 % less than Y (iii) 80 % less than Z, read off X, Y and Z from the graph.
(b) What is the range of ages for the middle 60 % of swimmers?
(c) Estimate the quartile ages from the Ogive.

Answers

Table SK1_4.1   Cumulative frequencies aged 'Less than' each upper age limit for members of the swimming team

Ages	Ages less than	Frequency	Cumulative Frequency	% Cumulative Frequency
17 < 19	19	14	14	28.0 %
19 < 21	21	19	33	66.0 %
21 < 23	23	11	44	88.0 %
23 < 25	25	4	48	96.0 %
25 < 27	27	1	49	98.0 %
27 < 29	29	1	50	100.0 %

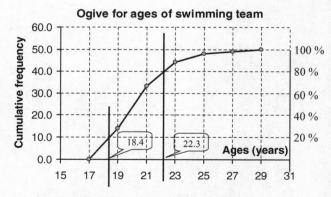

Figure SK1_4.1   Cumulative frequency graph for ages of swimming team.

X = 18.4, 20 % of swimmers are aged 18.4 or younger. Y = 20.7 years. Z = 22.3 years, 80 % of swimmers are aged 22.2 years or less. The middle 60 % are aged between 18.4 and 22.3 years.

(a) $X = 18.4$, 20 % of swimmers are aged 18.4 or younger. $Y = 20.7$ years. $Z = 22.3$ years, 80 % of swimmers are aged 22.2 years or less.

(b) The middle 60 % are aged between 18.4 and 22.3 years.

(c) From the Ogive, at the 25 % mark, $Q_1 = 18.6$; at the 50 % mark $Q_2 = 20.2$; at the 75 % mark, $Q_3 = 21.7$.

Skill Development Exercise SK1_5: Charts, tables, histograms and Ogives

Sort the data for the hours worked per week by the 49 staff in Table 1.15 into

(a) a stem-and-leaf plot.

(b) a frequency distribution table, with intervals 15 < 20 hours; 20 < 25 hours etc.
    Use Excel to plot, from the frequency distribution table (b)

(c) A histogram.

(d) An Ogive.

(e) From the Ogive, read off the number of hours worked by (i) 25 % or fewer (ii) 50 % or fewer (iii) 75 % or fewer of the group of 49 staff.

In each case give verbal descriptions of the data

Table 1.15   Hours worked in a given week by 49 staff | File: 49 Hours Worked |

20.0	37.3	54.2	25.3	59.6	24.5	29.7
18.0	38.8	42.1	39.5	56.8	16.9	28.5
45.5	42.0	39.5	42.6	40.0	44.2	40.1
44.0	56.4	30.2	20.0	22.7	37.8	23.4
26.0	20.2	36.1	18.3	19.7	36.8	26.5
24.0	23.4	15.4	20.0	38.9	42.1	24.1
41.0	18.5	21.3	22.6	37.2	42.9	17.9

Answers

(a)   Table SK1_5.1   Stem-and-leaf: Stem in units of 10 with increments of 10 Leaf Unit = 1.0

Stem	Leaves
(10's)	(1's)
1	5 6 7 8 8 8 9
2	0 0 0 0 1 2 2 3 3 4 4 4 4 5 6 6 8 9
3	0 6 6 7 7 7 8 8 9 9
4	0 0 1 2 2 2 2 2 4 4 5
5	4 6 6 9

(b)

Table SK1_5.2   Frequency distribution table for hours worked per week by 49 employees

Hours worked	15 < 20	20 < 25	25 < 30	30 < 35	35 < 40	40 < 45	45 < 50	50 < 55	55 < 60
Frequency	7	12	5	1	9	10	1	1	3

(c) Histogram from Frequency Table SK1_5.2

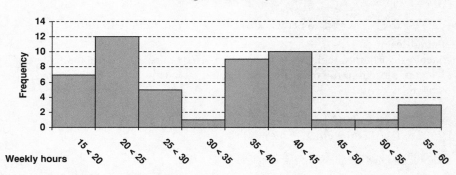

Figure SK1_5.1   Histogram for weekly hours worked by 49 employees.

There are 2 main groups here: one group works 30 hours or less per week while the other works between 35 and 45 hours per week

(d)

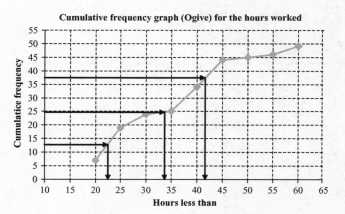

Figure SK1_5.2   Ogive for weekly hours worked by 49 employees.

(e) From the graph (i) 25 % of the group work 22.5 hours or less per week (iii) 50 % work 32 hours or less per week and (iii) 75 % work 42 hours or less per week. That means 25 % work 42 hours or more per week.

Progress Exercises 1.4: Histograms and Ogives

1. See question 1, PE1.2 Q1. For the data given in Table PE1.1 Q1 (File: No show).
   (a) Plot a histogram. Would you consider the data to be normally distributed?
   (b) Plot an Ogive. From the Ogive estimate the number of 'no shows' for the lower 50 % of flights
2. See question 2 PE1.2. For the data in Table PE1.2 Q2: (File: Exam marks).
   (a) Plot a histogram. Is this data approximately normally distributed?
   (b) Plot an Ogive. Calculate the maximum mark for the lower 30 % of results and the minimum mark for the upper 10 % of results.
3. See question 3 PE1.2, Table PE1.2 Q3, gives the duration of calls that exceeded one minute: (file: Mobile calls).
   (a) Plot a histogram. State any assumptions made. Describe the distribution of this data
   (b) Plot an Ogive: Calculate (i) the maximum duration for the lower 25 % of calls (ii) the minimum duration for the upper 25 % of calls (iii) the two limits within which the middle 50 % of calls lasted.
4. Is it possible to plot histograms for the data given in PE1.2 questions 4, 5 and 6? Give explanations for your answers.
5. The 'neighbour statistics' summarises certain social and economic statistics that relate to lifestyles in 'neighbourhoods' in the UK. Distance travelled to work is one such statistic. Table PE1.4 Q5 gives the distance travelled to work by the economically active inhabitants of neighbourhoods Brentwood (B 005A) and London (Lon 001A).

Table PE1.4 Q5   Distance travelled to work for (i) Brentwood (B) (ii) a London neighbourhood

File: Distance to work

Distance Travelled to Work (UV35)	B 005A	Lon 001A
All People (economically active 16–74)	1006	1056
Works mainly at or from home	101	105
Less than 2km	94	539
2km to less than 5km	160	237
5km to less than 10km	57	80
10km to less than 20km	159	16
20km to less than 30km	85	9
30km to less than 40km	285	4
40km to less than 60km	22	13
60km and over	10	21
No fixed place of work	30	32
Working outside the UK	3	0
Working at offshore installation	0	0

For each neighbourhood
   (a) plot (i) a histogram and (ii) an Ogive for those who travel to work (i.e. exclude those who work from home, no fixed place, etc.).
   (b) From the histogram, describe the distribution of distances travelled to work.
   (c) From the Ogive calculate the distance to work for (i) the 25 % that travel the shortest distance (ii) the 25 % that travel the longest distance (iii) the two distances within which the middle 50 % of workers travel.

6. The percentile points for income before and after tax are given in Table PE1.4 Q6 below.

File: Tax_income 90_91 03_04

Table PE1.4 Q6   Percentile points for total income before and after tax

Percentile points for total income before and after tax 1990–91 to 2003–04									
Taxpayers only									Amounts: £
Percentile point	1	5	10	25	50	75	90	95	99
Total income before tax									
1990–91	3190	3890	4650	:	10 600	:	23 200	30 100	57 200
2003–04	4820	5850	7000	10 100	16 000	25 100	37 100	50 600	111 000
Total income after tax									
1990–91	3150	3720	4300	:	8 980	:	18 700	23 300	40 400
2003–04	4800	5730	6790	9 290	14 000	21 100	30 700	39 000	77 000

: not given.

Source: Survey of Personal Incomes 2003–04 Table updated April 2006. Crown copyright material is reproduced with the permission of the Controller of HMSO and the Queen's Printer for Scotland.

(a) Describe the information contained in Table PE1.4 Q6.
(b) State the total income limit for the lower (i) 10 % (ii) 90 % of incomes before and after tax for the years 1990–91 and 2003–04.
(c) Plot Ogives for income before and after tax in 2003–04.

1.7   Line and Lorenz graphs

Line graphs are used extensively in various applications to show the relationship between two or more variables.

1.7.1   Lorenz graphs

On one particular homepage for the Office for National Statistics (ONS) it was reported that '1 % of the population owns 23 % of the wealth'. This unequal distribution of wealth is a well-known phenomenon. Unequal distribution of proportions (or percentages) arises in many other areas such as taxation where a small proportion of the population contribute a large proportion of tax revenue: in sales revenue where a large number of goods account for a small proportion of sales revenue. Such disparities in proportions are illustrated graphically by Lorenz curves.

To plot a Lorenz curve, first calculate the cumulative percentage frequency for each variable. Then plot an XY graph with the cumulative percentage frequency for one variable on the vertical and the cumulative percentage frequency for the other variable on the horizontal axis.

The data in Table 1.16 was given as the percentage cumulative frequencies on the NSO home page. The Lorenz curve is plotted in Figure 1.18:

Table 1.16   Percentage of population (x) that own percentage of marketable wealth (y) in UK

% most wealthy (x)	1	5	10	20	50
% of wealth owned (y)	23	43	56	74	94

Source: Inland Revenue on personal wealth.

Note: The distribution of people's marketable wealth relates to all adults in the UK. Estimates of wealth are derived from the Inland Revenue estimates from the estate multiplier method and ONS national accounts balance sheet method.

Estimates for individual years should be treated with caution as they are affected by sampling error and pattern of deaths that year.

Reading the Lorenz curve. If wealth was equally distributed, then 1 % of the population would own 1 % of the wealth, 20 % own 20 % of the wealth . . . 100 % of the population would own 100 % of the wealth. This situation is represented in Figure 1.18 by the line joining the points $(0, 0)$ and $(100, 100)$ – it is called the 45° line or line of equal distribution. The extent to which the Lorenz curve deviates from the 45° line is an indication of the disparity between proportions. For example, to see how much the most wealthy 30 % actually own draw a vertical through $x = 30$ until it cuts the Lorenz curve: it cuts the curve at the point at $y = 81$. So in reality (in 2002) the most wealthy 30 % of the population own 81 % of the wealth!

The Gini coefficient

The extent to which the Lorenz curve deviates from the 45° line is measured by the Gini coefficient. The Gini coefficient ranges from 0 to 100.

The Gini coefficient is zero for perfect equality when the Lorenz curve coincides with the 45° line.

The Gini coefficient is 100 for total inequality. For example when 0.1 % (rounds to 0 %) of the population own 100 % of the wealth. In this case the Lorenz curve would be the shape of the upper triangle in Figure 1.18.

The value of the Gini coefficient is the area enclosed between the Lorenz curve and the 45° line calculated as a percentage of the triangular area. In Figure 1.18, the Gini coefficient is approximately $\frac{35}{50} \times 100 = 70$. This value differs substantially from zero, indicating a substantial departure from equality of distribution.

Further useful information on wealth and other examples of unequal distribution of proportions such as the inequality of household income are available on the www.statistics.gov.uk.

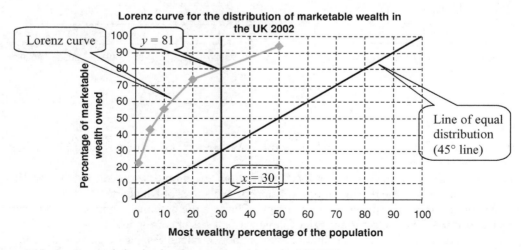

Figure 1.18   Lorenz graph for the distribution of wealth in the UK 2002.

1.7.2   XY (scatter) plots and line graphs

These graphs are used to show trends and relationship between two variables. Several trends may be compared by plotting line graphs on the same diagram. For example, the average private household size in Ireland for selected years from 1926 to 2002 are given in Table 1.17:

Table 1.17   Average household sizes in Ireland from 1996 to 2002 | File: Households 1 |

x = year	1926	1936	1946	1961	1966	1971	1979	1981	1986	1991	1996	2002
y = average household size	4.48	4.31	4.16	3.97	4.01	3.93	3.72	3.66	3.53	3.34	3.14	2.94

A Scatter plot is a plot of the (x, y) data points. In this example, let $x =$ the year and $y =$ household size. The method for plotting line graph manually is illustrated for the data in Table 1.17.

Step 1: Scale each axis. Check the smallest and largest number to be plotted on each axis. Choose a sensible scale to include all points and leave space to write in titles and labels.

The x-axis: since years range from 1926 to 2002, the x-axis must span a period of $2002 - 1926 = 76$ years. A sensible choice of scale could range from 1920 to 2010.

The y-axis: since the y values range from 2.94 to 4.48, a scale ranging from 2 to 5 would be sensible.

Step 2: Plot the points. The scatter plot is given in Figure 1.19(a).

Step 3: Join the points to get the corresponding **line graph**.

Step 4: Give a verbal description of the information displayed in the graph. From this graph the general trend is for the household size to drop steadily (with the exception of a brief stationary period from 1960–65) from an average of 4.5 in 1926 to less than three in 2002.

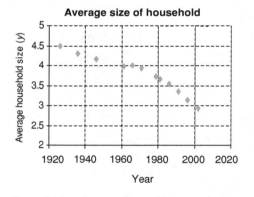

Figure 1.19a   Scatter plot and Line graph   Figure 1.19b   for average household size 1926 to 2002.

XY(scatter) and line graphs in Excel

To plot line ordinary XY graphs in Excel, enter the x and y points into an Excel worksheet, either as two columns or two rows.

Click on the graph icon, then select XY(Scatter) from the drop-down menu (Step 1 of 4): see Figure 1.20. Then follow the remainder of the 4 steps as explained for bar charts.

To plot two or more trends on the same graph. Click on the graph icon and **select** 'XY scatter' graph at step 1 of 4. Then select the x-values along with all sets of corresponding y-values. The only requirement here is that all sets of data must have common x-values. See Figure 1.23a and 1.23b.

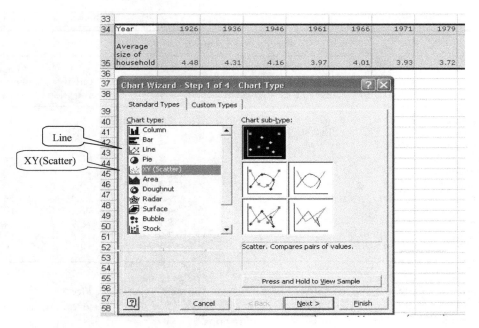

33								
34	Year	1926	1936	1946	1961	1966	1971	1979
35	Average size of household	4.48	4.31	4.16	3.97	4.01	3.93	3.72

Figure 1.20 Selecting scatter plots in Excel.

Line graphs in Excel: For data with **equally spaced *x*-values only** select 'Line' graph in step 1 of 4. The *x*-labels are displayed on the graph.

Progress Exercises 1.5: Line graphs

1. The numbers (in thousands) employed in three sectors in Ireland are given in Table PE1.5 Q1:

Table PE1.5 Q1   Numbers (in thousands) employed   File: Employment 1

Economic Sector (NACE Rev.1)	Mar– May 98	Mar– May 99	Mar– May 00	Mar– May 01	Mar– May 02	Mar– May 03	Mar– May 04	Mar– May 05
Agriculture, Forestry & Fishing	136	137.3	132.9	122.5	124	116.6	117	113.7
Transport, Storage & Communication	87	96.2	101.3	111.1	111.7	112.1	113.2	118.2
Health	113.9	120.3	133	144	159.3	169.9	177	188

Source: CSO.

On the same diagram plot the numbers employed in the three sectors from 1998 to 2005. Give a verbal description of the trends.
2. The unemployment rates (expressed as percentages) by highest qualification for regions in the UK for years 2000/01 and 2001/02 is given Table PE1.5 Q2:

Table PE1.5 Q2   Percentage unemployed by highest qualification   File: Qualif unemployed

	(i) No qualification	(ii) Degree or equivalent	(iii) GCE A Level or equivalent	(iv) GCSE grades A*−C or equivalent	(v) Other qualification	Total Unemployed (thousands)
				Percentages unemployed		
United Kingdom	8.9	2.7	3.8	5.5	6.9	1409
North East	11.9	not given	4.1	9.3	8.7	74
North West	8.7	not given	3	5.5	8.3	153
Yorkshire and the Humber	9	2.9	4.2	5.5	8.3	128
East Midlands	8		3.1	4.4	4.4	83
West Midlands	10.2	2.5	3.8	5.5	7.4	141
East	5.5	2	3	4.4	6.2	107
London	12	3.9	6.1	8.1	8.9	248
South East	7.1	2.7	2	3.8	4.8	153
South West	7.2	2.5	3.1	4.3	4.1	87

Source: Labour Force Survey, Office for National Statistics.

(a) On the same diagram, plot line graphs for the percentages unemployed for
(i) 'No qualification' (ii) 'Degree or equivalent' (iii) 'GCE A level or equivalent' (iv) 'GCSE grades A* − C level or equivalent' for all regions.

(b) Describe and compare the unemployment rates for the groups (i) to (iv) in (a).

(c) Plot 'other qualification' on the graph in (a). Compare the unemployment for this group with those in (a).

3. Table PE1.5 Q3 gives an extract from the distribution of marketable wealth series C produced by the Office for National Statistics:

Table PE1.5 Q3   Distribution of marketable personal wealth   File: Personal wealth extract

Year	Percentages of Population with less than				Total adult population
	£5000	£15 000	£50 000	£100 000	Thousands
1976	60	90	99	100	40 486
1978	52	83	98	99	40 824
1980	49	77	97	99	41 356
1982	46	71	96	99	41 937
1984	41	64	93	98	42 765
1986	40	61	90	97	43 322
1988	34	51	86	94	43 822
1990	35	49	78	94	44 284
1992	27	45	78	93	44 765
1994	26	43	78	92	45 007
1996	28	43	75	91	45 191
1998	27	39	68	89	45 654
2000	29	40	65	86	47 828

Source: Office for National Statistics. Crown copyright material is reproduced with the permission of the Controller of HMSO.

(a) Plot a line graph for the percentage of the population with less than £15 000 over the years 1976 to 2002. Describe the trend displayed in the graph.

(b) On the same diagram, plot the graphs for the percentage of the population with less than £5000 and less than £100 000 over the years 1976 to 2002. Give a verbal comparison of the trends between 1976 and 2002.

4. **See question 3.** For the years 1976 and 2000 (File: Personal wealth extract)

   (a) Plot a graph for the percentages of the population with less than £5000, £15 000, £50 000 and £100 000 of marketable wealth.

   (b) From the graph estimate the percentage (to the nearest 5 %) of the population with less than (i) £40 000 (ii) £75 000 of marketable wealth.

   (c) Calculate the numbers of adults with less than (i) £40 000 (ii) £75 000 of marketable wealth.

5. The percentage of the population that own 1 %, 2 %, 5 %, 10 %, 25 % and 50 % of the wealth is given in Table PE1.5 Q5:

File: Personal wealth all

Table PE1.5 Q5   Percentages of the population that own given percentages of the wealth

Year	Percentages of Wealth Owned						Coefficient Gini
	1 %	2 %	5 %	10 %	25 %	50 %	
1990	18	24	35	47	71	93	64
2002	24	31	45	57	75	94	71
2003[1]	21	28	40	53	72	93	67

[1]provisional.
Source: Office for National Statistics. Crown copyright material is reproduced with the permission of the Controller of HMSO.

On the same diagram, plot a Lorenz curve for 2002 and 2003. Using the curves explain the values given in Table PE1.5 Q5 for the Gini coefficients.

6. The average weekly earnings and hours worked by males and females in the basic metals industries in the EU are given in Table PE1.5 Q6:

File: Basic metal

Table PE1.5 Q6   Average weekly earnings and hours worked in Basic metal industries

Basic Metals (NACE 27)	Average Earnings Per Week (Euro)		Average Hours Worked Per Week (Number)	
	Male	Female	Male	Female
2002Q4	662.07	349.33	45.0	34.8
2003Q4	707.96	360.21	49.4	35.0
2004Q4	643.21	363.66	45.8	33.3
2005Q4	681.99	403.50	45.8	33.9

Source: Eurostat.

(a) Plot a graph comparing the trends in male and female for the (i) average hours worked per week (ii) average weekly earnings. Give a verbal description of the trends.

(b) Calculate the average hourly pay for males and females from 2002 to 2005 (i.e., average pay per week divided by average hours worked). Plot a graph comparing the trends. How does this graph compare to those in (a)?

7. The percentage of all household income that is shared by given percentages of the population is given in Table PE1.5 Q7:

Table PE1.5 Q7   Percentage of households on percentage of household income

Percentage of households	25 %	40 %	60 %	80 %	100 %
Percentage of household income less than . . .	5 %	15 %	30 %	60 %	100 %

(a) Plot a Lorenz graph for the data.
(b) From the graph estimate the Gini coefficient. Give a verbal description on the distribution of household incomes.

8. Table PE1.5 Q8 displays the calculations of cumulative percentages for a Lorenz graph for the number and cost of prescriptions in regional pharmacy,

File: Prescriptions

Table PE1.5 Q8   Calculation of percentage cumulative frequencies for a Lorenz curve

Cost of prescription	Number of prescriptions	Cumulative frequencies	Prescriptions % Cum. freq.	Total value of prescript.	Cumulative frequency	Values % Cum. freq.
0 < 20	436	436	43.6	4 530	??	4.4
20 < 50	350	786	??	16 750	??	??
50 < 100	105	891	??	7 480	28 760	??
100 < 500	55	??	94.6	8 040	??	35.9
500 < 1000	32	??	97.8	31 640	68 440	66.7
over 1000	22	1000	100.0	34 100	102 540	100.0

(a) Fill in the blank cells, marked ??, in Table PE1.5 Q8.
(b) Plot a Lorenz graph and explain what the curve represents. Describe the distribution of cost over the number of prescriptions.

9. The use of computers and/or internet access for the years 1998 to 2000 for Australia is given in Table PE1.5 Q9:

Table PE1.5 Q9   Household access to computers or the Internet (Australia)

Percentages	Households with access to a computer at home			Households with access to the Internet at home		
	Nov-98	Nov-99	Nov-00	Nov-98	Nov-99	Nov-00
Household income						
$0–$49 999	34	33	37	10	12	21
$50 000 or more	69	71	77	34	43	57
Households						
With children under 18 years	67	69	74	25	35	48
Without children under 18 years	36	39	46	15	20	32
Region						
Metropolitan areas	50	53	59	22	30	40
Other areas	43	44	52	13	17	32
Total	**47**	**50**	**56**	**19**	**25**	**37**

Proportions are of all households in each category.
Source: Use of the Internet by Householders, Australia, November 2000 (Cat. no. 8147.0).

(a) On the same diagram, plot line graphs to display the percentages of households that have access to a computer and households that have access to the internet for income groups. Describe the trends.
(b) Plot graphs to display the trends for households (i) with children under 18 years and those without (ii) in metropolitan and other areas. Describe these trends.

1.8 Misleading graphs

In the previous sections we have seen that graphs offer an excellent means to summarise and convey information at a glance. However graphs may give misleading information, usually because the reader does not read the titles, labels and units carefully. Three brief examples are given below:

1. Pictograms: Abandoned puppies and dogs present a serious problem in terms of cost, public health, cruelty to animals, etc. for local councils, particularly in the period after Christmas. One council ran publicity drives with specially designed posters (Figure 1.21) for children, in an attempt to reduce the numbers of abandoned puppies. On these posters the height of each puppy was proportional to the frequency or number abandoned. Between 2004 and 2006, the number abandoned dropped from 500 to 250 – yet, on the poster, the drop looks much greater. The reason for this misrepresentation is that all the dimensions of the puppy are halved– not just the height.

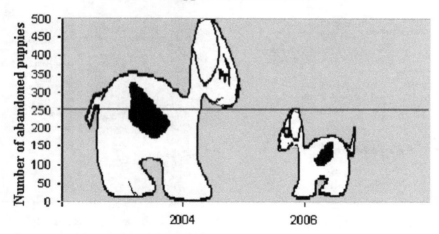

Figure 1.21   Pictogram of the number of abandoned puppies in 2004 and 2006.

2. Line graphs can give false impressions about rates of increase or decrease. For example, the graphs in Figure 1.22, give the impression that the increase in the value of imports is greater than exports between 2000 and 2006. Careful reading of the vertical axis will reveal that the value of imports has increased by £ 20m while exports have increases by £ 40m.

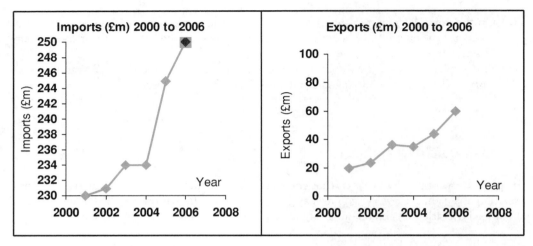

Figure 1.22   Imports and exports between 2000 and 2006.

3. Read units very carefully! In Figure 1.23a (birth rates for the Irish population from 1989 to 2004) it would appear at first glance that births outside marriage increased very rapidly between 1900 and 2000 and remain considerable higher than total births! Careful reading of the rates will reveal that the total birth rate is given in births per 1000 of the population, but the birth rate outside of marriage is a percentage of total births. Figure 1.23b gives both birth rates per 1000 of the population

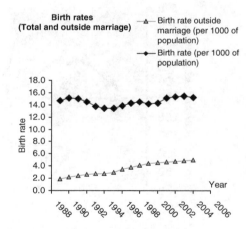

Figure 1.23a   Birth rates from 1989 to 2004;

Figure 1.23b   Birth rates from 1989 to 2004
Source: CSO.

Note: in any statistical report read and check the definition of any broad general terms used, such as 'housewife' 'wealthy', fertility', 'unemployment', 'single parent', 'divorces', etc.

Summary and overview of data collection, charts and graphs

The collection of representative, unbiased data is of fundamental importance to any study in statistics. Collecting representative data is not as simple as it might first appear. An outline of survey design and data collection methods is given in sections 1.1 and 1.2, followed by directions to some reliable sources of data in section 1.3. **The levels of data measurement** may be summarised as follows:

Ratio (data has a natural zero, measurements can be compared)
Interval (data has an arbitrary zero, measurements can be compared)
Ordinal (ordered or ranked data, equality, greater/less than)
Nominal (group classifications of data)

The level of data measurement determines the type of chart/graph and statistical analysis that may be carried out on the data. When the **raw data** is collected it must be sorted in order to read information from it. Sorting data into stem-and-leaf plots and frequency distribution tables gives some initial information. Information is conveyed visually through charts and graphs as summarised below.

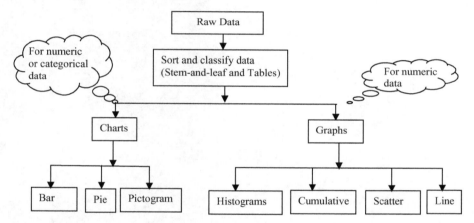

Figure 1.24   Summary of charts and graphs for categorical and numeric data.

While Microsoft **Excel** and other software produce accurate and colourful graphs quickly and easily, the reader should study the methods involved and plot simple graphs manually in order to enhance understanding and become skilled at 'reading' information accurately from graphs. When presenting and reading graphs the following are essential: titles, labels and units for horizontal and vertical axes must be stated; terminology must be explained; sources of data, when and how the data was collected along with any unusual circumstances must be given.

Descriptive Statistics

2

Chapter Objectives

Having carefully studied this chapter and completed the exercises you should be able to do the following

- Describe the measures of central tendency and dispersion
- Use formulae to calculate the summary statistics and measures of dispersion from raw data and grouped data
- Explain why the mean differs from the median when data is skewed
- Calculate the coefficients of skewness and coefficient of variation.
- Calculate and draw box plots. Describe the distribution of data from the box plots
- Calculate weighted averages
- Decide which statistics best describe a set of data
- Recognise when data is skewed from graphs and from calculated statistics
- Use calculator and Excel to calculate the summary statistics and measures of dispersion
- Give verbal description of a set of data in terms of summary statistics and measures of dispersion

2.1 Summary statistics for raw data: Mean, quartiles and mode

In Chapter 1, tables and charts/graphs provided visual information about data characteristics. However, it is necessary to be able to describe data numerically, for example by a value that is typical of the bulk of the data. Such a figure, calculated from the data is called **a summary statistic**. This is usually a mean (or average value), a median or a mode. In the following sections the methods and formulae for calculating means, medians, modes and other statistics will be explained.

2.1.1 The mean or average value for a set of raw data

The mean or average value per item (or datum) is $\dfrac{\text{The total value of all the data}}{\text{The number of data}}$

For example, the average age for a group of children whose ages are given in the list (referred to as Data A) below is calculated as follows:

Data A: ages $= 7, 8, 3, 5, 5, 6, 7, 4, 9$ years

The average age $= \dfrac{Total\ years\ for\ the\ entire\ group}{Number\ of\ children\ in\ the\ group}$

For data A, the average age $= \dfrac{7+8+3+5+5+6+7+4+9}{9} = \dfrac{54}{9} = 6.00$ years

Formula for the mean (or average)

The first step towards deriving a 'formula' for calculating the average is to describe, in words, how the calculation was carried out: it is then easy to generalise and finally to write this general description as a 'formula' that can be applied to any set of data. Formulae are much easier to remember and apply correctly if described verbally.

In general terms, the mean age for the group given in Data A was calculated by adding all the ages, then dividing the total by the number of individuals in the group: Let x_i represents an individual age where the subscript (i) refers to the position of that age in the list.

For data A, $x_1 = 7$, $x_2 = 8$, $x_3 = 3$, $x_4 = 5$, $x_5 = 5$, $x_6 = 6$, $x_7 = 7$, $x_8 = 4$, $x_9 = 9$

Then the mean or average age $= \dfrac{x_1 + x_2 + x_3 + x_4 + \cdots\cdots\cdots\cdots + x_9}{9} = \dfrac{54}{9} = 6$

The summation or addition of the 9 numbers may be represented by the symbol $\sum\limits_{i=1}^{i=n}$, as follows

$$\frac{x_1 + x_2 + x_3 + x_4 + \cdots\cdots\cdots\cdots + x_9}{9} = \frac{\sum\limits_{i=1}^{i=9} x_i}{9}$$

Σ: The uppercase Greek letter Sigma indicates addition (summation)

$i = 1$ at the bottom of Σ and $i = 9$ at the top of Σ means 'add each successive number in the list from the first (x_1) to the ninth (x_9) inclusive'

If the sum is written simply as $\sum x$ it is assumed that all the numbers, from the first to the last, are included in the summation.

Hence, in general, the calculation of the mean value for a list of N values is given by the formula:

$$\text{Mean or average value,} \quad \mu = \frac{x_1 + x_2 + x_3 + x_4 + \cdots\cdots\cdots\cdots + x_{n-1} + x_n}{N} = \frac{\sum\limits_{i=1}^{i=N} x_i}{N}$$

> μ is called 'mu', the symbol for the population mean

> N is the total number of items in the population

Note: you will come across two symbols for the mean or average (even though the calculations are exactly the same for each):

$\bar{x}$, 'x-bar' is used when the mean value has been calculated from sample data

μ, pronounced 'mu' (the lowercase Greek letter) indicates that the mean value has been calculated from population data.

In this chapter the symbol μ is used for the mean. In later chapters on statistical inference it will be necessary to use both symbols as appropriate to indicate whether the mean was calculated from sample data or population data.

Formula: mean value for raw data

$$\mu = \frac{x_1 + x_2 + x_3 + x_4 + \cdots\cdots\cdots\cdots + x_{N-1} + x_N}{N} = \frac{\sum\limits_{i=1}^{i=N} x_i}{N} \qquad (2.1)$$

2.1.2 The median, quartiles and percentiles

When a set of data is arranged in ascending order, the values that divide the data into quarters are called quartiles:

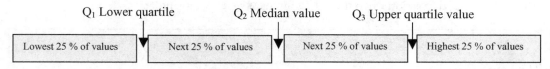

Q_1 Lower quartile Q_2 Median value Q_3 Upper quartile value

| Lowest 25 % of values | Next 25 % of values | Next 25 % of values | Highest 25 % of values |

Figure 2.1   Quartiles for a set of data that is arrayed (ordered) in ascending order.

The median is **the value of the middle item** when the data is arranged in either ascending or descending order (usually ascending)

> Ordered data is sometimes referred to as arrayed data

Median for an odd number of data

For example, find the median for Data A

First array the data in ascending order, then divide the ordered data in half. The median is the value of the item that is half way through the ordered data.

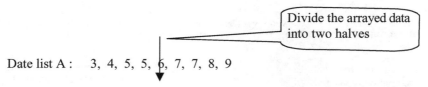

Divide the arrayed data into two halves

Date list A :   3, 4, 5, 5, 6, 7, 7, 8, 9

In this example, the middle item has the value six. Hence the median age for this group of children is six years.

The median is also called the **second quartile** (Q_2).

Verbally:
50 % of the children are aged six or younger and 50 % are aged six or older.

Median for an even number of data

If we have an even number of items the median is the average of the middle two values.

e.g.   Suppose we have numbers   3, 5, 5, 7, 8, 50

Divide the arrayed data into 2 halves

$$Q_2 = \text{Median} = \frac{5+7}{2} = 6$$

Definition

The Median or second quartile, Q_2, is the **value of the middle item** in an ordered set of data.
Q_2 is the value of the item that is at or nearest the $0.5(N+1)$th position in the ordered data   (2.2)

The **Lower Quartile** (Q_1) is the value of the item that is *one-quarter (25 %) of the way through the ordered data*. If N is the number of items in the raw data then Q_1 is the value of the item at or nearest to the $0.25(N+1)$th position when the data is arranged in ascending order

The **Upper Quartile** (Q_3) is the value of the item that is *three-quarters (75 %) of the way through the ordered data*. If N is the number of items in the raw data then Q_3 is the value of the item at or nearest to the $0.75(N+1)$th position when the data is arranged in ascending order

---

Definitions:

The Lower Quartile
Q_1 is the item whose position in the ordered list is at or nearest to the $0.25(N+1)$th   (2.3)

The Upper Quartile
Q_3 is the item whose position in the ordered list is at or nearest to the $0.75(N+1)$th   (2.4)

For example, for the ordered Data A, the upper and lower quartiles are shown graphically as follows:

$$Q_1 \quad Q_2 \quad Q_3$$
Date list A: 3, 4, 5, 5, 6, 7, 7, 8, 9

$$\downarrow \qquad \downarrow \qquad \downarrow$$

$$Q_1 = \frac{4+5}{2} = 4.5 \qquad : Q_2 = 6 : \qquad Q_3 = \frac{7+8}{2} = 7.5$$

Note: when the data set is small, the quartiles are readily calculated from the ordered data by inspection. However, use formulae (2.2), (2.3) or (2.4) when the data set is large.

Percentiles

Percentiles divide the arrayed data in hundredths.

Percentiles are used to describe distributions of populations by age, income, etc.

For example, Table 2.1, published by the Office for National Statistics (ONS) gives income before and after tax at selected percentile points (for taxpayers only) in 2003/04. The table reads as follows: 1 % of taxpayers have an income of £ 4820 or less before tax; 5 % of taxpayers have an income of £ 5850 or less before tax ... 99 % of taxpayers have an income of £ 111 000 or less before tax

Table 2.1   Income before and after tax 2003–04   File: Tax_income 03_04

Percentile point	1	5	10	25	50	75	90	95	99
Income before tax 2003–04	4820	5850	7000	10 100	16 000	25 100	37 100	50 600	111 000
Income after tax 2003–04	4800	5730	6790	9 290	14 000	21 100	30 700	39 000	77 000

Source: Survey of Personal Incomes 2003–04 Last updated April 2006.

In fact the median is the 50th percentile: 50 % of the taxpayers have an income of £ 14 000 or less after tax.

2.1.3   The mode

The **mode** is simply the value that occurs most often.

In data list A: 3, 4, 5, 5, 6, 7, 7, 8, 9 there is no single mode. The number 5 occurs twice but so does the number 7. So we could describe this data as being bi-modal, that is having two modes

Note: For numeric data the following problems arise

(i) There may be no mode or there may be several.
(ii) The mode may be a value near the beginning, middle or end of the data! In other words it can be anywhere and therefore may not representative.

However, in some applications such, as surveys that use nominal or ordinal data, it may be useful to make statements such as 'most people who read the Financial Times work in banks!'

Worked Example 2.1: Calculation of mean, median, quartiles and mode from raw data

Calculate the mean, median, quartiles and mode for the data given in (a) and (b) below. Give verbal explanations of the answers.

(a) Data B is a list of the hourly rates of pay for the 9 members of staff employed in a newsagent.
    Data B: 9, 7, 9, 50, 7, 7, 5, 8, 6
(b) A company has staff of 49 persons. The hours worked by each member of staff in a given week are given in Table 1.15 (reproduced below for convenience):

File: 49 Hours Worked

Table 1.15   Hours worked in a given week by 49 staff

20.0	37.3	54.2	25.3	59.6	24.5	29.7
18.0	38.8	42.1	39.5	56.8	16.9	28.5
45.5	42.0	39.5	42.6	40.0	44.2	40.1
44.0	56.4	30.2	20.0	22.7	37.8	23.4
26.0	20.2	36.1	18.3	19.7	36.8	26.5
24.0	23.4	15.4	20.0	38.9	42.1	24.1
41.0	18.5	21.3	22.6	37.2	42.9	17.9

Solution

(a) **The mean**: since there are 9 items to be averaged, $N = 9$

$$\mu = \frac{\sum_{i=1}^{N} x_i}{N} = \frac{x_1 + x_2 + x_3 + x_4 + x_5 + x_6 + x_7 + x_8 + x_9}{5}$$

The average hourly rate $= \dfrac{9 + 7 + 9 + 50 + 7 + 7 + 5 + 8 + 6}{9}$

$$= \frac{108}{9}$$

$$= 12.00$$

The average hourly rate of pay is 12.

Calculate the median and quartiles from raw data

(a) Data set B is small; hence quartiles are calculated immediately from the ordered data
    The dividing line falls on value 7: this is the median value or median hourly rate of pay
    Graphically, when data B is ordered we have:

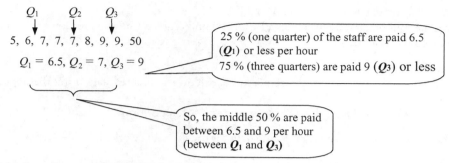

$Q_1 \quad Q_2 \quad Q_3$

5, 6, 7, 7, 7, 8, 9, 9, 50

$Q_1 = 6.5, Q_2 = 7, Q_3 = 9$

25 % (one quarter) of the staff are paid 6.5 (Q_1) or less per hour
75 % (three quarters) are paid 9 (Q_3) or less

So, the middle 50 % are paid between 6.5 and 9 per hour (between Q_1 and Q_3)

(b) **Calculate the mean from raw data**
Calculate the total number of hours worked by the entire staff in the given week. Then divide this total by the number of staff, 49, to find the average number of hours worked.

$$\mu = \frac{20.0 + 18.0 + 45.5 + \cdots + 24.1 + 17.9}{49} = \frac{1592.5}{49} = 32.5 \text{ hours}$$

The average number of hours worked per person in the week is 32.5

(c) **Calculate the median (Q_2) and quartiles (Q_1 and Q_3) from raw data by inspection**
The raw data for the 49-weekly hours worked is ordered in ascending order in Table 2.1:

Table 2.1   Hours worked by the 49 employees in ascending order

15.4	20.0	23.4	26.5	37.3	40.1	44.0
16.9	20.0	23.4	28.5	37.8	41.0	44.2
17.9	20.0	24.0	29.7	38.8	42.0	45.5
18.0	20.2	24.1	30.2	38.9	42.1	54.2
18.3	21.3	24.5	36.1	39.5	42.1	56.4
18.5	22.6	25.3	36.8	39.5	42.6	56.8
19.7	22.7	26.0	37.2	40.0	42.9	59.6

$Q_2 = 30.2$

Divide the arrayed data into two halves: the median is the value of the middle datum:

$\longleftarrow$ 24 items $\longrightarrow$ 25th item $\longleftarrow$ 24 items $\longrightarrow$
(middle)

Median: since there are an odd number of data, the median is the value of the 25th item in the ordered data: $Q_2 = 30.2$. See Figure 2.1b.
The lower quartile is approximately the median of the lower 24 items; so its value is the average of the 12th and 13th item: $Q_1 = \dfrac{21.3 + 22.6}{2} = 21.45$ hrs. See Figure 2.1b.
The upper quartile is approximately the median of the upper 24 items; so its value is the average of the 36th and 37th item: $Q_3 = \dfrac{41.0 + 42.0}{2} = 41.5$ hours. See Figure 2.1b.

Calculate the quartiles by formulae (2.2), (2.3) and (2.3)

For a large amount of data, use formulae (2.2), (2.3) and (2.4) to find the position of the quartiles in the arrayed list as follows:

Step 1: Array the data

Step 2: Calculate the **position** of the required quartile in the ordered data using the appropriate formula.

Step 3: Calculate the **value** of the item at or nearest to that position.

Apply the three steps in this method to the data for the 49 hours worked, Table 1.15

Step 1: The data is arrayed in Table 2.1 and Table 2.2

Step 2: Find the **position** of the quartile in the ordered data

The median

Step 2: Calculate the position of the second quartile by formula (2.2). Q_2 is the value of the item that is at or nearest the $0.5(n+1) = 0.5(49+1)$th $= 25$th position in the ordered list

Step 3: Calculate the value of the Median. The median number of hours worked is 30.2, (see Table 2.1) – this means that 50 % of employees work 30.2 hours or less per week; 50 % work 30.2 hours per week or more

The lower quartile, Q_1

Step 2: Calculate the position of the lower quartile by formula (2.3): The lower quartile is the value of the item that is closest to the $0.25(n+1) = 0.2(49+1) = 12.5$th position in the ordered list

Step 3: Calculate the value of Q_1, the average of the 12th and 13th item in Table 2.2.

$$Q_1 = \frac{21.3 + 22.6}{2} = 21.45$$

The upper quartile, Q_3

Step 2: Calculate the position of the upper quartile by formula (2.4): the upper quartile is the value of the item that is closest to the $0.75(n+1) = 0.75(49+1) = 37.5$th position in the ordered list

Table 2.2   Quartiles from ordered data for 49 employees

Mode = 20 15.4	20.0	23.4	26.5	37.3	40.1	44.0
16.9	20.0	23.4	28.5	37.8	41.0	44.2
17.9	20.0	24.0	29.7	38.8	42.0	45.5
18.0	20.2	24.1	30.2	38.9	42.1	54.2
18.3	21.3	24.5	36.1	39.5	42.1	56.4
18.5	22.6	25.3	36.8	39.5	42.6	56.8
19.7	22.7	26.0	37.2	40.0	42.9	59.6

$$Q_1 = \frac{21.3 + 22.6}{2}$$
$$Q_1 = 21.45$$

$$Q_3 = \frac{41.0 + 42.0}{2}$$
$$Q_3 = 41.5$$

So, the middle 50 % of staff work between 21.45 and 41.5 hours

Step 3: Calculate the value of Q₃, the average of the 37th and 38th item in Table 2.2,

$$Q_3 = \frac{41.0 + 42.0}{2} = 41.5$$

Calculate the mode from raw data

(a) In data B, the value that occurs most often is 7. It occurs three times, so the mode is 7.
(b) From the ordered data for Table 2.2, we see that the mode is 20 – it occurs 3 times.

2.1.4 Why not use the mean value to represent any set of data?

A summary statistic is representative if it is fairly typical of most of the data.
    Consider the following two examples:

Data list A: 3, 4, 5, 5, 6, 7, 7, 8, 9:

Mean: $\mu = 6$ Median, $Q_2 = 6$ Bimodal, 5 and 7

All summary statistics are representative

Data list B: 5, 6, 7, 7, 7, 8, 9, 9, 50

The mean is **NOT** representative of the bulk of the data. It is larger than most of the values and much smaller than 50

Mean: $\mu = 12$: Median $= 7$: Mode $= 7$

In Data list B the mean is not representative. All of the data, with the exception of one datum have values between 5 and 9, but the mean value is 12. The mean value is unrepresentative because it is calculated from all the data, including extreme values.
    Careful examination of the calculation of the mean for Data B reveals that the extreme value, 50, contributed almost as much as all the other values combined:

$$\frac{(9 + 7 + 9 + 7 + 7 + 5 + 8 + 6) + 50}{9} = \frac{(58) + 50}{9} = \frac{108}{9} = 12.00$$

The value 50 is called an '**outlier**'; it is very different from the rest of the data. Data in which the majority of values are close to each other but a small number of values that are extremely high or extremely low (outliers) are said to be **skewed**. Graphically skewed data has a long tail extending towards the outlier. When data is symmetrical distributed there is reasonable symmetry about a central value; see Figure 2.2 below.
    If data is skewed the median is more representative of the bulk of the data than the mean as it is not affected by extreme values.
    Note: if the outlier(s) are low value(s), the value of the mean will be smaller than the median, and the data would be described as 'skewed to the left'.

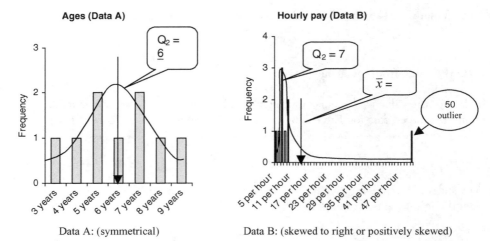

Figure 2.2   Symmetrically distributed data (A) and skewed data (B).

2.2   Summary statistics for grouped data: Mean, quartiles and mode

Most published statistics are presented in the form of Tables. For example, the age distribution for those employed in farming, fisheries and forestry in the years 1991 and 1996 is given in Table 2.3:

Table 2.3   Numbers employed in farming, fishing and forestry

Age group	1991	1996
15–19	4,585	2,826
20–24	11,872	9,319
25–34	27,171	24,492
35–44	31,299	28,210
45–54	31,626	30,902
55–64	33,477	25,846
65 and over	23,519	19,030
Totals	163,549	140,625

Source: CSO.

In this case it would be impractical to list all the raw data (304,174 data) – it would take several pages and give little or no information. It would make even less sense to attempt to calculate averages or array the data in order to calculate quartiles. Therefore it is necessary to develop formulae for the calculation of means and quartiles from grouped data; these formulae are similar to those used for raw data, but with adjustments for 'grouping'.

{ A
  s
  s
  u
  m
  e }
Grouping assumption: It is assumed that the values within each interval vary uniformly between the lowest and highest values for the interval. Hence the average value of the data in any interval is the mid-interval value: the mid-interval value is used to represent the group numerically.

In this section the calculation of the summary statistics – mean, quartiles and mode for grouped data will be demonstrated in Worked Examples 2.2 to 2.6.

2.2.1 Calculation of mean for grouped data

Preliminary:

The frequency distribution table for the hours worked by the 49 staff was given in Table SK1_5.2. When dealing with grouped data it is important to describe the contents of the table 'in words', as illustrated below.

Table SK1_5.2   Frequency distribution table for hours worked by 49 employees

Hours worked	15 < 20	20 < 25	25 < 30	30 < 35	35 < 40	40 < 45	45 < 50	50 < 55	55 < 60
Frequency	7	12	5	1	9	10	1	1	3

7 people worked from 15 up to but not including 20 hours

10 people worked from 40 up to but not including 45 hours

When data are summarised in a frequency distribution table a certain amount of detail is lost. For example, in Table SK1_5.2 the first group reads 'seven people worked from15 up to but not including 20 hours' If the raw data were not available then it would be impossible to say whether all 7 worked exactly 15 hours or 19.99 hours or various hours between these. Since the 'grouping assumption' may not be satisfied, statistics calculated from grouped data are usually approximate.

The mid interval value is the average value of the end points for the interval:

$$\text{Mid interval value} = \frac{15 + 20}{2} = \frac{35}{2} = 17.5$$

Steps in calculating the mean for grouped data

$$\text{The mean value was defined earlier as } \frac{\text{The overall total value of all the data}}{\text{The number of data (items)}}$$

In **Table SK1_5.2**, the number of hours worked by each person in the interval '15 < 20' is assumed to be the mid-interval value, 17.5. The total hours worked by the seven people in this interval is $17.5 \times 7 = 122.5$. Similarly the total hours worked by those in each of the remaining intervals in the table is the mid-interval value multiplied by the number of individuals (frequency) in that interval. The overall total for all 49 employees is calculated by summing the interval totals. Finally, calculate the mean value by dividing this overall total value by the number of employees.

The method is summarised in the following steps:

Step 1: Calculate the mid-interval value (x_i) for each interval

Step 2: Calculate the total for each interval ($f_i \times x_i$)

Step 3: Calculate the overall total by summing the interval totals $\sum (f_i \times x_i)$

Step 4: Divide the overall total by the number of items or individuals, $\sum_{all} f_i$

These steps are illustrated for the data in Table SK1_6.2 in Worked Example 2.2:

Worked Example 2.2: Calculation of the mean for grouped data

(a) Calculate the mean for the grouped data given in Table SK1_6.2 for the hours worked by the 49 employees
(b) Deduce the formulae for calculating the mean for grouped data by steps 1 to 4 above
(c) Compare the values of mean calculated from raw and grouped data. Explain why these values are different

Solution

(a) **Step 1:** Calculate the mid-interval value (x_i) for each interval
Refer to Table 2.4 for the following.

Step 2: Calculate the total for each interval ($f_i \times x_i$) given in column four, Table 2.4.

Table 2.4 Calculation of the mean from the frequency distribution table for the 49 employees

Intervals (hours)	Mid-value x_i	Frequency f_i	Sub-group totals = freq × mid-value	$f_i \times x_i$
15 < 20	17.5	7	7 × 17.5 =	122.5
20 < 25	22.5	12	12 × 22.5 =	270
25 < 30	27.5	5	5 × 17.5 =	137.5
30 < 35	32.5	1	1 × 32.5 =	32.5
35 < 40	37.5	9	9 × 37.5 =	337.5
40 < 45	42.5	10	10 × 42.5 =	425
45 < 50	47.5	1	1 × 47.5 =	47.5
50 < 55	52.5	1	1 × 52.5 =	52.5
55 < 60	57.5	3	3 × 57.5 =	172.5
Totals		$\sum_{all} f_i = 49$	$\sum f_i x_i =$	1597.5

f_i is the number of people (frequency) in interval i

x_i is the mid-interval value for interval i.

$f x_i$ total for interval 1.

This is the overall total calculated by summing the interval totals:

Step 3: Calculate the overall total by summing the interval totals ($f_i \times x_i$)

Step 4: Divide the overall total by the number of items or individuals, $\sum\limits_{all} f_i$

Hence, mean $= \dfrac{\text{Total hours worked by the entire staff}}{\text{Total no. of people on the staff}} = \dfrac{f_1 x_1 + f_2 x_2 + f_3 x_3 + \cdots\cdots\cdots + f_9 x_9}{f_1 + f_2 + f_3 + \cdots\cdots\cdots + f_9}$

$= \dfrac{1597.5}{49} = 32.602$

> **Note:** 32.500 was the exact average calculated from the raw data in Worked Example 2.1

(b) Formula 2.5, generalises the calculation of the mean for the nine intervals above to any set of data divided into L intervals.

Formula: the mean for grouped data

$$\mu = \frac{f_1 x_1 + f_2 x_2 + f_3 x_3 + \cdots\cdots\cdots + f_L x_L}{f_1 + f_2 + f_3 + \cdots\cdots\cdots + f_L} = \frac{\sum f_i x_i}{\sum f_i} = \frac{\sum f_i x_i}{N} \qquad (2.5)$$

where $L =$ the number of intervals and $N = \sum f_i$

(c) Why are the answers different when the mean calculated from raw data and grouped data?
The exact mean value was 32.500 calculated from the raw data in Worked Example 2.1. The value calculated from grouped data was 32.605. The difference arises because we assume that data within each interval is distributed uniformly throughout the interval. This is not always the case. When data is not uniformly distributed from the beginning to the end of the interval, the mid-interval value may be different from the average hours worked in group.

Hence in almost all cases, the mean, as well as any other statistics calculated from grouped data are **approximate** and will differ slightly from those calculated from raw data.

2.2.2 Quartiles, percentiles for grouped data

Quartiles may be estimated from grouped data by using

1. the frequency distribution table.
2. the Ogive (cumulative frequency graph).

Estimate the median and quartiles from the frequency distribution table

When data is sorted into a frequency table the data is ordered from the lowest to the highest values in blocks or intervals. Approximate estimates of the quartiles are made by identifying the intervals containing the items that are 25 %, 50 % and 75 % of the way through the ordered data. More accurate estimates are made by calculating how far into an interval each of these points are positioned and the value of the data at that point, using formulae (2.6), (2.7) and (2.8) given in Appendix B, (assuming the uniform distribution of data within intervals). The use of the Ogive to estimate quartiles is the graphical equivalent of this method.

Worked Example 2.3: Estimate the median and quartiles from the frequency table

(a) Identify the intervals that contain the lower quartile, the middle quartile (or median) and the upper quartile for the frequency distribution in Table SK1_5.2 (hours worked by the 49 staff).

(b) Estimate each quartile approximately

Solution

(a) The calculations based on the data in Table SK1_5.2 are shown in Table 2.5.

The lower quartile is the value of the item that is positioned 25 % of the way through the ordered data, i.e. at or nearest to $0.25(49 + 1)$th $= 12.5$th item in the ordered data. From the cumulative frequency column in Table 2.5 the 12.5th item is in the interval $(20 < 25)$. Since there are seven items in the first interval a further 5.5 items are required to reach the lower quartile value: this would be just less than half of the 12 items in this group, hence the value of Q_1 is 22.5 approximately See Table 2.5.

The middle quartile or median is the value positioned half way through the ordered data, i.e. at or nearest the item in the $0.5(49 + 1)$th $= 25$th position. The 25th item is in interval $(30 < 35)$. Since there is only one item in this group, use the mid-interval value as an estimate of the median, so 32.5, as indicated in Table 2.5.

The upper quartile is the value positioned 75 % through the ordered data, that is at or nearest to the $0.75(49 + 1)$th $= 37.5$th item. This item is in the interval $(40 < 45)$. To reach the 37.5th item requires the first 3.5 out of the 10 items in this group – approximately 1/3 of the way into the interval $(40 < 45)$. Hence the value of the upper quartile (Q_3) is 42 approximately (see Table 2.5).

Table 2.5   Identifying the intervals containing the median and quartiles

Intervals (hours)	Frequency: f_i	Less than	Cumulative frequency	
15 < 20	7	< 20	7	Q_1 the 12.5th item is in this interval: The value of Q_1 is 22.5 approximately
20 < 25	12	< 25	19	
25 < 30	5	< 30	24	Q_2 (median), the 25th item is in this interval: The value of Q_2 is 32.5 approximately
30 < 35	1	< 35	25	
35 < 40	9	< 40	34	
40 < 45	10	< 45	44	
45 < 50	1	< 50	45	Q_3 the 37.5th item is in this interval; the value of Q_3 is 42 approximately
50 < 55	1	< 55	46	
55 < 60	3	< 60	49	

Compare the results for the raw data calculated in Worked Example 2.1. $Q_1 = 21.45$; $Q_2 = 30.2$, $Q_3 = 41.5$.

Worked Example 2.4: Estimate the median and quartiles graphically from the Ogive

Estimate the median and quartiles graphically from the ogive for the data in Table 2.5 for the 49 weekly hours worked.

Solution

The Ogive was calculated in Skill Development Exercise SK1_5 and is reproduced below in Figure Sk1_5.2.

Draw horizontal lines through the 25 %, 50 %, 75 % cumulative frequency points from the vertical axis on the right-hand side of the Ogive to cut the Ogive at points (Q_1, 25 %), (Q_2, 50 %) and (Q_3, 75 %). To read off the x-coordinates, i.e. the values of Q_1, Q_2 and Q_3 drop perpendicular lines from these points onto the horizontal axis.

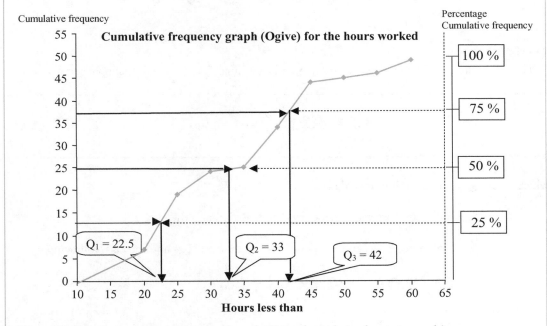

Figure SK1_5.2 Estimates of the quartiles from the Ogive (cumulative frequency graph).

Note: the same results are obtained by drawing horizontal lines through the points $y = \frac{1}{4} \times 50$, $y = \frac{1}{2} \times 50$, $y = \frac{3}{4} \times 50$ on the cumulative frequency axis on the left hand side of Figure SK1_5.2, then reading the values of the quartiles from the horizontal axis.

2.2.3 Mode for grouped data

The mode for grouped data is estimated from the frequency table (for hours worked by 49 employees) or estimated graphically from the histogram, as illustrated in Figure 2.3. (The histogram was plotted in Skill Development Exercise, SK1_5.)

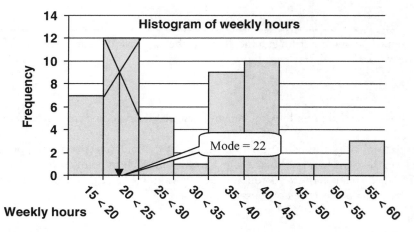

Figure 2.3   Mode estimated from the histogram.

The exact value of the mode may be calculated by formula (2.9) (derived by simple geometry!):

$$\text{Mode} = \text{L}_{\text{mode}} + \left(\frac{\ell_i}{\ell_i + \ell_2} \right) (w) \qquad (2.9)$$

where: L_{mode} = lower limit of modal class.

ℓ_1 = difference between frequency of modal class and the class below it.

ℓ_2 = difference between the frequency of the modal class and the class following it.

W = width of the interval of the modal class.

Worked Example 2.5: Calculate the mode from grouped data

Determine the mode for the data in Table 2.4 (frequency distribution table for the hours worked by 49 employees):

(a) Graphically from the histogram.
(b) By calculation.

Solution

(a) Select the group with the highest frequency; here it is the 20–25 hours interval. Draw lines from the corner points at the top of the modal group diagonally across to the top of the groups on either side as shown in Figure 1.17. The mode is 22.

(b) By calculation, using equation (1.7), for the 90-incomes data,

$$L_{mode} = 20, \ell_1 = 12 - 7 = 5, \ell_2 = 12 - 2 = 5, w = 5$$

$$Mode = 20 + \left(\frac{5}{5+7}\right)(5) = 22.08$$

The mode from the raw data, was 20.5.

Skill Development Exercise SK2_1: Mean and quartiles grouped data

(a) Calculate the summary statistics – mean, median, quartiles, mode- for the ages of the swimming team from the following table:

Table 1.9   Ages of swimmers available for competitions   File: Swimmers

Ages of swimmers									
18.8	18.7	20.2	17.4	29.6	23.7	19.6	18.4	20.6	25.5
17.9	24.2	20.9	19.2	23.1	21.5	21.9	20.7	22.3	21.7
20.5	21.6	20.4	20.2	19.5	19.9	19.9	20.8	17.2	19.1
18.1	19.2	18.8	18.8	21.2	24.3	22.3	22	18.8	21.7
18.1	20.9	21.1	17.9	18.6	18.8	20.1	20.9	20.1	22.4

(i) the raw data.
(ii) the grouped data given in the frequency table, Table SK1_2.2.

Answers

(a) From raw data: mean = 20.582; median, $Q_2 = 20.30$; $Q_1 = 18.80$; $Q_3 = 21.68$; mode = 18.8.
(b) Grouped data: mean = 20.48; median, $Q_2 = 20.16$; $Q_1 = 18.79$; $Q_3 = 21.82$; mode = 19.8.

2.2.4   Weighted averages

Recall equation (2.1) where the average value for a set of data was described as

$$\frac{\text{The total value of all the data}}{\text{The number of data}} = \frac{\sum x_i}{N}.$$

Such an average is referred to as a simple 'unweighted' average because the 'total value of all the data' was a simple sum of the individual values, such as the age in years of each child in data A.

In other applications the average value for a set of data is a weighted average. The following example demonstrates the difference between a simple average and a weighted average.

Consider five brands of wine W, X, Y, A-Reserve, and A-Vintage where the prices per bottle are given as: Wine W (1.20); Wine X (1.10); Wine Y (5.70); Wine A-Reserve (35.00); Wine A-Vintage (150.00). Three different 'crates of wine' are made up as follows:

Crate 1: one bottle of each wine

Crate 2: eight bottles of each wine

Crate 3: 123, 62, 32, 2 and 1 bottles of wines W, X, Y, A Reserve and A Vintage respectively.

The average price per bottle in each of the three crates is calculated as follows:

Crate 1: the crate consists of only one bottle of each wine then the overall average price per bottle is the simple average of the individual prices:

$$\frac{1.20 + 1.10 + 5.70 + 35.00 + 150.00}{5} = 38.6$$

Crate 2: if the crate consists of the same number of each wine then the overall average price per bottle is again 38.6.

In this example there are eight bottles of each wine, the average price is

$$\frac{8 \times 1.20 + 8 \times 1.10 + 8 \times 5.70 + 8 \times 35.00 + 8 \times 150.00}{8 \times 5}$$

$$= \frac{\cancel{8} \times (1.20 + 1.10 + 5.70 + 35.00 + 150.00)}{\cancel{8} \times 5} = 38.6$$

Crate 3: since the number of bottles of brands W, X, Y, A-Reserve and A-Vintage is 123, 62, 32, 2 and 1 respectively the average price per bottle is calculated as

$$\frac{123 \times 1.20 + 62 \times 1.10 + 32 \times 5.70 + 2 \times 35.00 + 1 \times 150.00}{200} = 2.81$$

Note: this is simply the total price for all the wines divided by the number of bottles!

In this calculation, the price of each wine is 'weighted' by the number of bottles of that wine in the crate; hence the overall average price per bottle is described as the **weighted average.** The definition of a wieghted average is given in equation (2.10) as follows:

$$\mu = \frac{w_1 x_1 + w_2 x_2 + w_3 x_3 + \cdots\cdots w_N x_N}{w_1 + w_2 + w_3 + \cdots\cdots w_N} = \frac{\sum w_i x_i}{\sum w_i} \qquad (2.10)$$

w_i are called the weights: the weights reflect the relative importance of each x_i.

Note: the calculation of the average price per bottle in crate 2 is also a 'weighted average' where all the weights are equal.

Worked Example 2.6: Calculation of weighted averages

In a statistic examination the overall marks calculated as follows: 50 % for the written paper; 30 % are for the practical examination and 20 % for an assignment.

Calculate the overall mark for a student who is awarded the following marks for each component: 31 % (paper); 84 % (practical) and 32 % (assignment).

Solution

From (2.19) the weighted average is $\mu = \dfrac{w_1x_1 + w_2x_2 + w_3x_3 + \cdots\cdots w_Nx_N}{w_1 + w_2 + w_3 + \cdots\cdots w_N} = \dfrac{\sum w_i x_i}{\sum w_i}$

The break down of the overall marks gives the weights. Hence

	Weights (w_i)	Student's marks (%) (x_i)	$w_1 \times x_i =$	
Paper	50	31	$50 \times 31 =$	1550
Practical	30	84	$30 \times 84 =$	2520
Assignment	20	32	$20 \times 32 =$	640
Totals	100			4710

$\sum w_i$

$\sum w_i x_i$

$\mu = \dfrac{\sum w_i x_i}{\sum w_i} = \dfrac{4710}{100} = 47.10$

Weighed averages are used are used extensively in the calculation of economic indicators. A well known economic indicator is the consumer price index, CPI. This is a weighted average of prices of a fixed basket of consumer goods and services. The weight used for each good/service is proportional to the average amount of that good/service purchased by all households in the country as determined by the Household Budget Survey.

For further information look up government websites such as www.CSO.ie; www.statistics.gov.uk; www.eurostat.org

Recall: the mean for grouped in equation (2.5):

$$\mu = \frac{f_1x_1 + f_2x_2 + f_3x_3 + \cdots\cdots\cdots + f_Nx_N}{f_1 + f_2 + f_3 + \cdots\cdots\cdots + f_N} = \frac{\sum f_i x_i}{\sum f_i}$$

This is in fact, a weighted mean where frequencies, f_i are the weights: This makes sense as intervals with higher frequencies make a greater contribution to the value of the mean, than those with lower frequencies.

A summary of the descriptive statistics is given in Table 2.6.

Progress Exercises 2.1: Summary statistics for raw and grouped data

1. Calculate the mean, median, mode and quartiles for the following data.
   16, 14, 18, 16, 108, 17, 18, 19, 16, 17, 16

   How do the calculated values of the mean, median and mode indicate that the data is skewed? Which statistic best describes the data? Give reasons for your choice.

2. In an interview of 40 randomly selected employees, data on their average weekly expenditure in the staff canteen was obtained, see Table PE2.1 Q2.

Table PE2.1 Q2 Average weekly expenditure in staff canteen (£)						File: Canteen	
23	0	20	18	30	25	10	35
10	15	0	2	22	25	15	23
30	15	10	5	15	15	13	15
0	5	15	10	9	15	18	15
25	20	24	25	21	4	32	22

(a) Sort the data into a stem-and-leaf plot.
(b) Calculate the values of the mean, median, mode and quartiles.
   Give a verbal description of expenditure patterns in the staff canteen.
3. See question 2: the weekly canteen expenditure in Table PE2.1 Q2.
   (a) Summarise the data in a frequency distribution table, using intervals $0 < 10$; $10 < 20$; $20 < 30$ and $30 < 40$.
   (b) Calculate the mean from the table.
   (c) Estimate the median and quartiles from the table.
   Describe the results above verbally.
   Explain why the values of the mean, quartiles and mode are different from those in question 2
4. See question 3: the weekly canteen expenditure in Table PE2.1 Q2.
   (a) Plot an Ogive.
   (b) From the Ogive estimate the median and quartiles.
   (c) From the Ogive estimate the two values between which the middle 80 % of expenditures fall.
5. See question 1, Progress Exercises 1.2 and 1.4. Data for the number of 'no shows' for scheduled flights given in Table PE1.2 Q1 (File: No show).
   (a) For the raw data, calculate the mean, mode and quartiles.
   (b) For the data summarised in the frequency distribution table, calculate the mean, mode and approximate values for the quartiles.
   Explain the results in (a) and (b) verbally. Do the values of the mean and median indicate symmetry or skewness of the data?
6. See question 2, Progress Exercises 1.2 and 1.4. Data for the exam marks for 60 students given in Table PE1.2 Q2 (File: Exams).
   (a) For the raw data, calculate the mean, mode and quartiles.
   (b) For the data summarised in the frequency distribution table (question 2, PE1.2), calculate the mean, mode and approximate values for the quartiles.
   Explain the results in (a) and (b) verbally. Do the values of the mean and median indicate symmetry or skewness of the data?
7. See question 3, Progress Exercises 1.2 and 1.4. Data for the duration of mobile calls is given in Table PE1.2 Q3 (file: Mobile calls).
   (a) For the raw data, calculate the mean, mode and quartiles.
   (b) For the data summarised in the frequency distribution table, calculate the mean, mode and approximate values for the quartiles.
   Explain the results in (a) and (b) verbally. Do the values of the mean and median indicate symmetry or skewness of the data?
8. A survey is carried out on the wear-miles for two brands of tyre: the number on miles travelled before each tyre had to be replaced is summarised in the Table PE2.1 Q8.

File: Tyres A and B

Table PE2.1 Q8   Wear miles for Brand A and Brand B tyres

Miles	Brand A	Brand B
less than 5000	10	6
5000 < 10 000	6	4
10 000 < 20 000	15	15
20 000 or more	19	55
Totals	50	80

(a) Draw a suitable chart to compare the wear-miles of the two brands.

(b) Calculate the mean life (wear-miles) for each brand of tyre. Give approximate values for the medians.

Write a brief report on the performance of Brand A and Brand B tyres using the results of (a) and (b) above.

9. The distribution of median and mean income before and after tax by age of taxpayer is given in Table PE2.1 Q9. (File: Tax and age)

File: Tax and age

Table PE2.1 Q9   Distribution of tax and income by age group

Distribution of median and mean income by age range and sex, 2002–03. Taxpayers only
Numbers thousands; Amounts £

Age	No. of taxpayers	Median income before tax	Median tax	Mean income before tax	Mean tax
Under 20	553	7 810	453	8 790	725
20–24	2 060	11 500	1200	12 700	1510
25–29	2 600	16 500	2210	19 000	2940
30–34	3 080	18 800	2590	23 100	4030
35–39	3 320	19 400	2640	25 600	4770
40–44	3 190	19 300	2600	26 600	5160
45–49	2 860	19 300	2650	26 700	5220
50–54	2 720	18 000	2460	25 100	4790
55–59	2 550	15 700	2040	22 900	4270
60–64	1 780	12 700	1470	18 300	3100
65–69	1 420	12 400	1010	16 700	2310
70–74	1 120	12 100	883	16 100	2080
75 and over	1 640	11 800	775	15 300	1780
All ranges	28 900	15 800	1950	21 600	3760

Source: Survey of Personal Incomes 2002–03 (see paragraph 2 of notes on Personal Incomes).
Table updated October 2004.

(a) Explain why the median income is lower than the mean incomes for all age groups.
(b) Calculate the total tax paid by each age group.
(c) Explain why the difference between the mean and median incomes is smallest for the two youngest age groups and greatest for age groups over 30.
(d) Plot line graphs (in Excel if available) to illustrate the distribution of
    (i) numbers of tax payers by age
    (ii) mean and median income by age
    (iii) numbers of taxpayers and mean tax by age
(e) Use Excel or otherwise to confirm the mean tax for all ages is approximately £3,760 as given in Table 2.1 Q9.
Write a brief report, based on the results above, describing the distribution of income and tax by age of taxpayer.

10. See question 5, Progress Exercises 1.4, Table PE1.4 Q5 on distance travelled to work by residents of two 'neighbourhoods' (file Distance to work).
    (a) Calculate the mean and median distance travelled to work by the residents of each neighbourhood. State any assumptions made.
    (b) Explain why the mean and median distance for each neighbourhood differ.
    (Exclude those who do not travel to work, i.e. those who work from home, no fixed place, etc.)

11. Define a weighted average. Explain why the overall average is not always the average of individual averages.
    An airline baggage handling division has estimated the average weights (Kg.) of baggage as follows:
    Holdalls (4 Kg); Small cases (9 Kg.); Large cases (12 Kg);
    Calculate the average weight per bag for a party that check in 12 holdalls, 8 small cases and 20 large cases.

12. Income before and after tax in 2004–5 by quintile group for all households is given in Table PE2.1 Q12:

File: Tax on income 04_05

Table PE2.1 Q12   Income before and after tax in 2004–5 by quintile group for all households

	Bottom	2nd	3rd	4th	Top	Overall
Original income	4 277	11 196	21 575	34 464	66 332	27 569
Final income	13 254	17 515	22 553	28 315	47 410	25 810

Source: ONS.

(a) Plot a bar chart to compare income before and after tax by quintile group
(b) Describe verbally a 'quintile group'. Hence describe, in words, the income before and after tax for each quintile group in Table PE1.2 Q12.

Table 2.6 Summary of the calculation and uses of summary statistics for raw and grouped data

Statistic	Calculations		When Used
	Raw Data	Grouped Data	
Mean ($\bar{x}$)	$\dfrac{\sum x_i}{n}$ (2.1)	$\dfrac{\sum f_i x_i}{\sum f_i}$ (2.5)	When data is evenly spread or symmetrical
Weighted average		$\dfrac{\sum w_i x_i}{\sum w_i}$ (2.10)	Use to calculate when data averages and sample sizes are given
Median (Q_2)	Get the value of the which is halfway through the ordered data (2.2)	Get the value of the 50th percent item measured from the Ogive, or calculate it (2.6)*	Use for any data, but particularly when data is skewed
Mode	Value which occurs most often	Estimate from the modal class in histogram or calculate it (2.9)*	Useful to quote for nominal or ordinal data.
Lower Quartile (Q_1)	The value of the item which is one quarter way through the ordered data (2.3)	The value of 25th percent item from Ogive or calculate it (2.7)*	Any data set
Upper Quartile (Q_3)	The value of item which is three quarters way the ordered data (2.4)	The value of 75th percent item from Ogive or calculate it (2.8)*	Any data set
Percentiles	The value if the item below which a given percentage of all values lie	The value of xth percent item from Ogive	Any data set

*use of these formulae is optional.

2.3 Measures of dispersion for raw data. Variance, QD, IQR

In the previous section a single numeric value, called a summary statistic, was used as a typical value to represent a set of data. However a summary statistic gives no indication about the dispersion of values within a set of data, as the following example will demonstrate.

A Tour operator is charged with planning activities for two groups of holidaymakers. One of the main considerations is age. On inquiry she is told that the average age (and the median age) for each group is 50. So she assumes that the individuals in the group are aged in or around 50 and proceeds to plan activities accordingly.

When the groups of tourists finally arrive she discovers that the ages of the individuals in each of the group are:

Group P: 48, 50, 52, 51, 49

Group Q: 2, 88, 76, 31, 64, 39, 50

Obviously, the average, or even the median age describes only aspect of the data: it gives a typical representative value but gives no indication whether the values of the data are close or very scattered. Another 'statistic' is required to give a measure of the spread or dispersion of values within a set of data.

Remember: **A 'statistic'** is a sample characteristic

The following statistics are the most common measures of dispersion:

- Range
- Variance
- Standard deviation
- Semi-interquartile range
- Quartile deviation

2.3.1 The range

The range is the difference between the highest and lowest value in the data set

Group P: Range $= 52 - 48 = 4$

Group Q: Range $= 88 - 2 = 86$

For group P, a mean of 50 and a range of four indicates that all the values in this group are close to 50.

For group Q a mean of 50 and a range of 86 indicates that at least two ages are very different from each other – but there is no indication whether these are just two outliers or whether all the data within the group are very different from each other! So the problem with the range is that it considers the highest and lowest values only.

2.3.2 Variance and standard deviation

A measure of dispersion from the mean value

The sum of the deviations for each data from the mean value might seem an obvious measure of dispersion. Consider the groups P and Q, each of which had a mean age of 50. The calculation of deviations from the mean and the sum of the deviations is given in Table 2.7.

Table 2.7   Calculation of the deviations of each value from the mean value

Group P x_i	$x_i - 50$	Group Q	x_i	$x_i - 50$	
48	−2		2	−48	**Remember**: When numbers have different signs subtract, but give the answer with the sign of the biggest number.
50	0		88	38	
52	2		76	26	
51	1		31	−19	
49	−1		64	14	
Total	0		39	−11	
			50	0	
		Total		0	

As you can see, the problem with this idea is that the sum of the deviations from the mean is zero in each case. In fact this is always the case: $\sum (x_i - \mu) = 0$ or $\sum (x_i - \bar{x}) = 0$.

The mean is a pivotal value, in the sense that the sum of the deviations of data above the mean value will balance the sum of the deviations of data values below the mean.

Therefore, to overcome the problem of positive and negative values totalling to zero, each value is squared before adding – see Table 2.8 below.

Table 2.8   Calculation of squared deviations from the mean for ages of tourists Group P and group Q

Group P			Group Q		
x_i	$(x_i - 50)$	$(x_i - \mu)^2 = (x - 50)^2$	x_i	$(x_i - 50)$	$(x_i - \mu)^2 = (x_i - 50)^2$
48	−2	$(-2)^2 = 4$	2	−48	$(-48)^2 = 2304$
50	0	$(0)^2 = 0$	88	38	$(38)^2 = 1444$
52	2	$(2)^2 = 4$	76	26	$(26)^2 = 676$
51	1	$(1)^2 = 1$	31	−19	$(-19)^2 = 361$
49	−1	$(-1)^2 = 1$	64	14	$(14)^2 = 196$
Total	0	10	39	−11	$(-11)^2 = 121$
			50	0	$(0)^2 = 0$
			Total	0	5102

Adding each $(x_i - \mu)^2$ gives the sum, $\sum (x_i - \mu)^2 = 10$ for group **P**

Adding each $(x_i - \mu)^2$ gives the sum, $\sum (x_i - \mu)^2 = 5102$ for group **Q**

Group **P**: $\sum (x_i - \mu)^2 = 10$ Group **Q**: $\sum (x_i - \mu)^2 = 5102$

The sum of the squared deviations from the mean certainly does indicate a larger variation of data from the mean value in Group Q compared to group P:

One final consideration: the number of data within each group is different.

If we are to use this formula to compare variation within several sets of data, the number of data must be taken into consideration.

This is achieved quite simply by dividing the sum $\sum (x_i - \mu)^2$ by the number of data in the set:

$$\frac{\sum (x_i - \mu)^2}{N}$$

This is the average value of the squared deviation (sometimes called the mean squared deviation). This statistic is called the **variance**:

Group **P**: Variance $= \dfrac{\sum (x_i - \mu)^2}{N} = \dfrac{10}{5} = 2$

Group **Q**: Variance $= \dfrac{\sum (x_i - \mu)^2}{N} = \dfrac{5102}{7} = 728.857$

The formula for the variance for any group on N data (items) is defined as follows.

$$\text{Variance for raw data: } \sigma^2 = \frac{\sum (x_i - \mu)^2}{N} \qquad (2.11)$$

σ^2: sigma squared' is the symbol for variance

So, **Variance, σ^2,** is the sum of the squared deviations from the mean value divided by the number of data

Standard Deviation: standard deviation is the square root of the variance

Group P: Standard deviation $\sigma = \sqrt{2} = 1.414$

Group Q: Standard deviation $\sigma = \sqrt{728.857} = 26.997$

σ: 'sigma' is the symbol for standard deviation

$$\text{Standard deviation } \sigma = \sqrt{\frac{\sum (x_i - \mu)^2}{N}} \qquad (2.12)$$

You will come across two symbols for variance, σ^2 and s^2

Population variance: σ^2 refers to variance calculated from population data by formula (2.11) above where the population size is referred to by the uppercase letter N.

Sample variance: s^2 refers to variance calculated from sample data by the formula $s^2 = \dfrac{\sum\limits_{i=1}^{n} (x_i - \bar{x})^2}{n - 1}$, where the lowercase letter, n is the sample size.

$$\text{Variance calculated from sample data: } s^2 = \frac{\sum\limits_{i=1}^{n} (x_i - \bar{x})^2}{n - 1} \text{ where } n \text{ is the sample size.} \qquad (2.13)$$

Further discussion on the use of s^2 is deferred to Chapter 6.

Worked Example 2.7: Variance and standard deviation from raw data (Swimmers)

Calculate the variance and standard deviation for the raw data in Table 1.9, reproduced below

Table 1.9 Ages of swimmers available for competitions | File: Swimmers |

				Ages of swimmers					
18.8	18.7	20.2	17.4	29.6	23.7	19.6	18.4	20.6	25.5
17.9	24.2	20.9	19.2	23.1	21.5	21.9	20.7	22.3	21.7
20.5	21.6	20.4	20.2	19.5	19.9	19.9	20.8	17.2	19.1
18.1	19.2	18.8	18.8	21.2	24.3	22.3	22	18.8	21.7
18.1	20.9	21.1	17.9	18.6	18.8	20.1	20.9	20.1	22.4

Solution Use the formula (2.10) $\sigma^2 = \dfrac{\sum (x_i - \mu)^2}{N}$

Step 1: calculate the mean value: this was calculated in Skill Development Exercise SK2_1 where $\mu = 20.582$

Step 2: follow the method outlined in Table 2.8: *for each value* subtract the mean (20.582) and then square the answer (this example works through the data column by column).

Step 3: Sum the squared deviations calculated in Step 2

$$\sum (x - \mu)^2 = (18.8 - 20.582)^2 + (17.9 - 20.582)^2 + \cdots\cdots\cdots + (22.4 - 20.582)^2$$

$$= 254.094.$$

This is the top line in formula (2.11).

Step 4: Divide the sum the squared deviations in Step 3 by the number of data N: $N = 50$

$$\sigma^2 = \frac{(18.8 - 20.582)^2 + (17.9 - 20.582)^2 + \cdots\cdots\cdots + (22.4 - 20.582)^2}{50}$$

$$= \frac{\sum (x_i - \mu)^2}{N} = 5.0619$$

Step 5: Standard deviation is the square root of variance, hence $\sigma = \sqrt{\sigma^2} = 2.2499$

2.3.3   Other measures of dispersion

The calculation of the mean and variance includes every item in a set of data. Hence, extreme high values or extreme low values (i.e. skewed data) will have a disproportionate influence the overall value of variance. In section 2.1 the median was found to be a more representative summary statistic for such data. In this section measures of dispersion that involve the quartiles will be shown to be more representative as measures of dispersion. These measures are the semi-interquartile range and quartile deviation, given by formulae (2.14) and (2.15).

The semi-interquartile range (or interquartile range) is the difference between the upper and lower quartile.

The quartile deviation is the semi-interquartile, divided by 2.

$$\textbf{Semi-interquartile range } \textbf{IQR} = Q_3 - Q_1 \qquad (2.14)$$

$$\textbf{Quartile deviation } \textbf{QD} = \frac{Q_3 - Q_1}{2} \qquad (2.15)$$

Worked Example 2.8: *IQR* and *QD* for raw data

Calculate the interquartile range and quartile deviation for data sets P and Q. Use the values of the median and quartiles to describe the dispersion of the data verbally.

Solution

Group:	Q_1	Q_2	Semi-interquartile range $IQR = Q_3 - Q_1$	Quartile deviation $QD = \frac{Q_3 - Q_1}{2}$
P: 48, 49, 50, 51, 52	48.5	51.5	$51.5 - 48.5 = 3$	1.5
Q: 2, 31, 39, 50, 64, 76, 88	31	76	$76 - 31 = 45$	22.5

Skill Development Exercise SK2_2: Measures of dispersion for raw data

The data given in (a) and (b) where

(a) The hourly rate of pay for employees in the newsagents: Data B: 9, 7, 9, 50, 7, 7, 5, 8, 6.
(b) The hours worked by each member of staff in a given week in a small company given in Table 1.15 (reproduced below).

Table 1.15   Hours worked in a given week by 49 staff

20.0	37.3	54.2	25.3	59.6	24.5	29.7
18.0	38.8	42.1	39.5	56.8	16.9	28.5
45.5	42.0	39.5	42.6	40.0	44.2	40.1
44.0	56.4	30.2	20.0	22.7	37.8	23.4
26.0	20.2	36.1	18.3	19.7	36.8	26.5
24.0	23.4	15.4	20.0	38.9	42.1	24.1
41.0	18.5	21.3	22.6	37.2	42.9	17.9

Calculate
(i) range; (ii) variance; (iii) standard deviation; (iv) semi-interquartile range; (v) quartile deviation.

Comment on the difference in the values of standard deviation and quartile deviation. Suggest possible reasons for the differences.

Answers
(a) Data B: 9, 7, 9, 50, 7, 7, 5, 8, 6 pay per hour for employees in the newsagent
   (i) Range = largest − smallest = $50 - 5 = 45$
   (ii) $\sigma^2 = \dfrac{\sum (x_i - \mu)^2}{N} = \dfrac{1638}{9} = 182$
   (iii) Standard deviation $\sigma = \sqrt{\dfrac{\sum (x_i - \mu)^2}{N}} = \sqrt{182} = 13.4907$

(iv) The quartiles were calculated in Worked Example 2.1: $Q_1 = 6.5$, $Q_3 = 9$
Hence, the interquartile range **IQR** $= 9 - 6.5 = 2.5$

(v) Quartile deviation **QD** $= \dfrac{Q_3 - Q_1}{2} = \dfrac{2.5}{2} = 1.25$

(a) Calculations from raw data, Table 1.15, hours worked by 49 staff and from the results of Skill development exercise SK1_5

(i) Range $= 59.6 - 15.4 = 44.2$

(ii) Working row by row through the ordered data (see SK1_5)

$$\sigma^2 = \frac{\sum\limits_{i=1}^{i=N} (x_i - \mu)^2}{N}$$

$$= \frac{(15.4 - 32.5)^2 + (20.0 - 32.5)^2 + \cdots\cdots\cdots (42.9 - 32.5)^2 + (59.6 - 32.5)^2}{49}$$

$$= \frac{292.41 + 156.25 + \cdots\cdots\cdots 108.16 + 734.41}{49}$$

$$= \frac{6751.30}{49}$$

$$= 137.78$$

> This is a lot of tedious arithmetic! Using methods for 'grouped' data in section 2.4 will help!

(iii) $\sigma = \sqrt{\dfrac{\sum\limits_{i=1}^{N} (x_i - \mu)^2}{N}} = \sqrt{137.78} = 11.738$

(iv) $\text{IQR} = Q_3 - Q_1 = 41.5 - 21.45 = 20.05$

(v) Quartile deviation $QD = \dfrac{Q_3 - Q_1}{2} = \dfrac{20.05}{2} = 10.025$

Comment on the difference in the values for variance and quartile deviation
For data B, standard deviation $= 13.49$ while quartile deviation $= 1.25$. The value of the standard deviation is large due to the presence of the outlier, 50. The effect of the value 50 is obvious from a careful examination of the calculations for variance, given in Table SK2_2.1.

Table SK2_2.1   Calculation of variance for Data B, hourly rate in newsagent (raw data)

> The value 50 contributes more to the measure of variance than all of the other data

x_i	$(x_i - \text{mean})$	$(x_i - \text{mean})^2$
9	$(9 - 12) = -3$	9
7	$(7 - 12) = -5$	25
9	$(9 - 12) = -3$	9
50	$(50 - 12) = 38$	1444
7	$(7 - 12) = -5$	25
7	$(7 - 12) = -5$	25
5	$(5 - 12) = -7$	49
8	$(8 - 12) = -4$	16
6	$(6 - 12) = -6$	36
Totals 108	0	1638

(b) Similarly, for the 49 weekly hours, the standard deviation = 11.038 while the quartile deviation is 10.025. In this case the higher values: 54.2, 56.4, 56.8, and 59.6 are not too different from the rest of the data, hence raising the value of the standard deviation slightly.

In general the quartile deviation is a better indication of dispersion than standard deviation when data is very skewed.

2.4 Measures of dispersion for grouped data. Variance, QD, IQR

The calculation of variance for the 50 swimmers in Worked Example 2.8 and the hours worked by the 49 staff in Skill Development Exercise, SK2_2 is very tedious. Variance and other measures of dispersion may be calculated from grouped data, using formulae (2.16), (2.17) and (2.18) as demonstrated in the following Worked Examples.

2.4.1 The formula for variance for grouped data:

The formula for calculating variance from raw data $\sigma^2 = \dfrac{\sum (x_i - \mu)^2}{N}$, given in equation (2.11)

The corresponding formula for the calculation of variance for grouped data is given in (2.16).

$$\sigma^2 = \frac{\sum f_i (x_i - \mu)^2}{\sum f_i} = \frac{\sum f_i (x_i - \mu)^2}{N}, \text{ where } N = \sum_{All} f_i \tag{2.16}$$

The standard deviation is the square root of the variance.

Standard deviation for grouped data:

$$\sigma = \sqrt{\frac{\sum f_i (x_i - \mu)^2}{\sum f_i}} = \sqrt{\frac{\sum f_i (x_i - \mu)^2}{N}} \text{ where } N = \sum_{All} f_i \tag{2.17}$$

There are several alternative formulae for variance – sometimes computationally more convenient than (2.16); one of the alternative formulae is given in (2.18).

Alternative formula for variance for grouped data:

$$\sigma^2 = \frac{1}{N} \sum f_i x_i^2 - \left(\frac{\sum f_i x_i}{N}\right)^2 \text{ where } N = \sum_{All} f_i \tag{2.18}$$

You will be asked to prove the equivalence of these formulae in the theoretical exercises.

Worked Example 2.9: Variance, standard deviation, *IQR* and *QD* for grouped data

The ages of members of the swimming team is given as grouped data in Table SK1_2.2, Skill Development Exercise SK2_1. For this data calculate

(a) the variance and standard range
(b) the semi-interquartile deviation and quartile deviation

Compare the answers with those for the raw data in Worked Example 2.6.

Solution

(a) The calculations for variance for grouped data are set out in Table 2.9

Table 2.9   Calculation of mean and variance for ages of swimming team from grouped data.

Ages	Mid-interval value x_i	Frequency f_i	$f_i x_i$	$(x_i - \text{mean})$	$(x_i - \text{mean})^2$	$f_i(x_i - \text{mean})^2$	$f_i(x_i)^2$
$17 < 19$	18.0	14	252	−2.5	6.1504	86.1056	4 536
$19 < 21$	20.0	19	380	−0.5	0.2304	4.3776	7 600
$21 < 23$	22.0	11	242	1.5	2.3104	25.4144	5 324
$23 < 25$	24.0	4	96	3.5	12.3904	49.5616	2 304
$25 < 27$	26.0	1	26	5.5	30.4704	30.4704	676
$27 < 29$	28.0	1	28	7.5	56.5504	56.5504	784
Totals		50	1024			252.48	21 224

$\sum f_i = 50$ $\sum f_i x_i = 1024$ $\sum f_i(x_i - \text{mean})^2 = 252.48$ $\sum f_i x_i^2 = 21\ 224$

Method: Use formula (2.14) where variance for grouped data $\sigma^2 = \dfrac{\sum f_i(x_i - \text{mean})^2}{\sum f_i}$

referring to Table 2.9 for all calculations

Step 1: Calculate the mid-interval values- see column 2.

Step 2 Calculate the mean: $\mu = \dfrac{\sum f_i x_i}{\sum f_i} = \dfrac{1024}{50} = 20.48$ (See columns 3 and 4).

Step 3: Calculate the deviation of each mid-interval value from the mean (column 5), then square each (column 6). Finally, multiply each squared deviation by the frequency to give the total value per interval.

Step 4: Sum of the squared deviations in column 6:

$$\sum f_i(x_i - \text{mean})^2 = 252.48.$$

Step 5: Divide the answer in Step 4 by the number of items $N = \sum f_i = 50$

Hence $\sigma^2 = \dfrac{\sum f_i(x_i - mean)^2}{\sum f_i} = \dfrac{252.48}{50} = 5.0496$

> Recall, $\sigma^2 = 5.0619$ in
> W.E. 2.6 for raw data

Step 6: Standard deviation, $\sigma = \sqrt{\dfrac{\sum f_i(x_i - mean)^2}{\sum f_i}} = \sqrt{5.0496} = 2.2471$

> Recall, $\sigma = 2.2499$ in
> W.E. 2.6 for raw data

OR, alternatively, by formula (2.16) the variance and standard deviation are calculated as follows:

$$\sigma^2 = \frac{1}{N}\sum f_i x_i^2 - \left(\frac{\sum f_i x_i}{N}\right)^2 \text{ where } N = \sum_{All} f_i \text{ and } \mu = \frac{\sum f_i x_i}{\sum f_i} = 20.48 \text{ (above)}$$

$$\sigma = \sqrt{\frac{1}{50}(21\,224) - (20.48)^2} = 2.2471$$

(b) Interquartile range and Quartile deviation for grouped data

The quartiles for the grouped data were calculated in Skill Development Exercise SK1_3 Hence the inter-quartile range and quartile deviation are calculated as follows

Q_1	Q_3	Inter-quartile range $IQR = Q_3 - Q_1$	Quartile deviation: $QD = \dfrac{Q_3 - Q_1}{2}$
18.6	21.7	$21.7 - 18.6 = 3.1$	1.55

2.5 Use of calculator for descriptive statistics

There are numerous brands and versions of calculator. Most scientific calculators have inbuilt statistical functions for descriptive statistics and regression. In general, the calculator will produce descriptive statistics in a mode called the 'SD mode' (Standard Deviation) for Casio calculators and 'Statxmode' for Shape calculator. Most calculators will also do calculations for regression (see Chapter 3).

It is not possible to give step by step instructions on how to enter data into the calculator as this varies, depending on the calculator – but it is straightforward and instructions illustrated by simple examples are given in the manual supplied with your calculator.

Having entered the data, the average value may be read out by pressing the key for $\bar{x}$ or μ as directed; the standard deviation for population or sample data is again read out by pressing the appropriate key. If variance is required, read out the standard deviation first; then square the standard deviation using the 'x^2' key on the calculator.

Note: on means and standard deviation from calculators
The mean of the given data is calculated by the same formula, whether the data is from a population or a sample. Hence the calculator uses just one symbol μ or $\bar{x}$.

The calculator does not use different symbols for population and sample size: n is simply the number of data.

Recall that the standard deviation is calculated by different formulae, depending on whether the data is from a population or a sample (see section 2.3.2).

Depending on the brand and version of calculator, the population variance may be symbolised as σ^2 or $[x\sigma_n]^2$: and is calculated by the formula $\dfrac{\sum(x-\mu)^2}{N}$ where μ is the population mean

The sample variance may be referred to as s^2, s_x^2 or $[x\sigma_{n-1}]^2$ and is calculated by the formula $\dfrac{\sum(x-mean)^2}{n-1}$ where $\bar{x}$ is the sample mean.

Progress Exercises 2.2: Measures of dispersion for raw and grouped data

---

1. See question 1, PE2.1. Calculate the variance (σ^2 and s^2), standard deviation (σ and s), interquartile range and quartile deviation for the following data:
   16, 14, 18, 16, 108, 17, 18, 19, 16, 17 and 16.
   Explain why the standard deviation is greater than the quartile deviation.
2. See question 2, PE2.1 Q2. (File Canteen): the weekly expenditure in the canteen
   Use the grouped data from Table PE2.1 Q2 to calculate the variance (σ^2 and s^2), standard deviation (σ and s), interquartile range and quartile deviation for the weekly expenditure in the canteen.
   Explain which measure of dispersion best represents the dispersion of in the data.
3. The heights of five individuals are given in Table PE2.2 Q3

Table PE2.2 Q3   Heights of 5 individuals given in (a) meters (b) centimetres

Height	meter	cm
	3.6	360
	2.5	250
	3.1	310
	3.8	380
	2.9	290

Calculate the mean, variance and standard deviation for heights that are measured in (a) meters; (b) centimetres.

Explain why the statistics; mean, variance and standard deviation are different when measurements are given in centimetres rather than meters.

4. See question 2, PE2.1. Data for the exam marks for 60 students given in Table PE1.2 Q2 (File: Exams).
   Calculate the variance, standard deviation, interquartile range and quartile deviation for (a) the raw data; (b) the data summarised in the frequency distribution table.
   Describe the data using the results in (a) and (b) and question 6, PE2.1.
5. See question 1, PE2.1. Data for the number of 'no shows' for scheduled flights given in Table PE2.1 Q1 (File: No show).

Calculate the variance (σ^2 and s^2), standard deviation (σ and s), interquartile range and quartile deviation for (a) the raw data; (b) the data summarised in the frequency distribution table. **Describe** the data using the results in (a) and (b) and question 5, PE2.1.

Is there any indication of extreme values in the data set from the relative sizes of the standard deviation and quartile deviation?

6. See question 3, PE2.1. Data for the duration of mobile calls is given in Table PE2.1 Q3, file: Mobile calls.

Calculate the variance (σ^2 and s^2), standard deviation (σ and s), interquartile range and quartile deviation for (a) the raw data (b) the data summarised in the frequency distribution table (assume an upper limit $18.0 < 19.5$).

Explain why the variance for the raw data is greater than that for the grouped data.

Describe the data using the results in (a) and (b) and question 7, PE2.1 Do the values of the standard deviation and quartile deviation indicate the skewness of this data?

7. See question 8 PE2.1, data for wear miles of tyres given Table PE2.1 Q8

Table PE2.1 Q8   Wear miles for Brand A and Brand B tyres

Miles	Brand A	Brand B
less than 5000	10	6
5000 < 10 000	6	4
10 000 < 20 000	15	15
20 000 or more	19	55
Totals	50	80

(a) Calculate the variance and standard deviation for wear miles for Brand A and brand B.
(b) Calculate the quartile deviation for wear miles for Brand A and brand B.
Explain why the variance for brand B is so large.

8. See question 10, PE2.1,[6] the distance travelled to work by residents of two neighbourhood', Brentwood and London (File: Distance to work).

For the distance travelled to work by residents of each neighbourhood (exclude those who do not travel to work, i.e. those who work from home, no fixed place, etc.) calculate
(a) the mean and median.
(b) the population variance and standard deviation.
Give possible explanation(s) why the results in (a) and (b) are different for Brentwood and London.

2.6   Other descriptive statistics. CV, skewness and box plots

In question 3, PE2.2, the value of variance (and standard deviation) was different when measurements were given in meters and centimetres, see Table 2.10. Hence the units of measurement must be stated when quoting statistics such as averages, variance, standard deviation, quartile deviation, etc.

Table 2.10 Variance, standard deviation and mean for
the same data with measurements in different units

Height	Meters	cm
	3.6	360
	2.5	250
	3.1	310
	3.8	380
	2.9	290
Variance	0.2216	2216
Standard deviation	0.4707	47.0744
Mean	3.18	318

Comparisons between different sets of data are enhanced by using statistics that are independent of units: two such descriptive statistics are the coefficient of variation and the coefficient of skewness.

2.6.1 The coefficient of variation

The coefficient of variation is independent of the units of measurements and is defined as

$$\text{Coefficient of variation: } CV = \frac{\sigma}{\mu} \times 100 \qquad (2.18)$$

For example, the coefficient of variation for the data in Table 2.10 is calculated as follows.

When measurements are recorded in meters is $CV = \dfrac{0.470744}{3.18} \times 100 = 14.80327$.

When measurements are recorded in centimetres is $CV = \dfrac{47.0744}{318} \times 100 = 14.80327$.

The coefficient of variation is also useful for comparing two sets of data when (i) the means are close but the variances are different or (ii) the means are different but the variances are close,

2.6.2 Skewness

When data is symmetrically distributed, the mean, median and mode are equal

If data is 'skewed' the relative positions of mean, median and mode (in grouped data) are illustrated in Figure 2.4:

Pearson's coefficient of skewness is a statistic that gives a measure of the degree of skewness in a set of data

$$\text{Pearson's Coefficient of skewness, } sk = \frac{3(\text{mean} - \text{median})}{\text{standard deviation}} \qquad (2.19)$$

Note: Pearson's coefficient of skewness is independent of the **units of measurement.**

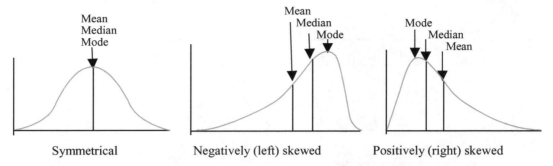

Figure 2.4   Relative positions of mean and median in symmetrical and skewed data.

To illustrate, the coefficient is zero for perfectly symmetrical data because the mean is equal to the median.

$$sk = \frac{3(\text{mean} - \text{median})}{\text{standard deviation}} = \frac{3(0)}{\text{std.deviation}} = 0$$

A negative value of Pearson's coefficient indicates skewness to the left where the mean is less than the median due to the presence of extreme small value(s). A positive coefficient indicates skewness to the right where the mean is greater than the median due to the presence of extreme large value(s).

Worked Example 2.10: Calculate Pearson's coefficient of skewness

Calculate Pearson's coefficient of skewness from the grouped data given in Table SK1_2.1 (frequency distribution for the ages in swimming team).

Solution

From Skill Development SK2_1 the summary statistics for the groped data in Table SK1_2.1 were calculated: $\mu = 20.48$; $Q_2 = 20.158$, $\sigma = 2.2471$

Pearson's Coefficient of Skewness, $sk = \dfrac{3(\text{mean} - \text{median})}{\text{standard deviation}} = \dfrac{3(20.48 - 20.158)}{2.2471} = 0.4299$

The coefficient is positive indicating the presence of at least on extreme high value.

2.6.3   Box plots

A box plot is a graphical display of the symmetry or skewness of a set of data. A box plot consists of the median at the centre of the 'box' with the quartiles as the respective lower and upper ends of the box. The whiskers extend from the lower end of the box to the minimum data value and from the upper end to the maximum data value, provided the minimum and maximum values are within one and a half inter-quartile ranges ($1.5 \times IQR$) from either end of the box. Data values between $1.5 \times IQR$ and

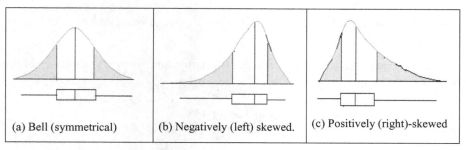

| (a) Bell (symmetrical) | (b) Negatively (left) skewed. | (c) Positively (right)-skewed |

Figure 2.5   Box plots depict the symmetry/skewness of a set of data.

$3 \times IQR$ units from the end of the box are called outliers and represented by a '*'. If a data value is further than $3 \times IQR$ units from the end of the box it is an extreme outlier and represented by a 'o'.

(**Note:** there are different conventions for representing 'whiskers). A summary of the way in which the distribution of the data is depicted by box plots is given in Figure 2.6. The calculation of box plots is illustrated in Worked Example 2.10.

Worked Example 2.10: Calculation of a box plot

Plot the box plot for the data for swimmers (see Skill Development SK2_1).

Solution

Form Skill Development Exercise SK2_1, for raw data, the minimum = 17.2, $Q_1 = 18.8$, $Q_2 = 20.3$, $Q_3 = 21.68$, mean = 20.58 and maximum = 29.6.

Plot the 'box' using the values of $Q_1 = 18.8$, $Q_2 = 20.3$ and $Q_3 = 21.68$ as shown in Figure 2.7.

Before plotting the 'whiskers' check whether the minimum and maximum are within $1.5 \times IQR$ units from the respective ends of the box.

In this example the value of the interquartile range is $IQR = Q_1 - Q_3 = 21.675 - 18.8 = 1.875$. Hence $1.5 \times IQR = 1.5 \times 1.875 = 2.8125$.

The difference between Q_1 and the minimum data value is $18.8 - 17.2 = 1.6$. This is within '$1.5 \times IQR$' units, hence draw the lower whisker.

The difference between Q_2 and the maximum value is $29.6 - 20.3 = 9.3$. This is greater than '$1.5 \times IQR$' units. The maximum value for the upper whisker is $Q_2 + 1.5 \times IQR = 21.675 + 2.8125 = 24.4875$. Hence draw the upper whisker (length 2.8125) from the upper end of the box to the point 24.4875. Represent the maximum value of 29.6 by a *.

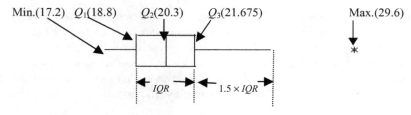

Figure 2.6   Construction of a box plot for age of swimmers.

Skill Development Exercise SK2_3: Dispersion and other statistics for grouped data

Calculate the measures of dispersion, coefficient of variation and Pearson's coefficient of skewness from the raw data for

(a) the hours worked by the 49 employees in Table 1.15.
(b) the daily sales of MP3 players given in Table 1.4.

Answers

(a)		(b)	
mean	32.5	Variance $\sigma^2 =$	163.2099
variance	137.7816	Std. dev. $\sigma =$	12.77536
std dev	11.73804	$CV =$	39.64768
Q_1	22.6	$Sk =$	0.287011
Q_2	30.2	$IQR =$	16.5
Q_3	41	$Q.D =$	8.25
IQR	18.4		
QD	9.2		
CV	36.11705		
Sk	0.587832		

Progress Exercises 2.3: Dispersion and other statistics for grouped data

1. For each of the sets of data, C, D and E

$$C : 48, 50, 52, 51, 49$$

$$D : 1, 3, 5, 4, 2$$

$$E : 1001, 1003, 1005, 1004, 1002$$

   (a) Calculate the range, mean, median, variance and coefficients of variation and skewness.
       Explain why the range and the variance are the same for each set, but the means are different.
   (b) Explain why the values of the CV are different for each set.
   (c) For each data set, give an explanation for values of the coefficient of skewness.
2. See question 2, PE2.1 where Table PE2.1 Q2 gives the weekly expenditure in the canteen for a sample of 40 employees (hence use s, the sample standard deviation etc.), (File: Canteen) calculate, from the raw data
   (a) coefficient of variation (CV); (b) Pearson's coefficient of skewness; (c) the box plot.
   Explain each of the results.

3. See question 5, PE2.2. Data for the number of 'no shows' for a sample of scheduled flights given in Table PE1.2 Q1, File: No show.
   Calculate, from the raw data
   (a) coefficient of variation (CV); (b) Pearson's coefficient of skewness; (c) the box plot. Explain each of the results.

4. See question 4, PE2.2. Data for the exam marks for 60 students (a population) given in Table PE1.2 Q2, (File: Exams).
   Calculate, from the raw data
   (a) coefficient of variation (CV); (b) Pearson's coefficient of skewness; (c) the box plot. Explain each of the results.

5. See question 6, PE2.2. Data for the duration of a sample of mobile calls is given in Table PE1.2 Q3, File: Mobile calls.
   Calculate, from the raw data
   (a) coefficient of variation (CV); (b) Pearson's coefficient of skewness; (c) the box plot. Explain each of the results.

6. See question 7 PE2.2, data for wear miles for samples of tyres (Tyres A and B), Table PE2.1 Q8
   Calculate (for each brand)
   (a) coefficient of variation (CV); (b) Pearson's coefficient of skewness.
   Explain how the results indicate the skewness for the wear miles for Brand B.

7. Theoretical exercise: prove that the coefficient of variation is independent of units.
   Hint: let x_i represent the original units, then ax_i represents the same measurements scaled by a constant, a.

2.7 Descriptive statistics in Excel

In this chapter descriptive statistics were described for both raw and grouped data. However, for large data sets manual calculations were long and tedious. Microsoft Excel has inbuilt statistical functions that will produce most of the descriptive statistics from raw data on the click of the mouse: the basic formula function may be used to carry out calculations for grouped data from tables.

2.7.1 Descriptive statistics in Excel

2.7.1.1 Calculation of the mean in Excel

Take the example of the MP3 players given in Table 1.4. Enter the data into an Excel spreadsheet, see Figure 2.7. To calculate the average value for this data, place the cursor in the cell where the answer is required: cell A8 was used in this example. Click on the f_x button on the menu bar.

The first drop-down menu allows you to select the category 'Statistical' followed by the statistical function that you require: in this case we select AVERAGE.

Having selected 'AVERAGE' you are requested to enter the data to be averaged: the easiest way to do this is to use the cursor to highlight the data, as shown in Figure 2.8a. Alternatively you may enter the range of cell addresses containing the data. In this case the data is in cells ranging from A2 to F6: enter this as 'A2:F6' (blank cells don't count).

The average value of 32.2222 is displayed on the dialogue box, even before clicking on OK. When you click OK the answer is given in cell A8 on the worksheet.

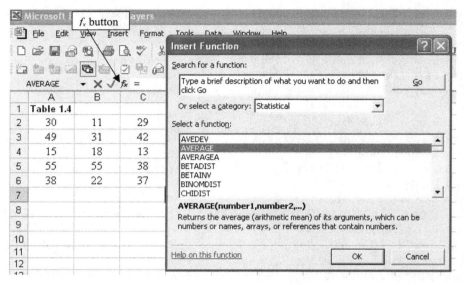

Figure 2.7 Calculating the average in Excel: drop-down menus.

The command or formula for calculating the average is displayed in the menu bar in Figure 2.8a, simultaneously with the drop down menus. Hence, the average could have been obtained by placing the cursor in cell A8 (or wherever the answer is required) and typing '= average(A2:F6)' (in either upper or lowercase letters) as shown on the menu bar in Figure 2.8b.

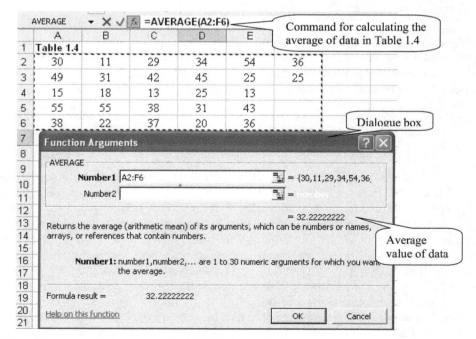

Figure 2.8a Calculating the average in Excel: Highlighting the data.

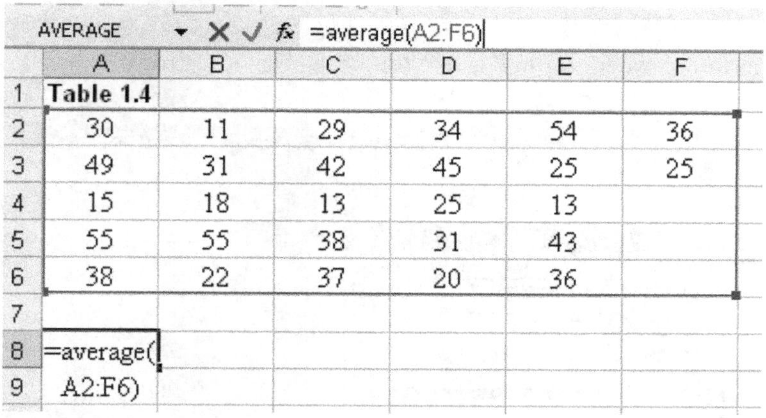

Figure 2.8b   Formula for average in Excel.

Other statistics are readily calculated in Excel by selecting the appropriate function in the statistical category or by typing the corresponding formula.

Variance in Excel

It was mentioned earlier that there are two formulae for calculating variance, depending on whether population or sample data were used. In Excel there are four functions for variance: VAR, VARA, VARP and VARPA. We will require two of these: VAR calculates the sample variance, s^2 and VARP calculates the population variance σ^2. The description of the selected function is given directly below the 'Select a function' window. See Figure 2.9:

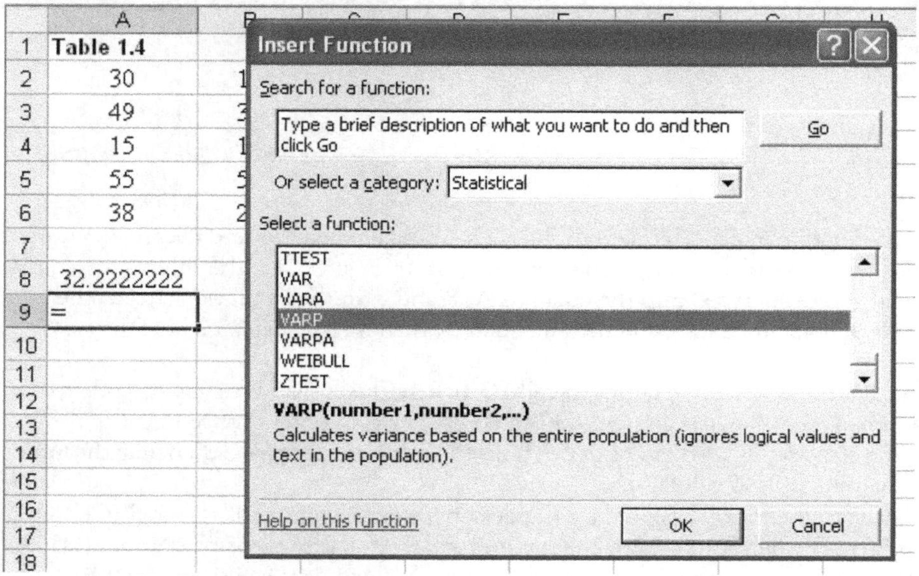

Figure 2.9   Calculating population variance in Excel.

Quartiles in Excel

Again, follow the drop-down menus in Figure 2.10: the median is quartile 2.

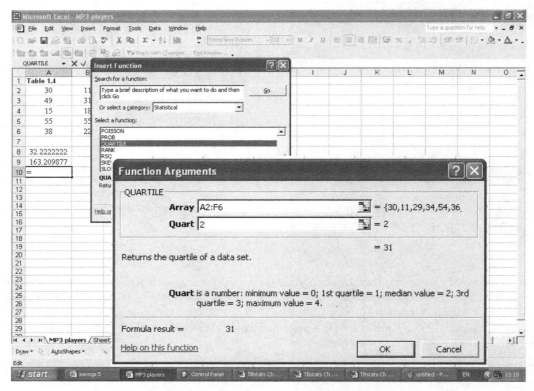

Figure 2.10   Calculating the median in Excel.

Grouped data: use Excel's formula function for calculations of statistics from grouped data

The formula function in Excel along with the 'auto-fill' facility may be used to carry out the arithmetic for grouped data that is arranged in rows or columns. In Figure 2.11, the intervals, frequencies and mid-interval values are typed into the columns A, B and C in the Excel worksheet. The calculations for the mean and variance shown in the remainder of the table are explained as follows:

1. To calculate $f_1 x_1$ place the cursor in cell D3 (where the answer is required) and type = B3*C3: this formula multiplies the contents of cell B3 (f_1) by the contents of cell C3 (x_1).
2. Place the cursor at the corner of this cell until the black cross appears: then drag the formula down the remaining cells in column D.
3. (a) Calculate the sum $\sum f_i x_i$: place the cursor where the answer is required (Cell D12): either click on the $\sum$ button on the menu bar and follow instructions or type = sum(D3;D11). Also (b) similarly, calculate the sum of the frequencies, $\sum_{f_i}$ giving the answer in cell B12.

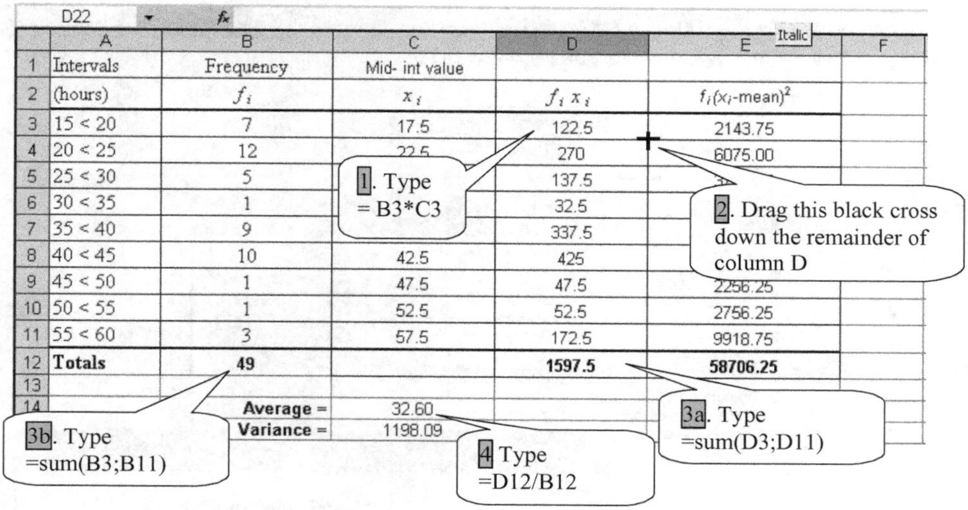

Figure 2.11   Calculation of the average value for grouped data using the Excel formula facility.

4. Finally, the average value is $\sum f_i x_i$ divided by $\sum f_i$. Enter the formula in cell C14 by typing $=$ D12/B12.

Using the formula function to calculate variance will be left as an exercise to the reader.

Data analysis in Excel

Excel will produce a complete set of summary statistics using 'data analysis'.

To demonstrate, arrange the data for the MP3 players in a single column in an Excel worksheet.

Select 'data analysis' from the menu bar, then select 'descriptive statistics' from the data analysis menu. In the Descriptive Statistics dialogue box highlight the data or enter the address for the data in the 'Input Range' box. Click 'labels' if the label was included in the input range. Into the 'Output Range' enter the cell address for the top left hand corner of the region in the worksheet where you wish to display the descriptive statistics (cell B1 in this example). Finally click on summary statistics. See Figure 2.12a.

The results of the data analysis for descriptive statistics are given below:

Note: (i) Kurtosis is a summary statistic that compares height or peak of the sample data with a Normal, bell shaped curve. A kurtosis value of zero indicates the data is bell shaped like the Normal curve. A negative kurtosis indicates a flatter distribution than the Normal while a positive kurtosis indicates a more peaked distribution than Normal. (ii) In Excel, the formula used to calculate skewness is

$$\frac{n}{(n-1)(n-2)} \sum \left(\frac{x_i - \bar{x}}{s}\right)^3$$

Descriptive statistics in Minitab

Minitab is one of the many statistical software packages on the market. Minitab will be introduced later in the text (Chapter 8) for statistical inference. However Minitab will also produce descriptive statistics with very useful graphical summaries. See section 8.4.2, Figure 8.11.

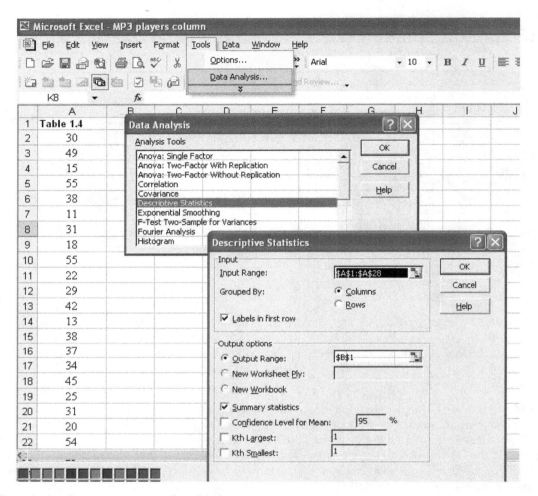

Figure 2.12a   Descriptive statistics for MP3 players.

Table 1.4		Explanations
Mean	32.22222222	... $\bar{x} = \dfrac{\sum x_i}{n}$
Standard Error	2.505454638	... $\dfrac{s}{\sqrt{n}}$, see Chapter 6
Median	31	
Mode	25	
Standard Deviation	13.01872419	... $s = \sqrt{\dfrac{\sum (x_i - \bar{x})^2}{n-1}}$
Sample Variance	169.4871795	
Kurtosis	−0.77975764	... s^2
Skewness	0.145858112	
Range	44	
Minimum	11	
Maximum	55	
Sum	870	
Count	27	

Figure 2.12b   Descriptive statistics for MP3 players.

Progress Exercises 2.4: Descriptive statistics in Excel (or otherwise)

In questions 1 to 5 below calculate the summary statistics and measures of dispersion, CV and skewness from (a) the raw data (b) grouped data.

1. Data for the number of 'no shows' for scheduled flights (File: No show).
2. Data for the exam marks for 60 students given in (File: Exams).
3. Data for the duration of mobile calls is given in (File: Mobile calls).
4. Data for wear miles of tyres given Table PE2.2.
5. For the data from neighbour statistics in File: Distance to work; calculate the summary statistics and measures of dispersion, CV and skewness. Describe the results as you would to a non-statistician.
6. The alcoholic drink released for home consumption in the UK for the years 1990/91 and 2003/04 is given in Table PE2.4 Q6:

File: Alcohol home

Table PE2.4 Q6   Alcoholic drink for home consumption

	Thousand hectolitres			
	Beer	Cider & perry	Wine	Spirits
1990/91	64 541	3777	7 342	987
1991/92	62 975	3912	7 391	888
1992/93	60 462	4162	7 737	863
1993/94	59 372	4593	8 117	851
1994/95	59 420	4916	8 434	887
1995/96	60 400	5581	8 920	814
1996/97	59 545	5677	9 997	823
1997/98	60 483	5536	10 092	810
1998/99	58 761	5603	10 476	840
1999/00	59 979	6291	12 118	926
2000/01	56 139	5943	13 047	937
2001/02	58 763	5900	14 795	978
2002/03	59 791	5872	12 757	1158
2003/04	60 594	5966	12 643	1210

Source: HM Customs and Excise.

(a) Calculate the summary statistics and measures of dispersion, CV and coefficient of skewness for each category of drink: Beer, Cider and perry, Wine and Spirits.
Explain how the statistics for each category describes consumption over the years.
(b) In Excel (or otherwise) plot graphs for the consumption of each beverage over the years. Do the graphs confirm the statistics in (a)?

7. The imports of crude petroleum by China for the years 2002 to 2005 is given in Table PE2.4 Q7

File: China imports 1

Table PE2.4 Q7   Imports of petroleum by China from 2002 to 2005

| China – Imports of merchandise trade, value, Petroleum, crude, Million US dollars | | | |
	2002	2003	2004	2005
Jan	660	1770	2363	2301
Feb	669	1537	2519	3144
Mar	800	1765	2267	3701
Apr	1138	1807	2504	4656
May	1044	1227	2454	3883
Jun	1229	1545	3063	3987
Jul	1131	1415	2622	4242
Aug	1251	1449	2657	3607
Sep	1217	2119	3160	4775
Oct	1070	1460	2856	4856
Nov	1523	1663	3715	4295
Dec	1029	2068	3734	4413

Source: UNESCAP: Source(s): China–NBS Official Communication.

(a) Plot graphs to compare the imports for each year 2002 to 2005.
(b) For each year calculate the mean, median, standard deviation, CV and coefficient of skewness for the value of imports.
(c) Explain how the statistics in (b) relate to the graphs in (a).

8. Anecdotal evidence suggests that in recent years air traffic has grown much faster than traditional means of transport such as rail. Statistics for freight and passenger travel in Malaysia is given in Table PE2.4 Q8:

File: Malaysia rail

Table PE2.4 Q8   Freight and passenger travel by rail from 2002 to 2005

| Malaysia – Railway freight, Total, Million ton-kilometres | | | | | | | | | | | |
	Jan	Feb	Mar	Apr	May	Jun	Jul	Aug	Sep	Oct	Nov	Dec
2002	98	78	93	86	84	89	92	88	97	96	95	81
2003	91	63	76	77	75	65	80	75	71	64	70	81
2004	71	68	75	76	82	84	87	93	97	93	89	102
2005	102	84	101	98	102	100	98	103	103	103	90	95

Table PE2.4 Q8   Freight and passenger travel by rail from 2002 to 2005 (Cont.)

	Jan	Feb	Mar	Apr	May	Jun	Jul	Aug	Sep	Oct	Nov	Dec

Malaysia – Railway passenger travel, Total, Million passenger-kilometres

	Jan	Feb	Mar	Apr	May	Jun	Jul	Aug	Sep	Oct	Nov	Dec
2002	76	100	93	78	100	100	87	87	103	83	92	140
2003	85	81	85	61	73	86	74	85	94	81	92	134
2004	103	81	96	81	94	111	80	105	78	78	112	133
2005	87	91	99	84	94	114	92	104	93	82	122	135

Source: UNESCAP: Printed on: 23 July 2006 01:02:26, Source(s): Malaysia–DoS Official Communication.

(a) Plot graphs to compare (i) the volume of freight (ii) passenger transport for each year 2002 to 2005.
(b) For each year calculate the mean, median, standard deviation.
(c) Explain how the statistics in (b) relate to the graphs in (a).
Does this data for Malaysia support the anecdotal evidence?
9. The income tax payable: by annual income[1], 2003/04[2] United Kingdom is given in Table PE2.1 Q12:

File: Tax vs income

Table PE2.1 Q12   Income tax payable by annual income 2003/04

Income	Number of taxpayers (millions)	Total tax payable (£ million)	Average rate of tax payable (percentages)	Average amount of tax payable (£)
£4615–£4999	0.5	10	0.4	20
£5000–£7499	3.3	510	2.5	160
£7500–£9999	3.8	1 990	6.0	520
£10 000–£14 999	6.5	8 270	10.3	1 280
£15 000–£19 999	5.1	11 660	13.3	2 310
£20 000–£29 999	6.3	23 670	15.5	3 780
£30 000–£49 999	3.7	25 220	18.3	6 790
£50 000–£99 999	1.3	21 950	25.9	17 210
£100 000 and over	0.4	30 810	34.0	76 080
All incomes	30.7	124 090	18.0	4 040

1 Total income of the individual for income tax purposes including earned and investment income. Figures relate to taxpayers only. 2 Based on projections in line with the April 2003 Budget.
Source: Inland Revenue.

(a) Confirm that the mean tax paid per individual over all incomes is *approximately* £4,040.
(b) Calculate the lower quartile income, upper quartile income and median income of taxpayers from the Table PE2.1 Q12.
(c) Plot an Ogive; hence estimate the median and quartiles from the Ogive.
(d) Explain the different values of the median and mean.
In questions 10 to 12, large sets of data are not shown in the text but are given in Excel files.

10. **Excel File**: Cholesterol. Total blood cholesterol is the sum of LDL (bad cholesterol) and HDL (good cholesterol). The Excel file 'cholesterol' records the LDL and HDL cholesterol for all adults in a rural community in units of mmol/L (as part of research on lifestyles). Individuals are considered high risk if their LDL exceeds 6.5 and low risk when LDL is less than 5.2: LDL levels between 5.2 and 6.6 are 'borderline'. On the other hand, when HDL is above 1.0, the individual is considered to be low risk but high risk if their HDL is less than 0.9: HDL levels between 0.9 and 1.0 are 'borderline'.

Calculate the descriptive statistics for LDL and HDL cholesterol. Compare the LDL and HDL cholesterol levels.

11. (a) **Excel file**: 'Savings 5'

To encourage school children to develop the habit of saving money five pilot groups are selected. Rather than spending pocket money on 'junk' food during the school day, the children are asked to 'save' the money instead. The amounts saved per week during the school year are summarised in the Excel file 'savings 5'.

For each of the five groups, give the descriptive statistics for their savings. Compare the savings patterns of the five groups

(b) **Excel file: 'Sick days'**

This file gives the number of sick days taken annually for a random sample of 30 employees from a manufacturing company (A) and construction company (B). Calculate the summary statistics and measures of dispersion for each sample. Use these statistics to compare the sick days taken by employees in the two companies.

12. **Excel file:** Hours worked_UK_2000

The data in this file was collected in a labour force and summarises the distribution of usual weekly hours of work by gender in Spring 2000.

Based on the data for 0 to 59 usual hours worked per week,

(a) plot a graph to display the usual hours worked by males and females. What is the modal hours worked by each gender?

(b) Calculate the (i) average (ii) variance (population) and (iii) coefficient of variation for the hours worked by males and females.

Briefly describe the difference between weekly hours worked for males and females.

Summary and overview of summary statistics and measures of dispersion

The objective in this chapter is to calculate statistics to describe a set of data numerically in terms of a representative value, a measure of dispersion and a measure of skewness.

The mean value or average value is the most commonly used representative value. Because the calculation of the mean includes all the data its value is influenced by extreme data rendering it unrepresentative for the bulk of the data. In such circumstances the median should be used. The mode may be used to describe categorical data. The formulae for the calculation of summary statistics from both raw and grouped data are summarised in Table 2.6.

Measures of dispersion are necessary in the description of a set of data since the representative values (mean, median or mode) give no indication of the spread of values. The most common measure of dispersion is variance: the average of the sum of squared deviation of each datum from the mean value.

Table 2.11   Summary of measures of dispersion and other descriptive statistics

Measure	Raw data		Grouped data		When used
Range	Highest value – lowest value		(Upper limit of last interval) – (lower limit of first interval)		Seldom used except to quote such as 'incomes range from 15 to 50'.
Variance	$\sum \dfrac{(x_i - \mu)^2}{N}$	(2.11)	$\dfrac{\sum f_i (x_i - \mu)^2}{\sum f_i}$	(2.16)	Widely used, when data is symmetrically distributed
Standard deviation	$\sqrt{\dfrac{\sum (x_i - \mu)^2}{N}}$	(2.12)	$\sqrt{\dfrac{\sum f_i (x_i - \mu)^2}{\sum f_i}}$	(2.17)	
Semi-inter-quartile range	$Q_3 - Q_1$ where Q_1 and Q_3 are calculated from raw data	(2.14)	$Q_3 - Q_1$ Q_1 and Q_3 are calculated from the ogive or frequency distribution table	(2.14)	Semi-interquartile range or quartile deviation is used for skewed data. Variance is inflated by outliers in skewed data
Quartile deviation	$\dfrac{Q_3 - Q_1}{2}$	(2.15)	$\dfrac{Q_3 - Q_1}{2}$	(2.15)	
Coefficient of variation	$CV = \dfrac{\sigma}{\mu} \times 100$			(2.18)	Use to give a measure of dispersion relative to numeric values in data
Pearson's coefficient of skewness	$\dfrac{3(\text{mean} - \text{median})}{\text{standard deviation}}$			(2.19)	Use to give an indication of skewness

Like the mean, the variance is effected by extreme values in skewed data; therefore other measures of dispersion, such as quartile deviation should be used. The formulae for the calculation of measures of dispersion from both raw and grouped data are summarised in Table 2.11. Pearson's coefficient is just one of the statistics that indicate the degree of skewness in a set of data. Box plots graphically display the symmetry/skewness for a set of data.

REGRESSION AND CORRELATION: INTRODUCTION

Chapter Objectives

At the end of this chapter you should be able to do the following

- Plot scatter diagrams and describe the relationship between the variables
- Explain the term 'least squares'
- Calculate the equation of the least-squares line from formulae and calculator
- Plot the least-squares line
- Verbally describe the slope and intercept
- State the limitations on the use of least-squares line
- Use Excel to plot scatter diagrams with the trend line and its equation
- Explain and calculate the coefficient of determination (R^2) as the proportion of variation explained by the least-squares line
- Explain that the values R^2 must be in the range $0 \leq R^2 \leq 1$
- Explain that $\sqrt{R^2} = \pm r$ where r is the correlation coefficient and interpret r
- Calculate the correlation coefficient and coefficient of determination by formulae, by calculator and Excel
- Calculate and interpret the rank correlation coefficient
- Use regression and correlation to describe and interpret the type and strength of the relationship between two variables

3.1 Introduction to regression: Scatter plots and lines

The word 'regression' was first used by the geneticist, Francis Galton who noted that tall fathers had shorter sons and short fathers had taller sons. He described this phenomenon as regression where heights tended to 'regress' towards the mean height for the population.

In every discipline from genetics and the sciences to business, economics and the social sciences we are interested in whether a relationship exists between two (or more) variables and in describing the nature and strength of the relationship. The simplest relationship is a linear relationship. In this chapter the formula or equation relating the two variables will be derived from the sample data by a method called 'the method of least squares'. The strength of the linear relationship is quantified by a statistic called the correlation coefficient.

3.1.1 Scatter plots

For an overview of whether a relationship exists or not, plot the points on an XY graph, called a **scatter plot or scatter diagram**. For example, the scatter plots in Figure 3.1 give savings vs. income for incomes ranging from € 2000 to € 300 000. Savings (y) are plotted on the vertical axis against income (x) on the horizontal axis.

Information from Scatter plots

(a) **Type of relationship**: an overview of the type of relationship between the variables, i.e. whether it is linear, a curve of some sort or whether there is no relationship.

(b) **Strength of the relationship**; when the (x, y) points trace out a very definite line or curve it is referred to as a 'strong' relationship. Otherwise the trend is described as 'moderate' such as the trend in Figure 3.2 or 'weak' if the points are so scattered that it is difficult to discern any particular trend.

(c) **Positive trends** where y increases as x increases, the trend is upward sloping. **Negative trends** where y decreases as x increases, the trend is downward sloping.

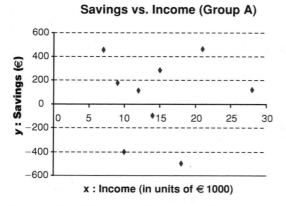

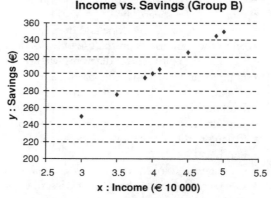

Figure 3.1(a) No discernible relationship between savings and income.

Figure 3.1(b) A perfect positive linear relationship between savings and income.

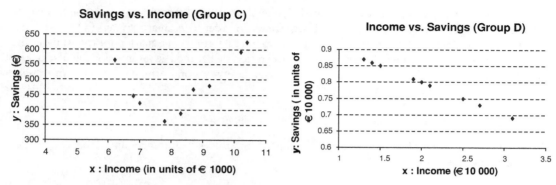

Figure 3.1(c) A curvilinear relationship between savings and income.

Figure 3.1(d) A perfect negative linear relationship between savings and income

Figure 3.1 Scatter plots for savings for various income ranges.

File: Savings vs. income

The graphs in Figure 3.1 represent the relationships ranging from no relationship at all in Figure 3.1(a) to perfect linear relationships in 3.1(b) and 3.1(d). Figure 3.1(c) is an example of a relationship that it is not linear: it is called a curvilinear relationship.

Linear trends. The examples given in Figure 3.1 were extreme cases: the points either fell on a perfect straight line or a curve or were completely random. In practise, in the case of 'real' data, the scatter plot will be more or less linear. For example, consider the annual savings for randomly selected individuals on incomes ranging from € 20 000 to € 110 000 given in Table 3.1:

Table 3.1   Savings vs. Income 1

File: Savings vs. income 1

$x =$ Income ($\times$ € 10 000)	2	3	4	5	6	7	8	9	10	11
$y =$ Savings (€)	240	250	364	350	472	422	410	495	687	750

The scatter plot for savings (y) against income (x) is given In Figure 3.2.

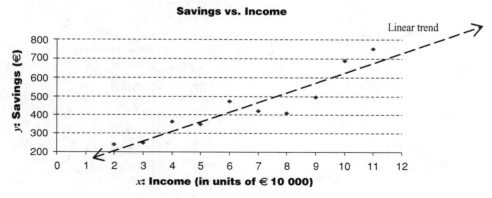

Figure 3.2   A linear trend between Savings (y) against Income (x).

The scatter plot in Figure 3.2 reveals that the general tendency for savings to increase as income increases is described by a straight line or a linear trend. In order to estimate savings (y) for incomes other than those given in the data it would be useful to have a formula or equation for the straight line relating savings (y) to income (x). In this chapter the rationale behind the method for calculating the equation of the 'best fit' line will be explained. Hence a brief revision of the mathematics of straight lines is essential before proceeding any further.

3.1.2   Review of the mathematics of straight lines

The equation of a straight line may be defined uniquely in terms of its intercept and slope: in statistics this is usually written as $y = a + bx$

- x is called the **independent variable** and is plotted on the horizontal axis. For example, $x =$ income in Figures 3.1 and 3.2.
- In regression analysis the independent variable, x may be referred to as the explanatory variable or the predictor variable or the exogenous variable (in economics).
- y is called the **dependent variable** and is plotted on the vertical axis. For example, $y =$ savings in Figures 3.1 and 3.2.
- In regression analysis the dependent variable, y may be referred to as the response variable or the regressor variable or the endogenous variable (in economics).
- In the equation of the line $y = a + bx$, a is described as **the intercept** and it is the point where the line crosses the vertical or y-axis.
- In the equation of the line $y = a + bx$, b is the described as **the slope** or **slant** of the line: slope is the change in y per unit increase in x. The slope is measured by taking any section of the line, such as A to B in Figure 3.3, measuring the change in height and the change in horizontal distance then
- Slope, $b = \dfrac{\text{change in vertical height }(y)}{\text{change in horizontal distance }(x)} = \dfrac{\Delta y}{\Delta x}$
- A slope of b means that an increase of one unit along the horizontal (x-axis) is accompanied by a change of b units along the vertical axis.

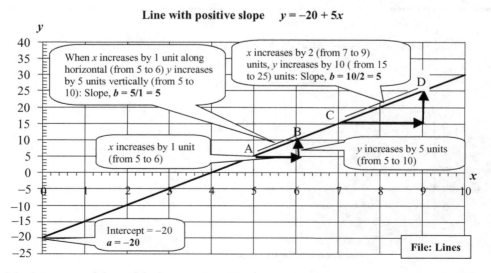

Figure 3.3   Intercept and slope of the line $y = 5x - 20$.

To plot the line $y = 5x - 20$ first calculate at least teo points: for example Table 3.2 gives the calculated values of y when $x = 0, 1, 2 \ldots 9$; plot these points then draw the line.

Table 3.2   Selected points for the line $y = 5x - 20$

x	0	1	2	3	4	5	6	7	8	9
y	−20	−15	−10	−5	0	5	10	15	20	25

3.1.2.1   An example of a straight line with positive slope

$a = -20$ (intercept)

$b = 5$ (slope)

The equation of the line in Table 3.2 is $y = -20 + 5x$. This line cuts the y axis at $y = -20$: the slope is 5. To 'see' what is meant by saying the slope is 5, sections of the line 'A to B' and 'C to D' are highlighted showing that for any section an increase of one unit along the horizontal (x) is accompanied by an increase of 5 units along the vertical (y).

3.1.2.2   Straight line with negative slope

The equation of the line in Figure 3.4 is $y = 16 - 4x$. The line has a negative slope, $b = -4$. Therefore y decreases by 4 units for each unit increases in x. The intercept in this case is $a = 16$.

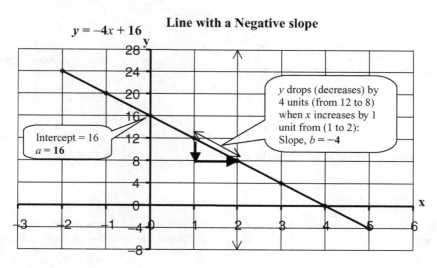

Figure 3.4   Intercept and slope of the line $y = 16 - 4x$.

Table 3.3   Selected points for the line $y = 16 - 4x$:

x	−2	−1	0	1	2	3	4	5
y	24	20	16	12	8	4	0	−4

3.2 The least-squares line. Criteria and equation for the best fit line

The least-squares line is the standard 'best' fit line that can be fitted to a set of data. In this section the criterion for a best fit line is explained. Then, based on this criterion, formulae are derived from which the equation of the least-squares line is calculated.

3.2.1 What is a 'best-fit' line?

If we were to rely on individual judgement to draw a line through a set of points such as those in Table 3.4 there would be numerous possibilities as illustrated in Figure 3.5:

Table 3.4   Rate of pay per hour for 4 randomly selected employees   File: Rates of pay

x: Hours worked per day	3	5	7	9
y: Rate of pay per hour (£)	7	6	7	40

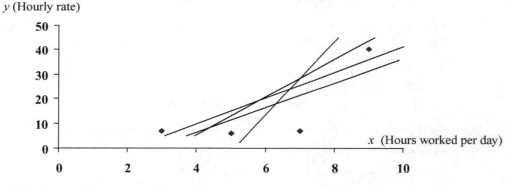

Figure 3.5   Some of the many possible trend lines.

This fairly extreme example illustrates the necessity for a 'standard, best fit' line.

Criteria for a best-fit line

To begin, we assume that the equation of the best fit line is: $y = a + bx$, where the objective is to find the values of a (intercept) and b (slope) so that this line is the standard 'best fit' line for the data.

- It would seem reasonable that a best-fit line should be as close as possible to all the points and well balanced between them. One way of measuring the closeness of a point to the line is to measure the vertical distance between the point and the line: This idea is illustrated for one point (x_4, y_4) in the scatter plot in Figure 3.6. The vertical distance between the point (x_4, y_4) and the corresponding point on the line is $d_4 = y_4 - \hat{y}_4$.

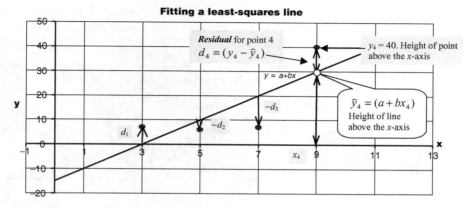

Figure 3.6   Residuals $d_1, -d_2, -d_3, +d_4$ for the 'best fit' line: $y = a + bx$.

Notes:

y_4 is the height of that **point** above the x-axis along the vertical line through $x = x_4$.

$\hat{y}_4$ is the height of the **line** above the x-axis along the vertical line through $x = x_4$.

The height of the line at $x = x_4$ is $\hat{y}_4$ where $\hat{y}_4$ is calculated by substituting $x = x_4$ into the equation of the line: Hence $\hat{y}_4 = a + bx_4$.

$\hat{y}$ is called 'y–hat'

Residuals

The vertical distance of any point (x_i, y_i) from the 'best fit' line is given as $d_i = y_i - \hat{y}_i$. d_i is frequently called a **'residual'**.

Residuals are positive for data points above the line; residuals are negative for points below the line.

A possible criterion for a best-fit line would be that the sum of the residuals

$$\sum d_i = + d_1 - d_2 - d_3 + d_4,$$

should be as small as possible.

But, for the 'best-fit' line this sum is always precisely zero: $\sum d_i = + d_1 - d_2 - d_3 + d_4 = 0$.

In other words, the best-fit line is so perfectly balanced between all the data points that the sum of the residuals above the line is the same as the sum of the residuals below it, giving a net sum of zero. This property, where the sum of positive and negative deviations is zero, arises frequently in statistics (remember deviations from the mean $\sum (x_i - mean) = 0$) and is generally dealt with by **squaring before** adding. **Hence, the criterion for the best fit line** is that the sum of the squared residuals, $\sum (d_i)^2$ should be a minimum as stated in (3.1):

$$\sum (d_i)^2 = (d_1)^2 + (-d_2)^2 + (-d_3)^2 + (d_4)^2 = (d_1)^2 + (d_2)^2 + (d_3)^2 + (d_4)^2 \qquad \text{(3.1)}$$

The criterion for the best fit line is that the sum of the squared residuals, $\sum (d_i)^2$ should be a minimum. Hence the name **'least-squares line'**.

Note: $\sum (d_i)^2 = (d_1)^2 + (d_2)^2 + (d_3)^2 + (d_4)^2$ is called a 'Sum of Squares'.

Based on the criterion in (3.1), formulae (3.2) and (3.3) are derived mathematically (see Appendix C). These formulae are used to calculate the slope b and intercept, a for the least-squares line for any set of (x, y) points.

$$\text{Slope: } b = \frac{n\left(\Sigma xy\right) - \left(\Sigma x\right)\left(\Sigma y\right)}{n\left(\Sigma x^2\right) - \left(\Sigma x\right)^2} \tag{3.2}$$

$$\text{Intercept: } a = \frac{\Sigma y - b\,\Sigma x}{n} \tag{3.3}$$

Hence the equation of the least-squares line $y = a + bx$.

where n = number of points.

Σx = sum of the x co-ordinates. Σx^2 = sum of x^2 (first square each x, then add)

Σy = sum of the y co-ordinates. Σxy = sum of the $(x \times y)$ values for each point.

Notes:

- Σx^2 is **not** the same as $(\Sigma x)^2$
- In regression, y is called the **dependent variable or response**: x is called the **independent variable,** the **predictor,** the **regressor** or the **explanatory variable.**
- There are several other formulae for a and b. These are listed at the end of the chapter.
- The values of a and b may be calculated directly (see Worked Example 3.1), using the calculator, Excel or Minitab or other statistical packages.
- There are alternative formulae for a and b: some of them are given in (3.2a), (3.2b) and (3.3a).

$$b = \frac{\Sigma\left(x - \bar{x}\right)\left(y - \bar{y}\right)}{\Sigma\left(x - \bar{x}\right)^2} \tag{3.2a}$$

$$b = \frac{\Sigma xy - \frac{\Sigma x \Sigma y}{n}}{\Sigma x^2 - \frac{\left(\Sigma x\right)^2}{n}} \tag{3.2b}$$

$$a = \bar{y} - b\bar{x} \tag{3.3a}$$

Limitations on the use and interpretation of a regression line

- Changes in y are *not necessarily caused* by changes in x. Changes in each variable may be caused by other factors. The variable, x may be just one of several variables that explain changes in y.
- The equation of the least-squares line is **only valid for the range of values of x** from which the least-squares line was calculated. There is no guarantee that the trend will continue outside this range.

Remember that the symbol, $\hat{y}$ (called y-hat) represents the value of y estimated by the equation of the least-squares line.

Worked Example 3.1 is a simple example used to demonstrate the calculation of the least-squares line and comments on its interpretation and limitations on its use.

Worked Example 3.1: Calculate and plot a least-squares line

The data in Table 3.4 gives the hourly rate of pay and hours worked per day for four randomly selected employees in a large company.

Table 3.4   Hourly rate (£) vs. hours per day for 4 randomly selected employees

x (hours per day)	3	5	7	9	File: Rates of pay
y (rate per hour, £)	7	6	7	40	

  (i) Plot the data on a scatter diagram and comment on the trend.

 (ii) Calculate the slope (*b*) and intercept (*a*) of the least-squares line: $y = a + bx$ and plot this line on the scatter diagram.

(iii) Give a verbal explanation of the slope and intercept of least-squares line. Comment on the limitations on the use of the least-squares line.

(iv) Use the equation of the least-squares line to **estimate** the hourly rate of pay for those who work 3, 5, 7 and 9 hours per day.

 (v) Use the results in (iv) to verify that the sum of the residuals is zero:

$$\sum d_i = +d_1 - d_2 - d_3 + d_4 = 0: \text{ that is, } \sum d_i = \sum (y_i - \hat{y}_i) = 0.$$

(vi) Can you assume that changes in hourly rate are explained completly by the number of hours worked per day?

(vii) Show that the point $(\bar{x}, \bar{y})$ lies on the least-squares line.

Solution

  (i) The scatter plot is given in Figure 3.7a shows a weak trend, that appears to be more curvilinear than linear.

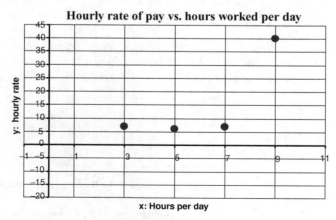

Figure 3.7a   Scatter diagram for hourly rate (£) vs. hours per day for 4 randomly selected employees.

(ii) Calculate the equation of the least-squares line: $y = a + bx$ using formula (3.2) and (3.3). To evaluate b and a the following summations must be calculated from the data then substituted into (3.2) and (3.3). Calculations are set out in Table 3.5.

$\sum x =$ sum of the x co-ordinates (column 1), $\sum x = 24$,
$\sum y =$ sum of the y co-ordinates (column 2), $\sum y = 60$,
$\sum x^2$, the sum of x^2 (column 3), $\sum x^2 = 164$
$\sum xy =$ sum of the $(x \times y)$ values for each point (column 4) $\sum xy = 460$.
n: the number of data points, $n = 4$

Table 3.5   Calculations of sums for intercept and slope of least-squares line

x: hours per day	y: rate per hour	$x \times y$	x^2
3	7	$(3)(7) = 21$	9
5	6	$(5)(6) = 30$	25
7	7	$(7)(7) = 49$	49
9	40	$(9)(40) = 360$	81
$\sum x = 24$	$\sum y = 60$	$\sum xy = 460$	$\sum x^2 = 164$

To calculation of slope, b, by substituting the summations in Table 3.5 into formula (3.2)

$$b = \frac{n(\Sigma xy) - (\Sigma x)(\Sigma y)}{n(\Sigma x^2) - (\Sigma x)^2} = \frac{4(460) - (24)(60)}{4(164) - (24)^2} = \frac{1840 - 1440}{656 - 576} = \frac{400}{80} = 5$$

Calculate intercept, a, from formula (3.3):
Now that the value of b is known, substitute this into formula (3.3) to calculate a

$$a = \frac{\Sigma y - b\Sigma x}{n} = \frac{60 - 5(24)}{4} = \frac{-60}{4} = -15$$

Substitute $a = -15$ and $b = 5$ into general equation for the least-squares line: $y = a + bx$
Hence **y = −15 + 5x is the equation of the least-squares line** for the data in Table 3.4.

To plot the line: calculate at least 2 points (x, y) from the equation. As a general rule calculate y for the lowest and highest value of x in the data and also for some intermediate value. Plot these points on the scatter diagram; join the points to obtain the least-squares line.

For example take $x = 0$, $x = 4$, $x = 9$
When $x = 0$, $\hat{y} = -15 + 5(0) = -15$
When $x = 4$, $\hat{y} = -15 + 5(4) = 5$
When $x = 9$, $\hat{y} = -15 + 5(9) = 30$

Plot these 3 points on the scatter diagram in Figure 3.7b. The least-squares line is the line joining the calculated points.

(iii) **The slope b is 5**. This means that rate per hour increases by £ 5 when the daily hours worked increases by one.

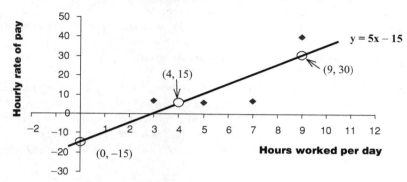

Figure 3.7b The least-squares line for hourly rate (£) vs. hours per day for 4 randomly selected employees.

Comment: This statement applies to those who work between three and nine hours daily only. It cannot be assumed that the trend continues for those who work less than three or more than nine hours daily.

The intercept $a = -15$. This means those who work no hours ($x = 0$) daily would earn $-£15$ per hour!

Comment: A rate of $-£15$ per hour makes no sense. This example reinforces the rule that the equation for the least-squares line is only valid for the range of x-values given in the data: $x = 3$ to 9.

(iv) The estimated values of y for $x = 3, 5, 7$ and 9 from the equation of the Least-squares line are given in Table 3.6. The sum of the residuals (in last column) is $(7 - 4 - 13 + 10) = 0$.

Table 3.6 Calculations showing $\sum(y_i - \hat{y}) = 0$ for least-squares line

x: hours per day	y: rate per hour	$\hat{y}_i = -15 + 5x_i$	$y_i - \hat{y}_i = $ residual	Residual
3	7	$-15 + 5(3) = 0$	$7 - 0 = 7$	7
5	6	$-15 + 5(5) = 10$	$6 - 10 = -4$	-4
7	7	$-15 + 5(7) = 20$	$7 - 20 = -13$	-13
9	40	$-15 + 5(9) = 30$	$40 - 30 = 10$	10
$\sum x = 24$	$\sum y = 60$		$\sum(y_i - \hat{y}) = 0$	0

(v) No. Changes in hourly rate may be due to numerous factors such as type of employment, qualifications etc.

(vi) From Table 3.6, $\sum x = 24$ and $\sum y = 60$.

Hence $\bar{x} = \dfrac{\sum x}{n} = \dfrac{24}{4} = 6$ and $\bar{y} = \dfrac{\sum y}{n} = \dfrac{60}{4} = 15$

Substitute $\bar{x} = 6$ and $\bar{y} = 15$ for x and y into the equation of the least-squares line

$$y = -15 + 5x$$
$$15 = -15 + 5(6)$$
$$15 = 15$$

The point ($\bar{x} = 6$ and $\bar{y} = 15$) 'satisfies the equation', therefore it is on the line

We say the point 'satisfies the equation' if, on substitution of the values of x and y the *Left-hand side = Right hand side*. This is true in this example.

In fact the mean values of a set of data, $(\bar{x}, \bar{y})$ will always lie on the least-squares line.

Important: the equations for y vs. x and x vs. y are different

The equation for the least-squares line for y vs. x was derived by minimising the squared deviations of the y-values from the line. The equation for the least-squares line for x vs. y would be derived by minimising the squared deviations of the x-values from the line. These two lines will have different equations. It will be left as an exercise to the reader in Progress Exercises 3.1, question 8 to show that least-squares line for hours worked per week (x) vs. hourly rate (y) in Table 3.4 is $x = 0.1199y + 4.2014$!

Skill Development Exercise SK3_1: Calculating a least-squares line

Calculate and plot the least-squares line on the scatter plot for the data given in Table 3.1 (File: Savings vs. income 1).

Table 3.1 Savings vs. Income

x = Income ($\times$ € 10 000)	2	3	4	5	6	7	8	9	10	11
y = Savings (€)	240	250	364	350	272	422	410	495	687	750

Answer

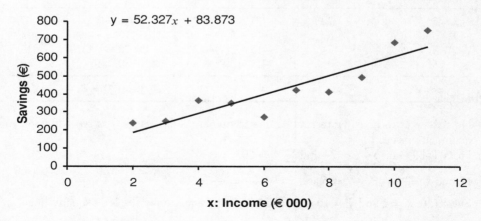

Figure SK3_1.1   Scatter plot and least-squares line for Savings vs. Income 1.

Skill Development Exercise SK3_2: Calculate a least-squares line profit vs. cars serviced

A garage offers a special rate for servicing cars that are due for the NCT test. The number of cars serviced and the profit in £ (hundreds) made by the garage is given for 5 randomly selected weeks.

Table SK3_2.1   Profit vs. number of cars serviced

x: Cars serviced	8	9	5	12	16
y: Profit (£ 00)	8	18	10	22	22

(a) Calculate the least-squares line for profit against units serviced (y against x).
(b) Plot the least-squares line on the scatter diagram.
(c) Verbally describe the slope and intercept. Comment on the limitations imposed on the use of the least-squares line.
(d) Use the equation of the line to estimate profit if 10 cars are serviced in one week. Comment on the scatter about the line.
    Would you consider this estimate of profit to be exact?
(e) Use the equation of the line to estimate profit if 8: 9: 5: 12 and 16 in one week. Hence calculate the residuals and show that the sum of the residuals is zero.

Answers

(a) The equation of the least squares line, $\hat{y} = -5.6 + 2x$.
(b)

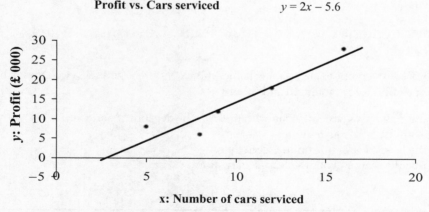

Figure SK3_2.1   Scatter plot and least-squares line for Profit vs. Number of cars serviced.

(c) **Slope** is the rate of change, that is, the change in y per unit change in. In this problem, slope is the rate at which profit increases for each additional car serviced. Therefore, the engineer makes a profit of two hundred pounds (£ 200) for each additional car serviced.

Intercept is the value of y when x is zero. The intercept of $-5.6 \times £100 = -£560$ could represent negative profit (a loss!). This may represent fixed costs, the cost of purchasing raw materials etc. before any cars are serviced.

(d) When 10 cars serviced in one week:

$$x = 10 \rightarrow y = -5.6 + 2(10) = 14.4 \text{ (that is } £1440)$$

In Figure SK3_2.1 the trend is fairly strong, the data points quite close to the least-squares line. Hence the estimated profits should be within a small margin of error of the true profit.

(e) $\hat{y}$, the values of y estimated by the equation of the least-squares line are given in column 6 in Table SK3_2.2.

The residuals are calculated in column 7. The sum of the residuals is zero.

Table SK3_2.2   Sums for calculation of slope and intercept of least-square line

				col. 5	col. 6	col. 7
x: Cars serviced	y: profit (£00)	$x \times y$	x^2	y^2	$\hat{y}$	$y - \hat{y}$
8	6	48	64	36	10.4	−4.40
9	12	108	81	144	12.4	−0.40
5	8	40	25	64	4.4	3.60
12	18	216	144	324	18.4	−0.40
16	28	448	256	784	26.4	1.60
Totals	**50.00**	**72.00**	**860**	**570**	**1352**	**0.00**

Skill Development Exercise SK3_3: Same least-squares line, more scatter

Table SK3_3.1 is the profit made by a mechanic who reconditions and sells two types of bicycle For **each type of bike**: (a) mountain and (b) city-jet

(i) Plot the data on a scatter diagram with profit as the dependent variable
(ii) Calculate the least-squares line
(iii) Plot the least-squares line on the scatter plot
(iv) Describe the scatter about the least-squares line

Table SK3_3.1   Profit on sale of reconditioned bikes

	Number sold	1	3	5	9	12
(a) Mountain	Profit €	−300	700	420	500	1600
(b) City-jet	Profit €	−55	228	452	945	1351

Answers are summarised on the graphs below.

The equations of the least-squares lines are almost identical, but the data points are scattered about the line in (a) but close to or on the line in (b).

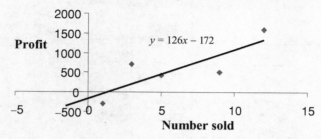

Figure SK3_3.1   Profit vs. number of mountain bikes sold.

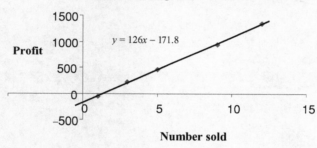

Figure SK3_3.2   Profit vs. number of city jet bikes sold.

3.3   Excel: XY (scatter) plots, the least-squares lines and formulae

Plot the XY (*scatter*) diagram for the data in Table 3.4 with chart title and a title for each axes in Excel as described in Chapter 1. To add the least-squares line and its equation of the scatter plot follow the steps set out below.

1. Point the cursor at any one of the data points and right-click the mouse; the 'Format data series' dialogue box appears.
2. Select 'Add Trendline' to bring up the 'Add Trendline' dialogue box, as illustrated in Figure 3.8.
3. Select the type of trend that best fits the data on the scatter diagram – for the present we will select 'linear', but note the other possibilities for curvilinear trends for future reference.
4. Click on 'Options' tab.
   The 'options' menu gives you several options, Figure 3.9.

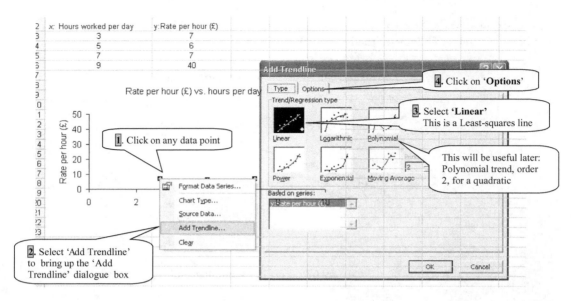

Figure 3.8   Right click on any data point → Add Trendline → Options.

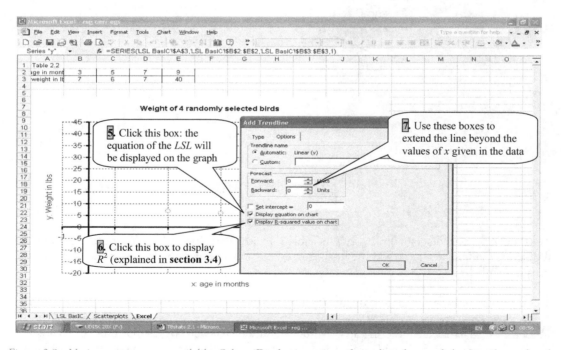

Figure 3.9   Various options are available. Select 'Display equation of trend' and extend the *Line* forward and backward.

For this example select the following:

5. Select 'display equation on chart'. This is the equation of the least-squares line for the data that Excel will plot on the scatter plot for values of x value given in the data.

6. Select display the 'R-squared value on chart'. The statistic R^2 will be explained in **section 3.4**.

7. To extend the line from either end enter the number units required in the boxes 'forward' or 'backward'.

The Least-squares line and its equation is given Figure 3.10:

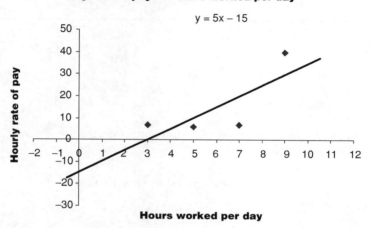

Figure 3.10   Excel: the least-squares line (linear trend line) and its equation is displayed on the scatter plot.

Skill Development Exercise SK3_4 in Excel: Savings vs. income Figure 3.1

Tables SK3_4.1 to SK3_4.4 below (File: Savings vs. income) give the data for savings vs. income in Figure 3.1 (a) to (d).

(a) For each set of data use Excel to plot the scatter diagram and the least-squares line with its equation.

(b) Fit a quadratic trend (called a polynomial, order 2) to the data in Table SK3_4.3. How does the quadratic trend compare with the linear trend?

Table SK3_4.1   Data for Figure 3.1(a)

File: Savings vs. income

Income (× 1000)	14	18	28	7	9	12	10	15	21
Savings	−100	−500	120	453	172	108	−400	280	459

Table SK3_4.2   Data for Figure 3.1(b)

Income (× 10 000)	3	3.5	3.9	4.0	4.1	4.5	4.9	5.0
Savings	250	275	295	300	305	325	345	350

Table SK3_4.3   Data for Figure 3.1(c)

Income (× 10 000)	6.2	6.8	7.0	7.8	8.3	9.2	8.7	10.2	10.4
Savings	565	445	422	361	387	478	468	592	624

Table SK3_4.4   Data for Figure 3.1(d)

Income (× 100 000)	1.3	1.4	1.5	1.9	2	2.1	2.5	2.7	3.1
Savings (× 1000)	0.87	0.86	0.85	0.81	0.8	0.79	0.75	0.73	0.69

Answers

(a) 3.1(a) $y = -0.0006x + 65.787$
   3.1(b) $y = 50x + 100$
   3.1(c) $y = 29.436x + 238.45$
   3.1(d) $y = -0.1x + 1$
(b) The quadratic trend is a better fit (closer to all the points) than the linear trend $y = 42.307x^2 - 678.04x + 3113.2$: Figure SK3_4.1.

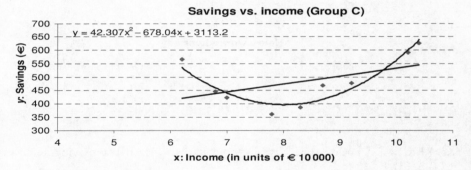

Figure SK3_4.1   Quadratic trend for savings vs. income (c).

Progress Exercises 3.1: Scatter plots and least-squares lines

It is recommended, at this introductory stage, that the reader attempt all these questions manually. Use Excel to confirm the answers.

1. Explain the terms 'slope' and intercept' of a straight line.
   State the value of slope and intercept and whether the line is upward sloping or downward sloping for each of the following equations
   (a) $y = 2 + 4x$ (b) $y = 8x - 5$ (c) $2y = 6x - 4$ (d) $48 - 4x = 4y$.
2. For each of the lines in question 1 above, evaluate y when $x = 0$ and $x = 10$. Hence plot each line.
3. Saving (S) is thought to be linearly related to income (Y is income in units of £ 000). The least-squares line, $S = -4 + 0.5Y$ was calculated from data for incomes ranging from 10 to 25 (£ 000).

(a) Plot the least-squares line for savings (S) against income (Y).

(b) Describe the intercept and slope in terms of income and savings.

(c) Is it possible to estimate savings for any income from this equation?

4. The following data gives the price of used cars for cars aged 1, 2, 3, 4, 5 and 6 years:

Table PE3.1 Q4   Value of cars against age

File: Car value

Age of car (years)	1	2	3	4	5	6
Value (£000)	9.5	9.3	6.0	6.5	4.7	5.2

(a) Plot the data on a scatter diagram.

(b) The least-squares line was calculated as $y = 10.35 - 0.99x$. Plot the line on the scatter plot. Estimate the value of a car that is 4.5 years old. Do you consider the estimate to be the actual value of the car? Give reasons for your answer.

(c) Is it reasonable to use this equation to predict the price of any car at any age?

5. The CSO published the number of new goods vehicles licensed for the first time during the months January to October 2001 – see Table PE3.1 Q5:

Table PE3.1 Q5   New goods vehicles licensed for the first from January to October 2001

Month (x)	Jan	Feb.	Mar.	Apr.	May	Jun.	Jul.	Aug.	Sept.	Oct.
New goods vehicles (,000) (y)	2.8	3.7	3.7	3.0	3.1	2.6	2.8	2.3	1.9	2.1

Source: CSO Ireland.

(a) Plot a scatter diagram for new goods vehicle against time

(b) Fit a least-squares line, assuming that $x = 1$ in January, $x = 2$ in February etc. and $\sum x = 55$, $\sum y = 28$, $\sum xy = 140.5$, $\sum x^2 = 385$, $\sum y^2 = 81.74$.

Based on the results above describe the trend in new goods vehicles licensed during the first 10 months of 2001

6. The percentage of smokers among the various age groups in Great Britain is given in Table PE3.1 Q6:

Table PE3.1 Q6   Prevalence of smoking in the adult population 2004   File: Smoking 2004

Great Britain		Percentages	
Ages	Mid-interval age	Men	Women
16–19	17.5	23	25
20–24	22	36	29
25–34	30	35	28
35–49	42	31	28
50–59	55	26	22
60 and over	70	15	14

Source: General Household Survey, Office for National Statistics.

(a) Plot a scatter diagram for the percentage of men and women that smoke against age (use the mid-interval age).

Describe the trend in each scatter diagram

(b) Omitting the first age group (16–19) in Table PE3.1 Q6 for men and women plot a scatter diagrams for (i) men against age and for (ii) women against age.

Calculate the equation of the least-squares lines for (i) men weight age and for (ii) women height age.

Give a verbal description of the least-squares lines.

Can you conclude that smoking is caused by age?

7. Table PE3.1 Q7 gives the percentages of adults drinking more than the recommended guidelines on at least one day last week (before the survey was taken) by age and sex in 2004.

Table PE3.1 Q7 Prevalence of drinking in the adult population 2004 | File: Drinking 2004 |

		Percentages	
Age	Mid-interval age	Men	Women
16–24	20	47	39
25–44	34.5	48	28
45–64	54.5	37	20
65 and over	74.5	20	5

Source: General Household Survey, Office for National Statistics.

Use the mid-interval ages in the following exercises.

(a) Plot a scatter diagrams for the percentages of (i) men against age and for (ii) women against age.

(b) Calculate the equation of the least-squares lines for (i) men and for (ii) women against age.

(c) Give a verbal description of the least-squares lines.

8. See Worked example 3.1, Table 3.4:

Table 3.4 Hourly rate (£) vs. hours per day for 4 randomly selected employees | File: Rates of pay |

| x (hours per day) | 3 | 5 | 7 | 9 |
| y (rate per hour, £) | 7 | 6 | 7 | 40 |

Plot the data for *x* vs. *y*.

Calculate the least-squares line for x: hourly rate (£) vs. y: hours per day. Plot this line on scatter diagram and give a verbal description of the least-squares line. Compare the equation of the line to that in WE 3.1. What do you conclude?

9. The heights of hand-crafted glass vases range from 5 to 22.5 cm. The heights and weights of six randomly selected vases are given in Table PE3.1 Q9:

File: Cut glass

Height (cm)	5	10	15	17.5	20	22.5
Weight (gm)	35	44.5	64	78.7	114	137.4

(a) Plot a scatter diagram for weights (y) against heights (x). Describe the relationship between weight and height.
(b) Calculate the equation of the least-squares line with weight as the response variable and height as the predictor variable.
   (i) Give a verbal description of the slope and intercept.
   (ii) Why is it nonsense to use the equation to estimate the weight of a vase that measures 1 cm high?
   (iii) State the limitations that apply to the use of this line when estimating weight from height.
(c) Would the equation of the least-squares line for height against weight differ from that in (b)? Confirm your answer by calculating the equation.
10. The numbers of males and females aged 25–34 and 55–59 that are in employment (in thousands) in all economic sectors is given in Table PE3.1 Q10

Table PE3.1 Q10   Numbers aged 25–34 in employment (,000)   **File: Gender**

	Year, Quarter	2005Q1	2005Q2	2005Q3	2005Q4	2006Q1
Age 25–34	Male	298	304.8	310.1	314.9	318.5
	Female	248.7	251.6	255.2	258.7	264.3
Age 55–59	Male	80.6	80.1	82	81.9	84.9
	Female	47.8	49.7	50.2	51.6	51.1

Source: CSO.

For each age group
(a) Plot a scatter diagram for the numbers of females (y) against the numbers of males (x) per quarter.
(b) Calculate the equation of the least-squares line for scatter plot in (a).
(c) Use the least-squares line in (b) to give a verbal description of the relationship between the numbers of males and females in employment.

3.4   Coefficient of determination and correlation

The equations for the least-squares lines in Figure SK3_2.1 and Figure SK3_2.2 reproduced below, were almost identical. Yet the points in Figure SK3_3.1 are quite scattered about the line indicating a relatively weaker relationship compared to that in Figure SK3_3.2.

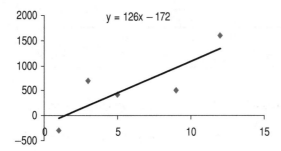

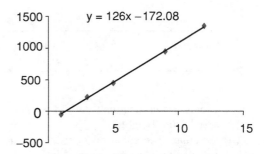

Figure SK3_3.1 Profits for mountain bikes vs. number of bikes sold. Moderately strong positive linear relationship between profit and numbers sold.

Figure SK3_3.2 Profits for City-Jet bikes vs. number of bikes sold. Very strong positive linear relationship between profit and numbers sold.

It is obvious that the least-squares line gives no indication regarding the strength of the relationship between the variables (i.e. the degree of scatter about the line); it is simply the 'best fit' line for *any* given set of points. It is for this reason that we require another statistic (recall a statistic is a sample characteristic) whose value will be indicative of the strength of the relationship.

The statistics that indicate the strength of a *linear* relationship are: **the coefficient of determination R^2** and its square root, the **correlation coefficient r.**

3.4.1 The coefficient of determination R^2 and correlation r

The value of R^2 will be shown to range from $R^2 = 0$, when there is no linear relationship between the variables to $R^2 = 1$ when there is a perfect linear relationship/trend. The correlation coefficient, r is defined as $r = \pm\sqrt{R^2}$, the value of r ranges from -1 to 1. Here again, a value of $r = 0$ indicates that there is no relationship between the variables; values of $r = -1$ or $r = +1$ indicate perfect linear trends; a negative sign indicates a downward sloping or decreasing trend and a positive sign indices an upward sloping or increasing trend.

The properties of **R^2 and r** are illustrated graphically in Figure 3.11 by four different scatter plots.

In graphs 3.11(a) $R^2 = 1$. All points lie on the least-squares line. Values of y estimated from the equation of the line are exactly the same as the observed values of y – so we say 'the least-squares line explains all the changes in y' by changes in x. The correlation coefficient is positive since the trend is positive (upward sloping): $r = \pm\sqrt{R^2} = \pm\sqrt{1} = +1$.

In the context of this example, the hourly rate of pay increases by £75 per hour for those who work between three and nine hours per day ($3 \leq x \leq 9$). For example, an accountant who works three hours per day is paid £100 per hour ($y = 3(75) - 125 = 100$) but an accountant who works five hours per day is paid £$(5 \times 75 - 125) = £250$ per hour. The hourly rate has increased by $2 \times £75$ above the rate for those who work three hours per day.

The scatter plots in **Figure 3.11(b)** shows a strong negative trend with the points close to the least-squares line: the calculated value of $R^2 = 0.9333$, which is close to 1. Hence values of y estimated by this line will be close to the observed values: the least-squares line explains most of the changes in y in terms of changes in x. The correlation coefficient, $r = -0.9661$, is negative, since the trend is downward sloping. In this example, the hourly rate of pay decreases by £8 per hour for each additional hour worked for those who work between one and five hours per day. A member of the catering staff who works two hours per day is paid £70 per hour ($y = -8(2) + 86 = 70$) while a person who works three hours per day is paid £62 per hour ($y = -8(3) + 86 = 62$). Those who work more hours per day are paid less per hour ($1 \leq x \leq 5$).

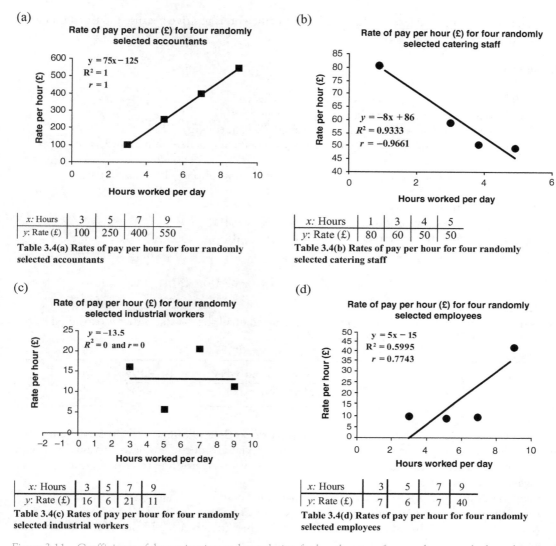

(a)

Rate of pay per hour (£) for four randomly selected accountants

$y = 75x - 125$
$R^2 = 1$
$r = 1$

x: Hours	3	5	7	9
y: Rate (£)	100	250	400	550

Table 3.4(a) Rates of pay per hour for four randomly selected accountants

(b)

Rate of pay per hour (£) for four randomly selected catering staff

$y = -8x + 86$
$R^2 = 0.9333$
$r = -0.9661$

x: Hours	1	3	4	5
y: Rate (£)	80	60	50	50

Table 3.4(b) Rates of pay per hour for four randomly selected catering staff

(c)

Rate of pay per hour (£) for four randomly selected industrial workers

$y = -13.5$
$R^2 = 0$ and $r = 0$

x: Hours	3	5	7	9
y: Rate (£)	16	6	21	11

Table 3.4(c) Rates of pay per hour for four randomly selected industrial workers

(d)

Rate of pay per hour (£) for four randomly selected employees

$y = 5x - 15$
$R^2 = 0.5995$
$r = 0.7743$

x: Hours	3	5	7	9
y: Rate (£)	7	6	7	40

Table 3.4(d) Rates of pay per hour for four randomly selected employees

Figure 3.11   Coefficients of determination and correlation for hourly rates of pay vs. hours worked per day.

File: Hourly rate

In graph **3.11(c),** there is no relationship between the variables; the equation of the horizontal line is $y = 13.5$. So, no matter what value x takes, the estimated value of y is always 13.5. Hence none of the changes (or variation) in y is explained by the least-squares line. $R^2 = 0$ and $r = 0$.

In this example, the least-squares lines estimates (or predicts) that individuals are all paid the same rate, £ 13.5 per hour, irrespective of how many hours they work per day ($3 \leq x \leq 9$).

The scatter plot in **Figure 3.11(d)** shows a weak relationship; the points are very scattered about the least-squares line. According to least-squares line the hourly rate increases as hours worked per day increase but the observed hourly rates of pay are different from those estimated by the line: hence the value of R^2 is small (0.5995). The correlation coefficient, is positive because the trend is upward sloping ($r = 0.7743$).

In summary

1. $\mathbf{0 \leq R^2 \leq 1}$: the value of R^2 ranges from 0 to 1 inclusive.
   $\mathbf{R^2} = 0$ when there is no linear relationship; see Figure 3.4(c)
   $\mathbf{R^2} = 1$ when there is a perfect linear relationship; see Figure 3.4(a)
2. The correlation coefficient is the square root of the coefficient of determination: $r = \pm\sqrt{R^2}$:

   $-1 \leq r \leq 1$: the value of the correlation coefficient, r, ranges from -1 to 1 inclusive.

   $r = -1$ for a perfect negative linear relationship
   $r = +1$ a perfect positive linear relationship. See Figure 3.4(a)
   $r = 0$ when there is no linear relationship. See Figure 3.4(c)

*3.4.2   The derivation of the coefficient of determination, R^2

* *This section is optional.*
R^2 is the proportion of the variation in y (from the mean of all the y-values) that is explained by the least-squares line; it is defined in equation (3.4).

$$R^2 = \frac{\text{Explained variation of y's from y-bar}}{\text{Total variation of y's from y-bar}} = \frac{\sum (\hat{y}_i - \bar{y})^2}{\sum (y_i - \bar{y})^2} \tag{3.4}$$

The **Explained variation,** also referred to as the **Regression Sum of Squares (RSS)** is the sum of the squared differences $(\hat{y}_i - \bar{y})^2$ the sum of the squared deviations of the estimated y values ($\hat{y}$) from y-bar. The variation is 'explained' in the sense that the least-squares line (regression line) explains changes in y in terms of x.

Total variation, $\sum (y_i - \bar{y})^2$ is also referred to as the **Total Sum of Squares (TSS)**.
TSS is the sum of the squared deviations of the observed y values from y-bar.
Worked Example 3.2 should serve to clarify the symbols: y (observed), $\hat{y}$ (estimated), $\bar{y}$ (mean) and terminology such as explained variation and total variation.

Worked Example 3.2: y (observed), $\hat{y}$ (estimated), $\bar{y}$ (mean), illustrated graphically:

The data in Table 3.4 gives the hourly rate of pay and hours worked per day for four randomly selected employees in a large company

Table 3.4   Hourly rate (£) vs. hours per day for four randomly selected employees | File: Hourly rate

| x (hours per day) | 3 | 5 | 7 | 9 |
| y (rate per hour, £) | 7 | 6 | 7 | 40 |

(a) In Excel, calculate the least-squares line for hourly rate as the dependent variable and hours worked as the explanatory variable. Plot the data and the least-squares line on a scatter plot.

(b) Calculate y-bar, $\bar{y}$. Plot the line $y = \bar{y}$ on the graph in (a).

(c) Show clearly the values of y_i, $\hat{y}_i$ for the fourth point in Table 3.4.

(d) On the graph indicate the total deviation, $(y - \bar{y})$; the explained deviation, $(\hat{y} - \bar{y})$ and the unexplained deviation, $(y - \hat{y})$ when $x = 9$.

Solution

(a) From Excel, the scatter plot with the least-squares line: $y = -15 + 5x$ is given in Figure 3.7, Worked Example 3.1.

(b) $\bar{y} = \dfrac{\sum y_i}{n} = \dfrac{7 + 6 + 7 + 40}{4} = 15$

The line $y = 15$ is a horizontal line that cuts the y axis at 15. This is illustrated graphically in Figure 3.11.

(c) For the observed point $(9, 40)$,

$y_i = 40$ the **observed** value when $x = 9$;

$\hat{y}_i$ is the value of y **estimated** by the least-squares line when $x = 9$.

$\hat{y}_i = a + b(x_i) = -15 + 5(9) = 30$.

Graphically it is the point $(x = 9, y = 30)$ on the least-squares line. See Figure 3.12.

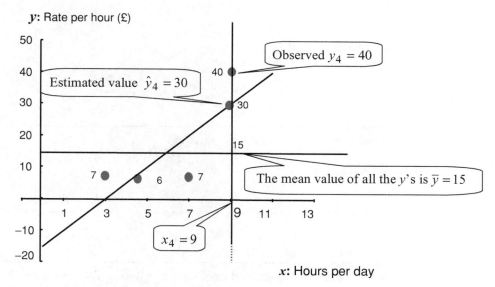

Figure 3.12   The points y_i $\hat{y}_i$ at $x = 9$ and the line $y = \bar{y}$, data in Table 3.4.

(d) Refer to Figure 3.13 $y_4 = 40$, $\hat{y}_4 = 30$, $\bar{y} = 15$, $x_4 = 9$

The total deviation is the difference between the observed values of y from the mean value of all the y's. Hence, when $x_4 = 9$, $y_4 - \bar{y} = 40 - 15 = 25$

The explained deviation is the difference, $(\hat{y}_4 - \bar{y})$ is $30 - 15 = 15$

The unexplained deviation is the difference between the observed value of y and the corresponding estimated value of y.

Hence the unexplained deviation $(\hat{y}_4 - y_4) = 40 - 30 = 10$

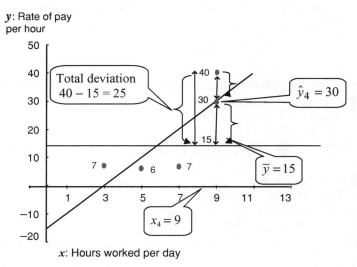

Figure 3.13   Rate per hour vs. Hours per day. Total deviation = Explained deviation + Unexplained deviation.

Note: 'variation' is 'deviation' squared
With the help of mathematics, you can prove that

Total variation = Explained variation + Unexplained variation

$$\sum (y_i - \bar{y})^2 = \sum (y_i - \hat{y}_i)^2 + \sum (\hat{y}_i - \bar{y})^2 \qquad (3.5)$$

$TSS = RSS + ESS$

Worked Example 3.3: Calculation of R^2 from formula (3.4)

The rates of pay per hour for four randomly selected employees are given in Table 3.4: reproduced here for convenience.

Table 3.4   Rate of pay per hour for four randomly selected employees

x: Hours worked per day	3	5	7	9
y: Rate of pay per hour (£)	7	6	7	40

Calculate the coefficient of determination using equation (3.4)

$$R^2 = \frac{\text{Explained variation}}{\text{Total variation}} = \frac{RSS}{TSS} = \frac{\sum (\hat{y}_i - \bar{y})^2}{\sum (y_i - \bar{y})^2}$$

Solution

Since $R^2 = \frac{RSS}{TSS}$ calculate (i) TSS (ii) RSS and finally (iii) $\frac{RSS}{TSS}$

(i) Total variation $(TSS) = \sum (y_i - \bar{y})^2$
    The calculations are carried out in Table 3.7.
    First calculate y-bar: $\bar{y} =$ mean value of the y-values,

$$\frac{\sum y_i}{n} = \frac{60}{4} = 15. \text{ See column 2}$$

Next calculate the deviation of each y-value from y-bar, then square. See Column 3.

Finally, sum the squared deviations to get the 'Total variation': $\sum (y_i - \bar{y}_i)^2 = 834$

Table 3.7 Calculation of explained and total variation for R^2

Column 1	Column 2	Column 3	Column 4	Column 5
x_i Hours worked per day	y_i Hourly rate (£)	Total variation (TSS) $(y_i - \bar{y})^2$	$\hat{y}$ $\hat{y}_i = 5x_i - 15$	Explained variation (RSS) $(\hat{y}_i - \bar{y})^2$
3	7	$(7 - 15)^2 = 64$	$5(3) - 15 = 0$	225
5	6	$(6 - 15)^2 = 81$	$5(5) - 15 = 10$	25
7	7	$(7 - 15)^2 = 64$	$5(7) - 15 = 20$	25
9	40	$(40 - 15)^2 = 625$	$5(9) - 15 = 30$	225
$\sum x_i = 24$	$\sum y_i = 60$	$\sum (y_i - \bar{y}_i)^2$ 834		$\sum (y_i - \bar{y}_i)^2 = 500$

(ii) Explained variation $(RSS) = \sum (\hat{y}_i - \bar{y})^2$
     Calculate $\hat{y}$ from the equation of the least-squares line: $\hat{y}_i = 5x_i - 15$ See column 4.
     Next calculate the deviation of each $\hat{y}$ (y-hat) from y-bar, then square. See column 5.
     Finally, sum the squared deviations to get 'Explained variation': $\sum (\hat{y}_i - \bar{y}_i)^2 = 500$

(iii) $R^2 = \dfrac{\text{Explained variation}}{\text{Total variation}} = \dfrac{RSS}{TSS} = \dfrac{\sum (\hat{y}_i - \bar{y})^2}{\sum (y_i - \bar{y})^2} = \dfrac{500}{834} = 0.5995$

The Coefficient of Determination $\times$ 100 is the percentage of the variation in y that is explained by the least-squares line. In this example, $R^2 \times 100 = 0.5995 \times 100 = 59.95\%$

In Progress Exercises 3.2 below you are required to calculate R^2 and explain the value of R^2 by referring to the graphs in Figure 3.11.

Progress Exercises 3.2: the Coefficient of Determination

---

1. The hourly rates of pay and hours worked per day for three randomly selected samples are given in Tables 3.4(a), (b) and (c).

Table 3.4(a)   Rates of pay per hour for four randomly selected accountants | File: Hourly rate |

x: Hours worked per day	3	5	7	9
y: Rate of pay per hour (£)	100	250	400	550

Table 3.4(b)   Rates of pay per hour for four randomly selected catering staff

x: Hours worked per day	1	3	4	5
y: Rate of pay per hour (£)	80	60	50	50

Table 3.4(c)   Rates of pay per hour for four randomly selected industrial workers

x: Hours worked per day	3	5	7	9
y: Rate of pay per hour (£)	16	6	21	11

For each data set in the tables above
(i) Calculate the coefficient of determination using formula (3.4). Calculate the correlation coefficient.
(ii) In Excel, plot the data for rate per hour against hours worked per day. Plot the least-squares line with its equation and the coefficient of determination. Give verbal explanations for the value of R^2 with reference to the graphs.

2. In Excel, plot the scatter diagram, linear trend, equation of the least-squares line and the coefficient of determination from the data for profit vs. sales of bikes in Table SK3_3.1.

Table SK3_3.1   Profit on sale of reconditioned bikes | File: Bikes |

	Number sold	1	3	5	9	12
(a) Mountain	Profit €	−300	700	420	500	1600
(b) City-jet	Profit €	−55	228	452	945	1351

3.4.3   Calculation of R^2 and r without reference to the least-squares line

Up to this point R^2 has been calculated directly from equation (3.4): to evaluate R^2 it has been necessary to calculate the equation of the least-squares line first. Many researchers are only concerned with the strength of the relationship between two (or more) variables and not the equation of the least-squares line.

There are several alternative formulae that may be used to calculate R^2 and r directly from the observed data points without reference to the least-squares line.

$$R^2 = \frac{\{n(\Sigma xy) - (\Sigma x)(\Sigma y)\}^2}{\{n(\Sigma x^2) - (\Sigma x)^2\} \times \{n(\Sigma y^2) - (\Sigma y)^2\}} \tag{3.6}$$

$$r = \frac{n(\Sigma xy) - (\Sigma x)(\Sigma y)}{\sqrt{\{n(\Sigma x^2) - (\Sigma x)^2\} \times \{n(\Sigma y^2) - (\Sigma y)^2\}}} \tag{3.7}$$

One of several alternative formulae is given in (3.7(b)):

$$r = \frac{\sum (x - \bar{x})(y - \bar{y})}{\sqrt{\sum (x - \bar{x})^2 \sum (y - \bar{y})^2}} \tag{3.7b}$$

Worked Example 3.4: Calculation of r from formula (3.7)

(a) Calculate the correlation coefficient for the hourly rates of pay and hours worked per day given in Table 3.4
(b) What is the percentage of the variation explained by the linear relationship (regression)?

Table 3.4   Hourly rate (£) vs. hours per day for four randomly selected employees

x (hours per day)	3	5	7	9
y (rate per hour, £)	7	6	7	40

Solution

(a) Formula (3.7) is given as $r = \dfrac{n(\Sigma xy) - (\Sigma x)(\Sigma y)}{\sqrt{\{n(\Sigma x^2) - (\Sigma x)^2\} \times \{n(\Sigma y^2) - (\Sigma y)^2\}}}$

The sums required for this formula are given in Table 3.8:

Table 3.8   Sums required for the calculation of r

Col. 1	Col. 2	Col. 3	Col. 4	Col. 5
x_i: Hours worked per day	y_i: Rate per hour	$x \times y$	x^2	y^2
3	7	(3)(7) = 21	9	49
5	6	(5)(6) = 30	25	36
7	7	(7)(7) = 49	49	49
9	40	(9)(40) = 360	81	1600
$\Sigma x = 24$	$\Sigma y = 60$	$\Sigma xy = 460$	$\Sigma x^2 = 164$	$\Sigma y^2 = 1734$

Hence substitute $\sum x = 24$, $\sum y = 60$, $\sum x^2 = 164$, $\sum xy = 460$, $\Sigma y^2 = 1734$ and $n = 4$ (the number of data points) into formula (3.7)

$$r = \frac{4(460) - (24)(60)}{\sqrt{\{4(164) - (24)^2\} \times \{4(1734 - (60)^2\}}}$$

$$= \frac{1840 - 1440}{\sqrt{\{656 - 576\} \times \{6936 - 3600\}}}$$

$$= \frac{400}{\sqrt{[80] \times [3336]}} \dots \text{don't forget to take the square root of the final answer in the denominator}$$

$$= \frac{400}{516.6043}$$

$$= 0.7743$$

(b) $R^2 = (r)^2 = (0.7734)^2 = 0.5995$

The percentage of the variation explained by the linear relationship (regression) is $R^2 \times 100 = 0.5995 \times 100 = 59.95\%$

Commonalities in the formulae for the slope and correlation

You may have noticed that the formula for the correlation coefficient, r and the slope of the least-squares line have certain commonalties, highlighted below:

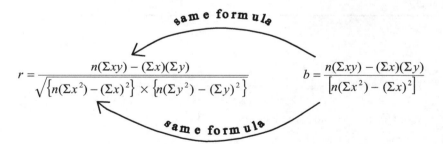

$$r = \frac{n(\Sigma xy) - (\Sigma x)(\Sigma y)}{\sqrt{\{n(\Sigma x^2) - (\Sigma x)^2\} \times \{n(\Sigma y^2) - (\Sigma y)^2\}}} \qquad b = \frac{n(\Sigma xy) - (\Sigma x)(\Sigma y)}{[n(\Sigma x^2) - (\Sigma x)^2]}$$

Figure 3.14   Comparison of formulae for b and r.

If the slope of the least-squares line has been calculated, the only additional 'sums' required to calculate r is $[n(\Sigma y^2) - (\Sigma y)^2]$.

For example, for the data in Table 3.4, the slope was calculated in Worked Example 3.1 as follows:

$$b = \frac{n(\Sigma xy) - (\Sigma x)(\Sigma y)}{n(\Sigma x^2) - (\Sigma x)^2} = \frac{4(460) - (24)(60)}{4(164) - (24)^2} = \frac{1840 - 1440}{656 - 576} = \frac{400}{80} = 5$$

To follow these with the calculation of r carry out one additional calculation:
$$n\sum y^2 - \left(\sum y\right)^2 = 4(1734) - (60)^2 = 3336 \text{ (See Table 3.8 for summations).}$$

$$\text{Hence } r = \frac{400}{\sqrt{\{80\} \times \{4(1734) - (60)^2\}}}$$

$$= \frac{400}{\sqrt{\{80\} \times \{3336\}}} = \frac{400}{516.6043} = 0.7743$$

Note: As mentioned at the beginning of this section the value of the slope (b) of a least-squares line gives no indication of the strength of the relationship (r) or vice-versa – even though there are commonalities in formulae used to calculate each.

The sign of the correlation coefficient indicates whether the line has a positive slope (upward sloping) or negative slope (downward sloping).

Important

1. **the value of the correlation coefficient for y against x is the same as the correlation coefficient for x against y.**
2. **the correlation coefficient is independent of units.**

Recall that the equation of the least-squares line is dependent on units. It will be left as an exercise for the reader to demonstrate this in some of the questions in PE3.3.

Skill Development Exercise SK3_5: Calculate r and R^2 by formula 3

Calculate the coefficient of determination and the correlation coefficient using formulae (3.4) and (3.5) for

(a) Savings against income data in Table 3.1.
(b) Profit vs. cars serviced in Table SK3_2.1.
(c) Hourly rates of pay vs. hours worked per day in Tables 3.4(a), (b) and (c).

Give a verbal description of your answer
Answer (a) $R^2 = 0.8155$; $r = 0.9031$ (b) $R^2 = 0.6870$; $r = 0.8289$; (c) see Figure 3.11.

Skill Development Exercise SK3_6: Excel, plots and linear trends

Use Excel (options under 'Add Trendline' to find the coefficients of determination and then the correlation coefficients for:

(a) Savings against income data in Table 3.1.
(b) Profit vs. cars serviced in Table SK3_2.1.

(c) The data for savings vs. income in Figure 3.1(a), (b), (c) and (d) given in Tables SK3_4.1, SK3_4.2, SK3_4.3, SK3_4.4 (File: Savings vs. income).

(d) The data for the mechanics profit for sales of bikes Table SK3_3.1 (File: Bikes).

Give a verbal description of your answer

Answers

(a) $R^2 = 0.8155; r = 0.9031$: strong positive correlation.

(b) $R^2 = 0.6870; r = 0.8289$: moderately strong positive correlation.

(c) $R^2 = 0; r = 0$: no linear relationship; $R^2 = 1; r = 1$; perfectly linear positive relationship: $R^2 = 0.2241; r = 0.4734$: weak positive linear relationship (recall the curvilinear scatter plot for this data in Figure 3.1(c): $R^2 = 1; r = -1$; perfectly linear negative relationship.

(d) Mountain bikes $R^2 = 0.6842$: moderately strong positive linear relationship; City-Jets $R^2 = 0.9992$: strong positive linear relationship.

3.5 Rank correlation. Calculate and interpret rank correlation

Formulae (3.4), (3.6) and (3.7) calculate the coefficient of determination and the correlation coefficient for ratio and interval data (continuous data). In many applications the strength of relationships for ordinal or ranked data is of interest. For example, in the present competitive climate many businesses send staff on training courses to update skills and improve productivity. If such courses are effective with a strong relationship between level of success on the course and improved performance at work. Suppose a group of eight sales staff are graded 1 to 8 on course performance on the completion of the training course (1 for highest etc.). Subsequently the annual sales for each salesperson are also ranked: 1 for highest sales down to 8 for lowest sales. Correlation for this ranked data is calculated using formula (3.8):

$$\text{Spearman's rank correlation coefficient: } r_s = 1 - \frac{6 \sum D^2}{n(n^2 - 1)} \qquad (3.8)$$

where D is the difference between paired ranks: n is the number of data pairs

The calculation of Spearman's rank correlation coefficient is illustrated in Worked Example 3.5:

Worked Example 3.5: Calculation of Spearman's rank correlation coefficient

The performance on a training course and value of annual sales is given for 8 sales people are ranked and given in Table 3.9.

Calculate Spearman's rank correlation coefficient.

Comment of the relationship between performance on course and sales value achieved.

Table 3.9   Course performance rank and sales value rank for 8 sales people (rank 1 = highest)

Staff member	Keane	Jones	White	O'Neill	Mc G	Green	Brown	Black
Course rank	1	2	3	4	5	6	7	8
Sales value rank	7	5	4	8	6	2	1	3

Solution

Use Spearman's rank correlation coefficient: $r_s = 1 - \dfrac{6 \sum D^2}{n(n^2 - 1)}$

Since the data is already ranked (if it not, then rank the data before proceeding further), the only calculation required is $\sum D^2$

D = difference between the course rank and the sales rank for an individual. Each difference, D is then squared. Finally, $\sum D^2$ is calculated and substituted into formula (3.8)

The calculations are set out in Table 3.10 below:

Table 3.10   Calculation of Spearman's rank correlation coefficient

Course rank	Sales value rank	D = Course rank − Sales rank	D^2
1	7	−6	$(-6)^2 = 36$
2	5	−3	$(-2)^2 = 9$
3	4	−1	$(-1)^2 = 1$
4	8	−4	$(-4)^2 = 16$
5	6	−1	$(-1)^2 = 1$
6	2	4	$(4)^2 = 16$
7	1	6	$(6)^2 = 36$
8	3	5	$(5)^2 = 25$
		$\sum D = 0$	$\sum D^2 = 140$

Substitute $n = 8$ (number of sales people) and $\sum D^2 = 140$ into formula (3.8)

$$r_s = 1 - \frac{6 \sum D^2}{n(n^2 - 1)}$$

$$= 1 - \frac{6(140)}{8(8^2 - 1)} = 1 - \frac{840}{504} = 1 - 1.6667 = -0.6667$$

There is a moderate negative relationship between the ranks awarded for course performance and the rank reflecting the sales value achieved. So, there is a moderately strong tendency for those who performed well on the training course to bring in lower annual sales!

Is the course counterproductive? Before coming to any conclusion, consider other factors, such as prior experience etc. Remember, the existence of a relationship does not necessarily imply causation!

The coefficient of determination, $(r_s)^2 = (-0.6667)^2 = 0.4444$.

For the further example on ranked correlation see question 7, PE3.3

3.6 Use the calculator for linear regression and correlation

In section 3.2, Microsoft Excel was used to produce scatter plots and the equation of the trend line together with the coefficient of determination. The equation of the least-squares line and the value of the correlation coefficient and much more are produced readily on most scientific calculators.

Most scientific calculators have inbuilt statistical functions for descriptive statistics and regression.

To activate the inbuilt functions for regression and correlation select the correct mode. For example, on most Casio calculators the mode is called *LR* mode and Stat*xy* on Sharpe) calculators. You must read the manufacturer's instruction manual for your own brand of calculator. Manuals will give an example to explain how to enter the (x, y) data points into the calculator and how to read out the slope, intercept and correlation coefficient of the least-squares line, and also the results of inbuilt calculations such as $\bar{x}, \bar{y}, \sum x, \sum y, \sum xy, \sum x^2, \sum y^2, \sigma_x, \sigma_y, s_x, s_y, n$.

3.6.1 Calculation of the equation of the least-squares line and the correlation coefficient!

Having entered the data into the calculator, the slope, intercept and correlation coefficient are displayed by pressing the appropriate keys. In one version of Casio, pressing the keys 'shift' followed by '7' gives the value of a (intercept): pressing 'shift' followed by '8' gives the value of b (slope) and 'shift' followed by '9' gives the value of r (the correlation coefficient). Look up the manual for your particular calculator to find out how to read of these statistics and then practise by entering the data given in Table 3.4.

Progress Exercises 3.3: Regression and Correlation

In the following exercises it is recommended that the reader plot graphs and do calculations manually for the smaller data sets but use Excel and calculator for the larger data sets.

1. The following data gives numbers of football jerseys purchased at several prices.

Table PE3.3 Q1   Numbers of jerseys purchased at selected prices | File: Jerseys |

Price per jersey € P	6	10	14	18	20
Numbers of jerseys demanded Q	142	185	96	44	40

(a) Plot the data (Q against P) on a scatter diagram. Calculate the correlation coefficient.
(b) The slope of the least-squares line were calculated as –9.6 and intercept as 231.7. Write down the equation of the least-squares line.
(c) Plot the least-squares line on the scatter plot in (a).
(d) Calculate the number of jerseys purchased when the price is € 15. Would you expect the actual number of jerseys purchased to be the same as the calculated (estimated) number when the price is € 15?

2. (a) Explain briefly how each of the following is used in statistics:
   (i) Least squares; (ii) Coefficient of determination; (iii) Scatter plots.
   (b) The number of apartments given as a percentage of all domestic dwelling sold since 2000 is given in Table PE3.3 Q2: $x =$ the number of years since 2000.

Table PE3.3 Q2   Apartments, as a % of all domestic dwellings

	2000	2001	2002	2003	2004
Years since 2000 (x)	0	1	2	3	4
% Apartments (y)	20	24	25	35	31

The equation of the least-squares line is $y = 20.4 + 1.65x$.
   (i) Plot the data and the least-squares line on a scatter diagram. Calculate the correlation coefficient.
   (ii) Give a verbal explanation for the slope of the least-squares line. Hence describe the trend displayed on the graph in (i).
   (iii) In 2003, 280,000 domestic dwelling were sold. Calculate the number of apartments sold.
   (iv) Does the above analysis explain the cause of this trend.

3. The CO_2 emissions in (Gg) per capita for four selected regions is given in Table PE3.3 Q3.

Table PE3.3 Q3 CO_2 emissions in (Gg) per capita for 1995 to 2002

CO2 emissions per capita (Gg)	1995	1996	1997	1998	1999	2000	2001	2002
Greece	7.50	7.66	8.03	8.43	8.32	8.76	8.95	8.93
Ireland	9.23	9.37	9.63	10.09	10.53	10.93	11.19	11.37
United Kingdom	9.77	10.06	9.61	9.61	9.39	9.42	9.62	9.31
United States	19.16	19.62	19.66	19.48	19.60	19.99	18.93	18.35

Source: UNFCC Greenhouse gases inventory database.

   (a) Calculate the correlation for the annual emissions of CO_2 for Greece and Ireland.
   (b) Calculate the correlation for the annual emissions of CO_2 for UK and US.
   In each case (a) and (b) can you conclude that the trend in annual emissions is similar for both countries?
4. (a) What is the smallest and largest value possible for a correlation coefficient? Plot simple graphs to illustrate your answers.
   (b) The consumer price index for health for the years 2002 to 2005 is given in Table PE3.3 Q4.

Table PE3.3 Q4 CPI Health Index 2002 to 2005				File: CPI health
Year	2002	2003	2004	2005
Year since 2000 (x)	2	3	4	5
CPI Health Index (y)	105.3	113.4	120.2	127.7

Source: CSO.

(i) Plot the data on a scatter diagram. Describe the trend in the CPI for health between the years 2002 to 2005. Would you expect this trend to continue indefinitely?

(ii) Calculate the least-squares line for the CPI index from 2002 to 2005.
If x = the number of years since 2000, hence $x = 2$ in 2002, $x = 3$ in 2003 etc., show that

$$\sum x = 14, \sum y = 466.6, \sum x^2 = 196, \sum y^2 = 217\,715.6, \sum xy = 6532.4$$

(iii) Calculate the correlation coefficient. Comment on the strength of the relationship.

(iv) Give a verbal description of the trend in the CPI for health with reference to (i), (ii) and (iii).

5. (a) Explain the terms: least squares; coefficient of determination.

(b) The Met. Office collates reports on weather conditions around the country. The mean hours of sunshine and temperature (°C) at selected weather stations are given in Table PE3.3 Q5.

Table PE3.3 Q5 Mean hours sunshine and temperature 2005		File: Weather
	Mean Temperature (y)	Hours of sunshine (x)
Shannon	11.5	4.04
Cork	10.2	4.30
Malin	10.5	3.93
Dublin	10.1	4.42
Casement	10.4	4.31
Valentia	11.4	3.94
Kilkenny	10.3	4.42
Belmullett	10.9	3.94
Connaught	9.1	4.93
Clones	9.9	3.9

Source: Met Eireann.

(a) Plot the data on a scatter diagram. Comment on the trend.

(b) Calculate the equation of the least-squares line and the correlation coefficient: Describe verbally the slope and intercept.

(c) From the equation of the line estimate the temperature for
(i) 4 hours sunshine; (ii) 10 hours sunshine; (iii) no sunshine.
Can you conclude that sunshine causes changes in temperature?

(d) Calculate the correlation coefficient for temperature as the dependant variable and sunshine as the independent variable. Compare your answer with that in (b).

6. The percentages of males and females who achieve five or more grades A to C in GCSE or SCE Standard grade in their last year of compulsory education in 2000/01 and 2001/02 is given in Table PE3.3 Q6:

Table PE3.3 Q6   Percentages of final year students who achieve five or more grades A to C

GCSE or SCE standard grade	5 or more grades A–C	
	% Males	% Females
United Kingdom	47.2	58.0
North East	40.5	50.9
North West	44.7	55.0
Yorkshire and the Humber	40.2	51.3
East Midlands	45.8	56.0
West Midlands	44.3	55.4
East	50.1	60.8
London	45.4	55.9
South East	51.5	61.4
South West	50.3	61.9

Source: Department for Education and Skills; National Assembly for Wales; Scottish Executive; Northern Ireland Department of Education.

File: Qualif_Gender

(a) Calculate the correlation coefficient. Explain what it means.

(b) Plot the data on a scatter diagram for % Females vs. % Males and calculate the equation of the least-squares line. Describe the slope verbally in terms of percentages of males and females who achieve five or more grades A to C. Are there any restrictions on the use of the regression equation?

(c) From the least-squares line, calculate the percentage of females who achieved 5 or more grades A-C when the percentage of males is 40, 55 and 75. Is it possible for the percentage males to exceed females?

7. (a) What is rank correlation?

(b) The quality of golf balls is thought to be related to price. A team of golfers grade the quality golf balls by giving grade 1 to top quality etc.
Prices are graded A (most expensive) to E (least expensive), as follows

Table PE3.3 Q7   Quality of golf balls against price

Price grade	A	C	B	C	D	E	E
Quality grade	7	5	6	4	3	1	2

(i) Assign numeric ranks to prices, giving rank 1 to A, rank 2 to B etc. The tied rankings, such as the two rank C's, occupy rank 3 and 4: hence assign the average of the tied ranks: $\frac{3+4}{2} = 3.5$ to each C.

(ii) Calculate Spearman's rank correlation coefficient. Give a verbal explanation of the result.

8. The incidence of lung disease and sulphur dioxide (SO_2) emissions for selected countries in 1995 is given in Table PE3.3 Q8:

Table PE3.3 Q8   Sulphur dioxide (SO_2) emissions and lung disease in 8 countries (1995)

	SO_2	Lung
Croatia	14.74	17.64
Finland	18.79	15.62
Germany	24.42	16.62
Greece	51.81	2.01
Ireland	44.71	10.73
Israel	51.6	4.92
Italy	23.26	18.53
Luxembourg	21.97	18.33

Source: World Health Organization database.

File: Lungs and sulphur 1

Where SO_2: Sulphur dioxide emissions, kg per capita per year. Lung disease are bronchitis/emphysema/asthma for all ages per 100 000
(a) Plot the data for lung disease against sulphur dioxide emissions.
(b) Calculate the equation of the least-squares line and the correlation coefficient.
(c) What can you conclude about the relationship between the given lung diseases and the level of sulphur dioxide emissions?

9. See question 8. The data on sulphur dioxide emissions and lung diseases for France and Denmark are added to the list in Table 3.3 Q9:

Table PE3.3 Q9   Sulphur dioxide (SO_2) emissions and lung disease in 10 countries (1995)

	SO_2	Lung
Croatia	14.74	17.64
Denmark	28.41	43.56
Finland	18.79	15.62
France	16.06	8.99
Germany	24.42	16.62
Greece	51.81	2.01
Ireland	44.71	10.73
Israel	51.6	4.92
Italy	23.26	18.53
Luxembourg	21.97	18.33

Source: World Health Organization database.

File: Lungs and sulphur 2

Plot the scatter diagram; calculate the equation of the least-squares line and the correlation coefficient. Compare these results with those in question 4.

10. See questions 8 and 9. The data on sulphur dioxide emissions and lung diseases for a further seven countries (marked by a *) are added to the list in Table PE3.3 Q10.

Table PE3.3 Q10 Sulphur dioxide (SO_2) emissions and lung disease in 17 countries (1995)

	SO_2	Lung	
Croatia	14.74	17.64	
Czech Republic	105.61	12.27	*
Denmark	28.41	43.56	
Estonia	82.48	15.78	*
Finland	18.79	15.62	
France	16.06	8.99	
Georgia	4.59	9.79	*
Germany	24.42	16.62	
Greece	51.81	2.01	*
Hungary	68.25	37.2	*
Iceland	89.41	15.7	*
Ireland	44.71	10.73	
Israel	51.6	4.92	*
Italy	23.26	18.53	
Latvia	23.74	17.3	*
Lithuania	25.9	34.75	*
Luxembourg	21.97	18.33	

Source: World Health Organization database.

File: Lungs and sulphur 3

Plot the scatter diagram; calculate the equation of the least-squares line and the correlation coefficient.

What do you conclude about the relationship between the given lung diseases and sulphur dioxide emissions from the results of the analysis in questions 7, 8 and 9?

11. The number of motor vehicle accidents for all ages per 100 000 of the population is given in Table PE3.3 Q11:

Table PE3.3 Q11 Motor vehicle accidents per 100 000 of the population

	2004	2003	2000	1995	1990	1980	1975
Greece	16.08	14.56	18.06	21.32	20.87	16.85	16.9
Iceland	6.77	7.47	11.55	9.82	9.77	9.54	18.08

Source: World Health Organization database.

File: Motors Greece and Iceland

(a) Plot the scatter diagram; calculate the equation of the least-squares line and the correlation coefficient, with Greece as the independent variable.
(b) Do you consider the data point for 1975 an outlier?
Remove the data point for 1975, and then repeat part. (a) Comment on the effect of the removal of the 1975 data.
(c) What inference may be deduced from the least-squares line and correlation calculated in (b)?
How would this inference differ from trend lines for motor vehicle accidents for each country plotted against years?

12. (a) (i) Describe the relationships between motor accidents in the Netherlands and the United Kingdom given in Table PE3.3 Q12 by using scatter plots initially and then by fitting a regression line and calculating the correlation coefficient.
    (ii) Plot line graphs for motor accidents in the Netherlands and the United Kingdom from 1975 to 2004 on the same diagram. Describe the results in (i) and (ii).

Table PE3.3 Q12   Motor vehicle accidents per 100 000 of the population   File: Motors accidents

	2004	2003	2000	1995	1990	1980	1975
Greece	16.08	14.56	18.06	21.32	20.87	16.85	16.9
Iceland	6.77	7.47	11.55	9.82	9.77	9.54	18.08
Netherlands	4.69	5.83	6.47	7.36	8.07	12.71	16.79
United Kingdom	5.42	5.7	5.59	5.87	9.09	11.21	11.91

Source: World Health Organization database.

(b) Fit regression lines and give the correlation coefficient for the motor vehicle accidents against year for each of the countries given in Table PE3.3 Q12
Note: let x = the number of years since 1975
y = Motor vehicle accidents per 100 000 of the population
Describe and compare the change in the number of accidents per 100 000 over the years for the four countries.

13. Life expectancy, incidence of motor accidents and lung diseases are given in Table PE3.3 Q13
    (a) (i) Plot a scatter diagram and calculate the least-squares line and correlation coefficient for lung disease against motor accidents.
        (ii) Remove the data for Ukraine, then repeat (i).
    Explain the results in (i) and (ii). Comment on the effect of the Ukraine data on the least-squares line.

Table PE3.3 Q13   Data fro selected countries 2000   File: Life expect

Country	Life expectancy	Motor accidents	Lung disease
Sweden	79.77	5.81	16.19
Switzerland	79.86	7.13	16.73
Tajikistan	62.44	4.37	30.85
TFYR Macedonia	72.01	5.37	22.27
Ukraine	67.76	10.96	45.69
United Kingdom	77.02	5.59	8.77

Source: WHO.

(b) (i) Plot a scatter diagram and calculate the least-squares line and correlation coefficient for life expectancy against motor accidents.
    (ii) Remove the data for Ukraine, then repeat (i).
Explain the results in (i) and (ii).

Comment on the effect of the Ukraine data on the least-squares line and the correlation coefficient.

14. Underage drinking is of major concern. The figures in Table PE3.3 Q14 is an extract from those released by the National Centre/NFER, Drug use smoking and drinking among young people in England 2003.

File: 13 years

Table PE3.3 Q14   Percentage of pupils aged 13 years who drank last week (before to the survey), by age and gender. England, 1992 to 2003

				percentages					
	1992	1994	1996	1998	1999	2000	2001	2002	2003
Boys	15	22	27	16	16	18	22	20	22
Girls	11	16	22	14	17	19	22	21	19

Source: NatCen/NFER Drug use, smoking and drinking among young people in England in 2003, Headline Figures.

(a) Plot line graphs for the percentage of boys and girls who drank alcohol last week. Describe the trends.
(b) Plot the data on a scatter diagram with the percentage of boys as the independent variable and the percentage of girls as the dependent variable.
Calculate the correlation coefficient.
Describe the relationship between the percentage of boys and girls aged 13 who drank last week.

15. Eurobarometer conducts opinion polls throughout the EU on a range of issues. A poll on the National Health Services for member states was conducted in 2003. The respondents were asked to select ONE statement from 1, 2, 3 or 4 in Table PE3.3 Q15 below that best described the health services in their own country. The survey results (in percentages) are given in Table PE3.3 Q15:

Table PE3.3 Q15   Opinion poll on health services (rates on scale 0 to 100) 2003   File: Health service

			D												UK
	B	DK	Total	GR	E	F	IRL	I	L	NL	A	P	FN	S	Total
1. Runs well	23.8	16.5	15.6	2.9	13.2	22.0	3.7	6.5	21.7	6.5	31.8	1.8	24.0	11.4	8.3
2. Minor changes needed	41.3	35.1	31.5	15.9	32.4	41.9	16.7	24.4	46.0	39.1	35.4	12.5	48.5	36.3	22.9
3. Fundamental changes needed	22.7	39.4	34.8	50.5	38.6	25.5	39.3	45.6	20.4	46.8	23.0	39.0	21.0	37.8	49.7
4. Completely rebuild system	5.2	6.1	11.1	27.6	12.1	7.0	32.9	19.9	8.2	6.8	4.3	41.4	3.2	10.6	15.8
5. Uncertain/don't know	7.0	2.8	6.9	3.1	3.7	3.6	7.5	3.5	3.8	0.8	5.5	5.4	3.1	3.8	3.3

Source: Eurobarometer.

(a) Calculate the correlation coefficient for 'Completely rebuild system' and ' Runs well'. Describe the relationship between opinions on 'Completely rebuild system' and ' Runs well'.

(b) Plot a scatter diagram for 'Completely rebuild system' against ' Runs well' and calculate the least-squares line. How does the scatter diagram relate to your answer in (a)?

(c) Calculate the correlation coefficient for the responses to the five questions between (i) Ireland (IRL) and Austria (A). (ii) Austria (A) and Finland (FN). Based on these results would you consider the opinions on the health services in groups (i) and (ii) to be similar? Explain your answer fully, using scatter plots if necessary.

16. See question 10, PE3.1 (File Gender)

For age groups (i) 25–34 (ii) 55–59 calculate the correlation coefficient for the numbers of females against the number of males in employment. With reference to the graphs and least-squares lines in question 10, PE3.1 and the correlation coefficients describe the relationship between the numbers of males and females in employment over the period 2005Q1 to 2006Q1.

3.7 Why bother with formulae?

Now that you have been introduced to Excel and the calculator you must be wondering why do we bother using all those tedious formulae. Some of the reasons are given below.

Examination of formulae enhances understanding: for example, the formula for the slope and intercept of the least-squares line uses ALL the data to fit the line as close as possible to ALL the points.

$$R^2 \text{ was defined in equation (3.4) as } R^2 = \frac{\text{Explained variation of } y\text{'s from } y\text{-bar}}{\text{Total variation of } y\text{'s from } y\text{-bar}}$$

$$= \frac{\sum (\hat{y}_i - \bar{y})^2}{\sum (y_i - \bar{y})^2}$$

- From the (3.4) we can 'see' that if every estimated value of y, $\hat{y}_i = \bar{y}$ then $\sum (\hat{y}_i - \bar{y})^2 = 0$ so $R^2 = \frac{0}{\sum (y_i - \bar{y})^2} = 0$. This is the situation when slope of the line, $b = 0$, there is no relationship between the variables.

- This is the situation when the line is a perfect fit for the observed data. If every $\hat{y}_i = y_i$ then $\sum (\hat{y}_i - \bar{y})^2 = \sum (y_i - \bar{y})^2$ giving $R^2 = \frac{\sum (y_i - \bar{y})^2}{\sum (y_i - \bar{y})^2} = 1$.

- Formula (3.7a) may be used to explain why the correlation for y vs. x is the same as that for x vs. y.

- This formula may also be used to show that r is independent of units.

$$r = \frac{\sum (x - \bar{x})(y - \bar{y})}{\sqrt{\sum (x - \bar{x})^2 \sum (y - \bar{y})^2}} \tag{3.5a}$$

The value of r is unchanged when x and y are interchanged.

Formulae may be used to explain the effect of changes in units for (x, y) data points on the equation of the least-squares line.

Suppose the time worked was recorded in minutes rather that hours, what effect would this have on regression coefficients a and b?

Again return to the formula (3.2) and (3.3) for a and b; replace every x by $60x$ (60 minutes in each hour)

Calculating slope:

$$b = \frac{n(\Sigma xy) - (\Sigma x)(\Sigma y)}{n(\Sigma x^2) - (\Sigma x)^2}$$

$$b_{new} = \frac{n(\Sigma 60xy) - (\Sigma 60x)(\Sigma y)}{n(\Sigma (60x)^2) - (\Sigma 60x)^2}$$

$$= \frac{60n(\Sigma xy) - 60(\Sigma x)(\Sigma y)}{3600n(\Sigma x^2) - 16(\Sigma x)^2}$$

$$= \frac{60\left[n(\Sigma xy) - (\Sigma x)(\Sigma y)\right]}{3600\left[n(\Sigma x^2) - (\Sigma x)^2\right]}$$

$$= \frac{1}{60}b$$

Calculating the intercept:

$$a = \frac{\Sigma y - b\Sigma x}{n}$$

$$a_{new} = \frac{\Sigma y - b_{new}\Sigma 60x}{n}$$

$$= \frac{\Sigma y - \frac{b}{60}60\Sigma x}{n}$$

$$= \frac{\Sigma y - b\Sigma x}{n}$$

$$a_{new} = a$$

Slope: the rate of change per minute is $\frac{1}{60} \times$ the rate of change per hour: the value of the new slope is $\frac{1}{60} \times$ the value of the original slope.

If you think about it, this makes sense: if the change in pay is £ b per hour, then it should be £ $\frac{1}{60} \times b$ per minutes.

Intercept: The intercept does not change.

Summary and overview of regression and correlation

The foregoing sections discussed how regression and correlation may be used to describe and interpret the strength and type of relationship that exists between two variables. The following points should be taken into consideration:

- The least-squares line is the line fitted to a set of data so the squared deviations of the *independent variable* y from the line is minimised.
- The least-squares line for y vs. x is different from the least-squares line for x against y.
- The coefficient of determination R^2 is the proportion of the variation in y explained by the least-squares line.
- The value of R^2 ranges from 0 to 1 inclusive.
- The correlation coefficient (r) is an index of the strength of a **linear relationship.** The value of the correlation coefficient ranges from -1 to $+1$ inclusive. **Note:** $r = \pm\sqrt{R^2}$.

- The correlation coefficient may be close to zero when the relationship is **not a linear relationship**. In Figure 3.1(c) the relationship was strong, but not linear the correlation coefficient was 0.4734.
- The correlation coefficient for y vs. x is the same as the correlation coefficient for x against y.
- When describing the relationship between the variables in regression any claims about causation can only be made after careful consideration of the context and as many other contributory factors as possible.
- Data sources should always be reviewed and the method of data collection.
- Be clear on the units that describe the data. For example, in Worked Example 3.1, the 'hours worked per day' the independent variable was given in units of £ 1000, £ 10 000 and £ 100 000.
- Outliers in the data need to be carefully checked – unusual values may be due to erroneous recording; special causes or they may be normal but rare occurrences.
- The presence of outliers can seriously distort the equation of the line and the correlation coefficient. For example, in WE3.1, if the point ($x = 9$, $y = 40$) hourly rate of £ 40 is included, then the equation is $y = 5x - 15$, $R^2 = 0.5995$, but if excluded, the equation is $y = 6.6667$ and $R^2 = 0$. See Figure 3.15 graphs (a) and (b) respectively.

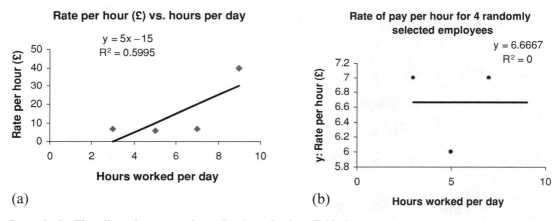

(a)   (b)

Figure 3.15   The effect of removing the outlier from the data, Table 3.4.

Other examples in relation to effect of an outlier is given in questions 11 and 13, PE3.3.

- Relationships may be spurious: for example, in PE3.3 the data in question 8 appeared to show a relationship between lung diseases and SO_2 emissions, but this may be coincidental for this group: the addition of data from other countries revealed no overall relationship.
- Clustered data can give misleading results.
- Sometimes a set of data may consist of different trends over different intervals. For example. Figure 3.16 is the plot and equation for the combined data in Tables SK3_4.1 to SK3_4.4 for savings vs. income (originally Figures 3.1(a), (b), (c) and (d)).

The coefficient of determination is 0.5177, or 51.77 % of the variation is explained by the least-squares line: $y = 182.39 + 29.014x$. Each increase of € 10,000 in income will result in an additional € 29 in savings!

REGRESSION AND CORRELATION: INTRODUCTION

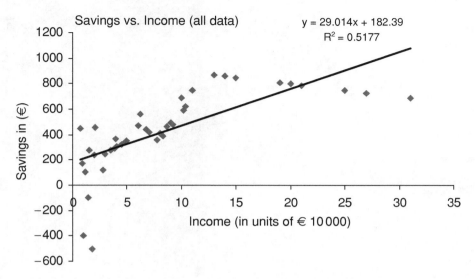

Figure 3.16   The totality of the data on savings vs. income from Figures 3.1 and 3.2.

Closer inspection of the graph reveals that for different income ranges in the combined data there are different trends:

1. For incomes from € 7000 to € 30 000 approximately, there is no relationship between savings and income: $y = 45.188x + 10.092$ and $R^2 = 0.0118, r = 0.1086$. See Figure 3.16(i).
2. For incomes from € 35 000 to € 110 000, there is a positive relationship between savings and income: $y = 45.522x + 123.58$. $R^2 = 0.7213, r = 0.8493$. See Figure 3.16(ii).
3. For incomes from € 130 000 to € 310 000, there is a perfect negative linear trend between savings and income: $y = 1000 - 10x, R^2 = 1, r = -1$. See Figure 3.16(iii).

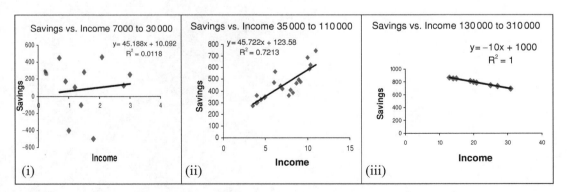

Figure 3.16 (i), (ii), (iii)   Data on Savings vs. income for three income ranges.

PROBABILITY

4

Chapter Objectives

Having carefully studied this chapter and completed the exercises you should be able to do the following

- Define and describe probability in terms of relative frequency and equally likely events
- Explain the axioms (self evident truths) of probability
- State and apply the counting rules for probability
- Define and describe independent events, mutually exclusive events and dependent events
- State the 'AND' and 'OR' rules
- Apply and solve problems based on the 'AND' and 'OR' rules
- Solve problem involving conditional probability
- Define a probability distribution
- Define and calculate joint, marginal and conditional probability distributions from a contingency table
- Derive and explain Bayes' Theorem
- Solve problems based on Bayes' Theorem

4.1   Introduction to probability

Probability is essentially a measure of the likelihood or chance of 'something' happening. Some examples to illustrate the use of probability in everyday language are given below:

1. 'There is a 50:50 chance that a head will turn up when you toss a coin';
2. 'The probability of getting an even number on one toss of a die is 0.5';

3. 'The odds of passing this exam are 3 to 1';
4. 'The odds of that horse winning are 100: 1';
5. 'There is a 30 % chance of a child developing mumps when more than two children in the class are already infected';
6. 'There is 1 chance in 3 of getting a "1 or a 6" on one roll of a die';
7. 'There is a 0.001% chance of an allergic reaction to that medication';
8. 'There is 1 chance in 10 000 that the new PC will breakdown in the first 2 years', etc., etc.

If 'something' is unlikely to happen, such as allergic reaction to medication or a PC breaking down, the probability is very small – on the other hand, when something is very likely to happen the probability is large. But, in the context of probability what is 'large' or 'small'? The first step in answering this question is to clearly define 'probability' and how it is measured. Then, armed with a clear definition, it is possible to develop the study of probability further, deducing methods and rules for calculating the probability for more complex events, ultimately leading to the theory of sampling and statistical inference. BUT a word of warning – at all stages in the study of probability it is vitally important to be able to state, *in words*, the situation in question and what its probability means.

4.1.1 Terminology

The 'something' that happens by chance is called an **Event**.

An experiment is a repeatable process, circumstance or action that causes an event to happen. Examples of experiments are given above: in 1, the experiment is the tossing of a coin. In 2 and 6 the experiment is the tossing of a die.

A trial is a single execution of an experiment, such as a single toss of a die.

An outcome is the result of a single trial such as a six on a die.

An event consists of a single outcome or one more outcomes, such as the event of an even number showing when a die is tossed: this event comprises the outcomes {2, 4, 6}.

The list of all possible **outcomes** of an experiment is called the '**sample space**' of the experiment, such as {1, 2, 3, 4, 5, 6}, the list of all possible numbers that show when a die is tossed.

An event space is the list of all outcomes that define the event, such as the list of even numbers.

Note: outcomes are also referred to as 'sample points'.

The following Worked Examples should clarify these ideas.

Worked Example 4.1: Terminology: experiment, events and sample space

An experiment entails tossing two coins. Draw a diagram to illustrate

(a) the sample space.
(b) the event space for (i) Heads on both coins (ii) Heads on one coin and Tails on the other coin (iii) Heads on neither coin (i.e. two Tails) (iv) at least one coin shows Heads.

Solution

(a) In Figure 4.1, the experiment is the process of tossing two coins; there are a total of four outcomes: HH (Heads on the first and second coin); HT (Heads on the first coin and Tails on the second coin); TH (Tails on the first coin and Heads on the second coin); TT (Tails on the first coin and on the second coin);

Hence the **S**ample space, **S** is the set {HH; HT; TH; HH}.

(b) Let the upper case letter **A** represent the 'event of two Heads'. The event space $A = \{HH\}$ so this event is a single outcome from S. **B** represents the 'event of one Head'. $B = \{HT, TH\}$, since two outcomes result in Heads showing on one of the coins. **C** represents the 'event of no Heads'. $C = \{TT\}$. **D** represents 'at least one coin shows Heads': 'at least one Head' means Heads on one or more of the coins. Hence $D = \{HT, TH, HH\}$.

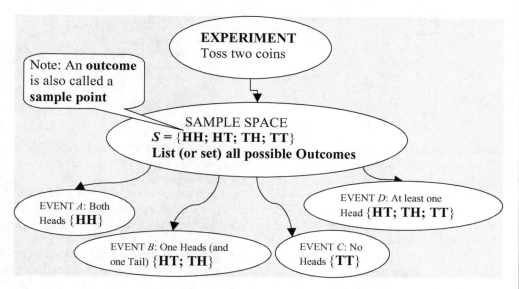

Figure 4.1   Experiment: toss two coins.

As already stated, an event can be any subset of one or more outcomes from the sample space. In Figure 4.1, Event D contains three outcomes: event B contains two outcomes: events A and C contain only a single outcome each.

Worked Example 4.2: Sample space and event spaces

An experiment consists of 1 roll of a fair die. Write out the sample space for the experiment; the event spaces for even numbers and for odd numbers.

Solution

The sample space, S, is the **set** of all possible outcomes: $S = \{1, 2, 3, 4, 5, 6\}$ Let A be the event of an odd number, hence the event space for A is $\{1, 3, 5\}$. Similarly the event space for B: an even number is $\{2, 4, 6\}$.

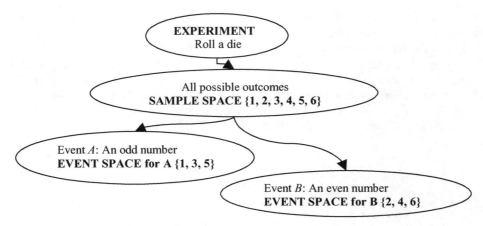

Figure 4.2   Experiment; roll a die; sample space; event space.

4.1.2   A review of some essential maths for sets

Sets are a mathematical tool that is useful in explaining some concepts and rules in probability; hence a quick review on the essentials is necessary at this stage. Sets are represented graphically by diagrams such as those in Figure 4.2. Such diagrams are called **Venn Diagrams**.

A set is a well defined collection of distinct items, numbers, or variables. A set is symbolised by an upper case letter, such as S, A, B above. The members of the set are sometimes referred to as the **elements** or **sample points** of the set. If the elements of a set are listed, the list is normally enclosed in curly brackets. The number of elements in a set, A, is written $\#A$.

When two or more sets (say A and B) are combined to include all the elements that are in either or both sets (with common elements appearing once only) the combined set is also a set called '**A union B**', written as '$A \cup B$', sometimes called 'A OR B', or even '$A + B$'.

The elements that are common to both sets constitute a set that is called '**A intersection B**', written as $A \cap B$ or 'A AND B' or simply AB.

The sample space, S, is referred to as the set of all possible outcomes of an experiment.

The events A and $\bar{A}$ are called **complementary events**: The event $\bar{A}$ is the set of all outcomes in the sample space, S that are not in A.

$\bar{A}$ is called 'Not A'.

These ideas are illustrated in Worked Example 4.3 below.

Worked Example 4.3: Some essential basics on sets:

Let A be the set of spades and B be the set of Aces in a deck of 52 cards.

(a) List the sets A, and B and $\bar{A}$
(b) Sketch the given sets A, B on the same diagram, showing common elements in the overlap area.
(c) Write out the elements of the set $A \cup B$: label this set on the sketch in (b)
(d) Write out the set $A \cap B$: label this set on the sketch in (b)
(e) State the number of elements in A, B, $A \cup B$ and $A \cap B$
(f) Use this example to explain why $\#(A \cup B) = \#A + \#B - \#AB$

Solution

(a) $A = \{2\spadesuit\, 3\spadesuit,\, 4\spadesuit,\, 5\spadesuit,\, 6\spadesuit,\, 7\spadesuit,\, 8\spadesuit,\, 9\spadesuit,\, 10\spadesuit,\, J\spadesuit,\, Q\spadesuit,\, K\spadesuit,\, A\spadesuit\}$:
$B = \{A\spadesuit,\, A\diamondsuit,\, A\heartsuit,\, A\clubsuit\}$
$\bar{A} = \{13\ \text{Clubs};\ 13\ \text{Hearts};\ 13\ \text{Diamonds}\}$ – i.e. all cards except Spades.

(b) A sketch of the set A AND B is given in Figure 4.3, where the common element ($A\spadesuit$, the Ace of spades) appears once only and is placed in the overlap area (shaded) of the 2 sets.

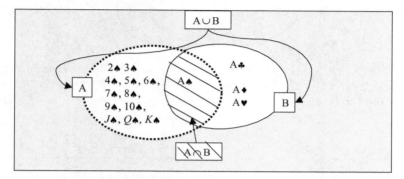

Figure 4.3   Venn diagram representation of the Union and intersection of sets.

(c) '$A \cup B$' is a list of every element within the boundary of the sets A and B in Figure 4.3. Hence
$A \cup B = \{2\spadesuit\, 3\spadesuit,\, 4\spadesuit,\, 5\spadesuit,\, 6\spadesuit,\, 7\spadesuit,\, 8\spadesuit,\, 9\spadesuit,\, 10\spadesuit,\, J\spadesuit,\, Q\spadesuit,\, K\spadesuit,\, A\spadesuit\ A\diamondsuit,\, A\heartsuit,\, A\clubsuit\}$:

(d) $A \cap B$ is the list (or set) of elements common to both A and B.
$A \cap B = \{A\spadesuit\}$, as given in part (b) above.

(e) Counting the number of elements in the sets
$\#A = 13$; $\#B = 4$; $\#(A \cup B) = 16$; $\#(A \cap B) = 1$.

(f) The number of elements in the set $(A \cup B)$ may be counted as the sum of the elements in each of the three sections in Figure 4.3 $\#(A \cup B) = 12 + 1 + 3 = 16$. This number may also be counted as the sum of elements in sets A and B minus the number in the overlap, $(A \cap B)$. The reason for subtracting the number of elements in $(A \cap B)$ is that these are counted twice when the elements in sets A and B are added. $\#(A \cup B) = \#A + \#B - \#AB = 13 + 4 - 1 = 16$.

Skill Development Exercise SK4_1: Some essential basics on sets

Given the sets $A = \{2, 4, 6, 8, 11\}$, $B = \{4, 6, 7, 8, 9\}$, $C = \{1, 2, 8, 10, 12\}$

(a) Sketch the given sets A, B on the same diagram, showing common elements in the overlap area.

(b) Write out the set $A \cup B$: Indicate this set on the sketch in (a).

(c) Write out the set $A \cap B$: Indicate this set on the sketch in (a).

(d) State the number of elements in A, B, $A \cup B$ and $A \cap B$.

(e) Use the above example to demonstrate that $\#(A \cup B) = \#A + \#B - \#AB$.

(f) Write out and sketch the set $\{A \cap B \cap C\}$.

(g) Write out and sketch the set $\{A \cup B \cup C\}$.

(h) Use the above example to demonstrate that
$$\#\{A \cup B \cup C\} = \#A + \#B + \#C - \#\{A \cap B\} - \#\{B \cap C\} - \#\{A \cap C\} + \#\{A \cap B \cap C\}.$$

Answers

(a)

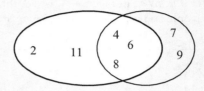

(b) $A \cup B = \{2, 4, 6, 7, 8, 9, 11\}$ (c) $A \cap B = \{4, 6, 8\}$ (d) $\#A = 5$: $\#B = 5$: $\#A \cup B = 7$: $\#A \cap B = 3$ (e) $\#(A \cup B) = \#A + \#B - \#AB \rightarrow 7 = 5 + 5 - 3$ (f) $\{A \cap B \cap C\} = (8)$ (g) $\{A \cup B \cup C\} = \{1, 2, 4, 6, 7, 8, 9, 10, 11, 12\}$ (h) $10 = 5 + 5 + 5 - 3 - 1 - 2 + 1.$

Progress Exercises 4.1: Review some essential maths for sets

---

1. Define the following terms, illustrating your answer with an appropriate example:
   (a) An experiment.
   (b) An outcome.
   (c) An event.
   (d) A sample space.
   (e) A sample point.
   (f) An event space.
2. Cards are selected from a deck of 52 cards. Write out and sketch the sets for the following. State the number of elements in each set.
   (a) The sample space.
   (b) The event space for red cards.
   (c) The event space for red Aces.
   (d) The event space for red cards and Aces.
3. A die and a coin are tossed. Write out and sketch the sets for the following.
   (a) The sample space.
   (b) The event space for a Head on the coin and an even number on the die.
   (c) The event space for a Head on the coin and a 1 or a 6 on the die.
4. Given the sets $S = \{1, 2, 3, 4, 5, 6, 7, 9, 12, 18\}$, $A = \{4, 9, 12, 16, 18\}$, $B = \{1, 3, 5, 9, 12\}$ and $C = \{2, 6, 7, 9, 12\}$, write out and sketch the following sets (note: S in the sample space):
   (a) $\overline{A}$
   (b) $\overline{B}$
   (c) $A \cup B$
   (d) $A \cap B$
   (e) $\overline{A} \cup B$
   (f) $\overline{A} \cap B$
   (g) $\{A \cap B \cap C\}$
   (h) $\{A \cup B \cup C\}$.
   State the number of elements in each set.

5. Two dice are rolled. Write out and sketch the following
  (a) The set of all possible outcomes.
  (b) The set of outcomes for which the numbers are equal.
  (c) The set of outcomes whose total is an even number or both numbers are equal.
  (d) The set of outcomes whose first or second number is unity.
  (e) The event space for outcomes where unity turns up on at least one face.
6. A bank plans to conduct a survey on two groups of customer: I: customers who use online banking and II: customers who use cheques. If $\#I = 4900$, $\#II = 850$, $\#(I \cap II) = 450$ calculate the number of customers:
  (a) who use online banking.
  (b) who use online banking only.
  (c) who cheques only.
  (d) who use online banking OR cheque book.

4.1.3 Definitions of probability

In the introduction, probability was described as the likelihood that an event will occur. For example, we may be interested in the probability or likelihood that a certain combination of numbers will turn in a lotto. There are three ways of assigning probability to an event:

1. Theoretical probability (classical probability).
2. Relative frequency or empirical probability.
3. Intuitive or subjective probability.

1. Theoretical probability also called classical or 'a priori' (before the fact):
This is probability assigned to events without making any observations or measurements. It is used when all outcomes are known, particularly when all outcomes are equally likely. The probability of any event, A, when all outcomes are equally likely is written as $P(A)$, where

$$P(A) = \frac{\text{number of outcomes which result in event A}}{\text{total number of all possible outcomes}} = \frac{\#A}{\#S} \qquad (4.1)$$

where $\#A$ and $\#S$ are the numbers of elements in the event and sample space respectively.

Worked Example 4.4: Probabilities for equally likely outcomes

(a) Two fair coins are tossed. Use formula (4.1) to calculate the probability of the following events
    (i) Two Heads (ii) One Head (iii) No Head (iv) At least one Head
(b) A fair die is rolled once. Write out the probability of (i) an even number (ii) an odd number
    (iii) a 1 or a 6

Solution

(a) In Worked Example 4.1, when two coins are tossed the sample and event spaces were
$S = \{HH, HT, TH, TT\}; \#S = 4$ (sample space)
$A = \{HH\}; \#A = 1. B = \{HT, TH\}; \#B = 2. C = \{TT\}; \#C = 1.$
$D = \{HT, TH, HH\}; \#D = 3.$ (event spaces)

Hence by formula (4.1)

(i) $P(\text{Two Heads}) = P(A) = \dfrac{\#A}{\#S} = \dfrac{1}{4}$

(ii) $P(\text{One Head}) = P(B) = \dfrac{\#B}{\#S} = \dfrac{2}{4}$

(iii) $P(\text{No Heads}) = P(C) = \dfrac{\#C}{\#S} = \dfrac{1}{4}$

(iv) $P(\text{At least one Head}) = P(D) = \dfrac{\#D}{\#S} = \dfrac{3}{4}$

(b) In Worked Example 4.2, for one roll of a die, the sample space and event spaces for odd and even number are

$S = \{1, 2, 3, 4, 5, 6\}; A = \{1, 3, 5\}$ and $B = \{2, 4, 6\}$ respectively
Also, if C is the event '1 OR 6', then $C = \{1, 6\}, \#C = 2$
By formula (4.1)

(i) $P(\text{Odd number turns up}): P(A) = \dfrac{\#A}{\#S} = \dfrac{3}{6} = \dfrac{1}{2}$

(ii) $P(\text{Even number turns up}): P(B) = \dfrac{\#B}{\#S} = \dfrac{3}{6} = \dfrac{1}{2}$

(iii) $P(\text{1 or 6 turns up}): P(C) = \dfrac{\#C}{\#S} = \dfrac{2}{6} = \dfrac{1}{3}$

2. Relative frequency or empirical probability: This type of probability is estimated from a large number of observations. The probability of an event is estimated by the relative frequency with which the event occurs and is calculated by the equation:

$$P(A) = \frac{\text{Number of times A happened}}{\text{Total number of times A was observed}} = \frac{f_A}{f_T} \qquad (4.2)$$

A practical difficulty with this measure of probability is that a large number of observations are required in order to obtain a measure of probability that allows for randomness and the various circumstances in which the event may occur. For example, suppose we attempt to estimate the probability that scheduled flights on a particular route are on time.

Let A be the event of arriving on time. Records for one week show that 18 out of the 24 were on time: hence the probability is estimated as $P(A) = \dfrac{18}{24} = 0.75$.

However, this probability is based on one particular week only: perhaps this was a holiday week and traffic was particularly heavy: perhaps there was a strike of some sort. A more reliable estimate of probability would require a much larger number of flights over a longer period of time so that various circumstances that affect air traffic are considered. A more reliable estimate of probability is given by taking 500 records spanning a one year period. If 450 out of 500 were late, then $P(A) = \dfrac{450}{500} = 0.9$.

3. Intuitive or subjective probability
This is the third type of probability, important in business and even in betting. It is a subjective estimate made by an individual or group of individuals on the likelihood of an event happening based on knowledge, experience and intuition. Since this probability is a subjective estimate there are no definitions or rules for its calculation. For example, a retailer states that s/he is 90 % certain that a certain good will be sold out within 2 days; a stockbroker states that s/he is 99.5 % certain that a particular stock will double in value within a year, etc.

4.1.4   Different ways of stating the probability of an event

4.1.4.1   Proportions or percentages

Probability is essentially the proportion of times an event is likely to happen. When a probability (proportion) is multiplied by 100, the probabilities may be quoted as a percentage.

For example, the chance of a head turning up when a coin is tossed is $\dfrac{1}{2}$ or we could say there is a 50 % chance ($\dfrac{1}{2} \times 100$ %) of getting Heads.

Quoting probabilities in terms of percentages is often more meaningful: for example stating that 'there is a 60 % chance that student T will pass the exam' instead of saying 'the probability of student T passing is 0.6'

But remember, when doing calculations, percentages must always be written as proportions by putting the percentage over 100 – so 60 % is $\dfrac{60}{100}$ when it comes to calculations!

4.1.4.2   Odds and probability

'The odds of passing this exam are 3 to 1'; this is another way of saying there are 4 chances; 3 for passing and 1 for failing, so in terms of probability

$$\text{The probability of passing} = \frac{3}{4} = 0.75$$

Warning about betting: in betting 'odds' has a different meaning – they refer to monetary rewards set by bookies, casinos, etc. so that they will make a profit and therefore cannot be converted to probabilities. 'The odds of that horse winning are 100 : 1'; means that when you place a bet of € 1 you may win € 100 or loose € 1.

NOTE: In problems on probability it is extremely important that you state everything clearly and logically in WORDS before attempting any calculations. You will see this in section 4.2.

4.1.5   The axioms of probability

The axioms of probability are the 'self evident truths' that set out the fundamental properties of probability and are stated As follows:

Axiom 1. Probability is a measure which lies between 0 and 1, inclusive. $0 \leq P(A) \leq 1$.

Axiom 2. For any event A, the probability of event A happening plus the probability of event A not happening is one. $P(A) + P(\overline{A}) = 1 \ldots$ where $\overline{A}$ is the complement of **A**.

The following Worked Example illustrates these ideas.

Worked Example 4.5: Axioms of probability

A mid-term paper in mathematics for economics was taken by 200 students, 175 students passed. Use this example to demonstrate that

(a) The probability of any event can never be less than 0 or greater then 1. This is axiom 1.
(b) The probability of passing plus the probability of failing is 1, This is axiom 2.
(c) The probability of failing (not passing) is the complement of the probability of passing.

Solution

(a) Let A be the event that a student passes:

By formula (4.2), the largest value of A occurs when all 200 pass, hence $P(A_{\max}) = \dfrac{200}{200} = 1$

The smallest value A occurs when no student passes, hence $P(A_{\min}) = \dfrac{0}{200} = 0$

So, if an event never happens its probability is 0. If an event always happens its probability is 1. This is **Axiom 1**: $0 \leq P(A) \leq 1$.

Therefore, *in problem solving, if you calculate a probability that is greater than one or less than zero, the probability that your answer is right is zero, the probability that your answer is wrong is one.*

(b) In the example above, 175 students out of 200 pass the examination, hence 25 do not pass. The two events are

A: pass the exam: hence $P(A) = \dfrac{f_A}{f_S} = \dfrac{175}{200}$

Not A: does not pass, in other words fails, hence $P(\overline{A}) = \dfrac{f_{\overline{A}}}{f_S} = \dfrac{25}{200}$

Hence, to show that $P(A) + P(\overline{A}) = 1$, substitute the probabilities calculated above

$$\frac{175}{200} + \frac{25}{200} = \frac{200}{200} = 1$$

This is **Axiom 2**: $P(A) + P(\overline{A}) = 1$.

Axiom 1: $0 \leq P(A) \leq 1$	(4.3)
Axiom 2: $P(A) + P(\overline{A}) = 1$	(4.4)

(c) If a student does not pass then the only other possible outcome is a fail.
Hence $\overline{A}$, the complement of **A** (**passing**) is **not passing (i.e., failing)**.

Progress Exercises 4.2: Introduction to probability

In the probability calculations give answers rounded correct to four decimal places.

1. What is probability?
   Describe how probability is assigned using the following approaches: a priori; relative frequency and subjective.
2. If two die are tossed calculate the probability that the two numbers that turn up are (a) both six; (b) the same; (c) both greater than 3.
3. Records, over a period of weeks, found that it had rained on 315 days out of 400.
   (a) Write down the probability that (i) it rains (ii) does not rain.
   (b) State the axioms of probability. Use this example to explain these axioms.
4. The following data is an extract from the 2001 census of Northern Ireland.

Table PE4.2   Extract census NI 2001   File: Census NI 2001

District	Population	Male	Female	Persons living in households	Persons living in communal establishments	Area (hectares)
South Antrim	99 810	49 334	50 476	97 721	2089	71 314
South Down	104 658	52 096	52 562	103 089	1569	129 801

   (a) If an individual is selected at random from all districts what is the *probability* that the individual is (i) from South Antrim (ii) male (iii) living in a household.
   (b) For the South Down populations calculate the *percentage* of (i) males (ii) females (iii) persons living in households.
      Write down the population density (persons per hectare).
   (c) For the South Antrim population calculate the *probability* that a randomly selected person (i) is male (ii) is female (iii) living in a household.
      Write down the population density (persons per hectare).
   (d) For the population living in communal establishments, calculate the *proportion* who live in (i) South Down (ii) South Antrim.
5. (Refer to question 2, Progress Exercise 4.1). One card is selected from a deck of 52 cards. Calculate the probability that the selected card is (a); a red card; (b) an Ace; (c) a red card AND an Ace (d) a red card OR an Ace.
6. (Refer to question 3, progress exercises 4.1). An experiment involves tossing a die and a coin.
   (a) State the sample space for the experiment.
   (b) State the event spaces for (i) Heads on the coin and an even number on the die (ii) Heads on the coin and 'a 1 or a 6' on the die.
   (c) Calculate the probability that the outcome is (i) Heads on the coin and an even number on the die (ii) Heads on the coin and 'a 1 or a 6' on the die.
7. The questions in a quiz fall into the following six categories, numbered 1 to 6:
   1. *General knowledge*. 2. *History*. 3. *20th century*. 4. *Literature and art*. 5. *Sport*. 6. *Science*
   A player rolls two dice: the category from which the question is selected is given by the smaller of the two numbers (or the common number if both are equal) that show on the dice. For

example, if the dice show a 3 and a 6, then the question is selected from category 3, 20th century.

Calculate the probability that a player is asked a question from each of the categories.

8. What is the probability that you will pass an examination on introductory probability on completion of this chapter? Is this subjective probability?

9. A company has six male and four female applicants for vacancies.
   (a) If two applicants are randomly selected what is the probability that (i) both are male; (ii) one is male and the other is female.
   (b) If three applicants are randomly selected what is the probability that (i) all are male; (ii) one is male and two are female.

10. Following extensive geological surveys, an oil company will proceed to drill for oil if there is a good chance that the size of the oil well is commercially profitable and the site is workable. As a result of the surveys on two sites the chances of finding commercially profitable oil wells are estimated as 80 % and 95 % respectively while the chances that the sites are workable are estimated as 94 % and 58 % respectively. Which site should the company consider and why?

4.2 Multiplication and addition rules for probability

In Worked Example 4.4 the probability of simple events (tossing coins and dice, etc.) was calculated by the definition (4.1); $P(A) = \dfrac{\#A}{\#S}$, by counting the number of outcomes in the sample space, $\#S$ and event space, $\#A$. For more realistic events, this method becomes unwieldy so it is necessary to deduce rules and methods for calculating probability. These methods are based on the multiplication and addition rules and are referred to as the 'AND' and 'OR' rules.

Note the terms 'AND' and 'OR' were described earlier in section 4.1.2 on sets.

4.2.1 'AND' rules (multiplication rules)

The use of AND features in everyday statements such as 'we will swim in the sea if it is summer' AND 'the air temperature is over 25°C'. So, to go swimming in the sea both events must be simultaneously true 'summer' AND 'over 25°C'.

When we say the event 'A AND B' happens we mean that both A and B happen together at the same time. The probability that both A and B happen at the same time is written as $P(A \text{ AND } B)$; $P(A \cap B)$ or $P(AB)$. The rules for calculating $P(AB)$ depend on whether the events A and B are **independent, mutually exclusive** or **dependent**.

The terms '**independent events**', '**mutually exclusive events**' or '**dependent events**' are explained with the AND rules below.

Rule 1: $P(AB) = P(A).P(B)$ for independent events (4.5)

Events are said to be **independent** when the outcome of one event has no effect on the outcome of the other event. For example, if an experiment involves tossing 2 coins, then whatever turns up on the first coin has no effect on the outcome for the second coin. The probability of Heads or Tails on the second coin is the same irrespective of whether the first coin showed Heads or Tails.

For example, if A = Heads on the first coin and B = Heads on the second coin, then since A and B are **independent** the probability of getting two Heads is calculated by the rule

$$P(AB) = P(A).P(B) = \frac{1}{2}\frac{1}{2} = \frac{1}{4}$$

This answer agrees with the calculation based on samples spaces in Worked Example 4.4.

This rule is stated as 'the probability that A and B happen together at the same time is the product of the two separate probabilities' (hence the name 'multiplication rule').

$P(AB) = P(A).P(B)$ for
Independent events

Rule 2 $P(AB) = 0$ for mutually exclusive events (4.6)

Events are said to be **mutually exclusive if the** occurrence of one event excludes or prevents the other occurring.

For example, consider an experiment of tossing one coin and let the events A = **Heads**, B = **Tails**. If Heads show when the coin is tossed, Tails cannot possibly show at the same time. The events A and B are described as **mutually exclusive**.

The probability of mutually exclusive events happening together at the same time is zero. P (**Heads** and **Tails**) = 0: $P(AB) = 0$.

$P(AB) = 0$ for
Mutually Exclusive events

Recall: If an event never happens, its probability is zero.

Rule 3 $P(A B) = P(A) P(B/A)$ or $P(A B) = P(B) P(A/B)$, for dependent events (4.7)

Events are said to be **dependent** when the occurrence of one event has an effect on the other. As an example, consider the experiment of drawing two cards from a pack, where the first card is *not replaced* before drawing the second. If the probability of getting two Aces is required, then let A = an Ace on the first drawn card, B = an Ace on the second.

The probability that the first card is an Ace is $P(A) = \frac{4}{52}$.

There are three Aces left in a total of 51 cards. The probability that the second card is an Ace is $\frac{3}{51}$: the probability that the second card is an Ace has been changed by the occurrence of the first event. This second probability is calculated *on the condition* that the first card drawn was an Ace (otherwise we don't end up with two Aces!) and is written $P(B/A)$.

Hence $P(B/A) = \frac{3}{51}$.

Finally, the probability that the first card AND the second are Aces is calculated

$$P(A B) = P(A)P(B/A) = \frac{4}{52}\frac{3}{51}.$$

$P(A B) = P(A) P(B/A)$
for **dependent events**

Note: If the events are independent, then A has no effect on B so $P(B/A)$ is the same as $P(B)$ Hence $P(AB) = P(A)P(B/A) = P(AB) = P(A)P(B)$ which is the AND rule for independent events.

Summary of the AND Rules

1. $P(AB) = P(A)P(B)$, when A and B are independent   (4.5)

2. $P(AB) = 0$ when A and B are mutually exclusive   (4.6)

3. $P(AB) = P(A)P(B/A)$ or $P(AB) = P(B)P(A/B)$
                                      when A and B are dependent   (4.7)

Hence it follows that when A and B are independent then $P(A/B) = P(A)$ and $P(B/A) = P(B)$

4.2.2   The 'OR' rules (addition rules)

In everyday language 'OR' is used in statements such as 'If it is summer OR the temperature (air) is over 25°C we will go swimming in the sea'. This means that we will go swimming in the sea if any of the following statements are true:

(a) 'it is over 25°C'...Call this event A.
(b) 'it is summer'...Call this event B.
(c) 'it is over 25°C **and** summer'...This is the event AB.

So if the event 'A OR B' happens then 'A happens or B happens or AB happen'.
   The meaning of 'OR' in this statement is the same as that used in defining the set $A \cup B$ in section 4.1.2.

4.2.2.1   The 'OR' rules for two events

Rule1: $P(A \cup B) = P(A) + P(B)$ for mutually exclusive events   (4.8)
Rule2: $P(A \cup B) = P(A) + P(B) - P(AB)$ for non-mutually exclusive events.   (4.9)

These rules will be explained through Worked Example 4.6.

Worked Example 4.6: Deduce the 'OR rule' when events are mutually exclusive

One card is selected from a deck of cards.

(a) Calculate the probability that the card is a Spade OR a Heart by counting the number of outcomes in the sample space and event space and using (4.1).
(b) Use this example to deduce **Rule 1** for mutually exclusive events $P(A \cup B) = P(A) + P(B)$

Solution

The sketch of the sample space for $\{A \cup B\}$ is in Figure 4.4. Note: these sets do not overlap and are called 'disjoint' sets.

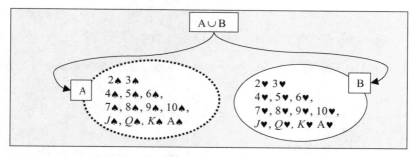

Figure 4.4   The Set 'Spades OR Hearts'.

Let the event A = the card is a Spade, hence #A = 13.

Let the event B = the card is a Heart, hence #B = 13.

(a) There are 26 cards that are either Spades or Hearts, hence $\#\{A \cup B\} = 26$
The sample space is the set of 52 cards, hence $\#S = 52$
Use formula (4.1) to calculate this probability.

$$P(A\ OR\ B) = \frac{\text{number of outcomes that result in A OR B}}{\text{Total number of all possible outcomes}} = \frac{\#\{A\ OR\ B\}}{\#S}$$

Substituting into (4.1)

$$P(\text{Spade or a Heart}) = \frac{\text{number of cards that are Spades or Hearts}}{\text{Total number of cards}} = \frac{26}{52}$$

(b) This same result is obtained by adding the individual probabilities (often referred to as the addition rule for mutually exclusive events)
$\#\{A\ OR\ B\} = \#\{A\} + \#\{B\}$ ... since the sets are disjoint, see Figure 4.4
$\dfrac{\#\{A\ OR\ B\}}{\#S} = \dfrac{\#\{A\}}{\#S} + \dfrac{\#\{B\}}{\#S}$ ... dividing both sides by $\#S$.
$P(A \cup B) = P(A) + P(B)$ ... from the classical definition of probability
$\dfrac{26}{52} = \dfrac{13}{52} + \dfrac{13}{52}$ ... demonstrating the rule.

Worked Example 4.7: Deduce the 'OR rule' for events that are not mutually exclusive

One card is selected from a deck of cards.

(a) Calculate the probability that the card is a Spade or an Ace by counting the number of outcomes in the sample space and event space and using (4.1).
(b) Use this example to deduce **Rule 2** for non-mutually exclusive events
$P(A \cup B) = P(A) + P(B) - P(AB).$

Solution

In worked Example 4.3, the event space for $A \cup B$ (a Spade OR an A) was sketched and is reproduced below.

Hence $A \cup B = \{2\spadesuit\ 3\spadesuit, 4\spadesuit, 5\spadesuit, 6\spadesuit, 7\spadesuit, 8\spadesuit, 9\spadesuit, 10\spadesuit, J\spadesuit, Q\spadesuit, K\spadesuit, A\spadesuit\ A\diamondsuit, A\spadesuit, A\clubsuit\}$:

(a) $\#\{A \cup B\} = 16$, the number of cards that are Spades OR Aces

The sample space is the set of 52 cards, hence $\#S = 52$

Substituting into (4.1) above

$$P(\text{Spade OR an A}) = \frac{\text{number of cards that are 'Spades OR Aces'}}{\text{Total number of cards}} = \frac{16}{52}$$

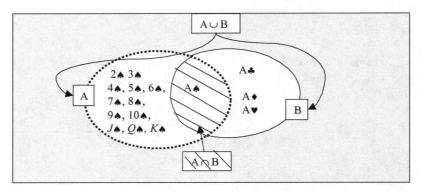

Figure 4.3 (reproduced): Union and intersection of sets.

(b) To write down a general rule for calculating $P(A\,\text{OR}\,B)$ consider how the number of elements in the set $\{A \cup B\}$ were counted, since this is the numerator in formula (4.1) .

$$P(A\,\text{OR}\,B) = \frac{\text{number of outcomes that result in A OR B}}{\text{Total number of all possible outcomes}} = \frac{\#\{A\,\text{OR}\,B\}}{\#S}$$

When the number of Spades, $\#(A)$ is added to the number of Aces, $\#(B)$ the Ace of Spades, $\#(A \cap B)$ is counted twice and so this number must be subtracted to give the number of cards in the set 'Spades OR Aces'. See Figure 4.3 above.

Hence $\#(A \cup B) = \#(A) + \#(B) - \#(A \cap B)$

To convert these numbers to probabilities divide both sides of this last equation by $\#(S) = 52$, the number of elements in the sample space

$$\frac{\#(A \cup B)}{\#(S)} = \frac{\#(A) + \#(B) - \#(A \cap B)}{\#(S)}$$

$$= \frac{\#(A)}{\#(S)} + \frac{\#(B)}{\#(S)} - \frac{\#(A \cap B)}{\#(S)} \left.\begin{array}{c} \\ \\ \\ \end{array}\right\} \quad \boxed{P(A \cup B) = P(A) + P(B) - P(AB).}$$

$$= \frac{13}{52} + \frac{4}{52} - \frac{1}{52} = \frac{26}{52}$$

Hence

$P(A \cup B) = P(A) + P(B) - P(AB)$. This is the OR rule for two events that are not mutually exclusive.

The addition rules for 2 events, A and B, are summarised as follows:

Summary for the OR Rules (also called the addition rules)

1. $P(A \text{ OR } B) = P(A) + P(B)$ when A and B are mutually exclusive   (4.8)

2. $P(A \text{ OR } B) = P(A) + P(B) - P(AB)$ when A and B are not mutually exclusive   (4.9)

Skill Development Exercise SK4_2: The OR (Addition) rule

Use sets to prove the OR rules

(a) $P(A \cup B) = P(A) + P(B) - P(AB)$.

(b) $P\{A \cup B \cup C\} = P(A) + P(B) + P(C) - P\{AB\} - P\{BC\} - P\{AC\} + P\{ABC\}$

Demonstrate the OR rule for sets A, B and C

where $A = \{2, 4, 6, 8, 11\}$, $B = \{4, 6, 7, 8, 9\}$, $C = \{1, 2, 8, 10, 12\}$

4.2.3 Tree diagrams

In Worked Example 4.1 when two coins were tossed it was easy to list the four outcomes {HH, HT, TH, TT}. However, if 3 coins or even 4 coins were tossed then it might not be so easy to list all the outcomes. A tree diagram can be very useful to 'see' and list all the possible outcomes in these situations.

See Figure 4.5 for two coins: begin at the 'Start' box, where one coin is tossed. There are two outcomes, Heads or Tails. The probability of each outcome is written on the line (branch!) leading to that particular outcome as shown on Figure 4.5. The second coin is tossed. The outcomes for the second coin are Heads or Tails and again the probability of each outcome is written on the lines leading from the first outcome to the second as shown on Figure 4.5. Each final outcome is given by the sequence of outcomes along each branch from the start box through to the end. For example, in Figure 4.5, the first branch goes from 'H on coin 1 to H on coin 2' written as HH. Similarly the final outcomes from the remaining branches are HT, TH, TT. The probability of a final outcome is the product of the two separate probabilities for outcomes along that branch.

Note: In a sequential step-by-step experiment, such as that in Figure 4.5, the number of final outcomes may be calculated by the following 'counting rule':

Counting rule 1: If an experiment is executed in sequence of k steps with n_1 possible outcomes at step 1; n_2 possible outcomes at step 2 ..... n_k possible outcomes at step k then the number of final outcomes is $(n_1)(n_1)\ldots(n_k)$.

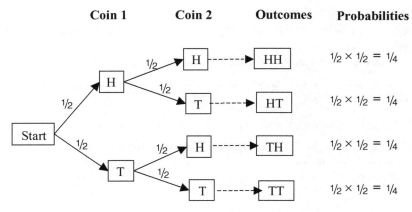

Figure 4.5   Tree diagram for tossing 2 coins.

In Figure 4.5, the experiment involves two intermediate steps, each with two possible outcomes. The number of final outcomes for the experiment is $(n_1)(n_1) = (2)(2) = 4$.

Worked Example 4.8: Probability calculations using 'AND' and 'OR' rules

An experiment involves tossing two coins. Calculate the probability that the outcome is (a) 2 Heads; (b) 1 Head; (c) No Heads, by each of the following methods:

Method I Direct calculations using $P(A) = \dfrac{\#A}{\#S}$

Method II Apply the appropriate 'AND' and/or 'OR' rules

Solution

Method I: These calculations were carried out in Worked Example 4.4

(i) $P(\text{Two Heads}) = P(A) = \dfrac{\#A}{\#S} = \dfrac{1}{4}$

(ii) $P(\text{One Head}) = P(B) = \dfrac{\#B}{\#S} = \dfrac{2}{4}$

(iii) $P(\text{No Heads}) = P(C) = \dfrac{\#C}{\#S} = \dfrac{1}{4}$

Method II: Apply the appropriate 'AND' and/or 'OR' rules

(a) Probability calculations are always easier if you work through the method step by step, stating the events 'in words' at the outset.
    When two coins are tossed,
    Step 1: In WORDS; 2 Heads can only turn up if the first coin shows Heads **AND** the second coin shows Heads at the same time.

Step 2: Write the verbal statement in Step 1 as an equation

2 Heads = Heads on the first **AND** Heads on the second coin

Step 3: Take the probability of both sides

$P(2\ \text{Heads}) = P(\text{Heads on the first } \textbf{AND} \text{ Heads on the second coin})$

Step 4: Apply the appropriate rule(s)

The coins are independent, so 'AND' means 'multiply'

$P(2\ \text{Heads}) = P(\text{Heads on the first coin}) \times P(\text{Heads on the second coin})$

Step 5: Calculate the probabilities

$$P(\text{HH}) = \frac{1}{2} \times \frac{1}{2} = \frac{1}{4}$$

(b) **Method II (rules).** When 2 coins are tossed

Step 1: In WORDS; the outcome of one Heads happens when the coins show Heads on the first and Tails on the second OR Tails on the first and Heads on the second coin.

Step 2: One Head = 'Heads and Tails' OR 'Tails and Heads'

Step 3: Take the probability of both sides

$P(\text{One Head}) = P(\text{HT OR TH})$

Step 4: The events HT and TH are mutually exclusive, since, if the outcome HT occurs then TH cannot happen at the same time. Hence the 'OR' means a straightforward addition

$P(\text{One Head when 2 coins are tossed}) = P(\text{HT}) + P(\text{TH})$

Step 5: $P(\text{One Head when 2 coins are tossed}) \rightarrow \dfrac{1}{4} + \dfrac{1}{4} = \dfrac{1}{2}$

(c) 'No Heads' when two coins are tossed

Step 1: In WORDS; No Heads means neither the first coin nor the second coin (so both coins show Tails)

Step 2: No Heads = 'Tails and Tails'

Step 3: Take the probability of both sides

$P(\text{No Heads}) = P(\text{'Tails AND Tails'})$

Step 4: The coins are independent, so AND means multiply

$P(\text{No Heads}) = P(\text{Tails}) \times P(\text{Tails})$

$P(0\ \text{Heads}) = \dfrac{1}{2} \times \dfrac{1}{2} = \dfrac{1}{4}$

Skill Development Exercise SK4_3: 'AND' and 'OR' and basic probability

Three coins are thrown.

Calculate the probability of (i) 3 Heads (ii) 2 Heads (iii) 1 Head (iv) No Heads by the following methods.

Method I Use the classical definition of probability given in (4.1).

Method II Applying the appropriate 'AND' and/or 'OR' rules.

Answer

Method I: Draw a Tree diagram similar to that given in Figure 4.5 but with an additional branch for the third coin.

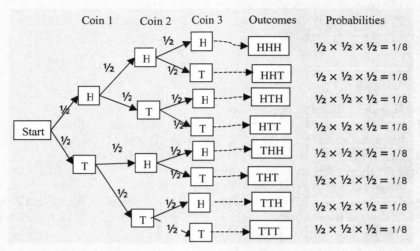

Figure SK4_3.1   Tossing three coins.

Refer to the tree diagram given in Figure SK4_3.1.
The sample space for three coins is
$S = \{HHH, HHT, HTH, HTT, THH, THT, TTH, TTT\}$

Method I: Direct use of sample and event spaces

(a) $P(3 \text{ Heads}) = \dfrac{\#E}{\#S} = \dfrac{1}{8}$, $E = \{HHH\}$, one outcome gives 3 Heads.

(b) $P(\text{Two Heads}) = \dfrac{\#E}{\#S} = \dfrac{3}{8}$, $E = \{HHT, HTH, THH\}$, three outcomes produce 2 Heads in various orders.

(c) $P(\text{One Head}) = \dfrac{3}{8}$, $E = \{HTT, THT, TTH\}$, three outcomes produce 1 Head in various orders.

(d) $P(\text{No Heads}) = P(TTT) = \dfrac{1}{8}$.

The same answers are obtained by Method II.

Worked Example 4.9: Application of the 'AND' and 'OR' rules

The probabilities that a PC and a printer function without fail for the first year of guarantee is 0.95 and 0.60 respectively. Assuming the probabilities are independent, calculate the probabilities that within the year

(a) both break down.
(b) the printer breaks down but the PC does not.
(c) only one of them breaks down.
(d) both function without fail.
(e) at least one of them functions without fail.

Solution

It may help to draw a tree diagram as in Figure 4.6.

Let A be the event that the PC functions; $\overline{A}$ is the event that the PC breaks down.

Let B be the event the printer functions: $\overline{B}$ is the event that the printer breaks down.

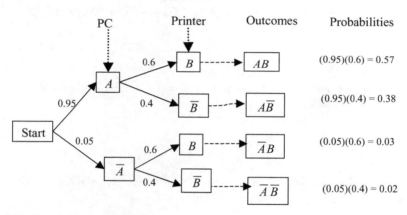

Figure 4.6 Tree diagram for PC and printer.

(a) Both break down means the PC breaks down 'AND' the printer breaks down
$P(\overline{A}\,\overline{B}) = P(\overline{A})P(\overline{B}) = (0.05)(0.4) = 0.02$.
There is a 2 % chance that both PC and printer will break down within the year.
(b) The printer breaks down 'AND' the PC functions.
$P(A\,\overline{B}) = P(A)P(\overline{B}) = (0.95)(0.40) = 0.38$.
There is a 38 % chance that the printer will break down but not the PC.
(c) **Step 1:** Only one breaks down means the printer breaks down but the PC functions 'OR' the printer functions and the PC breaks down
Step 2: Only one breaks down $\rightarrow A\overline{B}$ OR $\overline{A}B$
Step 3: $P(\text{Only one breaks down}) \rightarrow P(A\overline{B}$ OR $\overline{A}B)$
Step 4: $P(\text{Only one breaks down}) \rightarrow P(A\overline{B}) + P(\overline{A}B)$
Step 5: $P(\text{Only one breaks down}) = (0.95)(0.4) + (0.05)(0.6) = 0.41$

There is a 41 % chance that only one breaks down.
(d) Both function without fail means the PC functions 'AND' the printer functions

$$P(AB) = P(A)P(B) = (0.95)(0.60) = 0.57$$

There is a 57 % chance that both will function without fail for the year.

(e) There are *three methods* for solving this problem, Method 3 is the easiest. Also remember that 'at least one' means one or more'.

Method 1

At least one functions means
only the printer functions 'OR' only the PC functions 'OR' both function

At least one functions $\rightarrow \overline{A}B$ OR $A\overline{B}$ OR AB

P(At least one functions) $\rightarrow P(\overline{A}B$ OR $A\overline{B}$ OR $AB)$

P(At least one functions) $= P(\overline{A}B) + P(A\overline{B}) + P(AB)$
$$= 0.03 + 0.38 + 0.57 = 0.98$$

Method 2

$P(A \cup B)$ is another way, and simpler way, of dealing with 'at least one'
$$P(A \cup B) = P(A) + P(B) - P(AB) = 0.95 + 0.60 - -(0.95)(0.60) = 0.98$$

Method 3

Use the complementary events.

P(none function) + P(one or more functions) = 1

P(one or more functions) = 1 − P(none function) = 1 − (0.05)(0.04) = 0.98

4.2.4   Further counting rules: factorials; permutations; combinations

In simple probability calculations it was necessary to *count the number of outcomes*. For example, the classical definition of probability was given in (4.1) as

$$P(A) = \frac{\text{number of outcomes which result in event } A}{\text{total number of all possible outcomes}}$$

In more realistic probability calculations 'counting' can become complex. The counting rules are used in these situations. Counting rule 1 was already mentioned and the formula given in formula (4.10). A further three counting rules are given below: the application of these rules will be demonstrated in Worked Example 4.10.

Counting rule 1. If an experiment is executed in sequence of k steps with n_1 possible outcomes at step 1; n_2 possible outcomes at step 2.....n_k possible outcomes at step k then the number of final outcomes is

$$(n_1)(n_1) \ldots (n_k). \tag{4.10}$$

Counting rule 2. There are $n!$ arrangements of n different objects.
where n factorial is evaluated as

$$n! = (n)(n-1)(n-2)....(2)(1) \tag{4.11a}$$

Of these n objects, if n_1 are identical, n_2 are identical. . . . n_k are identical, then the number of different arrangements is

$$\frac{n!}{n_1!\, n_2!\, \ldots \ldots n_k!} \tag{4.11b}$$

Counting rule 3. There are $^n P_r$ permutations (arrangements) of n different objects, taken r at a time.

$$^n P_r = \frac{n!}{(n-r)!} \tag{4.12}$$

Counting rule 4. There are $^n C_r$ combinations of n different objects, taken r at a time where

$$^n C_r = \frac{n!}{(n-r)!\, r!} \tag{4.13}$$

Worked Example 4.10: Counting rules; permutations; combinations

(a) In a large building passwords are used to open office doors, filing cabinets, etc. How many different passwords are possible if
    (i) a four digit password is selected from the letters E, X, A, M;
    (ii) a four digit password is selected from the letters M, A, X, A;
    (iii) a four digit password is selected from digits 0 to 9 inclusive.
(b) A company plans to send a team to a sales conference.
    Calculate the number of different teams of three that may be selected from four available staff.
(c) Four employees are randomly selected from a panel of 12 females and 6 males for shift work. What is the probability that (i) no female is selected; (ii) two male and two females are selected? Explanations are given in Appendix 4.1.

Solutions

(a) (i) By counting rule 1, there are $4! = (4)(3)(2)(1) = 24$ different passwords.
        Explanation: there are four ways of choosing the first letter: three ways of choosing then second: two ways of choosing the third and only one way to choose the last letter.
        Calculator: Use the $n!$ key on the calculator to confirm your answer.
    (ii) By counting rule 2, (4.11b), the number of different arrangements is

$$\frac{4!}{2!} = \frac{(4)(3)(2)(1)}{(2)(1)} = (4)(3) = 12$$

    (iii) By counting rule 3, there are $^{10} P_4$ ways for choosing 4 numbers from 10 different numbers. Hence

$$^{10} P_4 = \frac{10!}{(10-4)!} = \frac{10!}{(6)!} = \frac{(10)(9)(8)(7)(6)(5)(4)(3)(2)(1)}{(6)(5)(4)(3)(2)(1)} = (10)(9)(8)(7) = 5040.$$

Calculator: Use the $^n P_r$ key on the calculator to confirm your answer.

(b) Let A, B, C and D represent the four individuals from which the team of three is selected.

Hence, by counting rule 4, there are ${}^4C_3 = \dfrac{4!}{(4-3)!\,3!} = \dfrac{(4)(3)(2)(1)}{(1)\,(3)(2)(1)} = 4$ different teams

Calculator: Use the ${}^nC_r$ key on the calculator to confirm your answer.

(c) (i) The panel consists of 12 females (F) and 6 males (M).

The probability that no females are selected in a team of four is stated as:

$P(F = 0 \text{ AND } M = 4)$

$= \dfrac{(\text{no. of ways of selecting no F from 12}) \times (\text{no. of ways of selecting four M from 6})}{(\text{no. of ways of selecting four people from a panel of 18})}$

$= \dfrac{{}^{12}C_0 \times {}^6C_4}{{}^{18}C_4} = \dfrac{1 \times 15}{3060} = 0.0049$

(ii) The required probability is stated as follows:

$P(F = 2 \text{ AND } M = 2)$

$= \dfrac{(\text{no. of ways of selecting two F from 12}) \times (\text{no. of ways of selecting two M from 6})}{(\text{no. of ways of selecting four people from a population of 18})}$

$= \dfrac{{}^{12}C_2 \times {}^6C_2}{{}^{18}C_4} = \dfrac{66 \times 15}{3060} = 0.3235$

Progress Exercises 4.3: Probability with rules for addition, multiplication and counting

In the probability calculations give answers rounded correct to four decimal places

1. Explain and illustrate the following terms with an appropriate example:
   (a) Independent events; (b) Dependent events; (c) Mutually exclusive events.
2. State, with reasons, whether the following events are independent events, dependent events or mutually exclusive events:
   (a) A six on the first throw of a dice followed by a six on the second throw.
   (b) Selecting a diamond followed by a club from a deck of cards.
   (c) The average daily temperature and sunshine in Melbourne (Australia).
   (d) The average daily temperature in Singapore and the season.
   (e) Studying for an examination and passing the examination.
3. A sales promotion offers a special price for PC with a printer. The probabilities that the PC and the printer function satisfactorily for three years are 0.84 and 0.72 respectively. What is the probability that at the end of three years that:
   (a) both will function satisfactorily.
   (b) only the printer will function satisfactorily.
   (c) at least one will function satisfactorily.
   State any assumptions made.
4. A bag contains 10 golf balls: 3 white, 2 yellow and 5 orange. A golfer picks two balls at random (without replacement). Calculate the probability that the selected balls are
   (a) both orange; (b) both the same colour; (c) both different colours; (d) both not white.

5. In an internet café, one of the five PC's is out of order. Two customers arrive; each sits down at a PC. Calculate the probability that:
   (a) both customers select PC's that are working .
   (b) only one selects a PC that is working.
   (c) both select PC's that are out of order.

6. An firm estimates that 15 % of accounts have at least one error. If an auditor selects two accounts what is the probability that there are errors in:
   (a) both accounts; (b) one account; (c) neither account.

7. The national statistic office produce 'life tables': these tables estimate the numbers of persons who will survive to each given age in the table starting with a base population of 100 000. Table PE4.3 Q7 is an extract from the abridged life table for Scotland 2002–2004.

Table PE4.3 Q7 Extract from abridged life table for Scotland 2002–2004

Age x	Males	Females
0	100 000	100 000
1	99 413	99 572
5	99 293	99 476
10	99 235	99 421
15	99 154	99 355
20	98 817	99 180

Source: Office for National Statistics. Crown copyright material is reproduced with the permission of the Controller of HMSO.

(a) Calculate the probability that (i) a male aged 15 will survive to age 20?; (ii) a female aged 1 will survive to age 15?

(b) Calculate the probability of surviving from each given age to the next given age for (i) males; (ii) females. Comment on the difference between the survival rates.

8. State, with explanations, whether the following statements are true or false.
   (a) If two events are independent then $P(AB) = P(A)P(B)$.
   (b) If two events are independent then $P(A/B) = P(A)$.
   (c) If two events are dependent then $P(AB) = 0$.
   (d) $P(A \text{ OR } B) = P(A) + P(B)$ when events A and B are independent.

9. How many car registration number plates are possible if the number plate consists of three different letters, selected from $\{E, C, O\}$ followed by four different numbers selected from the numbers 0 to 9.

10. To cater for the growing number of customers, a telephone company prefixes a number of digits to existing phone numbers. Calculate the number of different telephone numbers that are possible in the following:
    (a) A four digit number selected from the digits 0 to 9 but the first digit cannot be 0.
    (b) The numbers in (a) with the prefix 5, 6.
    (c) A six digit number selected from the digits 0 to 9 but the first digit cannot be 0 or any numbers in (b).

11. A box contains 20 hand held torches, four of which are defective.
    If 4 torches are selected at random what is the probability that:
    (a) no defective torches are selected.
    (b) two defective torches are selected.

4.3   Joint, marginal and conditional probability

4.3.1   Calculating joint and marginal probabilities

The simplest way to illustrate the ideas of 'joint', marginal and conditional probabilities is through a contingency table. A contingency table (such as Table 4.1 below) is a 2-way cross-tabulation for every possible combination of levels for two variables: the data in contingency tables are always *count data*. For example, in Table 4.1 the variables are accounting and mathematics results graded in three levels; Merit, Pass and Fail. The data is the examination results for 500 students, distributed into cells according to the results in both subjects. For example, 120 students were awarded 'a Merit in accounting AND a Merit in mathematics'. In terms of probability this is described as a **joint event**. The row totals are the total number of students that were awarded a Merit, Pass or Fail in accounting respectively: The column totals are the numbers awarded each grade for mathematics. A marginal event refers to events associated with one variable only. For example, in Table 4.1, the event 'Pass in accounting' refers to all those who passed accounting, irrespective of how they performed in the mathematics test.

Table 4.1   Examination results in maths and accounting for 500 students

		Mathematics			Row Totals
		Merit	Pass	Fail	
Accounting	Merit	120	10	0	130
	Pass	100	190	30	320
	Fail	20	10	20	50
	Column Totals	240	210	50	500

Joint event: a Merit in mathematics **AND** in accounting

Marginal event a Pass in accounting

Total of 500 students

Marginal event Fail mathematics

If the numbers are large, the frequencies in the contingency table may be converted to probabilities using the relative frequency definition: $P(A) = \dfrac{f_A}{f_T}$, where the overall total $f_T = 500$ (in Table 4.1) and f_A is the frequency of event A.

In Table 4.2, the **joint probabilities** are calculated by dividing each joint frequency in the body of the table by the f_T: the totality of these probabilities constitutes a **'joint probability' distribution***.

The **marginal probabilities** are calculated by dividing the marginal frequencies by the total frequency, f_T: The marginal probabilities in the column margin constitute the **'marginal probability distribution'** for accounting grades: The marginal probabilities in the row margin constitute the 'marginal probability distribution' for mathematics grades*. See Worked Example 4.11.

***A probability distribution** is a list of every possible outcome with the corresponding probability.

Worked Example 4.11: Joint and marginal probability distributions

Using the data given in the contingency table, Table 4.1

(a) Calculate all the joint probabilities for mathematics and accounting grades. Hence state the probability that a student is awarded (i) a Merit in mathematics and a Merit in accounting; (ii) a 'Pass OR a Merit in mathematics' and a Fail in accounting.
What is the joint probability distribution?

(b) Write out the probabilities of being awarded a Merit, a Pass and a Fail respectively in (i) mathematics; (ii) accounting.
What are the marginal probability distributions for (i) mathematics grades (ii) accounting grades?

Solution

(a) The joint probabilities for mathematics and accounting grades are given in Table 4.2

Table 4.2   Joint and marginal probabilities from Table 4.1

		Mathematics			
Joint probability P (Merit in maths **AND** Merit in accounting) $= 0.24$		Merit	Pass	Fail	Row Totals
Accounting	Merit	$\frac{120}{500} = 0.24$	$\frac{10}{500} = 0.02$	$\frac{0}{500} = 0$	$\frac{130}{500} = 0.26$
	Pass	$\frac{100}{500} = 0.20$	$\frac{190}{500} = 0.38$	$\frac{30}{500} = 0.06$	$\frac{320}{500} = 0.64$
	Fail	$\frac{20}{500} = 0.04$	$\frac{10}{500} = 0.02$	$\frac{20}{500} = 0.04$	$\frac{50}{500} = 0.10$
Column Totals		$\frac{240}{500} = 0.48$	$\frac{210}{500} = 0.42$	$\frac{50}{500} = 0.10$	1

Marginal probability P (Pass in accounting)

Total probability $=1$

Marginal probability P (Merit in maths) $= 0.48$

P {Pass **OR** a Merit in maths} **AND** {Fail in accounting}) $= \dfrac{20+10}{500} = 0.04 + 0.02$

(i) P (Merit in both mathematics **AND** in accounting) $= 0.24$, Table 4.2.

(ii) P ('Pass OR a Merit in mathematics' **AND** a Fail in accounting) $= 0.06$.

(b) The last row in Table 4.2 gives the probabilities of being awarded each grade in mathematics (regardless of accounting result) while the last column gives the probabilities of being awarded each grade in accounting (regardless of mathematics results). These probabilities are called the 'marginal' probabilities – simply because they appear in the margins of the contingency table.

The marginal probability distribution for mathematics grades is the list of all mathematics grades with the corresponding probabilities. The marginal probability distribution for accounting grades is the list of all accounting grades with the corresponding probabilities. Both marginal probability distributions are given below.

Marginal distribution for Mathematics			Marginal distribution for Accounting		
Merit	Pass	Fail	Merit	Pass	Fail
$\frac{240}{500} = 0.48$	$\frac{210}{500} = 0.42$	$\frac{50}{500} = 0.10$	$\frac{130}{500} = 0.26$	$\frac{320}{500} = 0.64$	$\frac{50}{500} = 0.10$

4.3.2 Conditional probabilities and conditional probability distributions

Recall: conditional probability $P(A/B)$, was introduced in section 4.2 in 'Rule 3', equation (4.7) for dependent events from $P(AB) = P(A)P(B/A) \ldots$

Rearrange this equation to solve for the conditional probability:

$$P(B/A) = \frac{P(AB)}{P(A)} \qquad (4.14)$$

Equation (4.14) is used to calculate conditional probability from the joint probability, $P(AB)$ and $P(A)$, the probability of the condition, A.

Conditional probabilities may be calculated from the contingency table by considering the row or column to which the condition refers as the reduced sample space. The required conditional probability is calculated from the reduced sample space. For example, for the data in Table 4.1, the probability that a student was awarded a Merit in mathematics when it is known that the student was awarded a Pass in accounting is a conditional probability where the reduced sample space is the set of students with a Pass in accounting. So, extract the group of students who were awarded a Pass in accounting. This group is extracted from row two of Table 4.1 and presented in Table 4.3:

Table 4.3   Extract from Table 4.1

	Merit	Pass	Fail	Total
Mathematics result → for students who were awarded a Pass in accounting	100	190	30	320

The probability that any one of these students was awarded a Merit (or any other grade) in mathematics is calculated as a simple relative frequency:

$$P(\text{a Merit in mathematics, given a Pass in accounting}) = \frac{100}{320}$$

But, this probability is conditional on the fact that student was awarded a Pass in accounting and hence must be written formally as $P(B/A) = \dfrac{P(AB)}{P(A)} = \dfrac{100}{320}$

... where A is the event: Pass in accounting and B is the event Merit in mathematics.

Note:

1. Reduced Sample Spaces and conditional probability

It is important to realise that when a condition is *given* this is additional information. This additional information is used to reduce the sample space from that of the overall table to the group to which the condition refers. For example, Table 4.3 is the reduced sample space from Table 4.1.

2. A Conditional Probability distribution is the set of probabilities for all outcomes that satisfy a given condition. See Worked Example 4.12.

Worked Example 4.12: Conditional probability and distributions

The following question refers to the data in Table 4.1.

(a) When it is known (given) that the student has a Pass grade in accounting calculate the probability that the student was awarded (i) a Merit in mathematics (ii) a Pass in mathematics.

(b) Write out the conditional probability distribution for students who were awarded a Pass grade in accounting from (i) the reduced sample space (ii) by calculating the conditional probabilities from the joint and marginal probabilities: $P(B/A) = \dfrac{P(AB)}{P(A)}$

(c) A student is known to have failed accounting, calculate the probability that a student has 'Merit or a Pass' in mathematics.

Solution

Let A = a Pass in accounting: B = a Merit in mathematics: C = a Pass in mathematics: D = a Fail in mathematics: F = a Fail in accounting

(a) (i) From Table 4.2, $P(A) = 0.64$, $P(AB) = 0.20$. Substitute into equation (4.8)

$$P(B/A) = \frac{P(AB)}{P(A)} = \frac{0.20}{0.64} = 0.3125$$

   (ii) From Table 4.2, $P(A) = 0.64$, $P(AC) = 0.38$. Substitute into equation (4.8)

$$P(C/A) = \frac{P(AC)}{P(A)} = \frac{0.38}{0.64} = 0.59375$$

(b) The conditional probability distribution may be calculated directly from the reduced sample space from Table 4.1 (see Table 4.3), for students who were awarded a Pass in accounting as follows.

Table 4.4   Conditional probability distribution from reduced sample space from Table 4.1

	Merit (B) (given a pass in accounting)	Pass (C) (given a pass in accounting)	Fail (D) (given a pass in accounting)	Total
Mathematics grade → for students who were awarded a Pass in accounting (A)	$\dfrac{100}{320} = 0.3125$	$\dfrac{190}{320} = 0.59375$	$\dfrac{30}{320} = 0.09375$	$\dfrac{320}{320} = 1$
	$P(B/A)$	$P(C/A)$	$P(D/A)$	

(c) Given that a student has a Pass in accounting, the conditional probabilities that the student was awarded a Merit in mathematics and a Pass in mathematics were calculated in part (a) above. The conditional probability for a Fail in mathematics is

$$P(D/A) = \frac{P(AD)}{P(A)} = \frac{0.06}{0.64} = 0.09375$$

Where $P(A) = 0.64$, $P(AD) = 0.06$ from Table 4.2.

Hence $P(B/A)$, $P(C/A)$ and $P(D/A)$ are the conditional probabilities for every possible mathematics grade for those who are known to have passed accounting, see Table 4.4, and hence constitute the conditional probability distribution.

(d) In WE 4.8, P (a Merit OR Pass in mathematics' AND Fail in accounting) = 0.06. Hence $P((B$ OR C) AND $F) = 0.06$; $P(F) = 0.10$. Substitute into equation (4.8)

$$P((B \text{ OR } C)/F) = \frac{P((B \text{ OR } C) \text{ AND } F)}{P(F)} = \frac{0.06}{0.10} = 0.6$$

Skill Development Exercise SK4_4: Conditional probability

From Table 4.1, calculate the probability that the student was awarded (a) a Merit in maths; (b) a Pass in maths; (c) a Merit or a Pass in maths when it is known (given) that the student has a Fail in accounting.

Solution

(a) $P(\text{Merit in maths}) = \dfrac{240}{500} = 0.48$

(b) $P(\text{Pass in maths}) = \dfrac{210}{500} = 0.42$

(c) $P(\text{Merit OR Pass in maths}) = \dfrac{450}{500} = 0.90$

Progress Exercises 4.4: Joint, marginal and conditional probabilities

In the probability calculations give answers rounded correct to four decimal places

1. The data in Table PE4.4_1 gives the number of parking fines issued in one month in three different areas by five traffic wardens, Tom, Ralph, Alf, Frank and Carol:

Table PE4.4 Q1   Number of parking fines issued by traffic wardens

Wardens	Main street	Shop street	Old Market	Row totals
T	120	300	210	630
R	150	240	150	540
A	210	210	180	600
F	270	210	240	720
C	90	150	270	510
Col. totals	840	1110	1050	3000

(a) Define and use the data in Table PE4.4 Q1 to give an example of the following terms: joint probability; marginal probability; conditional probability.

(b) Calculate the joint probability distribution for the data in Table PE4.4 Q1.

(c) Calculate the percentage of fines issued by each of the traffic wardens. How are these percentages related to the marginal probability distribution of fines issued by traffic wardens? Write out the marginal distribution.

(d) What is the probability that a fine was issued by warden F in the Old Market?

(e) *Given* that a fine was issued in Main St., what is the probability that it was issued by warden C?

(f) Given that a fine was issued by warden F, what is the probability that it was in Shop Street?

2. Companies select students for work experience on the basis of the results of a technical and a personal interview. The grades given F (weak), B (adequate), A (excellent) for a cohort of students are summarised in Table PE4.4 Q2 (subscripts; p for personal; T for technical).

Table PE4.4 Q2   Technical and personal interview results

| Grades for personal interview | Technical Grades | | | Row totals |
	A_T	B_T	F_T	
A_P	45	70	10	125
B_P	110	64	22	196
F_P	25	68	86	179
Column totals	180	202	118	500

(a) Write out the marginal probability distributions for the technical and personal interviews grades. Describe each distribution verbally.

(b) Write out the joint probability distribution for technical and personal grades.

(c) What percentage of students were awarded?
   (i) grade A for both interviews?
   (ii) grade A for the technical interview, *given* that s/he has already be awarded grade B for the personal interview?
   (iii) grade F for the personal interview and grade B for the technical interview?
   Convert the percentages calculated in (i) to (iii) above to probabilities, stating whether these are joint or conditional probabilities.

(d) Show, by calculation, that grade A in the personal interview is independent of grade A in the technical interview.

3. In a 'blind' experiment randomly selected patients are administered pain relief medications A, B or C. The patients are asked to rate the effectiveness of the medication as excellent, good or no effect. The results for 300 patients are summarised in Table PE4.4 Q3:

Table PE4.4 Q3   Pain relief

| Rate | Medication | | | Totals |
	A	B	C	
Excellent	34	65	23	122
Good	45	33	65	143
No effect	8	10	17	35
Totals	87	108	105	300

(a) Describe and give an example of a 'blind' experiment.
(b) Write out the probability distribution for the effectiveness of the medications.
(c) Calculate the probability that a randomly selected patient rated the medication as follows:
  (i) Excellent.
  (ii) no effect.
  (iii) if it is known that the patient was given C.
(d) What is the probability that a patient was given medication A if it is known that they rated the effect as excellent?

4. (a) Explain the following, with an appropriate example:
  (i) Independent events.
  (ii) Mutually exclusive events.
  (iii) Dependent events.
(b) An Internet café has eight PC's, 3 of which are faulty. On a randomly selected morning, three customers arrived when the café opened. If each customer selects a PC what is the probability that:
  (i) all three select a faulty PC.
  (ii) only the first customer selects a faulty PC.
  (iii) two of them select faulty PC's.
  (iv) none of them selects a faulty PC.

5. (a) Explain the following terms: independent events; conditional probability; probability distribution.
(b) Table PE4.4 Q5 below summarises the number of flights according to origin and minutes late on arrival in Singapore.

Table PE4.4 Q5   Late flight

| Minutes late | Origin of flight | | | | |
	London	Sydney	Bangkok	Paris	Row totals
5 but less than 15	19	15	21	12	67
15 but less than 30		43	16	14	99
30 but less than 45	10		17	22	84
45 or more	8	6		15	50
Column totals	63	99	75	63	

(i) Fill in the missing values in contingency Table PE4.4 Q5.
(ii) Write out the probability distributions for Origin of flight and Minutes late.
(iii) Calculate the probability that a flight from Sydney was less than 30 minutes late.
(iv) If it is know that a flight is from London, calculate the probability that it is at least 30 minutes late.
(v) Calculate the probability that a flight arrives from Paris and is between 15 and 30 minutes late.
(vi) Given that a flight arrives from Bangkok calculate the probability that it is less than 30 minutes late.
(vii) Calculate the average minutes late for flights from Paris.

6. Refer to the Table PE4.4 Q5.

An association of air travellers claim that flights from London are more likely to be late than flights from Sydney. Calculate the conditional probabilities that flights are at least 30 minutes late given the origin of the flight is (a) London; (b) Sydney.

Do your results confirm the views of the travellers association?

7. Registration records for five disciplines in a college give the numbers of male and female students as summarised in Table PE4.4 Q7:

Table PE4.4 Q7   Numbers of male and female students registered for five disciplines

Numbers	Discipline					Totals
	Business	Engineering	Science	Pharmacy	Humanities	
Male	214	412	102	60	50	838
Female	326	88	128	60	30	632
Totals	540	500	230	120	80	1470

(a) For each of the five disciplines given in Table PE4.4 Q7, calculate the conditional probabilities that the student is (i) male; (ii) female.

(b) Calculate the marginal probability distributions for (i) discipline, (ii) gender.

(c) Given that the student is (i) male; (ii) female calculate the conditional probability distributions that the student is studying each of the disciplines.

Based on the results of (a), (b) and (c) above (and any other appropriate probabilities) discuss the gender balance.

8. (a) A study is conducted to determine whether Vitamin C or Garlic (Odourless tablets) reduce the incidence of the common cold. Volunteers were randomly given Vitamin C, Garlic tablets or a placebo. The results after 3 months are summarised below:

Table PE4.4 Q8   Colds/no colds for three months

	Medication			Totals
	Vitamin C	Garlic	Placebo	
Cold	48	36	68	152
No cold	152	176	120	448
Totals	200	212	188	600

(b) Write out the probability distribution for the incidence of colds for each medication.

(c) Calculate the probability that a randomly selected volunteer
  (i) was given garlic and caught cold;
  (ii) did not catch a cold;
  (iii) caught a cold, if it is known that they were given vitamin C;
  (iv) was given the placebo, if it is known that they did not catch a cold.

4.4 Bayes' Rule

4.4.1 Conditional Probability and Bayes' Rule

The final application of conditional probability in this chapter is **Bayes' Rule**. The applications of Bayes' Rule in this section will examine the inference that can be drawn from tests for conditions that are rare (viruses, diseases; presence of performance enhancing drugs; errors in accounts; bugs in software, etc.) when tested by tests that are less than 100 % reliable. With Bayes' Rule we will calculate the probability that a rare condition is present given that the diagnostic test for the condition is positive

---

Worked Example 4.13: Conditional probability: Example of Bayes' Rule

An importer of smoke alarms has 3 suppliers: A_1; A_2 and A_3.
20 % of the smoke alarms are supplied by A_1; 70 % are supplied by A_2 and 10 % are supplied by A_3. From historical data it has been established that the percentage of faulty alarms from suppliers A_1; A_2 and A_3 is 12 %; 3 % and 2 % respectively.

(a) If the total number of alarms imported is 1000, present the above information in terms of frequencies. Calculate the total number of faulty alarms.
(b) Calculate the probability of selecting a faulty alarm What is a reduced sample space for faulty alarms?
(c) Given that a customer returns a faulty alarm, calculate the probability that the alarm was supplied by (i) A_1, (ii) A_2 (iii) A_3.

Solution

(a) Since there are 1000 alarms, the numbers imported from each supplier; A_1; A_2 and A_3, are 20 %, 70 % and 10 % of 1000 respectively. See row 1 of Table 4.5(a)

Table 4.5(a)   Smoke alarms

	Supplier			
	A_1	A_2	A_3	Total
Number of alarms from Supplier	200	700	100	1000

The number of faulty alarms from each supplier is 12 %, 3 % and 2 % of the respective number supplied in row 1. See row 2 Table 4.5(b).

$$\text{Number of faulty alarms} = \text{Number supplied} \times \frac{\text{stated percentage faulty}}{100}.$$

The total number of faulty alarms is 47.

Table 4.5(b) The number of faulty alarms from each supplier

	Supplier			
	A_1	A_2	A_3	Total
Number of alarms from Supplier A_i	200	700	100	1000
Percentage faulty, given Supplier A_i	12 %	3 %	2 %	
Number faulty from Supplier $A_i =$	$200 \times \dfrac{12}{100}$	$700 \times \dfrac{3}{100}$	$100 \times \dfrac{2}{100}$	
No. from Supplier $A_i \times$ percent faulty	$= 24$	$= 21$	$= 2$	47

(b) The probability of a faulty alarm is $\dfrac{\text{\# Faulty alarms}}{\text{\# Total alarms}} = \dfrac{47}{1000} = 0.047$

The condition: 'given that an alarm is faulty' reduces the sample space to the set of faulty alarms. See Table 4.5(c). The 'reduced sample space' will be used in the calculation of probabilities conditional on faulty alarms in part (c).

Table 4.5(c) Reduced sample space for faulty alarms

	Supplier				
	A_1	A_2	A_3	Total	
Number of alarms from Supplier	200	700	100	1000	Reduced sample space
Number faulty from Supplier	24	21	2	47	

(c) Given that the returned alarm is faulty, the probability that it was supplied by A_1, A_2 and A_3 is calculated from this reduced sample space.

(i) P(supplied by A_1, given it is faulty) $= \dfrac{24}{47} = 0.5106$;

(ii) P(supplied by A_2, given it is faulty) $= \dfrac{21}{47} = 0.4468$;

Most likely source of the faulty alarm is A_1

(iii) P(supplied by A_3, given it is faulty) $= \dfrac{2}{47} = 0.0426$.

4.4.2 Bayes' Rule: calculations

Worked Example 4.13 will be shown to be a particular case of **Bayes' Rule**.

Bayes' Rule may be explained in various ways, such as (i) the tabular approach, used above; (ii) the Venn diagram approach; (iii) use of tree diagrams.

The tabular approach, as used above in Table 4.5(b). This approach is generalised by writing the relative frequencies in Table 4.5(b) probabilities as given in Table 4.6.

The probability of selecting a faulty alarm is called the total probability

An alarm is faulty if it was supplied by (supplier 1 AND faulty) 'OR' (supplier 2 AND faulty) 'OR' (supplier 3 AND faulty)

Table 4.6   Probabilities calculated from Table 4.5 Smoke alarms

	Supplier			Total
	A_1	A_2	A_3	
Probability from Supplier A_i	$P(A_1) = 0.2$	$P(A_2) = 0.7$	$P(A_3) = 0.1$	1
Probability faulty, (given Supplier A_i)	$P(F/A_1) = \dfrac{12}{100}$	$P(F/A_2) = \dfrac{3}{100}$	$P(F/A_3) = \dfrac{1}{100}$	
Probability 'from Supplier A_i AND faulty'	$P(A_1)P(F/A_1) =$ $0.2 \times \dfrac{12}{100} = 0.024$ $= P(A_1 \text{ AND } F)$	$P(A_2)P(F/A_2) =$ $0.7 \times \dfrac{3}{100} = 0.021$ $= P(A_2 \text{ AND } F)$	$P(A_3)P(F/A_3) =$ $0.1 \times \dfrac{2}{100} = 0.002$ $= P(A_3 \text{ AND } F')$	$P(F) = 0.047$

$F = (A_1 \text{ AND } F) \text{ OR } (A_2 \text{ AND } F) \text{ OR } (A_3 \text{ AND } F') \dots$

Hence, taking probabilities

$P(F) = P(A_1 \text{ AND } F) + P(A_2 \text{ AND } F) + P(A_3 \text{ AND } F') \dots \text{addition rule (4.8)}$
$P(F) = P(A_1) P(F/A_1) + P(A_2) P(F/A_2) + P(A_3) P(F/A_3) \dots \text{multiplication rule (4.7)}$

This is the sum of the probabilities in row 3, Table 4.6

Calculate the 'revised' probability conditional on the given characteristic

That is calculate a. $P(A_1/F)$; b. $P(A_2/F)$; c. $P(A_3/F)$

These calculations are based on equation (4.14) $P(A/B) = \dfrac{P(AB)}{P(B)}$.

Hence $P(A_i/F) = \dfrac{P(A_i F)}{P(F)} = \dfrac{P(A_i)P(F/A_i)}{P(F)}, i = 1, 2, 3$ and $P(F)$ is given above.

Hence, divide each probability from row 3 in Table 4.6 by the total probability $P(F)$.

(a) $P(A_1/F) = \dfrac{P(A_1 F)}{P(F)} = \dfrac{0.024}{0.047} = 0.5146$

(b) $P(A_2/F) = \dfrac{P(A_2 F)}{P(F)} = \dfrac{0.021}{0.047} = 0.4468$

(c) $P(A_3/F) = \dfrac{P(A_3 F)}{P(F)} = \dfrac{0.002}{0.047} = 0.0426$

Note: the above probabilities were calculated from the reduced sample space, in Table 4.5(c).

Bayes' Rule is a generalisation of the conditional probabilities (a), (b) and (c)

$$P(A_i/F) = \frac{P(A_i F)}{P(F)} \dots \text{where } i = 1, \ 2 \text{ or } 3$$

$$P(A_i/F) = \frac{P(A_i)P(F/A_i)}{P(A_1)P(F/A_1) + P(A_2)P(F/A_2) + P(A_3)P(F/A_3)}$$

$$P(A_i/F) = \frac{P(A_i)P(F/A_i)}{\sum\limits_{j=1}^{3} P(A_j)P(F/A_j)}$$

This is 'Bayes' rule' for a population partitioned into three mutually exclusive events

Note: each numerator: $P(A_i)P(F/A_i)$ is one of the three probabilities
$P(A_1)P(F/A_1) + P(A_2)P(F/A_2) + P(A_3)P(F/A_3)$ in the denominator. The subscript j is used for the summation over all outcomes: i is used to indicate a specific outcome $P(A_i)P(F/A_i)$.

Worked Example 4.14: Venn diagram explanation for Bayes' Rule

---

Use Venn Diagrams to explain Bayes' Rule for Worked Example 4.13

Solution

Step 1: Start by writing down any information, probabilities, etc. given in the question. In this example these probabilities are: the percentage of alarms from each supplier: $P(A_1) = 0.20$, $P(A_2) = 0.70$ and $P(A_3) = 0.10$.
Also, when the supplier is known, then the percentage of faulty alarms was given. Hence the conditional probabilities of a Faulty alarm given the supplier are

$$P(F/A_1) = \frac{12}{100}, \; P(F/A_2) = \frac{3}{100} \; P(F/A_3) = \frac{2}{100}$$

Note: the complements of the above probabilities may be required occasionally: For example, $P(\overline{F}/A_1) = 1 - P(F/A_1) = \frac{88}{100}$.

Step 2: State the population:
In Worked Example 4.13, the population is the set of all smoke alarms supplied by the three suppliers A_1, A_2 and A_3. This population may be represented by the Venn diagram in Figure 4.7(a). The population is partitioned into three mutually exclusive and collectively exhaustive events: the alarms from suppliers A_1; A_2 and A_3 who supplied 20 %, 70 % and 10 % respectively of all alarms.

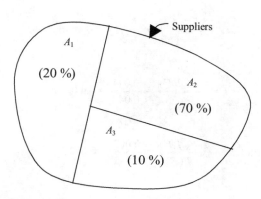

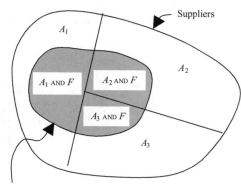

F: Faulty Alarms

Figure 4.7(a)   Venn diagram representation of the partitioned population.

Figure 4.7(b)   Venn diagram representation of the partitioned population and the sub-population of faulty alarms (F).

Step 3: **State the sub-population of interest and calculate the total probability.**

The sub-population of interest here is the set of faulty alarms, F. This sub-population is shown on the Venn diagram in Figure 4.7(b).

Calculate the 'total probability' of selecting an element from this sub-population. In this example, the total probability is the probability of selecting a Faulty alarm.

The event of a Faulty alarm occurs if any one of the following events occurs: 'An alarm is from $\{A_1$ AND Faulty$\}$ OR $\{A_2$ AND Faulty$\}$ OR $\{A_3$ AND Faulty$\}$' Write this statement in terms of probabilities:

$$P(\text{Faulty alarm}) = P((A_1 \text{ AND Faulty}) \text{ OR } (A_2 \text{ AND Faulty}) \text{ OR } (A_3 \text{ AND Faulty}))$$

$$P(F) = P(A_1 \text{ AND } F) + P(A_2 \text{ AND } F) + P(A_3 \text{ AND } F)$$

.... using addition (4.8) rule and letting F = Faulty.

Write out the joint probabilities in terms of known (or given) probabilities using the multiplication rule (4.7) for dependent events. Hence

$$P(F) = P(A_1)P(F/A_1) + P(A_2)P(F/A_2) + P(A_3)P(F/A_3)$$

Hence, in this example,

$$P(F) = (0.20)\left(\frac{12}{100}\right) + (0.70)\left(\frac{3}{100}\right) + (0.10)\left(\frac{2}{100}\right)$$
$$= 0.024 + 0.021 + 0.002 = 0.047$$

(The probability of selecting a faulty alarm is 0.047).

Step 4: **calculate the 'revised probabilities'** that an alarm was from a certain supplier, *given* it was found to be faulty. The revised probabilities are calculated from reduced sample space of faulty alarms in Figure 4.7(c). The required revised probabilities may be calculated from set of 'faulty alarms' as follows:

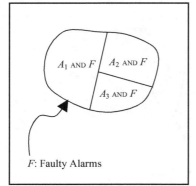

A_1 AND F A_2 AND F

A_3 AND F

F: Faulty Alarms

(a) $P(A_1/F) = \dfrac{P(A_1F)}{P(F)} = \dfrac{0.024}{0.047} = 0.5146$

(b) $P(A_2/F) = \dfrac{P(A_2F)}{P(F)} = \dfrac{0.021}{0.047} = 0.4468$

(c) $P(A_3/F) = \dfrac{P(A_3F)}{P(F)} = \dfrac{0.002}{0.047} = 0.0426$

Figure 4.7(c)   The Venn diagram representation of the sub-population (subset) of faulty alarms.

Skill Development Exercise SKD4_5: Tree diagrams and Bayes' Rule

Use a tree diagram to explain Bayes' Rule for Worked Example 4.13.

Solution

(outline solution only).

State the population (all alarms): **from the partitioned population state** the probabilities for each supplier are $P(A_1) = 0.20$, $P(A_2) = 0.70$ and $P(A_3) = 0.10$. See tree diagram Figure SK4_5.1 (this is the same as Table 4.6, row 1).

Given the supplier, the percentage of their product that is faulty is given by the next set of lines (branches) with probabilities:

$$P(F/A_1) = \frac{12}{100}, \ P(F/A_2) = \frac{3}{100} \ P(F/A_3) = \frac{1}{100}$$

See Figure SK4_5.1 (and row 2 in Table 4.6).

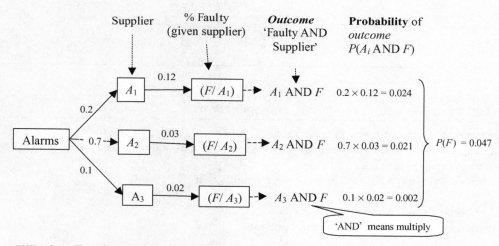

Figure SKD4_5.1   Tree diagram for calculating the probability of a faulty alarm.

Note: the percentage of faulty alarms is conditional or dependent on the supplier.

The Total probability of a faulty alarm is the sum of the probabilities for the three outcomes resulting in a faulty alarm.

The 'revised' probabilities are calculated from the equations:

$$P(A_i/F) = \frac{P(A_i F)}{P(F)} = \frac{P(A_i)P(F/A_i)}{P(F)}, i = 1, 2, 3$$

From these examples, it should be obvious that the use of various graphical tools to explain Bayes' Rule and calculated the 'revised' probabilities all involve essentially the same steps. These steps are summarised as follows:

Bayes' Rule: Summary of method

Step 1: State the base population. Partition the base population into n mutually exclusive events according to information given in the question and write down the probabilities: $P(A_j)$, $j = 1,2,\ldots n$.

Step 2: State the sub-population of interest (B). From the information in the question, write down the conditional probabilities: $P(B/A_j)$, $j = 1,2,\ldots n$.

Step 3: Calculate the total probability of selecting an item from the sub-population of interest:

$$P(B) = \sum_{j=1}^{n} P(A_j)P(B/A_j).$$

Step 4: Calculate the 'revised' probability conditional on B.

$$P(A_i/B) = \frac{P(A_i B)}{P(B)} = \frac{P(A_i)P(B/A_i)}{P(B)}$$

Bayes' Rule:
$$P(A_i/F) = \frac{P(A_i)P(F/A_i)}{\sum_{j=1}^{n} P(A_j)P(F/A_j)} \qquad (4.15)$$

Worked Example 4.15: Bayes' Rule, an application

It is known that 5 % of athletes take performance enhancing drugs. The test for such drugs gives a positive result in 98 % of cases when they are present, but it also gives a positive result in 4 % of cases when no drugs are present.

Given that a test is positive calculate the probability that

(a) performance enhancing drugs are present.
(b) performance enhancing drugs are not present.

Solution

Follow the step by step method given above, entering results into Table 4.7.

Step 1: The population is the population of athletes. This population is partitioned into athletes who take performance enhancers (A_1) and athletes who do not take performance enhancers (A_2). Hence the probabilities are $P(A_1) = 0.05$: $P(A_2) = 0.95$.

Step 2: The sub-population of interest (B) is the set of athletes who tested positive. The probability conditional on performance enhancers is $P(B/A_1) = \frac{98}{100}$, conditional on **no** performance enhancers $P(B/A_2) = \frac{4}{100}$.

Step 3: Calculate the total probability of a positive test:

$$P(B) = P(A_1)P(B/A_1) + P(A_2)P(B/A_2) = (0.05)(0.98) + (0.95)(0.04) = 0.087$$

Table 4.7   Calculations for Bayes' Rule: test for the presence of performance enhancing drugs

	Athletes		Total
	A_1 (takes dugs)	A_2 (takes no dugs)	
Probability A_i	$P(A_1) = 0.05$	$P(A_2) = 0.95$	1
Probability positive, given A_i	$P(B/A_1) = \dfrac{98}{100}$	$P(B/A_2) = \dfrac{4}{100}$	
Probability 'from A_i AND positive'	$P(A_1)P(B/A_1) =$ $0.05 \times \dfrac{98}{100} = 0.049$	$P(A_2)P(B/A_2) =$ $0.95 \times \dfrac{4}{100} = 0.038$	$P(B) = 0.087$

Step 4: Calculate the revised probability conditional for the test being positive

Use Bayes' Rule $P(A_i/B) = \dfrac{P(A_i B)}{P(B)} = \dfrac{P(A_i)P(B/A_i)}{P(B)}$ to calculate the probability that

(a) drugs are present, given a positive test $P(A_1/B) = \dfrac{0.049}{0.087} = 0.5632$

(b) drugs are NOT present, given a positive test $P(A_2/B) = \dfrac{0.038}{0.087} = 0.4368$

Comment: the test is obviously not reliable; given that the test is positive, there is only a 56 % chance that the athlete has taken drugs. Unfortunately, given that the test is positive there is a 44 % chance that the athlete did not take drugs!

You are asked to explain why this test in unreliable in SK4_4.

(see article on 'The Power of Bayes' Mathematics Today Vol. 41 No. 4 August 2005)

Worked Example 4.16: Bayes' Rules and unreliable tests

Apply the information in Worked Example 4.15 to a group of 1000 athletes.

Calculate the number of athletes that test positive when (i) drugs are present; (ii) drugs are not present.

Use the results (i) and (ii) to explain why this test for performance enhancing drugs is unreliable.

Answer

The answers to (i) and (ii) are given in Table 4.8:

Table 4.8   Bayes' Theorem for drug testing

	Athletes		Total
	A_1 (takes drugs)	A_2 (takes no drugs)	
Numbers (A_i)	50	950	1000
Percent positive, given A_i	98 %	4 %	
Number 'from A_i AND positive'	(i) $50 \times \dfrac{98}{100} = 49$	(ii) $950 \times \dfrac{4}{100} = 38$	87

The results show that the number of false positives is 38 out of the 87 positives from the group of 1000 athletes, while the number of true positives is only 49.

It is because of this large proportion of false positives that the test is unreliable. Hence, given that a test is positive, the probability the athlete has not taken drugs is $\left(\dfrac{38}{87}\right) = 0.4368$ (or a 43.68 % chance that the athlete has not taken drugs).

At the same time the probability that an athlete has taken drugs, given a test is positive is low:

$$\left(\frac{49}{87}\right) = 0.5632$$

Progress Exercise 4.5: Probability, including Bayes' Rule

In the probability calculations give answers rounded correct to four decimal places

1. (a) A company receives components from three different suppliers. The percentage of the company's total supply received from each supplier with their percentage defectives is given in Table PE4.5 Q1:

Table PE4.5 Q1   Quality of components from three suppliers

	Supplier A	Supplier B	Supplier C
Percentage of Total supplied	60 %	30 %	10 %
Percentage defectives	5 %	8 %	3 %

   (b) Use the data in Table PE4.5 Q1 to deduce Bayes' Rule
   Calculate the probability that
    (i) a randomly selected component is defective.
    (ii) given that a defective component is selected that it was supplied by A, by B, by C.
2. (a) State Bayes' Theorem.
   (b) A test is designed to detect a fault which occurs in 2.5 % of PC boards. The test is 94 % accurate, both for boards that have the fault and boards that do not have the fault. Calculate the probability that
     (i) a randomly selected board tests positive;
     (ii) the fault is present given that the test is positive;
     (iii) the fault is not present given that the test is positive.
   Would you consider this test satisfactory? Give reasons for your answer.
3. State Bayes' Rule.
   A procedure is developed to test for certain bacteria. If the bacteria are present the test will give a positive result in 98 % of cases, but in 4 % of cases the test is positive when the bacteria are not present. If the bacteria are present in 1.2 % of cases calculate the following probabilities:
   (a) the test is positive;
   (b) the test is negative;
   (c) the bacteria are present given the test is positive;
   (d) the bacteria are not present given the test is negative.

4. A program designed to scan emails for viruses is known to detect 97 % of infected emails. But the program also gives false alarms in 4 % of emails that do not contain the virus. If three emails in every 10 000 received by a certain company are infected, calculate the probability that the
   (a) scan detects a virus (scan is positive);
   (b) email contains a virus, given the scan is positive;
   (c) email does not contain a virus, given the scan is negative.
5. A retailer buys boxes of pre-packed grapes from three suppliers: 10 % of the total supply from A, 25 % from B and 65 % from C. Quality control have determined that 90 % of packs from supplier A are perfect, while 95 % and 82 % from B and C respectively are perfect.
   (a) Calculate the probability that a randomly selected pack of grapes is perfect.
   (b) A customer returns a mouldy pack of grapes. Calculate the probability that it was supplied by (i) A (ii) B (iii) C.
6. It is generally believed that there is a 95 % chance that a jury selected to try a criminal case will arrive at the correct verdict, that is, there is a 95 % chance that a guilty defendant is found guilty and a 95 % chance that an innocent defendant is found innocent. If 96 % of people brought to trial are guilty calculate the probability that
   (a) a defendant is found (i) guilty (ii) innocent.
   (b) Given that a defendant is found innocent by the jury, what is the probability that s/he is (i) guilty; (ii) innocent.
   (c) Given that a defendant is found guilty by the jury, what is the probability that s/he is (i) guilty; (ii) innocent.

Summary and overview of probability

At the outset, probability is not easy. It is difficult to visualise, terminology is strange and we give examples of tossing coins, dice and selecting cards. But gamblers were the first to realise the importance of studying the rules of chance. Industrialists realised that a large random variation in their products resulted on poor quality and lower profits. So, as a result of understanding probability gamblers, industrialists, investors etc. can make informed decisions and minimise risk. An important first step in the study of probability is to state the problem to be solved in words, then use the words 'AND' and OR as appropriate.

4.1   Introduction to probability

The discussion and explanations in probability require a large amount of terminology. It is important to be absolutely clear on the following:

An event: an experiment: outcomes: sample points. 'sample space'.

Since the use of **Sets** and **Venn diagrams**: help visualise and explain some basic concepts and ideas in probability a brief introduction to the essentials is given in the text.

The probability of an event is a measure of the likelihood that the event will happen. Probability may be assigned to an event by the following methods:

1. A priori (classical)

$$P(A) = \frac{\text{number of outcomes which result in event } A}{\text{total number of all possible outcomes}} = \frac{\#A}{\#S} \qquad (4.1)$$

2. Relative frequency

$$P(A) = \frac{\text{Number of times A happend}}{\text{Total number of times A was observed}} = \frac{f_A}{f_T} \qquad (4.2)$$

3. Subjective probability is assigned on the judgement of an individual (or group) based on prior knowledge and other information.

Having looked at the methods of assigning probability the following **Axioms** or self evident truths or properties about probability may be stated:

Axiom 1: $0 \leq P(A) \leq 1$ **(4.3)** states that the value of probability must be between 0 and 1 inclusive.

Axiom 2: $P(A) + P(\overline{A}) = 1$ **(4.4)** states that probability of A happening plus probability of A happening is unity.
($\overline{A}$ is the complement of A and called 'not A')

4.2 The addition and multiplication rules (AND and OR rules)

Summary of the AND rules

The event 'A AND B' means that A and B happen together at the same time. There are three AND rules depending on whether the events are independent, mutually exclusive or dependent.

1. $P(AB) = P(A) \, P(B)$, when A and B are independent   (4.5)
2. $P(AB) = 0$, when A and B are mutually exclusive   (4.6)
3. $P(AB) = P(A)P(B/A)$ or $P(AB) = P(B) \, P(A/B)$ when A and B are dependent   (4.7)

Summary for the OR rules (also called the addition rules)

'A OR B' means A happens OR B happens OR both (A AND B) happen.
The rules are as follows:

1. $P(A \, \text{OR} \, B) = P(A) + P(B)$ when A and B are mutually exclusive   (4.8)
2. $P(A \, \text{OR} \, B) = P(A) + P(B) - P(AB)$ when A and B are **not** mutually exclusive   (4.9)

Use of tree diagrams helps visualise the number of ways a sequence of events can happen and hence count the number of final outcomes, necessary for probability calculations.
The counting rules are used to count 'the number of ways' events can happen: this is essential in the calculation of probabilities using definition (4.1).

Counting rule 1: If an experiment is executed in sequence of k steps with n_1 possible outcomes at step 1; n_2 possible outcomes at step 2.....n_k possible outcomes at step k then the number of final outcomes is $(n_1) \, (n_1). \ldots . . (n_k)$.   (4.10)

Counting rule 2. There are $n!$ arrangements of n different objects.
n factorial is evaluated as $n! = (n)(n-1)(n-2) \ldots . (2)(1)$.   (4.11a)

Of these n objects, if n_1 are identical, n_2 are identical. ... n_k are identical, etc. then the number of arrangements is $\dfrac{n!}{n_1! \, n_2! \ldots \ldots n_k!}$

(4.11b)

Counting rule 3. There are $^n P_r$ permutations (arrangements) of n different objects, taken r at a time.

$$^n P_r = \frac{n!}{(n-r)!}$$

(4.12)

Counting rule 4. There are $^n C_r$ combinations of n different objects, taken r at a time where $^n C_r = \dfrac{n!}{(n-r)! r!}$

(4.13)

4.3 Joint, conditional and marginal probabilities

Conditional probability is the probability of an event calculated on the condition that the specified previous events have already happened.

$$P(B/A) = \frac{P(AB)}{P(A)}$$

(4.14)

Joint probability is the probability of an event satisfying both criteria simultaneously.

A probability distribution was defined as the list of every possible outcome with the corresponding probability.

Bayes' Rule: In many applications, a sample space may be partitioned into mutually exclusive and exhaustive events A_i hence giving all the probabilities $P(A_i)$. If further information in the form of conditional probabilities $P(B/A_i)$ is available, then the revised probabilities for A_i conditional on $B : P(A_i/B)$ may be calculated by Bayes' Rule.

$$P(A_i/F) = \frac{P(A_i)P(F/A_i)}{\displaystyle\sum_{j=1}^{n} P(A_j)P(F/A_j)}$$

(4.15)

Applications of Bayes' Rule feature in the testing for conditions that are rare (diseases; presence of performance enhancing drugs; errors in accounts; bugs in software, etc.) by tests that are less than 100 % reliable.

INTRODUCTION TO PROBABILITY DISTRIBUTIONS

5

Chapter Objectives

Having carefully studied this chapter and completed the exercises you should be able to do the following

- Define a random variable: state whether the random variable is discrete or continuous
- Define and describe the properties of a probability distribution
- Calculate probabilities for empirical and discrete uniform distributions
- State the assumptions and calculate probabilities for the Binomial distribution
- State the assumptions and calculate probabilities for the Poisson distribution
- State the properties of the Normal distribution
- Use Binomial, Poisson and Normal probability tables
- Calculate probabilities for Normally distributed random variables
- For Normally distributed data calculate the limit(s) within which a given percentage of values fall
- Use Excel to calculate probabilities for the Binomial, Poisson and Normal
- Derive and explain the formulae for mean and variance of a random variable in terms of expected values
- Define, explain and calculate the expected value and variance for empirical distributions
- State the mean and variance of the Binomial, Poisson and Normal random variables. Hence approximate the Binomial with the Poisson and Binomial and Poisson with Normal probabilities
- Calculate probabilities for the sums and differences of Normal, independent random variables

5.1 Introduction to probability distributions and random variables

The concept of a probability distribution was introduced briefly in Chapter 4 where it was described as a list of every possible outcome with the corresponding probability. The corresponding probabilities were calculated as simple relative frequencies. In Chapter 4 you may also have noted that probability calculations for even simple real life situations become complex very quickly! But, *under certain well defined conditions* it is possible to derive formulae (such as the Binomial formula introduced in 5.2) to calculate the probabilities. These formulae define 'some of which are introduced in this chapter'. Some of the terminology associated with theoretical probability distributions is explained below.

5.1.1 Random variable

In statistics the outcome of a random experiment is variable and determined by chance. The numeric value of the outcome is called a random variable. For example, when a die is thrown the face value that turns up may be any one of the numbers: 1, 2, 3, 4, 5 or 6. The face value that turns is a **random variable**. In other situations the outcome of a random experiment is not numeric, so a formula or rule is used to assign a numeric value to the each outcome. For example if three coins are tossed the outcomes are three Heads, two Heads and one Tail, one Head and two Tails or three Tails.

In this experiment, a possible rule for assigning a numeric value to each outcome is to count the number of Heads. Then the random variable may have values 3, 2, 1 or 0:

3 (when the outcome is three Heads)

2 (when the outcome is two Heads and one Tail)

1 (when the outcome is one Head and two Tails)

0 (when the outcome is three Tails).

> Hence the more general definition of a **random variable** is the rule that assigns a numeric value to each outcome of a random experiment.

The numeric values of a random variable may be discrete or continuous.

> **A discrete random variable** can assume a finite number of distinct values.

Examples of discrete random variables are given above in (a) tossing a die in which there were six distinct outcomes (b) throwing three coins in which there were four distinct outcomes.

> **A continuous random** variable can assume any value within a continuous interval.

As an example of a continuous random variable consider the time taken to process an online customer order. If it is known that all orders take between one to 20 minutes to process then the random variable may assume any value within this range. The number of outcomes depends on how precisely time is measured: if time is measured correct to two decimal places (hence ranging from 1.00 to 20.00 minutes), there are 2000 outcomes: if time is measured correct to four decimal places there are 200 000 outcomes, etc.

Notation: A random variable is referred to by an uppercase letter, X, Y, Z, etc.

The numeric values assumed by the random variable are referred to by the corresponding lowercase letter, x, y, z, etc.

Some examples of random variables:

(a) The number that turns up when a die is tossed.

X represents the number that turns when a die is tossed; the possible values of X are $x = 1, 2, 3, 4, 5, 6$.

(b) X represents the number of Heads that turn up when three coins are tossed; the possible values of X are $x = 0, 1, 2$ and 3.

(c) Y represents the number of defectives in a random sample of five: the possible values of Y are $y = 0, 1, 2, 3, 4$, and 5.

(d) W represents the number of arrivals in a queue for an ATM machine between 6 pm and 7 pm: $w = 0, 1, 2, 3.........$arrivals.

(e) T represents the time in minutes (correct to two decimal places) to process an online order: t is any time from $t = 0.00$ minutes to the maximum given time.

The probability that the random variable X will assume the value, x is written as $P(X = x)$ or simply $P(x)$ on a **trial** of a random experiment.

A **trial** of a random experiment is a single execution of the experiment, such as one toss of a die.

5.1.2   Probability distributions

Definitions

An empirical probability distribution is a list of every outcome of a random experiment with the corresponding probability.

A discrete probability distribution is the probability distribution of the discrete random variable. The outcomes can only assume separate or discrete values. The Uniform, Binomial and Poisson distributions are discrete distributions.

A continuous probability distribution is the probability distribution of a continuous random variable: the outcomes can be any value from of continuous interval. The Normal distribution is a continuous distribution.

The discrete uniform probability distribution models situations where each discrete outcome is equally likely to happen. For example, tossing a die: the outcomes of the experiment are $x = 1, 2, 3, 4, 5, 6$. All outcomes are equally likely; the probability of each outcome is $\frac{1}{6}$. The probability distribution is written out in Table 5.1:

Table 5.1   Discrete uniform distribution for the experiment of tossing a die

Outcome	$x =$	1	2	3	4	5	6
Probability	$P(x) =$	$\frac{1}{6}$	$\frac{1}{6}$	$\frac{1}{6}$	$\frac{1}{6}$	$\frac{1}{6}$	$\frac{1}{6}$

Note: $\sum P(x) = 1$: the sum of the probabilities must be unity. This is an essential property of all probability distributions.

Skill Development Exercise SK5_1: Discrete probability distributions

Write out the joint, marginal and conditional probability distributions for the data given in Table SK5_1.1. Confirm that $\sum P(x) = 1$ in each case.

Table SK5_1.1   Teachers, principals, gender

	Teachers	School Principals
Male	420	188
Female	1580	72

State the random variables: Suggest a rule for assigning values to the random variables.

Answer: The joint probability distribution is given in the body of Table SK5_1.2. $\sum P(x) = 0.19 + 0.08 + 0.70 + 0.03 = 1$. The marginal distributions are in the margins: in each case $\sum P(x) = 1$.

Table SK5_1.2   Joint and marginal probability distributions for teachers, principals, gender

	Teachers	School principals	Marginal distribution for gender
Male	0.19	0.08	0.27
Female	0.70	0.03	0.73
Marginal distribution for roles	0.88	0.12	1.00

The conditional distributions when a person is known to be (i) a teacher; (ii) a principal are given in Table SK5_1.3:

Table SK5_1.3   Conditional distributions for teachers, principals, gender

	(i) Given a person is a teacher	(ii) Given a person is a principal
Male	0.21	0.72
Female	0.79	0.28
Total	1	1

The conditional distributions when a person is known to be (iii) male (iv) female are given in Table SK5_1.4:

Table SK5_1.4   More conditional distributions for teachers, principals, gender

	Teacher	School principal	Total
(iii) Given a person is a Male	0.69	0.31	1
(iv) Given a person is a Female	0.96	0.04	1

5.1.3 Graphs of probability distributions

The outcomes (x) of the experiment are marked on the horizontal axis and probabilities (y) are plotted on the vertical axis. In Figure 5.1(a) vertical lines are drawn from the outcomes to the corresponding probability points (y) for clarity. A probability histogram depicts probability as area for discrete distributions. The outcomes are on the horizontal axis; each outcome is represented by an interval, nominally of unit width. Rectangles are drawn over each interval; the height of the rectangle is proportional to the probability of the outcome. See Figure 5.1(b). The sum of the areas of the 6 rectangles:

$$\text{Area} = 6 \left(\frac{1}{6} \times 1 \right) = 1.$$

Note: The area of a probability histogram must be unity.

In probability graphs area represents probability. For example, in Figure 5.1(b) the probability of a '2 OR 3' turning up when a die is thrown is represented by the area of the rectangles over outcome 2 and outcome 3.

$$P(2 \text{ OR } 3 \text{ on one throw}) = 2 \times \frac{1}{6} = \frac{2}{6} = \frac{1}{3}.$$

Uniform probability distribution

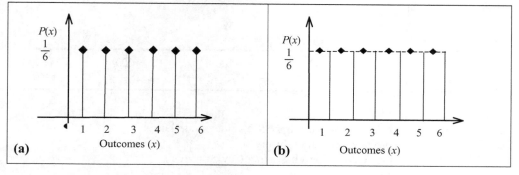

Figure 5.1   (a) Discrete uniform distribution for the experiment of tossing a die (b) Area = 1.

Summary of the properties of a probability distribution

1. The sum of all the probabilities is unity.
2. The total area under the probability histogram or probability curve is unity.

Note: Probability histograms are useful as a graphical tool to aid visualisation of certain ideas such as reading probabilities from tables and approximating discrete probabilities with continuous. But, discrete probabilities are readily calculated by formulae, such as the Binomial and Poisson in section 5.2. However, for continuous random variables (section 5.3) it will be necessary to calculate probabilities in terms of area under the probability curve.

Progress Exercises 5.1: Random variables and empirical distributions

1. What is a random variable?
   In each of the following experiments name the random variable and state the values it may assume and state for the number of outcomes in the sample space.
   (a) a die is thrown.
   (b) a card is selected from a box containing ten cards numbered 1 to 10.
   (c) two dice are thrown.
2. Define a random variable and give all the outcomes of the following experiments:
   (a) two coins are thrown.
   (b) a die and a coin are thrown.
   (c) a ball is selected from a box containing four white and six yellow balls.
3. Explain the difference between a discrete and a continuous random variable.
   State whether the random variables for each of the following is discrete or continuous.
   (a) the time taken to assemble a component.
   (b) the number of components assembled per hour.
   (c) the weight of a component.
   (d) the number of defective components in a sample of 20.
4. (a) Define and give an example of a discrete probability distribution.
   (b) Give the probability distributions for experiments (a), (b) and (c) in question two.
5. A retailer's sales records summarise the value of purchases made by customers on Mondays over a period of one year:

Table PE5.1 Q5   Value of purchases vs. number of customers

Value (£)	10 < 50	50 < 100	100 < 200	Over 200
Number of customers	4560	3982	2328	1140

(a) Write out the probability distribution for the value of purchases.
(b) Confirm that the probabilities sum to unity.
(c) Calculate the average value of purchases made by customers.
   State any assumptions made.

6. As part of statistical process control, the number of defective moulded plastic parts in samples of 100 parts is recorded. Over a period of time the following data was collected:

Table PE5.1 Q6   Number of defective parts in samples of size 100

Number of defectives	0	1	2	3	4	5	6	more than 6
Number of samples	80	125	64	21	15	3	5	4

(a) Write out the probability distribution for the number of defective parts per sample.
(b) Plot the probability histogram.
(c) Calculate the average number of defective parts per sample of 100.
(d) Calculate the proportion of defective parts for the entire data given in the Table PE5.1 Q6. State any assumptions made.

7. Examination results and part-time hours worked by full-time students are given in the table below:

Table PE5.1 Q7   Exam grades vs. hours worked (part time)

Exam result	Part-time hours work per week				Row totals
	0	1–5	6–10	>10	
Distinction	30	15	10	5	60
Credit	25	25	10	20	80
Pass	35	25	20	30	110
Fail	10	5	10	25	50
Column totals	100	70	50	80	300

(a) Calculate the joint probability distribution for results and hours worked.
(b) Calculate the marginal probability distribution for exam results.
(c) Calculate the marginal probability distribution for hours worked.
(d) Calculate the conditional probability distribution for results given that hours worked is greater than ten per week.

8. The probability distribution for the number of defects in new cars is as follows:

Table PE5.1 Q8   The probability distribution for the number of defects per new cars

X = Number of defects per car	0	1	2	3	4	5
Probability of X defects per car	0.10	0.24	0.30	0.15	0.10	p

(a) Calculate p.
(b) Calculate the proportion of cars that have less than two defects.
(c) Calculate the percentage of cars that have at least four defects.
(d) Calculate the average number of defects per car.

9. The problem of obesity among children is been a cause of concern. The probability distribution of weights of children aged between ten and 12 years is given in Table PE5.1 Q9.

Table PE5.1 Q9   The probability distribution for weights of children aged ten to 12 years

Y: weight in kilos	10 < 20	20 < 30	30 < 40	40 < 50	50 < 60	Over 60
Probability $P(y)$	0.08	0.22	0.30	x	0.09	0.12

(a) Calculate the value of x.
(b) What percentage of children weigh less than 40 kilos.
(c) Calculate the average weight of children in this group.
(d) In a group of 560 children, how many weigh more than 40 kilos?

5.2   The Binomial probability distribution

The Binomial formula has many applications such as opinion polls, statistical quality assurance, process control, etc.

The Binomial formula is the only probability formula that will be 'derived' from basics in this text. It is one of the easiest formulae to derive: it serves as an example to emphasise that probability distribution formulae are derived from clearly defined conditions (assumptions). Hence, the use of such formulae is valid only when the underlying assumptions are satisfied.

Unfortunately, deriving the Binomial does require some maths: a quick review of the maths for the Binomial is given in Appendix D.

5.2.1   Derivation of the binomial formula for sample sizes, $n = 2$ and $n = 3$

To explain how the Binomial formula is derived the conditions (assumptions) under which the probabilities are calculated. Under these conditions the probabilities for every outcome is calculate for sample size 2 (Worked Example 5.1) and size 3 (Worked Example 5.2). The results from both examples are generalised to explain the Binomial formula.

Assumptions for the Binomial

The Binomial formula may be used to calculate probabilities when the following conditions are true.

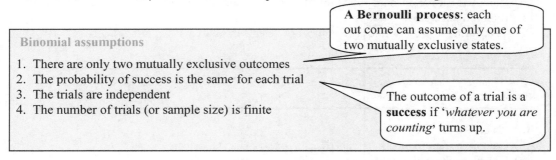

Binomial assumptions

1. There are only two mutually exclusive outcomes
2. The probability of success is the same for each trial
3. The trials are independent
4. The number of trials (or sample size) is finite

A Bernoulli process: each out come can assume only one of two mutually exclusive states.

The outcome of a trial is a **success** if '*whatever you are counting*' turns up.

These assumptions are satisfied in the experiment given in Worked Example 5.1.

Worked Example 5.1: Derivation of Binomial probabilities for $n = 2$

Two cards are selected from a box containing red and white cards; replacing the first card before selecting the next (this is called sampling with replacement).

(a) Show that this experiment satisfies the Binomial assumptions 1, 2, 3, 4 above.

(b) To generalise, let p = proportion of red cards and q = proportion of cards that are not red. Draw a tree diagram to show all possible outcomes for an experiment that selects a sample of two cards, with replacement. Then write out the probabilities that a sample of two cards contains (i) no red cards (ii) one red card (iii) two red cards in terms of p and q.

(c) Summarise the results in (b) by writing out the list of every possible outcome with the corresponding probability. Expand $(q + p)^2$ and show the successive terms are the same as the probabilities in (b).

(d) Calculate the probabilities in (b) when the sample is selected from a box containing six red and four white cards.

Solution

(a) Check assumptions

1. There are only two possible outcomes: a card is either red or white. The outcomes are mutually exclusive since a card cannot be both red and white.

2&3. Since sampling is with replacement, the probability of getting a red card is the same on each draw (or trial). The trials are independent since the numbers of red and white cards are the same for each trial: the outcome of one draw has no effect on the outcome of any other draw.

4. The sample size is 2, which is finite.

(b) In this example we are concerned with the 'number of red cards in a sample of size 2'.

(i) we are **counting red cards**.

(ii) Hence the random variable **X represents the number of red cards** in the sample, hence $x = 0, 1, 2$. All possible outcomes of the experiment are shown in the tree diagram in Figure 5.2.

(iii) **p = probability of selecting a red card**: q = probability of not selecting a red card (By Axiom 2: $p + q = 1$)

> p may also be described as the 'proportion' of red cards:
> $$p = \frac{\text{number of red cards}}{\text{total number of cards}} = \frac{6}{10}$$

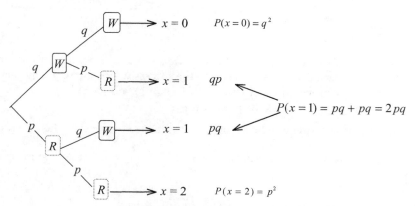

Figure 5.2   Tree diagram for the experiment of selecting two cards, with replacement.

The probabilities are calculated as follows:

(i) P (zero red cards) $= P$ (two white cards):
There is only way of selecting 'no red cards'
'the first is not red AND second card is not red':
Hence $P(x = 0) = q \times q = q^2$

> 'AND' means multiply

(ii) P (one red card) $\rightarrow P$ (one red card AND one white):
There are two ways of selecting one red card and one white card:
'a red first AND a white second' OR 'a white first AND a red second'

Hence $P(x = 1) = pq + qp = 2pq$

> 'OR' means add

(iii) P (two red cards): there is one outcome with two red cards:
'the first card is red AND second card is red'. Hence $P(x = 2) = p \times p = p^2$. The list of every possible outcome with the corresponding probability is summarised in Table 5.2:

Table 5.2   Binomial distribution where $X =$ number of reds in a sample size two

> Write out the list of every possible outcome

x	0	1	2
$P(x)$	$P(x = 0) = q^2$	$P(x = 1) = 2qp$	$P(x = 2) = p^2$
$(q + p)^2$	q^2	$2qp$	p^2
$(0.6 + 0.4)^2$	0.36	0.48	0.16

> (d)... substituting the values $p = 0.4$, $q = 0.6$

(c) Expanding $(q + p)^2$

$$(q + p)^2 = (q + p)(q + p)$$
$$= q(q + p) + p(q + p)$$
$$= q^2 + qp + pq + p^2$$
$$= q^2 + 2qp + p^2$$

> Multiplying out the brackets is called 'expanding'

The probabilities in row two of Table 5.2 (q^2, $2qp$, p^2) are identical to those obtained by multiplying out (expanding) $(q + p)^2$. Hence the successive terms of the Binomial $(q + p)^2$ give formulae for the probabilities of $P(x = 0)$, $P(x = 1)$ and $P(x = 2)$.
    (Note the value of x is same as the power on p).

(d) The probability of selecting a red card on any trial is $p = \dfrac{6}{10} = 0.6$ (the proportion of red cards). The probability of not selecting a red card is $q = \dfrac{4}{10} = 0.4$ (the proportion of non-red (white) cards).

See Appendix D for a review on the essential maths for the Binomial.

Worked Example 5.2: Derivation of Binomial probabilities for $n = 3$

Repeat the experiment in Worked Example 5.1 parts (b) to (d) for a sample size 3.

Solution

(b) The solution is outlined briefly as follows:
The tree diagram is given in Figure 5.3

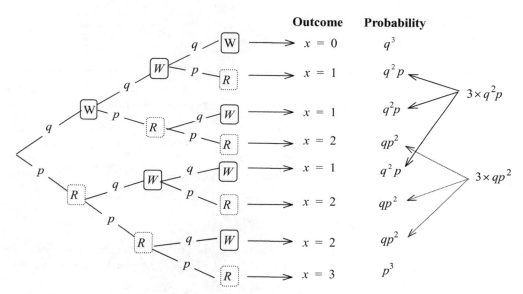

Figure 5.3   Tree diagram for the experiment of selecting three cards with replacement.

In Figure 5.3, it is evident that the probability of getting one red card in three trials is $q^2 p$: this probability is same irrespective of the order in which the red card is selected. Hence, the probability of one red card in a sample of size 3.

$P(x = 1) =$ (number of ways of selecting one red card) $\times$ (probability of one red card in one trial).

Hence, the probabilities for $x = 0, 1, 2, 3$ are given as follows (see Figure 5.3).

$P(x = 0) = 1 \times q^3$ (**1** outcome with 0 red cards: {WWW})

$P(x = 1) = 3 \times q^2 p$ (**3** outcomes with 1 red card: {RWW; WRW; WWR})

$P(x = 2) = 3 \times qp^2$ (**3** outcomes with 2 red cards: {RRW; RWR; WRR})

$P(x = 3) = 1 \times p^3$ (**1** outcome with 3 red cards: {RRR})

The results are summarised in Table 5.3:

Table 5.3   Binomial where $X =$ number of red cards in a sample of three

> Write out the list of every possible outcome

x	0	1	2	3
$P(x)$	$P(0) = q^3$	$P(1) = 3q^2p$	$P(2) = 3qp^2$	$P(3) = p^3$
$(q+p)^3$	q^3	$3\,q^2p$	$3\,qp^2$	p^3
$(0.6+0.4)^2$	0.216	0.423	0.288	0.064

> Substituting $p = 0.4$, $q = 0.6$

> ... the power on p is the same as the value of x

(c) Row three of Table 5.3 is given by the successive terms of the Binomial $(q + p)^3$

$$(q + p)^3 = (q + p)(q + p)^2$$

> 'expanding' $(q+p)^2$

$$= (q + p)(q^2 + 2qp + p^2)$$
$$= q(q^2 + 2qp + p^2) + p(q^2 + 2qp + p^2)$$
$$= q^3 + 2q^2p + qp^2 + pq^2 + 2qp^2 + p^3$$
$$= q^3 + 3q^2p + 3qp^2 + p^3$$

The terms in row three of Table 5.3 are identical to the probabilities in row two. Hence the probabilities of $P(x = 0)$, $P(x = 1)$, $P(x = 2)$ and $P(x = 3)$ are given by successive terms of Binomial $(q + p)^3$.

(d) See Row four of Table 5.3.

Terminology: In the above discussion the random variable was the number of red cards in three trials. Since red cards are being counted, if the outcome of a trial is a red card it is called a '**success**'. So, instead of saying 'the probability selecting one red card in a sample of three cards. . . .' we say 'the probability of one success in three trials . . .'

5.2.2   The Binomial formula for the probability of x successes in a sample of size n

The results from Worked Examples 5.1 and 5.2 for sample sizes $n = 2$ and 3 respectively are two examples the Binomial distribution for a sample of size $n = 2$ and 3.

It can be shown that for a sample size n the probabilities; $P(0)$, $P(1)$, $P(2) \ldots . P(x) \ldots P(n)$ are given by the successive terms of the expansion of $(q + p)^n$.

The only problem with this method is that expanding $(q + p)^n$ when the power, n is greater than three is time consuming. However, as explained in Appendix D, the expansion may be written out in general terms by the formulae given Table 5.4:

Table 5.4   Probability of x number of successes in sample of n

x	0	1	2	.....	x	.....
$P(x)$	$P(0)$	$P(1)$	$P(2)$		$P(x)$	
$(q + p)^n$	$^nC_0 p^0 q^n$	$^nC_1 p^1 q^{n-1}$	$^nC_2 p^2 q^{n-2}$	......	$^nC_x p^r q^{n-x}$	......

Where $n!$, called 'n factorial' is evaluated as $n! = (n)(n-1)(n-2)....3.2.1$
and $^nC_x = \dfrac{n!}{(n-x)!\ x!}$. (See Chapter 4, section 4.2.3).

Looking carefully at each column of Table 5.4 you should notice a pattern developing in successive terms, for example in a sample size n,

1 success $\rightarrow P(x = 1) = {}^nC_{[1]} p^{[1]} q^{n-[1]}$
2 successes $\rightarrow P(x = 2) = {}^nC_{[2]} p^{[2]} q^{n-[2]}$

....

x successes $\rightarrow P(x) = {}^nC_{[x]} p^{[x]} q^{n-[x]}$

... the power on p is the same as x, the number of successes

This last expression is called the Binomial probability distribution function which we will refer to as the 'Binomial formula'.

Binomial probability distribution function (Binomial formula)

$$P(x) = {}^nC_x p^x q^{n-x}$$

(5.1)

$P(x)$ is the probability of x successes in a sample size n

Explanation for $^nC_x = \dfrac{n!}{(n-x)!\ x!}$

If you recall, in Chapter 4, section 4.2.3, it was stated that when n objects consists of x identical objects of one type (the x successes) and the remaining $(n-x)$ of another type (non-successes or fails)

$\underbrace{S, S, S ...... S}_{x\ \text{Successes}}\ \ \underbrace{F, F, F ...... F}_{(n-x)\ \text{Fails}}$

The number of different arrangements by equation (4.11b) is $\dfrac{n!}{x!(n-x)!}$

Since there are only two groups of identical objects, this formula is identical to

$^nC_x = \dfrac{n!}{(n-x)!\ x!}$

^nC_x is sometimes written as $\binom{n}{x}$

For example, in Worked Example 5.3 there were three outcomes (or arrangements) that contained two red and one white card: {RRW; RWR; WRR}. This could be calculated by the formula,

$$^3C_2 = \frac{3!}{(3-2)!\ 2!} = \frac{3!}{(1)!\ 2!} = \frac{3 \times 2 \times 1}{(1)\ (2 \times 1)} = 3.$$

Remember, ^nC_x and $n!$ may be evaluated on most calculators.

Relaxing assumption two: trials are independent. In the examples above, the probability of a success was the same for each card selected because sampling from the small population of ten cards was *with replacement*.

In many applications you will see statements such as '4 % of components are defective' with no mention of population size or whether sampling is with or without replacement. As a general rule, if the sample size (n) is small compared to the population size (N) the probability of success in successive trials are sufficiently close to say 'trials are independent'. A guideline is that $\dfrac{n}{N} \times 100 \leq 5\,\%$.

To calculate Binomial probabilities you must be given (i) sample size, n (ii) the proportion of successes p must be given or else the data from which to calculate p must be given.

Since n and p are essential for Binomial probability calculations they are called the **parameters of the Binomial**: $P(x) = {}^nC_x\, p^x q^{n-x} = {}^nC_x\, p^x (1-p)^{n-x}$
(it follows that $q = 1 - p$).

The notation $B(n, p)$ means a **B**inomial probability distribution, sample size, n: proportion of successes p.

> **A parameter** is a key characteristic of a probability distribution. Its value must be known to calculate probabilities

Skill Development Exercise SK5_2: Binomial probabilities for sample size two

A building society audit revealed that 10 % of borrowers are classified as 'new' borrowers. In a publicity drive to attract more new borrowers, the society offers 'We'll pay your mortgage for one year' for two randomly select borrowers. Calculate the probability the the two borrowers are: (a) not 'new'; (b) only one is 'new'; (c) both 'new'.

Answer

Method
Step 1: Check that the assumptions for the Binomial are satisfied.
Step 2: State the random variable, X, hence p and n.
Step 3: Calculations: $n = 2$; $p = 0.1$, hence $q = 1 - p = 0.9$.

The probability distribution is

x (number of new borrowers in a sample of 2)	0	1	2
$P(x)$ (probability of x new borrowers in a sample of 2)	0.81	0.18	0.01

Alternatively: Since $n = 2$, write out the probability distribution for samples, size two by expanding $(q + p)^2$. Substitute $p = 0.1$: $q = 0.9$. Then read off the answers (a), (b) and (c).

x	0	1	2
$(q + p)^2$	q^2	$2qp$	p^2
$P(x)$	$P(0) = q^2 = 0.81$	$P(1) = 2qp = 0.18$	$P(2) = p^2 = 0.01$

Skill Development Exercise SK5_3: Binomial: multiple choice questions

An exam consists of four multiple choice questions.

If each question has only one correct answer calculate the probability that a student randomly selects the correct answer to (a) all four questions; (b) any three questions; (c) any two questions; (d) any one question; (e) none of the questions when there is a choice of (i) two; (ii) three; (iii) four possible answers.

Answers

four questions → sample size (or number of trials) $n = 4$. p is the proportion of correct answers on offer: these are given in (i), (ii) and (iii) below:

(i) When there is a choice of two answers per question: $p = \dfrac{1}{2}$, hence $q = \dfrac{1}{2}$.

(ii) When there is a choice of three answers to a question: $p = \dfrac{1}{3}$, hence $q = \dfrac{2}{3}$.

(iii) When there is a choice of four answers to a question: $p = \dfrac{1}{4} = 0.25$, hence $q = \dfrac{3}{4}$.

Substitute these values of p and q into (5.1) to calculate $P(0) \ldots P(4)$ for (i), (ii) and (iii)

Table SK5_3.1   Probability of x correct to four multiple choice questions for (i) two; (ii) three; (iii) four choices

x	0	1	2	3	4
$P(x)$ $(q + p)^4$	$P(0)$ ${}^4C_0 p^0 q^4$ $= q^4$	$P(1)$ ${}^4C_1 p^1 q^{4-1}$ $= 4pq^3$	$P(2)$ ${}^4C_2 p^2 q^{4-2}$ $= 6p^2 q^2$	$P(3)$ ${}^4C_3 p^3 q^{4-3}$ $= 4p^3 q^1$	$P(4)$ ${}^4C_4 p^4 q^{4-4}$ $= p^4$
$p = \dfrac{1}{2}, q = \dfrac{1}{2}$	0.0625	0.2500	0.3750	0.2500	0.0625
$p = \dfrac{1}{3}, q = \dfrac{2}{3}$	0.1975	0.3951	0.2963	0.0988	0.0123
$p = \dfrac{1}{4}, q = \dfrac{3}{4}$	0.3164	0.4219	0.2109	0.0469	0.0039
Comment	Probability that all answers are wrong	Probability that one answer is correct	Probability that two answers are correct	Probability that three answers are correct	Probability that all four answers are correct

Comment: probability of guessing all four correct decreases as the number of choices increases.

Note: there are advantages to using fractions when it comes to describing certain the results. For example, when there are four answers to choose from, the probability of getting the correct answer to all four questions is P (four correct) $= P$ (Q1 AND Q2 AND Q3 AND Q4 correct).

$$= \frac{1}{4} \times \frac{1}{4} \times \frac{1}{4} \times \frac{1}{4} = \left(\frac{1}{4}\right)^4 = \frac{1}{256} = 0.0039$$

Stating that there is a 'one chance in 256' of getting all four correct is easier to visualise than stating 'there is a probability of 0.0039' of getting all four correct.

Worked Example 5.3: Binomial probabilities on tax returns: $n = 5$ and $n = 20$

It is known that that one out of every five tax returns will contain errors and are classed as faulty.

(a) Calculate the probability that in the sample of five (i) none are faulty; (ii) at least one is faulty.
(b) Calculate the probability distribution for the sample of five tax returns. Hence write down the probability that at most two are faulty.
(c) An inspector randomly selects a sample of 20 tax returns. Calculate the probabilities that in the sample of 20 (i) seven are faulty; (ii) at most two are faulty.

Solution

Step 1: Start by checking the assumptions. Then state the random variable: X and hence assign values to p and q.
Check assumptions for the Binomial
1. Two mutually exclusive and exhaustive outcomes. A tax return is faulty or not faulty.
2. Trials are independent. Sampling is from a large batch, hence the probability of selecting a faulty return is approximately the same for each trial.
3. Fixed sample size. Sample size $n = 5$ in (a) and (b) and $n = 20$ in (c).

Step 2: State n and p:

Objective: count the **faulty** tax returns —— .. counting **faulty** returns

X = number of **faulty returns:** $p = P$ **(faulty)** $= \frac{1}{5} = 0.2.$ $q = 1 - p = 1 - 0.2 = 0.8$
Sample size $n = 5$ for part (a) $n = 20$ for part (b)

Step 3: Calculations

(a) (i) To calculate $P(0)$, by substituting $n = 5$, $x = 0$, $p = 0.2$, $q = 0.8$ into the Binomial formula: $P(x) = {}^n C_x p^x q^{n-x}$

Suggestion: Sometimes it is easier to substitute for one symbol at time and simplify. Hence substitute n first, followed by x and finally the values of p and q

$n = 5, p = 0.2$

$n = 5 \rightarrow : P(x) = {}^5 C_x p^x q^{5-x}$
$x = 0 \rightarrow P(0) = {}^5 C_0 p^0 q^{5-0} = 1 \times 1 \times q^5 = q^5$ .... since ${}^5 C_0$ and $p^0 = 1$
Finally, substitute the values of p and q (only q is required here)
$q = 0.8 \rightarrow P(0) = (0.8)^5 = 0.3277.$
There is a 33 % chance that a sample of five will contain no faulty tax returns.

(ii) 'At least one' means one or more.
   That is '1 OR 2 OR 3 OR 4 OR 5' faulty returns
   P (at least one faulty) $= P(1) + P(2) + P(3) + P(4) + P(5)$, by the addition rule.
   The direct method is to calculate the five probabilities.
   But there is a much easier way! The basic property of a probability distribution is that the sum of all the probabilities is one, hence

$$1 = P(0) + P(1) + P(2) + P(3) + P(4) + P(5)$$

$$1 - P(0) = P(1) + P(2) + P(3) + P(4) + P(5)$$

$$1 - 0.32768 = P(1) + P(2) + P(3) + P(4) + P(5)$$

$$0.67232 = P \text{ (at least one faulty)}$$

(b) The probability distribution requires the probability of each outcome. Calculate each probability following the steps outlined above. If you wish, use your calculator to evaluate ^nC_r as well as evaluating powers of p and q.
   The probability distribution is given in Table 5.5:

Table 5.5   The probability distribution of faulty tax returns

x	0	1	2	3	4	5
$P(x)$	$P(0)$	$P(1)$	$P(2)$	$P(3)$	$P(4)$	$P(5)$
$(q+p)^5$	$^5C_0 p^0 q^5$	$^5C_1 p^1 q^{5-1}$	$^5C_2 p^2 q^{5-2}$	$^5C_3 p^3 q^{5-3}$	$^5C_4 p^4 q^{5-4}$	$^5C_5 p^5 q^{5-5}$
	$= q^5$	$= 5pq^4$	$= 10p^2q^3$	$= 10p^3q^2$	$= 5p^4q$	$= p^5$
$(p = 0.2)$	0.3277	0.4096	0.2048	0.0512	0.0064	0.0003

'At most two faulty' means 'zero OR one OR two' faulty returns.
Hence P (at most two faulty) $= P(0) + P(1) + P(2) = 0.3277 + 0.4096 + 0.2048 = 0.9421$.

(c) (i) To calculate $P(2)$, first substitute $n = 20$, followed by $x = 2$ into (5.1)

$$n = 20 \rightarrow P(x) = {}^{20}C_x\, p^x q^{20-x}$$

$$x = 2 \rightarrow P(2) = {}^{20}C_2\, p^2 q^{20-2} = 190 \times p^2 \times q^{18} \ldots \text{since} {}^{20}C_2 = 190$$

$n = 20$
$p = 0.2$

Finally substitute the values of p and q

$$P(2) = 190(0.2)^2(0.8)^{18} = 0.1369.$$

There is a 13.69 % chance that a sample of 20 will contain two faulty tax returns
(ii) 'at most 2' means zero OR one OR two faulty tax returns
   P (at most two are faulty) $= P(0) + P(1) + P(2)$, by the addition rule.
   Calculate $P(0)$ and $P(1)$ as follows:

$$P(0) = {}^{20}C_0\, p^0 q^{20-0} = q^{20} = 0.0115$$

$$P(1) = {}^{20}C_1\, p^1 q^{20-1} = (20)pq^{19} = 0.0576$$

Hence
$$P \text{ (at most two are faulty)} = P(0) + P(1) + P(2) = 0.0115 + 0.0576 + 0.1369 = 0.2060$$

5.2.3 Discrete cumulative probabilities distributions and applications

In Chapter 1, a cumulative frequency was the sum of all frequencies up to and including the frequency for a given interval, say interval r: $\sum_{i=1}^{i=r} f_i$. Similarly, a cumulative probability is the sum of all the probabilities, up to and including the probability $P(x = r)$ of r successes. The cumulative probability is written as $P(x \le r) = \sum_{x=0}^{x=r} P(x)$.

A cumulative probability distribution is a list of every outcome x with the corresponding cumulative probability $P(x \le r): r = 0. 1. 2. 3. \ldots \ldots n$.

Cumulative probabilities are used in some statistical probability tables and in many software packages, such as Excel, Minitab and SPSS.

Cumulative Binomial probability tables

Table 5.6 is an extract from cumulative Binomial probability tables for samples size $n = 5$. These tables give the cumulative probabilities $P(x \le r) = \sum_{x=0}^{x=r} P(x)$ for various values of p: their use is demonstrated in Worked Example 5.4.

Table 5.6   Extract from cumulative Binomial probability tables for $n = 5$

| | | | | | $n = 5$ | | | | | |
r	$p = 0.01$	$p = 0.05$	$p = 0.10$	$p = 0.20$	$p = 0.50$	$p = 0.60$	$p = 0.80$	$p = 0.90$	$p = 0.95$	$p = 0.99$
0	0.9510	0.7738	0.5905	0.3277	0.0313	0.0102	0.0003	0	0	0
1	0.9990	0.9774	0.9185	0.7373	0.1875	0.0870	0.0067	0.0005	0	0
2	1	0.9988	0.9914	0.9421	0.5000	0.3174	0.0579	0.0086	0.0012	0
3	1	1	0.9995	0.9933	0.8125	0.6630	0.2627	0.0815	0.0226	0.0010
4	1	1	1	0.9997	0.9688	0.9222	0.6723	0.4095	0.2262	0.0490
5	1	1	1	1	1	1	1	1	1	1

Worked Example 5.4: Reading Binomial probabilities from tables

Refer to Worked Example 5.3(a) and (b) where $n = 5$, $p = 0.20$.

(a) Write down the cumulative probability distribution.
   Plot the cumulative probability distribution.
(b) Plot the probability histogram on the same diagram as the cumulative probability distribution in (a).
   Use the Binomial tables to write down the following probabilities:
   (i) At most two are faulty; (ii) None are faulty; (iii) Exactly three are faulty; (iv) At least one is faulty.

   Suggestion: it may help to use the probability histogram to 'see' how to write down the individual probabilities from the cumulative tables.

Solution

(a) The Binomial probabilities for x, the number of faulty tax returns ($n = 5$, $p = 0.20$), that were calculated in Worked Example 5.3(a) are given in column two of the table in Figure 5.4.

   The cumulative probabilities, $P(x \le r)$, are calculated by summing the individual probabilities for $x \le r$. See column 3, Figure 5.4.

r	$P(x = r)$	$P(x \le r)$
0	0.3277	0.3277
1	0.4096	0.7373
2	0.2048	0.9421
3	0.0512	0.9933
4	0.0064	0.9997
5	0.0003	1

Cumulative distribution distribution

Figure 5.4   Cumulative probability distribution.

The cumulative distribution is plotted with cumulative probabilities on the vertical axis against x on the horizontal axis.

(b) The probability histogram is plotted from the probability distribution given in Table 5.4. The cumulative probability distribution is superimposed on the graph with the histogram in Figure 5.5.

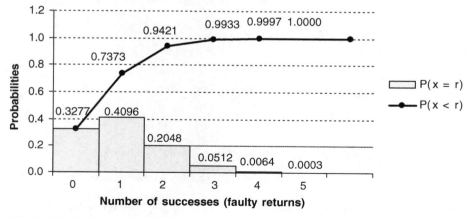

Figure 5.5   Probability histogram and cumulative distribution: $n = 5$: $p = 0.2$.

To read the Binomial probabilities in Worked Example 5.3 from the cumulative Binomial tables, view the column $p = 0.2$ in Table 5.6.

Table 5.7   Cumulative Binomial table 5.6 reproduced, highlighting the column, $p = 0.20$

				$n = 5$						
r	$p = 0.01$	$p = 0.05$	$p = 0.10$	$p = 0.20$	$p = 0.50$	$p = 0.60$	$p = 0.80$	$p = 0.90$	$p = 0.95$	$p = 0.99$
0	0.9510	0.7738	0.5905	0.3277	0.0313	0.0102	0.0003	0	0	0
1	0.9990	0.9774	0.9185	0.7373	0.1875	0.0870	0.0067	0.0005	0	0
2	1	0.9988	0.9914	0.9421	0.5000	0.3174	0.0579	0.0086	0.0012	0
3	1	1	0.9995	0.9933	0.8125	0.6630	0.2627	0.0815	0.0226	0.0010
4	1	1	1	0.9997	0.9688	0.9222	0.6723	0.4095	0.2262	0.0490
5	1	1	1	1	1	1	1	1	1	1

(i) From WE 5.3(b) 'At most two faulty' meant '0 OR 1 OR 2' faulty returns.
This probability was calculated $P(\text{at most two faulty}) = P(0) + P(1) + P(2) = 0.9421$
Now, use the Binomial tables. $P(x \leq 2)$ is the cumulative probability and it may be read directly from Table 5.6 $P(x \leq 2) = 0.9421$.

(ii) The probability that none are faulty is written as $P(0)$. Graphically this is represented as the area of the rectangle over $x = 0$ in Figure 5.5. But, $P(0) = \sum_{x=0}^{x=0} P(x) = P(x \leq 0)$, see Figure 5.5, Hence from cumulative Binomial tables, Table 5.6 $P(x \leq 0) = 0.3227$.

(iii) The probability 'exactly three' is represented by the area of the rectangle over $x = 3$. Referring to Figure 5.5, the area of the rectangle $x = 3$ may be described as 'the sum of the areas of the rectangles up to $x = 3$' minus 'the sum of the areas of the rectangles up to $x = 2$':

$$\sum_{x=0}^{x=3} P(x) - \sum_{x=0}^{x=2} P(x)$$

From the tables, the probability of exactly three is

$$\sum_{x=0}^{x=3} P(x) - \sum_{x=0}^{x=2} P(x) = 0.9932 - 0.9421 = 0.0511$$

(iv) From WE 5.3. 'At least one' means one or more.
$P(\text{at least one faulty}) = P(1) + P(2) + P(3) + P(4) + P(5)$
$P(\text{at least one faulty}) = 1 - P(0) = 1 - 0.3277 = 0.6723$
... where $P(0). = \sum_{x=0}^{x=0} P(x) = P(x \leq 0)$ from Table 5.6.

5.2.4   Binomial probability calculations in Excel

To illustrate probability calculations in Excel, Worked Example 5.3 with $n = 5$ and $p = 0.20$ will be used again.

Open an Excel worksheet and type values for X: 0, 1, 2, 3, 4, 5 in column A as shown in Figure 5.6. Place the cursor in cell B3 in the second column, and click f_x on the menu bar. In the statistical category, click on BINOMDIST (Binomial distribution) on the 'Select a function' dialogue box.

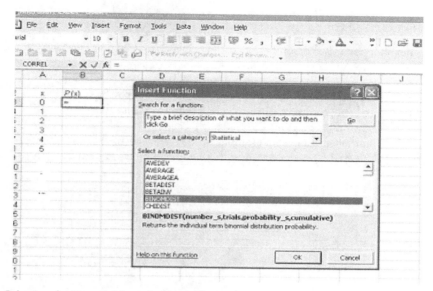

Figure 5.6   Selecting the Binomial from the 'Insert Function' dialogue box.

On the drop-down menu, enter the values of r, n, p in the first three boxes. In the fourth box type True if the cumulative probability is required: type False to calculate the individual probability.

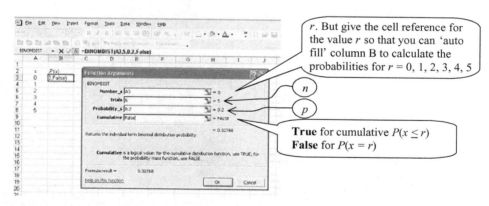

Figure 5.7   Entering the parameters for a Binomial calculation.

Click on OK: the value of the probability is stored in cell B3.
Now, auto fill the probabilities for $r = 1, 2, 3, 4$ and 5.

Note: use Excel to plot probability histograms and cumulative distributions using the methods described in Chapter 1.

Estimating p from data

If p (the proportion of successes) must be calculated from raw data, then the formula function in Excel may used to do the arithmetic. See Worked Example 5.5.

Worked Example 5.5: Binomial calculations in Excel or otherwise

A local butcher shop records the number of customers who pay for their purchases by cash over a period of ten days. The data is given in Table 5.8.

(a) From this data calculate the proportion, p of customers who pay by cash.
(b) Use the calculated value of p in (a) to calculate the probability that
    (i) at least 15 customers in a sample of 20 pay by cash;
    (ii) no more than 75 customers in a sample of 100 pay by cash;
    (iii) between 30 % and 50 % of customers in a sample of 60 do not pay by cash.

Table 5.8   Total customers and numbers of these who paid by cash over a ten day period   File: Cash

Day	Number of customers	Paid by cash
1	58	28
2	64	38
3	150	82
4	120	75
5	210	84
6	84	51
7	90	63
8	124	92
9	218	165
10	102	54
Total	**1220**	**732**

Solution

(a) The proportion of customers who paid cash

$$p = \frac{\text{Number who paid cash}}{\text{Total number of customers}} = \frac{732}{1220} = 0.60$$

(b) Use Excel, tables or otherwise to calculate
    (i) Sample size, $n = 20$, $x = 15$, $p = 0.6$. Hence $P(x \geq 15) = 0.1256$
    (ii) 'no more than 75' means 75 or less.
        Hence, with sample size, $n = 100$, $x = 75$, $P(x \leq 75) = 0.9994$
    (iii) 30 % of 60 is 18; 50 % of 60 is 30.
        $P(18 \leq x \leq 30)$ is the sum of the areas of rectangles 18 to 30 inclusive
        The proportion who did not pay cash is $1 - 0.60 = 0.40$.
        Hence for $n = 60$, $p = 0.4$,

$$P(18 \leq x \leq 30) = \sum_{x=0}^{x=30} P(x) - \sum_{x=0}^{x=17} P(x) = 0.9555 - 0.0413 = 0.9142.$$

Progress Exercises 5.2: Binomial probabilities

1. (See Worked Example 4.4) two coins are thrown (or a fair coin is thrown twice). Use the Binomial to calculate the probability that the outcome is (a) two Heads (b) one Head (c) no Heads.

2. (see Skill Development Exercise SK4_3). Three coins are thrown. Use the Binomial to calculate the probability that the outcome is (a) three Heads; (b) two Heads; (c) one Head; (d) no Heads.

3. (see Worked Example 4.10(c)) Four employees are randomly selected from a panel consisting of 12 females (F) and six males (M) for shift work. Assuming sampling is with replacement, calculate the probability that (a) there are no females; (b) there are two males and two females on the panel.

   Why are these answers (a) and (b) different from those of Worked Example 4.10(c)?

4. A county council has 520 employees. A survey concluded that 43 % of employees prefer to work flexi-time.

   (a) Calculate the number of employees that would prefer to work flexi-time.

   (b) Calculate the proportion of employees that would prefer to work flexi-time.

   (c) A random sample of 50 employees is required for a follow-up survey.
   (i) Describe the method you would use to select the random sample.
   (ii) Calculate the probability that the sample of 50 contains less than 18 employees who prefer to work flexi-time.

5. (a) State the assumptions for the Binomial distribution.

   (b) A quality control inspector selects a random sample of ten components from each incoming batch. The batch is returned to the supplier if the sample contains any defectives. If 8 % of components in the batch are defective, calculate the probability that the
   (i) sample contains no defectives;
   (ii) sample contains at least one defective;
   (iii) batch is accepted;
   (iv) batch is returned.

   Calculate the maximum percentage of defectives in a batch that will result in the batch being accepted 90 % of the time.

6. A hardware chain store has estimated that 20 % of credit card purchases were less than £ 20. A random sample of ten credit card purchases is selected.

   (a) What is the probability that the number of purchases less than £ 20 were
   (i) at most one; (ii) more than two.

   (b) What is the probability that the number of purchases valued £ 20 or more were (i) exactly three; (ii) at least three.

7. Ms G, a Green party candidate, gained 40 % of votes in a local election. Following the election, Ms G holds a series of public discussions on local issues. If she selects a random sample of ten constituents for a discussion panel, what is the probability that the sample will include (a) fewer than three; (b) at least two; (c) exactly five individuals who voted for her in the election.

8. A low fares airline has estimated that there is an 85 % chance of selling a seat on scheduled flights during the week prior to departure if a 'special offers' campaign is launched. If there are ten unsold seats available, what is the probability that (a) exactly seven; (b) more than seven will be sold; (c) what is the most likely number of seats that will be sold?

9. The manager of a car sales room claims that 10 % of customers who purchase a particular model of car will contact the sales room at least once with queries regarding the operation of the air conditioning and/or audio system within the first year. If a manager sells 50 cars, calculate

the probabilities that the number of customers that will contact the sales room about the air conditioning and/or audio system within the first year is 0, 1, 2, 3, 4, 5, 6, 7, 8, 9, 10 or more.

10. A PC service centre has 20 online support technicians. Over a period of time it has been observed that, on average, two technicians are absent from their stations at any given time. Calculate the probability that (a) at least one technician is absent; (b) at most three technicians are absent.

11. The prevalence of a particular disease among bovines is thought to be between 3 % and 8 %. What sample size should be selected if the researcher wishes to be 95 % certain that the sample contains at least one animal with this disease for (a) 3 % prevalence; (b) 8 % prevalence?

12. Hotel bookings record whether a booking was made online, by phone or otherwise. The data for 12 weeks is given in Table PE5.2 Q12.

Table PE5.2 Q12   Mode of booking hotel rooms over a 12 week period   File: Online

| Week | Mode of booking | | | Total |
	online	phone	other	
1	104	150	35	289
2	85	121	24	230
3	74	65	45	184
4	73	143	130	346
5	86	243	45	374
6	35	150	182	367
7	404	225	122	751
8	108	194	76	378
9	126	124	125	375
10	148	128	75	351
11	212	184	99	495
12	225	193	242	660
Total	1680	1920	1200	4800

(a) From this data calculate the proportions who booked (i) online; (ii) by phone; (iii) other. What proportion did not book online?

(b) The hotel randomly selects samples of guests for special offers. Calculate the probability that
   (i) at least ten guests in a sample of 20 booked online;
   (ii) exactly four guests in a sample of 20 did not book online;
   (iii) between ten and 20 guests in a sample of 50 booked by phone;
   (iv) 40 % of guests in a sample of 120 booked online.

5.3   The Poisson probability distribution

Binomial probability calculations require a finite sample size. There are many situations where a sample size does not feature. For example sample size does not feature in the calculation of the probability of x emergency calls per hour or x faults per km of cable, etc. In situations such as these (subject to given assumptions outlined below), a formula called the 'Poisson probability formula' will be used to calculate the probability of x occurrences of an event over a given interval of time or length (area, volume, time, etc.).

5.3.1   The Poisson probability distribution function and assumptions

The Poisson probability distribution function is named after Simeon Poisson who first proposed it in a book published in 1837. The Poisson formula is $P(x) = \dfrac{\lambda^x e^{-\lambda}}{x!}$. This 'Poisson probability formula' will be used to calculate probabilities in situations where a rare, random event occurs at a uniform rate, λ subject to the assumptions outlined in section 5.3.1.1.

Some examples of situations where the Poisson formula may be used are given below:

1. A certain type of cable has an average of 3 faults per km (mean $\lambda = 3$ per km). The Poisson formula may be used to calculate probabilities such as:
   (a) P(no faults in one km); (b) P(at least 2 faults in one km).

   > λ is the symbol λ is the mean (or average) rate of occurrence for a given interval.

2. A liquid contains an average of 12 radioactive particles per cubic meter: (mean $\lambda = 12$ per cubic meter).
   The Poisson formula may be used to calculate probabilities such as:
   (a) P(ten particles in one cubic meter); (b) P (at least 5 particles in one cubic meter).
3. Sheet metal has an average of two defects per sq m: (mean $\lambda = 2$ per square meter)
   Calculate the probability that there is at most 1 defect per sq. m.
4. An ambulance service receives an average of four calls in 1 hour during 'off peak' hours. ($\lambda = 4$ in 1 hour).
   Use the Poisson formula to calculate that during 'off peak' hours there are (a) three calls in one hour; (b) At most two calls in 1 hour; (c) one call in 30 minutes.

The Poisson formula may also be used to approximate Binomial probabilities, since the Poisson formula is simpler and easier formula to use than the Binomial.

Assumptions for the Poisson process

The Poisson formula is only valid when the conditions (assumptions) for a Poisson process are satisfied. These assumptions are explained in the context of example 4 above where calls for an ambulance service that arrive at the rate of four per hour. (λ = four calls in 1 hour).

Assumptions

1. The average rate at which an event occurs per interval is uniform. For the ambulance service the average rate was given as λ = four calls in 1 hour. Since this rate is uniform (during off peak hours) the rate may be halved to give an average of two calls in 30 minutes; doubled to give an average of eight calls in 2 hours, etc.
2. The number of calls is not influenced by what has happened in the previous (off peak) interval. So if there were no calls in 1 hour, this has no effect on the chance of x calls in the next hour. This property is often quoted as 'the Poisson has no memory!'
3. There is practically no chance of more than one call arriving in a very short time. For example, the probability of more than one call in (i) 1 minute is 0.0645 (ii) 1 second is 0.0110 (calculation of probabilities given in Worked Example 5.6(d)). As the interval becomes smaller, the probability of more than one call per interval approaches zero.

Poisson probability calculations

> The Poisson probability distribution function is given by the formula
>
> $$P(x) = \frac{\lambda^x e^{-\lambda}}{x!} \qquad (5.2)$$
>
> where λ is the mean rate at which X occurs in a given interval
> **Note**: some texts (and tables) use the symbol μ for the mean rate instead of λ.

> The number e

Brief notes on the number e

e is the symbol representing the number 2.71828... because its decimal part is unending (just like π is used to represent 3.14159...). To evaluate e^{power} use any scientific calculator, for example $e^{1.5} = 4.4817$. See Appendix E for further information on the importance of e.

To calculate probabilities using the Poisson formula you must be given (or the data to calculate) the parameter λ: $P(x) = \dfrac{\lambda^x e^{-\lambda}}{x!}$

λ is interval is sometimes called 'area of opportunity'.
Worked Example 5.5 will demonstrate Poisson probability calculations.

Worked Example 5.6: Poisson probability calculations

The average rate of calls for an ambulance service is four calls in 1 hour during 'off peak' hours. Calculate the probability that during 'off peak' hours the ambulance service receives

(a) three calls in hour;
(b) at most two calls in 1 hour;
(c) one call in 30 minutes;
(d) more than one call in (i) 1 minute (ii) 1 second.

Solution

(a) The average rate is given as $\lambda = \mathbf{4}$ **in 1 hour**.
Hence the probability of four calls in 1 hour is calculated by substituting Step 1: $x = 3$ Step 2: $\lambda = 4$ into the Poisson formula (5.2)

$$P(x) = \frac{\lambda^x e^{-\lambda}}{x!} \qquad (5.2)$$

Step 1: substitute $x = 3$ into (5.2) $P(x = 3) = \dfrac{\lambda^3 e^{-\lambda}}{3!}$
Step 2: substitute $\lambda = 4$ into the result of step 1 and calculate the probability:

$$P(x = 3/\lambda = 4) = \frac{4^3 e^{-4}}{3!} = \frac{(64)(0.01832)}{6} = 0.1954$$

(b) 'at most two' means '0 OR 1 OR 2'

$P(\text{at most two}) = P(0) + P(1) + P(2)$

Step 1: substitute the values of $x = 0$, 1 and 2 in the successive Poisson formulae

$$P(\text{at most two}) = \frac{\lambda^0 e^{-\lambda}}{0!} + \frac{\lambda^1 e^{-\lambda}}{1!} + \frac{\lambda^2 e^{-\lambda}}{2!}$$

Step 2: substitute $\lambda = 4$, hence evaluate the probability

$$P(\text{at most two}) = \frac{4^0 e^{-4}}{0!} + \frac{4^1 e^{-4}}{1!} + \frac{4^2 e^{-4}}{2!} = e^{-4}(1 + 4 + 8) = 0.2381$$

Remember $0! = 1$ and $e^0 = 1$ (any real number raised to the power zero is unity).

(c) We now require a probability for a 30 minute interval. Since the rate is assumed to be uniform, then four calls in 1 hour means that the rate for 30 minutes is two calls: $\lambda = 2$ in 30 minutes: Hence, the probability of one call in 30 minutes is

$$P(x = 1) = \frac{\lambda^1 e^{-\lambda}}{1!} = \frac{2^1 e^{-2}}{1!} = \frac{2(0.1353)}{1} = 0.2706 \ldots \text{substituting } x = 1 \text{ first, then } \lambda = 2$$

(d) The probability of more than one call; $P(x \geq 1) = 1 - P(x = 0)$

(i) For 1 minute $\lambda = \frac{4}{60}$ hence $P(x \geq 1) = 1 - \frac{\left(\frac{4}{60}\right)^0 e^{-\frac{4}{60}}}{0!} = 1 - 0.9355 = 0.0645$

(ii) For 1 second $\lambda = \frac{4}{60 \times 60}$ hence $P(x \geq 1) = 1 - \frac{\left(\frac{4}{360}\right)^0 e^{-\frac{4}{360}}}{0!} = 1 - 0.9890 = 0.0110$

5.3.2   Tables for cumulative Poisson probabilities Table 2, Appendix G

The cumulative probability Tables for the Poisson distribution gives the sum of all probabilities up to and including $P(x = r)$ for various values of λ: $\sum_{x=0}^{x=r} P(x \leq r) = \sum_{x=0}^{x=r} \frac{\lambda^x e^{-\lambda}}{x!}$.

The use of the cumulative probabilities tables will be demonstrated in Worked Example 5.7.

Worked Example 5.7: Poisson probability calculations

The average rate for calls to an ambulance service is four per hour during 'off peak' hours.

(a) Plot the probability histogram and cumulative probability distribution for $\lambda = 4$ on the same diagram.

(b) Use the extract of cumulative Poisson probability tables (Table 5.9) to calculate the probabilities of (i) three calls in 1 hour; (ii) At most two calls in 1 hour.

Solution

(a) The Poisson probabilities for $\lambda = 4$ are $P(0) = \frac{\lambda^0 e^{-\lambda}}{0!} = e^{-4} = 0.0183$, $P(1) = \frac{\lambda^2 e^{-\lambda}}{1!} = \frac{4^1 e^{-4}}{1!} = 0.0733$, $P(2) = \frac{\lambda^2 e^{-\lambda}}{2!} = \frac{4^2 e^{-4}}{2!} = 0.1465$, $p(3) = \frac{\lambda^3 e^{-\lambda}}{3!} = \frac{4^3 e^{-4}}{3!} = 0.1954$ etc.
See Figure 5.8.

Table 5.9   An extract of the Poisson cumulative probability for $\lambda = 4$ to $\lambda = 4.5$

r	$\lambda = 4.0$	$\lambda = 4.1$	$\lambda = 4.2$	$\lambda = 4.3$	$\lambda = 4.4$	$\lambda = 4.5$
0	0.0183	0.0166	0.0150	0.0136	0.0123	0.0111
1	0.0916	0.0845	0.0780	0.0719	0.0663	0.0611
2	0.2381	0.2238	0.2102	0.1974	0.1851	0.1736
3	0.4335	0.4142	0.3954	0.3772	0.3594	0.3423
4	0.6288	0.6093	0.5898	0.5704	0.5512	0.5321
5	0.7851	0.7693	0.7531	0.7367	0.7199	0.7029
6	0.8893	0.8786	0.8675	0.8558	0.8436	0.8311
7	0.9489	0.9427	0.9361	0.9290	0.9214	0.9134
8	0.9786	0.9755	0.9721	0.9683	0.9642	0.9597
9	0.9919	0.9905	0.9889	0.9871	0.9851	0.9829
10	0.9972	0.9966	0.9959	0.9952	0.9943	0.9933
11	0.9991	0.9989	0.9986	0.9983	0.9980	0.9976
12	0.9997	0.9997	0.9996	0.9995	0.9993	0.9992
13	0.9999	0.9999	0.9999	0.9998	0.9998	0.9997
14	1.0000	1.0000	1.0000	1.0000	0.9999	0.9999
15	1.0000	1.0000	1.0000	1.0000	1.0000	1.0000

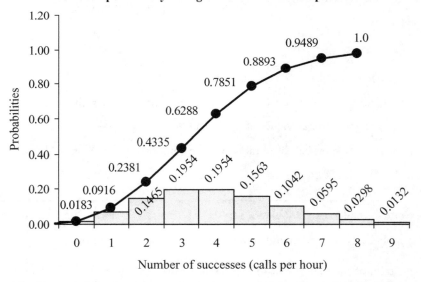

Figure 5.8   The probability histogram and cumulative probability distribution for $\lambda = 4$.

(b) (i) The probability of three calls per hour is $\sum_{x=0}^{x=3} P(x) - \sum_{x=0}^{x=2} P(x)$ that is, the sum of the areas of the rectangles up to and including $P(x = 3)$ – the sum of the areas of the rectangles up to and including $P(x = 2)$.

Hence $\sum_{x=0}^{x=3} P(x) - \sum_{x=0}^{x=2} P(x) = 0.4335 - 0.2381 = 0.1954$

(ii) 'at most two' means '0 OR 1 OR 2'
$P(\text{at most two}) = P(0) + P(1) + P(2)$
This is the cumulative probability for $r = 2$ and may be read directly from the tables:
$$\sum_{x=0}^{x=2} P(x) = 0.2381.$$

5.3.3 Poisson probability calculations in Excel

Set up an Excel worksheet as directed for the Binomial in Figure 5.6 but click on 'Poisson' in the 'Insert Function' menu. In the 'Function Arguments' menu, Figure 5.10, enter

1. the value for r the number of successes or the address for the r;
2. λ: the Mean for the Poisson; and
3. True for a cumulative or False for the individual probability (called a probability mass function).

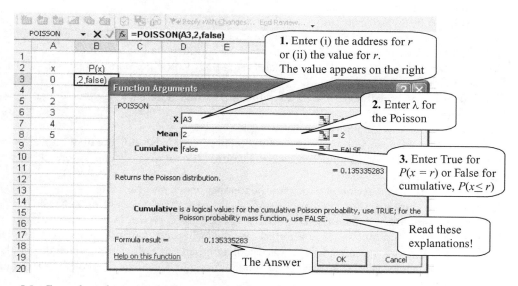

Figure 5.9   Enter the information for Poisson probability calculations.

Skill Development Exercise SK5_4: Use the calculator for Poisson probabilities

Use the calculator to calculate the probabilities listed in each of the examples 1, 2, 3 and 4 given in the introduction to section 5.3. Confirm your answers in Excel.

Answers

Example 1 (a) 0.0498; (b) 0.8009: Example 2 (a) 0.1048; (b) 0.9797
Example 3: 0.4060: Example 4 (a) 0.1353; (b) 0.3233; (c) 0.3679 (d) 0.1144

Progress Exercises 5.3: Poisson Probabilities

1. Evaluate the following correct to four decimal places

   (a) (i) e^2; (ii) $3e^{-2}$; (iii) $\dfrac{3\,e^{-2.5}}{3!}$; (iv) $\dfrac{4!e^{-5}}{1+e^{2.2}}$

   (b) Substitute (i) $x = 4$ into the formula $\dfrac{\lambda^x e^{-\lambda}}{x!}$; (ii) $\lambda = 2.4$ into $\dfrac{\lambda^4 e^{-\lambda}}{4!}$

2. Assuming a Poisson process calculate the following probabilities
   (a) Exactly two arrivals in 5 minutes at an ATM when the average rate is four in 5 minutes
   (b) At least three defects in a kilometre of telephone cable when the average rate is five defects per kilometre.
   (c) An angler catching more than three sea trout per day on a river where the average catch is 2.5 per angler per day.
   (d) An angler catching more than nine sea trout in three days on a river where the average catch per angler is 2.5 per day

3. The roads authority carried out surveys on the volume of traffic on minor roads. On one particular road an average of 96 vehicles pass a given point between the hours of 3.00 pm and 7.00 pm.
   (a) Under what conditions would the traffic flow be considered a Poisson process?
   (b) Calculate the average number of vehicles using the road (i) per hour (ii) per minute (iii) per 15 minutes.
   (c) Calculate the probability that (i) 25 vehicles pass in 1 hour (ii) one vehicle passes in 1 minute

4. State the assumptions for a Poisson process.
   The number of defects in the casing for flat screens is recorded during the final inspection of the finished product: For 12 randomly selected casings, the number of defects were: 0; 2; 0; 3; 2; 5; 2; 2; 1; 2; 1; 4.
   (a) Calculate the average and variance for the number of defects per casing.
   (b) Use the average in (a) to calculate the probability of at most three defects per casing.
   (c) Do you think the number of defects per casing is a Poisson process? Give reasons.

5. The organisers of a mountain rescue service are reviewing the demand for its services over week-ends in summer. From records, it was calculated that 120 calls were received during the 48 hour period between Friday midnight and Sunday midnight. Calculate the probability (during weekends) that
   (a) one call is received in a 1 hour period;
   (b) at most two call are received in a 1 hour period;
   (c) 12 calls are received in a 12 hour period;
   (d) more than 50 calls are received in a 24 hour period.

6. A local network in an organisation averages 2.8 down-times in four weeks. Calculate the probability that the number of down-times is
   (a) exactly one in a week;
   (b) none in a week;
   (c) more than two in a week;
   (d) 12 in six weeks;
   (e) between 10 and 15 in six weeks.

7. A particular brand of sail canvas is known to have an average of nine rough spots in a length of 10 m.

(a) Calculate the probability that there are less than five rough spots in a 10 m length.
(b) Calculate the probability distribution for the number of rough spots in 5 m lengths. Hence state which outcome is most likely.
(c) Confirm that probability of (i) one rough spot per m (ii) five rough spots per 5 m.

8. A car rental company reviews the staffing of their car rental kiosk in the airport arrivals terminal. Since the number of staff on duty depends on the demand, the number of cars rented per hour is recorded for 125 hourly intervals. The number of times (frequency) that 0, 1, 2....11, 12 cars were rented per hour is given in Table PE5.3 Q8:

Table PE5.3 Q8   Number of times that 0, 1, 2......, 11, 12 cars were rented per hour for 125 hours   File: Rented cars

Cars rented per hour	0	1	2	3	4	5	6	7	8	9	10	11	12
Frequency		1	3	4	10	15	20	24	19	13	10	4	1	1

(a) Calculate the mean number of rentals per hour.
(b) Assuming a Poisson process, use the mean number of cars rented per hour in (a) to calculate (i) the probability that 0, 1, 2, 3...., 11, 12 cars are rented per hour (ii) the number of times that 0, 1, 2, 3...., 11, 12 cars are rented per hour in a total of 125 hours. Hint: to calculate the number of times A happens use f_A use the definition of probability: $P(A) = \dfrac{f_A}{N}$.

Do you think the assumption of a Poisson process is correct?
(The number of times that you would expect 0, 1, 2, 3...., 11, 12 rentals per hour in a total of 125 hours, assuming a Poisson distribution with a given λ are called the expected frequencies.)

9. On average, two typos in every five pages are not detected by the spell checker in the first draft of a report. Calculate the probability that there will be (a) exactly two typos undetected in 5 pages (b) exactly four typos undetected in ten pages (c) at most 20 typos undetected in 50 pages.

10. A company records the number of absentees per day for a period of 80 days in Table PE5.3 Q10.

Table PE5.3 Q10   Absentees per day for a period of 80 days   File: Absentees per day

Absentees per day	0	1	2	3	4	5	6	7	8	9	10
Days	6	9	12	18	18	9	4	0	3	0	1

(a) Calculate the mean number of absentees per day.
(b) Use the average in (a) to calculate the Poisson probabilities for zero to ten absentees per day using the mean in (a). Hence calculate the expected number of days on which there are 0, 1, 2 .... ten absentees for an 80 day period. Do your results indicate that the absentee rate is modelled by a Poisson process?

5.4   The Normal probability distribution

The Normal curve was introduced in 1733 by De Moivre as an approximation to certain Binomial distributions (reprinted in 1738 in The Doctrines of Chance). However Hans Frederick Gauss gave a rigorous account of its properties in 1809.

The Normal probability is a continuous probability distribution and it is the most important probability in statistics. It will feature in the remainder of the text, not just in probability calculations but also in statistical inference.

5.4.1   What is the Normal probability distribution?

It has been long been recognised that large numbers of measurements, when sorted and plotted in a probability (relative frequency) histogram, tend to assume a bell-shaped form, similar to that given in Figure 10(a): the relative frequency (or probability) histogram for the time taken by most customers to complete transactions at an ATM machine (in this example the majority of customers take between two and 20 minutes).

Since time (T) is a **continuous random** variable it may be recorded to any degree of precision: if time is measured to the nearest $^1/_4$ minute there are 72 intervals between two and 20 minutes: this probability histogram in Figure 5.10(a). But if time is measured correct to two decimal places, (that is from $t = 2.00; 0.01; 0.02 \ldots 19.99, 20.00$ minutes), there are 1,800 intervals, hence the probability histogram will assume the shape of a continuous curve in Figure 5.10(b).

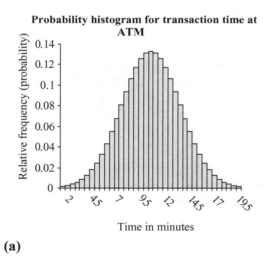

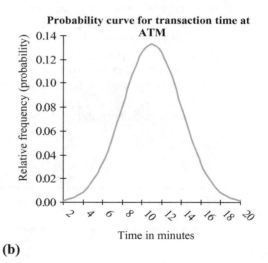

(a)   **(b)**

Figure 5.10(a)   Bell shaped probability histogram.   Figure 5.10(b)   Normal, bell shaped curve.

This symmetrical 'bell shaped' curve is the classic 'Normal' probability distribution curve: it is the most important continuous probability distribution in statistics. To highlight its importance, W.J. Youden (1900 to 1971) presented it in a manner similar to that in Figure 5.11.

THE
NORMAL
LAW OF ERROR
STANDS OUT IN THE
EXPERIENCE OF MANKIND
AS ONE OF THE BROADEST
GENERALIZATIONS OF NATURAL
PHILOSOPHY ♦ IT SERVES AS THE
GUIDING INSTRUMENT IN RESEARCHES
IN THE PHYSICAL AND SOCIAL SCIENCES AND
IN MEDICINE AGRICULTURE AND ENGINEERING ♦
IT IS AN INDISPENSABLE TOOL FOR THE ANALYSIS AND THE
INTERPRETATION OF THE BASIC DATA OBTAINED BY OBSERVATION AND EXPERIMENT

Figure 5.11   The Normal distribution presented by W.J. Youden.

5.4.2   The characteristics of the Normal curve

The equation of the Normal curve is given by:

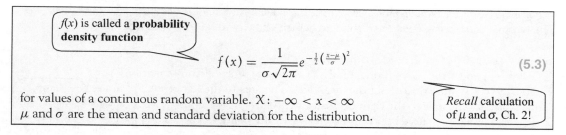

$f(x)$ is called a **probability density function**

$$f(x) = \frac{1}{\sigma\sqrt{2\pi}}e^{-\frac{1}{2}\left(\frac{x-\mu}{\sigma}\right)^2}$$

(5.3)

for values of a continuous random variable. $X: -\infty < x < \infty$
μ and σ are the mean and standard deviation for the distribution.

Recall calculation of μ and σ, Ch. 2!

The equation, $f(x)$ in (5.3) traces out a bell shaped curve, symmetrical about the mean μ:
The value of μ determines the location of the curve, see Figure 5.12(a): the by the value of its standard deviation determines the width of curve: See figure 5.12(b).

The area under this curve is always unity, a property of any probability distribution.

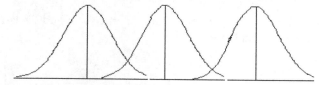

Figure 5.12(a)   Different means, same standard deviation.

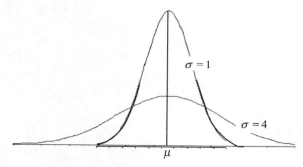

Figure 5.12(b)   Same mean, different standard deviations.

The probability that a random variable, X, has a value between $x = a$ and $x = b$ is given by the area under the curve between $x = a$ and $x = b$. Areas under curves are usually calculated by integration: $P(a \leq X \leq b) = \int_{x=a}^{x=b} f(x)dx$: see Figure 5.13.

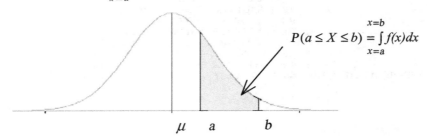

$$P(a \leq X \leq b) = \int_{x=a}^{x=b} f(x)dx$$

Figure 5.13 $P(a < x < b) =$ area under the curve (shaded) between $x = a$ and $x = b$.

Fortunately, for the Normal probability curve, it will not be necessary to use integration to calculate areas. The Normal curve has special characteristics which allow us to find the area from single set of tables.

Special properties of the Normal distribution

1. Total area under the curve is one. (This is true of any probability distribution).
2. The curve is symmetrical about the mean (μ). The area to the left of the mean is 0.5. The area to the right of the mean is 0.5.
3. The area under the curve between the mean and any point x depends on the number of standard deviations between x and μ.

> The number of standard deviations between x and μ is called the Z **score**.

   For example, in Figure 5.14,
   (a) The area between the mean and a point which is one standard deviation ($1 \times \sigma$) greater (or less) than the mean is 0.3413.
   (b) The area between the mean and a point which is two standard deviations ($2 \times \sigma$) greater (or less) than the mean is 0.4772.
   (c) The area between the mean and a point which is three standard deviations ($3 \times \sigma$) greater (or less) than the mean is 0.4986.

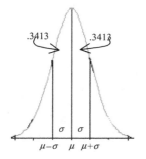

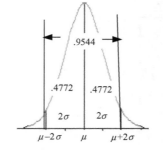

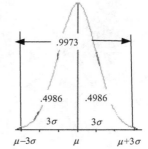

(a) 68.26 % of the total area lies between ($\mu - \sigma$) and ($\mu + \sigma$)

(b) 95.44 % of the total area lies between ($\mu - 2\sigma$) and ($\mu + 2\sigma$)

(c) 99.72 % of the total area lies between ($\mu - 3\sigma$) and ($\mu + 3\sigma$)

Figure 5.14   The area under the curve between x and μ depend on number of σ's between x and μ.

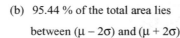

Use the Normal probability tables (Table 3, Appendix G) to determine areas under the Normal curve

Let Z = the number of standard deviations (σ's) between x and μ.
  In Figure 5.14 we say:

(a) 68.26 % of the area under the Normal curve lies between $Z = -1$ and $Z = 1$;
(b) 95.44 % of the area under the Normal curve lies between $Z = -2$ and $Z = 2$;
(c) 99.72 % of the area under the Normal curve lies between $Z = -3$ and $Z = 3$.

The Normal probability tables (an extract is given in Table 5.10) give the area under the Normal curve to the right of given values of Z (correct to two places of decimals) for values of Z ranging from $Z = 0.00$ and $Z = 4.00$ and selected higher values of Z.

Table 5.10   Extract from Normal probability distribution tables

Area in upper tail of the Normal probability distribution — Second decimal places for values of Z

$Z = \dfrac{x - \mu}{\sigma}$

	0.00	0.01	0.02	0.03	0.04	0.05	0.06	0.07	0.08
0.0	0.5000	0.4960	0.4920	0.4880	0.4840	0.4801	0.4761	0.4721	0.4681
0.1	0.4602	0.4562	0.4522	0.4483	0.4443	0.4404	0.4364	0.4325	0.4286
0.2	0.4207	0.4168	0.4129	0.4090	0.4052	0.4013	0.3974	0.3936	0.3897
0.3	0.3821	0.3783	0.3745	0.3707	0.3669	0.3632	0.3594	0.3557	0.3520
0.4	0.3446	0.3409	0.3372	0.3336	0.3300	0.3264	0.3228	0.3192	0.3156
0.5	0.3085	0.3050	0.3015	0.2981	0.2946	0.2912	0.2877	0.2843	0.2810
0.6	0.2743	0.2709	0.2676	0.2643	0.2611	0.2578	0.2546	0.2514	0.2483
0.7	0.2420	0.2389	0.2358	0.2327	0.2296	0.2266	0.2236	0.2206	0.2177
0.8	0.2119	0.2090	0.2061	0.2033	0.2005	0.1977	0.1949	0.1922	0.1894
0.9	0.1841	0.1814	0.1788	0.1762	0.1736	0.1711	0.1685	0.1660	0.1635
1.0	0.1587	0.1562	0.1539	0.1515	0.1492	0.1469	0.1446	0.1423	0.1401
1.1	0.1357	0.1335	0.1314	0.1292	0.1271	0.1251	0.1230	0.1210	0.1190
1.2	0.1151	0.1131	0.1112	0.1093	0.1075	0.1056	0.1038	0.1020	0.1003
1.3	0.0968	0.0951	0.0934	0.0918	0.0901	0.0885	0.0869	0.0853	0.0838
1.4	0.0808	0.0793	0.0778	0.0764	0.0749	0.0735	0.0721	0.0708	0.0694
1.5	0.0668	0.0655	0.0643	0.0630	0.0618	0.0606	0.0594	0.0582	0.0571
1.6	0.0548	0.0537	0.0526	0.0516	0.0505	0.0495	0.0485	0.0475	0.0465
1.7	0.0446	0.0436	0.0427	0.0418	0.0409	0.0401	0.0392	0.0384	0.0375
1.8	0.0359	0.0351	0.0344	0.0336	0.0329	0.0322	0.0314	0.0307	0.0301
1.9	0.0287	0.0281	0.0274	0.0268	0.0262	0.0256	0.0250	0.0244	0.0239
2.0	0.0228	0.0222	0.0217	0.0212	0.0207	0.0202	0.0197	0.0192	0.0188
2.1	0.0179	0.0174	0.0170	0.0166	0.0162	0.0158	0.0154	0.0150	0.0146
2.2	0.0139	0.0136	0.0132	0.0129	0.0125	0.0122	0.0119	0.0116	0.0113

$Z = 1.0$ → (0.1587)

Tail area

As an example to illustrate the use of the tables, consider Figure 5.14(a). To find the tail area to the right of $Z = 1.00$ look up $Z = 1.0$ (i.e. the value of Z up to first decimal place) in the left hand column of the tables: draw a horizontal line across from $Z = 1.0$ and a vertical line down through the column headed '0.00' as shown in Table 5.13. The tail area is given by the value at the intersection of the horizontal and vertical lines. Hence when $Z = 1.00$, the tail area is 0.1587 as shown in Figure 5.15:

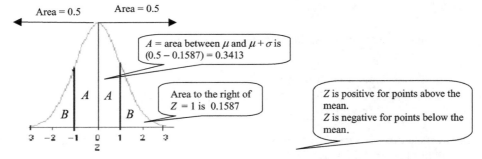

Figure 5.15   Area between μ and $\mu + \sigma$ (between $Z = 0$ and $Z = 1$).

With the upper tail area known, use the fact that the total area under the Normal curve is unity; the area to the left and right of the mean is 0.5 to calculate areas such as (i) the area to the left of $Z = 1$ and (ii) the area between $Z = 0$ and $Z = 1$ as follows.

Let B = the area to the right of $z = 1$. $B = 0.1587$
then

(i) the area to the left of $Z = 1$ is $1 - 0.1587 = 0.8413$ (area = $A + A + B$)
(ii) Let A = the area between $Z = 0$ and $Z = 1$. Hence area $A = 0.5 - 0.1587 = 0.3413$.

Note

Since the Normal curve is symmetric about the mean,

(i)   the area to the left of $Z = -1$ is the same as the area to the right of $Z = 1$ is .0.1587
(ii) the area between $Z = -1$ and $Z = 0$ is also equal to $A = 0.3413$ (area = B)
(iii) the area to the right of $Z = -1$ is the same as the area to the left of $Z = 1$. This area is 0.8413

(area = $A + A + B$)

Further examples on using Normal probability tables are given in Worked Example 5.8 and 5.9.

Worked Example 5.8: Using Normal Probability Tables

The time taken to complete a transaction at an ATM machine is normally distributed with a mean time of 11 minutes and standard deviation of 3 minutes. Use Normal probability tables to calculate the probability that a transaction will take between

(a) (i) 11 and 14 minutes; (ii) 8 and 14 minutes;
(b) (i) 11 and 17 minutes; (ii) 5 and 17 minutes;
(c) (i) 11 and 20 minutes; (ii) 2 and 20 minutes.

Solution

Since Z = the number of standard deviations between a point x and the mean, in this example where $\mu = 11$ and $\sigma = 3$ the values of Z for parts (a), (b) and (c) may be written down immediately, as shown in Figure 5.16:

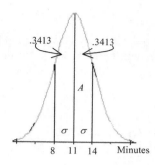 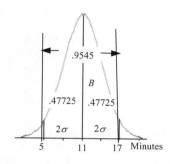 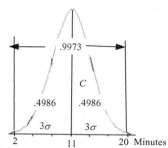

(a) 68.26 % of transactions take between $Z = -1$ and $Z = 1$.

(b) 95.44 % of transactions take between $Z = -2$ and $Z = 2$.

(c) 99.72 % of transactions take between $Z = -3$ and $Z = 3$.

Figure 5.16   The area under the curve between x and μ depend on number of σ's between x and μ.

The values of Z (or Z-score) are then used to look up the corresponding tail areas in the Normal probability distribution tables

(a) (i) Since $\mu = 11$ and $\sigma = 3$, then $x = 14$ is one standard deviation (three minutes) from the mean: hence $Z = 1.00$. From the Normal probability distribution tables (Table 5.9) the tail area from $Z = 1$ is 0.1587. Hence the area A is calculated $A = 0.5 - 0.1587 = 0.3413$. The probability that a transaction will take between 11 and 14 minutes is 0.3413.

   (ii) In Figure 5.16(a), point $x = 8$ is three minutes, one standard deviation below the mean; $Z = -1$

   Because of symmetry, areas equidistant on each side of the mean are equal. Since the area on the right is $A = 0.3413$, see (i), then the area to the left is also 0.3413, The total area between $Z = -1$ and $Z = +1$ is $2A = 2(0.3413) = 0.6826$.

   The probability that a transaction will take between eight and 14 minutes is 0.6826.

(b) (i) Since $\mu = 11$ and $\sigma = 3$, then $x = 17$ is two standard deviations (six minutes) from the mean: hence $Z = 2.00$. From the Normal probability distribution tables (Table 5.9) the tail area from $Z = 2$ is 0.0228. Hence, the area between the mean and $Z = 2$ is: $B = 0.5 - 0.0228 = 0.4772$.

   The probability that a transaction will take between 11 and 17 minutes is 0.4772.

   (ii) By symmetry, the total area between $Z = -2$ and $Z = 2$ is $2(0.4772) = 0.9544$.

   The probability that a transaction will take between five and 17 minutes is 0.9544.

(c) (i) Since $\mu = 11$ and $\sigma = 3$, then $x = 20$ is three standard deviations (nine minutes) from the mean: then $Z = 3.00$. From the Normal probability tables the tail area from $Z = 3$ is 0.0013.

   Hence, the area between the mean and $Z = 3$ is calculated as: $C = 0.5 - 0.0013 = 0.4987$.

   The probability that a transaction will take between 11 and 20 minutes is 0.4987.

   (ii) By symmetry, the total area between $Z = -3$ and $Z = 3$ is $2(0.4987) = 0.9974$.

   The probability that a transaction will take between two and 20 minutes is 0.9974.

Skill Development Exercise SK5_5: Use Normal Probability Tables

Flight times between London and Hong Kong are normally distributed with $\mu = 12.5$ hours, standard deviation $= 1.5$ hours. Calculate the probability that a flight will take (a) longer than 14 hours; (b) between 12.5 and 14 hours; (c) less than 14 hours; (d) less than 11 hours.

Answers

(a) P(flight time is longer than 14 hours): $= P(Z > 1.00)$: $= 0.1587$.
(b) P(flight time is between 12.5 and 14 hours) $= 0.3413$.
(c) P(flight time is less than 14 hours) $= 1 - 0.1587 = 0.8413$.
(d) P(flight time is less than 11 hours) $= 0.1587$.

5.4.3 A step-by-step method for calculating Normal probabilities

From the previous Worked Example, the values of x, μ and σ were such that it was possible to write down the Z-values immediately. However, in most problems, the Z-values must be calculated formula (5.4).

$$Z = \frac{x - \mu}{\sigma} \qquad (5.4)$$

Z is the number of standard deviations between x and μ

A step-by-step method for calculating Normal probabilities is set out, and illustrated in Worked Example 5.8:

Step 1. Sketch the normal curve, symmetrical about μ. Mark in the point x.
Shade in the area (probability) required.
Step 2. From x, calculate Z.
Step 3. Look up tables for the corresponding tail area in the Normal probability tables.
Step 4. Calculate the required area or probability.

Worked Example 5.9: A step-by-step method for calculating Normal probabilities

The time taken to process an email enquiry is normally distributed with a mean time of 500 sec. and a standard deviation of 10 sec.

What is the probability that a randomly selected email enquiry will be processed in
(a) more than 505 sec; (b) less than 485 sec; (c) between 485 and 505 sec.
State your results verbally.

Solution

(a)
Step 1. Sketch curve, mark in info.

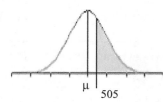

Step 2. Calculate Z
For $x = 505$
$$Z = \frac{x - \mu}{\sigma}$$
$$= \frac{505 - 500}{10} = 0.5$$

Step 3. Look up tables
When $= 0.5$,
area in the tail $= 0.3085$

Step 4. Calculate the required area (probability)
This is the required area
$P(x \geq 505 \text{ sec}) = 0.3085$

(b)
Step 1. Sketch curve, mark in info.

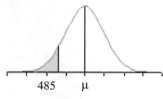

Step 2. Calculate Z
For $x = 485$
$$Z = \frac{x - \mu}{\sigma} = \frac{485 - 500}{10}$$
$$= -1.5$$

Step 3. Look up tables
When Z is negative, look up the tables for $Z = +1.5$
Tail area $= 0.0668$

Step 4. Calculate the required area (probability)
By symmetry, the area above $Z = 1.5$ is the same as the area below $Z = -1.5$
$P(x \leq 485 \text{ sec}) = 0.0668$

(c)
Step 1. Sketch curve; mark in info.

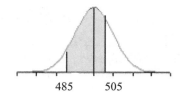

Step 2. Calculate Z
For $x = 505$, $Z = 0.5$
For $x = 485$, $Z = -1.5$

Step 3. Look up tables
$Z = 0.5$; tail area $= 0.3085$
$Z = 1.5$; tail area $= 0.0668$

Step 4. Calculate the required area (probability)
Subtract the tails areas from 1 to get the area in the middle
$1 - 0.0668 - 0.3085 = 0.6227$
$P(485 \leq x \leq 505) = 0.6247$

Method for calculating the limits that contain a given percentage of all values under the Normal curve

In worked Example 5.9 you were asked to calculate the probabilities that the time taken to process enquiries would be greater than a stated number of minutes, x (or less than or between two values of x). But, in some situations it might be useful to know that it takes between P and Q seconds to process, say 95 % of inquiries. Calculating the values of P and Q is possible when the times are Normally distributed, with μ and σ known. The method for calculating P and Q involves the four steps used in Worked Example 5.9, but *in the reverse order*. We start with probabilities that are given in the question and must work back to calculate the value(s) of x: $x = P$ and $x = Q$. The method for calculating the two limits, symmetrical about the mean, that contain a given percentage of all the data is set out as follows:

Step 1. Sketch a Normal curve marking and label the given area and tail areas.
Step 2. From the tail areas, look up tables to find Z.

Step 3. From the value of Z, calculate the number of units (d) between x and the mean. It will be shown that the number of units is $d = Z\sigma$.

Step 4. Calculate the values of P and Q: $Q = \mu + Z\sigma$ and $P = \mu - Z\sigma$.

Worked Example 5.10: Calculate the limits that contain a given percentage of all data

The time taken to process an email enquiry is normally distributed with a mean time of 500 sec. and a standard deviation of 10 sec.

(a) Between what **two times** (symmetrical about the mean) should (i) 95 %; (ii) 99 % of emails be processed? State your results verbally.
(b) Within what time should 94 % of emails be processed? (Or below what time, etc.)

State your results verbally.

Solution

(a) (i) **Step 1. Sketch a Normal curve.** Label the area in each tail as 0.025. Hence the area between the mean and each tail is 0.475

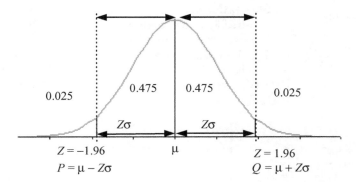

0.025   0.475   0.475   0.025

$Z\sigma$ $Z\sigma$

$Z = -1.96$ μ $Z = 1.96$
$P = \mu - Z\sigma$ $Q = \mu + Z\sigma$

Figure 5.17   Calculating the two limits for a Normal distribution within which 95 % of values fall.

Step 2. Since the area in each tail is 2.5 %. Look up the Normal tables for an area as close as possible to 0.025 (highlighted in Table 5.10). Follow the horizontal line to $Z = 1.9$ and the vertical line to 0.06. Hence $Z = 1.96$ for a tail area of 0.025.

    Alternatively, look up Table 5.11 percentage points of the Normal distribution. This table gives values of Z corresponding to selected tail areas (α) from $\alpha = 0.5$ to $\alpha = 0.000005$.

    From Table 5.11, when $\alpha = 0.025$, then $Z_\alpha = 1.96$.

Step 3. Since Z is the number of standard deviations from μ and each standard deviation is 10 seconds, then the point A is $1.96\,(10) = 19.6$ seconds greater than the mean.

Step 4. Finally:

Add the 19.6 to the mean to get the upper limit for x: $Q = 500 + 19.6 = 519.6$
Subtract 19.6 from the mean to get the lower limit for x: $P = 500 - 19.6 = 480.4$
In words: '95 % of email enquiries takes between 480.4 and 519.6 seconds to process'

(a) (ii) The calculation of the two limits that contain 99 % of all the values is the same as above except that there is 0.5 % of the area in each tail, therefore $Z = 2.5758$

$Q = 500 + 2.5758(10) = 525.758$
$P = 500 - 2.5758(10) = 477.4242$

In words: '99 % of email enquiries takes between 477.42 and 525.76 seconds to process'

(b) In words: since there is 94 % below x this leaves 6 % of the area in the upper tail, to find the value of Z corresponding to a tail area of 6 % you will have to use the Normal probability distribution tables: $Z = 1.56$. Hence $Q = 500 + 1.56(10) = 515.6$.

This result may be stated as '94 % of email enquiries are processed within 515.6 seconds'.

Table 5.11   Percentage point for the Normal probability distribution

Percentage point for the Normal probability distribution

α	Z_α	α	Z_α	α	Z_α	α	Z_α	α	Z_α
0.50	0.0000	0.050	1.6449	0.030	1.8808	0.020	2.0537	0.010	2.3263
0.45	0.1257	0.048	1.6646	0.029	1.8957	0.019	2.0749	0.009	2.3656
0.40	0.2533	0.046	1.6849	0.028	1.9110	0.018	2.0969	0.008	2.4089
0.35	0.3853	0.044	1.7060	0.027	1.9268	0.017	2.1201	0.007	2.4573
0.30	0.5244	0.042	1.7279	0.026	1.9431	0.016	2.1444	0.006	2.5121
0.25	0.6745	0.040	1.7507	0.025	1.9600	0.015	2.1701	0.005	2.5758
0.20	0.8416	0.038	1.7744	0.024	1.9774	0.014	2.1973	0.004	2.6521
0.15	1.0364	0.036	1.7991	0.023	1.9954	0.013	2.2262	0.003	2.7478
0.10	1.2816	0.034	1.8250	0.022	2.0141	0.012	2.2571	0.002	2.8782
0.05	1.6449	0.032	1.8522	0.021	2.0335	0.011	2.2904	0.001	3.0903

In general, if α represents the area in both tails of a Normal curve (that is, $\alpha/2$ is the area in each), then $(1 - \alpha)$ represents that area between the two tails, then

$(1 - \alpha)$ 100% of the area under the normal curve will fall between

$$\mu - Z_{\alpha/2} \text{ and } \mu + Z_{\alpha/2} \qquad (5.5)$$

For example 95 % of all values fall between $(\mu - 1.96\sigma)$ and $(\mu - 1.96\sigma)$

5.4.4 The standard Normal distribution

Step 2 in the method for calculating Normal probabilities required the calculation of the number of standard deviations (Z) between x and μ by the formula $Z = \dfrac{x - \mu}{\sigma}$.

The formula $Z = \dfrac{x - \mu}{\sigma}$ transforms (or rescales) any Normal probability distribution that has a mean μ and standard deviation σ to the standard Normal distribution.

The **standard Normal distribution** has a mean value, $\mu = 0$ and standard deviation, $\sigma = 1$. Hence the standard Normal tables (Z tables) are tables for the standard Normal distribution.

5.4.5 Sums or differences of Normal independent random variables

The distribution of the sums and differences of two normally distributed, independent random variables (NIRV) features in many applications, as illustrated in Worked Example 5.11 below. The distribution for the difference between NIRVs is fundamental to making inference about the difference between population means and proportions from Chapter 6 onwards. In this section the rules for writing down the distribution for the sum of two NIRVs or the difference between two NIRVs is stated, without proof and followed by a Worked Example.

Rule: Suppose the random variables:

X_1 is Normally distributed with mean μ_1 and standard deviation σ_1

X_2 is Normally distributed with mean μ_2 and standard deviation σ_2

> Then the sum of the random variables $X_1 + X_2$ is a random variable that is Normally distributed with mean: $\mu = (\mu_1 + \mu_2)$ and variance $\sigma^2 = \sigma_1^2 + \sigma_2^2$ (5.6)

i.e., standard deviation $\sigma = \sqrt{\sigma_1^2 + \sigma_2^2}$

> The difference between the random variables $X_1 - X_2$ is a random variable that is Normally distributed with mean: $\mu = (\mu_1 - \mu_2)$ and variance $\sigma^2 = \sigma_1^2 + \sigma_2^2$ (5.7)

Worked Example 5.11: Sums and differences of NIRV

A rural enterprise centre provides an advice service to small craft businesses on all aspects of the business, such as production methods marketing accounts etc. A candle making business incurs losses as a result of candle glasses being overfilled or under filled (hence poor quality) when coloured candle wax is poured into the glass tubes.

The wax is filled to heights that are normally distributed, with $\mu_1 = 25$ cm $\sigma_1 = 0.5$ cm, and the heights of the tubes are normally distributed, $\mu_2 = 26$ cm, $\sigma_2 = 0.4$ cm.

Calculate the probability that height of the gap between the top of the tube and the wax is between 0.5 cm and 1.50 cm.

Solution

The difference (gap) between the top of the wax and the top of the tube is Normally distributed. Let μ and σ be the mean and standard deviation of the 'gap'. Hence

$$\mu = (\mu_2 - \mu_1) = (26 - 25) = 1 \text{ cm}$$

$$\sigma^2 = \sigma_1^2 + \sigma_2^2 = (0.5)^2 + (0.4)^2 = 0.41. \text{ Hence } \sigma = \sqrt{\sigma_1^2 + \sigma_2^2} = 0.64$$

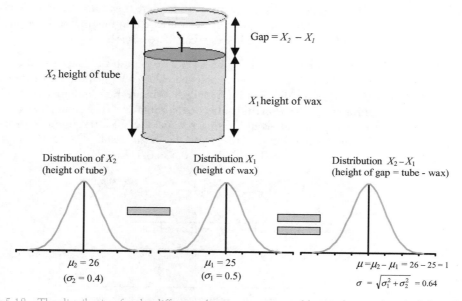

Gap $= X_2 - X_1$

X_2 height of tube

X_1 height of wax

Distribution of X_2
(height of tube)

Distribution X_1
(height of wax)

Distribution $X_2 - X_1$
(height of gap = tube - wax)

$\mu_2 = 26$
$(\sigma_2 = 0.4)$

$\mu_1 = 25$
$(\sigma_1 = 0.5)$

$\mu = \mu_2 - \mu_1 = 26 - 25 = 1$
$\sigma = \sqrt{\sigma_1^2 + \sigma_2^2} = 0.64$

Figure 5.18   The distribution for the difference between every possible pair $(x_1 - x_2)$ is also Normal.

With the probability distribution of the gap, X, known, the problem is reduced to calculating the usual Normal probabilities, $\mu = 1, \sigma = 0.64$.

Step 1: sketch the normal curve and mark in info given. Since the required area is symmetrical about the mean, calculate the area, A on the right hand side and multiply then answer by two.

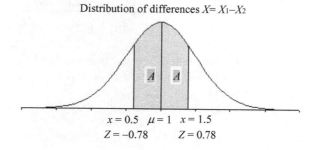

Distribution of differences $X = X_1 - X_2$

A A

$x = 0.5$ $\mu = 1$ $x = 1.5$
$Z = -0.78$ $Z = 0.78$

Step 2: $Z = \dfrac{x - \mu}{\sigma} = \dfrac{1.5 - 1.0}{0.64} = 0.78$

Step 3: Look up the normal tables; area in the tail is 0.2177.

Step 4: Hence $A = (0.5 - 0.2177) = 0.2823$. The required area $= 2 \times A = 0.5646$

$P(0.5 < \text{gap} < 1.5) = 0.5646$, the shaded area in the diagram above.

5.4.6 Normal probabilities in Excel

Excel offers four options on the Normal probability distribution as illustrated in Figure 5.19. These options will be explained with reference to Worked Example 5.8 where the transaction time is normally distributed with a mean time of 11 minutes and standard deviation of 3 minutes. In Excel, the probability that a transaction will take between 11 and 14 minutes is calculated as follows:

Option 1. Select NORMDIST as illustrated in Figure 5.19. This option produces the dialogue box given in Figure 5.20. In this dialogue box, enter the value of x (14), the mean (11) and the standard deviation (3) of the distribution. Finally, to obtain the cumulative probability for times up to 14 minutes, type 'true' in the bar headed '**cumulative**'. The cumulative probability is 0.84134774 is given in the dialogue box or, on clicking 'OK', the result is given in the cell in the worksheet where the cursor was placed. From this result, the required probability is calculated. $P(11 < x < 14) = 0.84134474 - 0.5 = 0.34134774$.

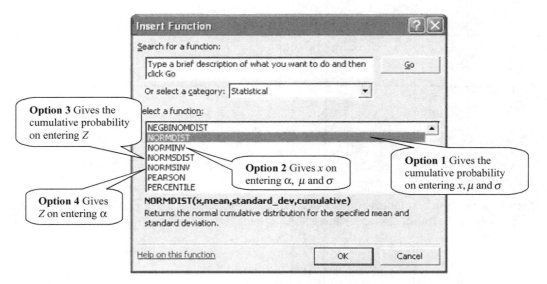

Figure 5.19   Normal probability functions in Excel: selecting 'NORMDIST'.

Option 2. This option is the inverse of option 1. To illustrate, enter the total area to the left of $x = 14$ (in Figure 5.20), along with the mean and standard deviation of the distribution in Figure 5.21. The value of x is returned.

Option 3. simply involves calculating, then entering the Z value to obtain the cumulative probability.

In **Option 4**, the Z value is returned when the total area to the left is entered.

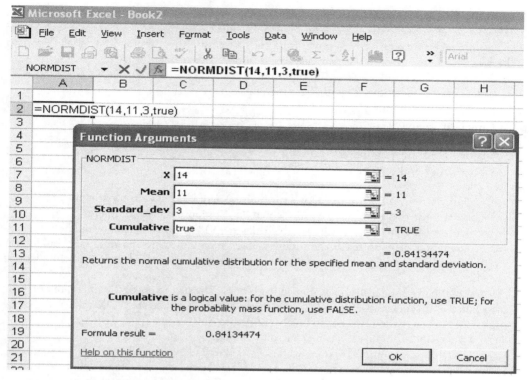

Figure 5.20   Dialogue box for 'NORMDIST'.

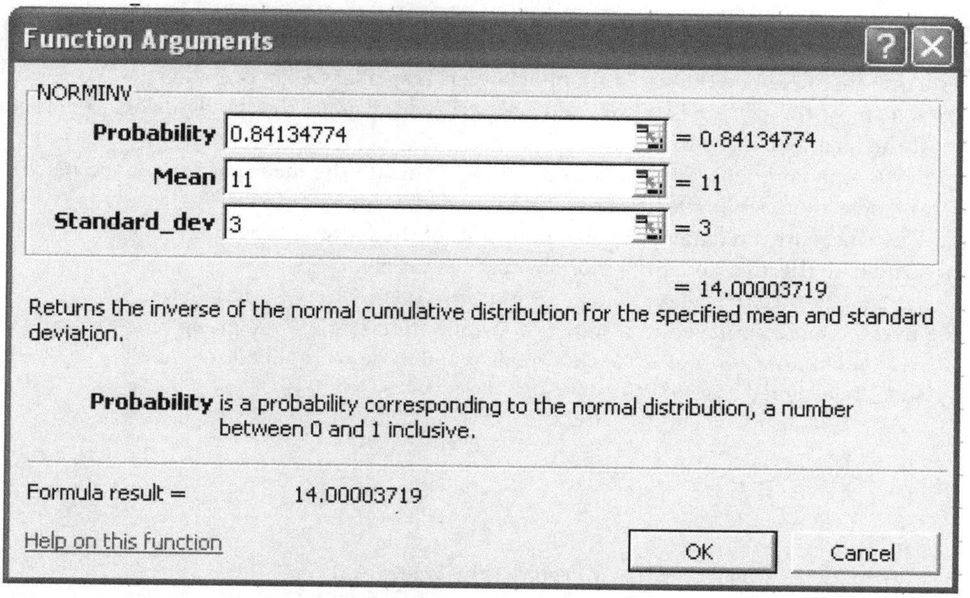

Figure 5.21   The value of x is returned for a given area to the left of x, μ and σ are entered.

Progress Exercises 5.4: Normal Probability Distribution

1. The time taken for a city bus to travel from the city centre to a suburban terminal, departing between 5.00 pm and 6.00 pm on weekdays, is normally distributed with a mean journey time of 36 minutes and standard deviation of six minutes.
    Calculate the percentage of journeys which take
    (a) longer that 42 minutes
    (b) less than 30 minutes
    (c) between 30 and 36 minutes
    (d) between 30 and 42 minutes
    (e) between 27 and 45 minutes.
2. (see question one).
    (a) Calculate the two time limits between which 90 % of journeys are complete.
    (b) Calculate the two time limits within which 95 % of journeys are complete.
    (c) Calculate the time limit within which the fastest 30 % of journeys are complete.
    (d) A commuter plans to reach the suburb at 6.15 p.m. Calculate the time he should allow for the bus journey if he is to be 95 % certain of arriving at the terminal at the planned time.
3. An automatic machine drills holes in wooden blocks. The diameters of the holes are normally distributed with a mean diameter of 550 mm and a standard deviation of 1.00 mm. What is the probability that a randomly selected block will have a hole with a diameter which is
    (a) (i) less than 550 mm (ii) between 548 and 552 mm
    (b) Calculate the two limits between which 99 % of diameters measure.
    Holes which are too small may be redrilled, but blocks with holes which are too large must be discarded. If a customer requires blocks with holes 549 mm and 551 mm calculate,
    (c) the percentage of blocks which must be redrilled
    (d) the percentage of blocks which must be discarded.
4. (a) See question three. Calculate the two limits between which 99 % of diameters measure. After maintenance the variation of diameters is smaller i.e., σ is smaller.
    (b) If no more than 5 % of blocks should be discarded, calculate the standard deviation required if the mean setting is to remain at 550 mm.
    (c) If the standard deviation is reduced to 0.50 calculate the mean setting for the machine if no more than 5 % of blocks should be discarded.
5. (a) Describe, with a diagram, the characteristics of the Normal curve.
    (b) Calculate the area under the Normal curve between
        (i) $Z = 1$ and $Z = 2.5$; (ii) $Z = -1$ and $Z = 1$; (iii) $Z = -1$ and $Z = 1.8$.
6. The average mark achieved in a statistics examination (for a large group of students) was 60 with a standard deviation of 12. Marks are assumed to be Normally distributed.
    If grades are assigned as follows:
        (i) $A \geq \mu + 2\sigma$
        (ii) $\mu + 2\sigma > B \geq \mu + \sigma$
        (iii) $\mu + \sigma > C \geq \mu - \sigma$
        (iv) $\mu - \sigma > D \geq \mu - 2\sigma$
        (v) $E \leq \mu - 2\sigma$
    (a) Calculate the range of marks for each of the grades A to E.
    (b) Calculate the percentage of students who are awarded each grade.
    (c) Calculate the number of students who are awarded each grade in a class of 124 students.

7. The heights of males in a given district are Normally distributed with a mean height of 70 inches and a standard deviation of 1.5 inches.
   (a) What percentage of males have heights (i) less than 72 inches; (ii) between 66 and 72 inches; (iii) more then 75 inches?
   (b) If 95 % of males measure between A and B inches, what are the values of A and B?
   (c) An anthropologist claims that only 10 % of males are more than C inches high. Calculate the value of C correct to two decimal places.

8. Calculated from data collected in a labour force survey, the average number of hours worked by women in paid employment is 32.4 hours per week with a standard deviation of 3.84 hours. Assume the number of hours worked is Normally distributed with the mean and standard deviation calculated above.
   (a) Calculate the proportion of women who work (i) more than 40 hours; (ii) between 35 and 40 hours; (iii) less than 30 hours per week.
   (b) If a researcher randomly selects a sample of 300 women in paid employment, how many of those selected work (i) less than 40 hours; (ii) between 35 and 40 hours; (iii) more than 30 hours per week?

9. (a) Sketch a normal curve marking in the values $Z = 0, 1, 2$. Indicate the number of standard deviations between each Z-value and the mean.
   (b) A manufacturer of a new brand of lithium battery claims that the mean life of the battery is 3,600 hours with a standard deviation of 250 hours.
      (i) What percentage of batteries will last for more than 3500 hours?
      (ii) What percentage of batteries will last more than 4000 hours?
      (iii) What percentage of batteries will last between 3500 and 4000 hours? If 800 batteries are supplied, how many should last between 3400 and 3800 hours?
      (iv) If it costs £ 10 to replace a battery under guarantee, how many hours should the manufacturer set as a minimum guaranteed life if the percentage of batteries to be replaced is 5 % or less?
      (v) Within what two limits should 95 % of all batteries last?

10. The average weight of adults in a population is Normally distributed with $\mu = 69$ kg and standard deviation $\sigma = 8.5$ kg.
    (a) What is the probability that a group of 12 adults weighs more than 900 kg?
    (b) Health and safety regulation require that the maximum load in certain lifts should not exceed 1,000 kg. (i) What is the probability that a group of 14 adults exceed 1000 kg? (ii) What is the maximum weight for 99 % of groups of (i) 14; (ii) 12? (iii) What is the maximum number allowed in the lift if health and safety regulations are not breached?

11. A company produces laptop computers tailored to customer specifications and guarantees to deliver the laptop within 48 hours. The times taken (in hours) for the four stages in the process from Order: production: QC and delivery are to process the order normally distributed with a means and standard deviations as follows:
    Order processed: $\mu = 4.3; \sigma = 3.6$ hours: **Production** $\mu = 19.5; \sigma = 6.2$ hours: **QC and other tests** $\mu = 9.6; \sigma = 2.2$ hours: **Delivery** $\mu = 6.4; \sigma = 1.8$ hours.
    (a) (i) Calculate the overall mean and variance for the time taken from receipt of order to delivery.
        (ii) Calculate the percentage of orders that are processed with the guaranteed time.
        (iii) Calculate the overall variance if 95 % of orders are processed within guarantee.

(b) Following a brainstorming session, the production team propose to reduce the production time to an average $\mu = 15.5$ hours; $\sigma = 2.5$ hours. Following this change (i) what percentage of orders should be delivered within guaranteed time?

(ii) If a batch of 58 orders, how many should be delivered within guaranteed time?

12. Snap-on rubber handles are made for aluminium walking poles. The external diameter of the poles are Normally distributed with $\mu = 2.12$ and $\sigma = 0.03$ and internal diameter of the handles are also Normally distributed with $\mu = 2.13$ and $\sigma = 0.04$. For a tight fit, the difference in diameters between the handles and poles should be should be within the range −0.15 to 0.18 cm. Calculate the percentage of pole/handle combinations that are within the tight fit range.

13. **Excel file: Cholesterol**   File: Cholesterol

Total blood cholesterol is the sum of LDL (bad cholesterol) and HDL (good cholesterol). The Excel file 'Cholesterol' records the LDL and HDL cholesterol for all adults in a small community in units of mmol/L. Individuals are considered high risk if their LDL exceeds 6.5 and low risk when LDL is less than 5.2; LDL levels between 5.2 and 6.5 are 'borderline'. On the other hand, when HDL is above 1.0, the individual is considered low risk and high risk if their HDL is less than 0.9; HDL levels between 0.9 and 1.0 are 'borderline'.

(a) Calculate the mean (μ) and variance (σ^2) for LDL and HDL cholesterol (see question 10(a) Progress Exercise 2.4).

(b) *Assuming normal a distribution* for levels of LDL, use the mean and variance calculated in (a) to calculate the probability that a randomly selected individual's level of LDL is in (i) the low risk range (ii) high risk range (iii) borderline range.

(c) Assuming a normal distribution for levels of HDL, use the mean and variance calculated in (a) to calculate the percentage of individuals whose levels of HDL is in (i) the low risk range; (ii) high risk range; (iii) borderline range.

(d) Assuming normal distributions for levels of LDL and HDL, use the means and variances calculated in (a) to calculate the percentage of individuals whose Total cholesterol (LDL + HDL) is (i) less than 5.2; (ii) greater than 6.5.

(e) Health professionals claim that one in five adults have total cholesterol levels above 6.47. Is this statement true for the adults in this community?

14. **Excel file**: 'Hours worked UK 2000'   File: Hours worked UK 2000

The Excel file gives the hours worked per week by males and females in the UK from data collected by the labour force survey spring 2000.

(a) Calculate the mean and variance for hours worked by (i) males (ii) females.
(See question 12, Progress Exercise 2.4).
Assuming Normal distributions (this assumption will be tested in later chapters).

(b) Use the results in (a) to calculate the probability that the weekly hours worked are (i) less than 20 for females (ii) between 35 and 45 hours for males.

(c) Write down the mean and variance for the difference between hours worked by males and females. Hence calculate the percentage of (i) males that work longer hours than females; (ii) females who work longer hours than males; (iii) males that work at least ten hours more than females per week.

15. **Excel file**: 'Savings 5'   File: Savings 5

To encourage school children to develop the habit of saving money five pilot groups are selected. Rather than spending pocket money on 'junk' food during the school day, the children are asked to 'save' the money instead. The amounts saved per week during the school year are summarised in the Excel file 'savings 5'.

(a) Calculate the mean and variance for each of the five pilot groups (see question 11(a)) Progress Exercise 2.4).
Assuming savings are normally distributed.

(b) Write down the mean and variance for the difference in amounts saved between group five and group one. What is the probability that a child in group one saves at least £ 10 more than those in group five?

(c) Write down the probability distribution for the total amount saved by all children in the combined five groups. What is the probability that a child will save (i) less than £ 25 (ii) between £ 25 and £ 30 (iii) more than £ 30.

5.5 Expected values (mathematical expectation)

Expected value. The expected value of a random variable is defined as its mean value. It could also be described as 'the value you would expect on average'. The formula for calculating expected values follows directly from the formula calculating the mean value for grouped data. The expected value is a very important mathematical tool in the theory of statistics. It is used in many applications such as the calculation of expected profits and losses and risk analysis. In this section expected values of parametric distributions such as the Binomial, Poisson and Normal will be used to explain how Binomial probabilities may be approximated by the Poisson and Normal.

5.5.1 Formula for mean and variance in terms of expected values

The mean value of a random variable

The formula for calculating the mean value or expected value for a random variable is based on (2.5), the mean for grouped data

$$\mu = \frac{\sum f_i x_i}{N} \qquad \text{where} \sum f_i = N$$

$$\mu = \sum \left(\frac{f_i}{N_i} x_i \right) \qquad \ldots \text{group } f_i \text{ with } N$$

$$\mu = \sum P(x_i)(x_i) \qquad \ldots \text{since } P(x_i) = \frac{f_i}{N} \text{ for large } N$$

$$\mu = \sum x_i P(x_i)$$

The expected value (average value) of the random variable, X is

$$\mu = E(X) = \sum x P(x) \tag{5.8}$$

The expected value of any function (or formula) of the random variable is defined as

$$E(g(X)) = \sum g(x) P(x) \tag{5.9}$$

where $g(X)$ is the function or formula in X; for example $E(X^2) = \sum x^2 P(x)$

The variance of a random variable is a measure of variation of its values about the mean, $E(X)$. The formula for variance in terms of expected values may be deduced from (2.14), the variance for grouped data as follows:

$$\text{variance} = \sum \left((x_i - \mu)^2 \frac{f_i}{N} \right)$$

$$= \sum \left((x_i - \mu)^2 P(x_i) \right) \ldots \text{ since } P(x_i) = \frac{f_i}{N} \text{ when } N \text{ is large}$$

$$V(X) = E(X - \mu)^2$$

Hence the formula for the variance of a random variable is:

$$V(X) = \sigma^2 = E(X - \mu)^2 \tag{5.10}$$

Or, an alternative formula is

$$V(X) = E(X^2) - [E(X)]^2 \text{ or } V(X) = E(X^2) - [\mu]^2 \tag{5.11}$$

Showing that formula (5.11) is a rearrangement of (5.10) will be left as an exercise.
See Appendix F for further material on expected values

Worked Example 5.12: Calculate expected value and variance of a random variable

An experiment involves throwing a fair die. Calculate the expected value and variance for the random variable X where X is the number that shows when the die is thrown. Give a verbal description of your answer.

Solution

To calculate expected values for a random variable we first require its probability distribution.

Step 1: list every possible outcome of the experiment and the corresponding probability (columns one and two, Table 5.12).

The Expected value is calculated in column three.

$$E(X) = \sum x P(x)$$

$$= 1 \left(\frac{1}{6} \right) + 2 \left(\frac{1}{6} \right) + 3 \left(\frac{1}{6} \right) + 4 \left(\frac{1}{6} \right) + 5 \left(\frac{1}{6} \right) + 6 \left(\frac{1}{6} \right)$$

$$= \left(\frac{1}{6} \right) (1 + 2 + 3 + 4 + 5 + 6)$$

$$= \left(\frac{1}{6} \right) (21)$$

$$= 3.5$$

In column 4, use the expected value, $E(X) = \mu = 3.5$ in the calculation of variance by formula (5.11).

Table 5.12 Calculation of expected values

x	$P(x)$	$xP(x)$	$(x-\mu)^2 P(x)$	$x^2 P(x)$
1	$\frac{1}{6}$	$1 \times \frac{1}{6} = 0.166667$	$(1-3.5)^2 = 1.041667$	$1^2 \times \frac{1}{6} = 0.166667$
2	$\frac{1}{6}$	$2 \times \frac{1}{6} = 0.333333$	$(2-3.5)^2 = 0.375000$	$2^2 \times \frac{1}{6} = 0.666667$
3	$\frac{1}{6}$	$3 \times \frac{1}{6} = 0.500000$	$(3-3.5)^2 = 0.041667$	$3^2 \times \frac{1}{6} = 1.500000$
4	$\frac{1}{6}$	$4 \times \frac{1}{6} = 0.666667$	$(4-3.5)^2 = 0.041667$	$4^2 \times \frac{1}{6} = 2.666667$
5	$\frac{1}{6}$	$5 \times \frac{1}{6} = 0.833333$	$(5-3.5)^2 = 0.375000$	$5^2 \times \frac{1}{6} = 4.166667$
6	$\frac{1}{6}$	$6 \times \frac{1}{6} = 1.000000$	$(6-3.5)^2 = 1.041667$	$6^2 \times \frac{1}{6} = 6.000000$
Totals	1	3.5	2.916667	15.16667

$$E(X) = \sum xP(x)$$

$$V(X) = \sum (x-\mu)^2 P(x)$$

$$E(X^2) = \sum x^2 P(x)$$

Alternatively, calculate variance by formula (5.11).

So, first calculate $E(X^2) = \sum x^2 P(x)$, column five: then, apply formula (5.11).

$$V(X) = E(X^2) - [E(X)]^2 = 15.16667 - [3.5]^2 = 2.91667$$

Verbal explanation: on each throw of a die numbers from one to six may turn up in any random order, such as 3, 4, 6, 3, 1, 3, 5, 4, 2 ...

Over a large number of trials, the average of these values is expected to be 3.5.

Expected values mean and variance for continuous random variables

For completeness, expressions for the mean and variance of continuous distributions are given:

The mean value:
$$E(X) = \sum xP(x) = \int_{-\infty}^{\infty} xf(x)dx \qquad (5.12)$$

The variance
$$V(X) = E(X-\mu)^2 = \int_{0}^{\infty} (x-\mu)^2 f(x)dx \qquad (5.13)$$

You will note that the summation ($\sum$) is replaced by the integral sign $\int$ and $P(x)$ is replaced by the probability density function $f(x)$.

The formulae for mean and variance of the Binomial and Poisson probability distributions may be derived using (5.8) and (5.10) or (5.11); the formulae for the mean and variance of the continuous Normal distribution is derived by equations (5.12) and (5.13). These formulae are given in Table 5.13.

The mean and variance for the Binomial, Poisson and Normal distributions are summarised in columns 2 and 3 in Table 5.13:

Table 5.13 Mean and variance of Binomial, Poisson and Normal probability distributions

Random Variable	Mean	Variance
Binomial (n,p)	np	npq
Poisson (λ)	λ	λ
Normal (μ, σ)	μ	σ^2

5.5.2 Approximating the Binomial and Poisson with the Normal

You may recall that Binomial probability calculations were tedious. An immediate application of expected values is the approximation of Binomial probabilities by (i) the Poisson and (ii) the Normal under the following conditions:

(i) If n is large and p is small ($n \geq 20$ and $p \leq 0.05$) the Binomial may be approximated with the Poisson. The approximation involves using the mean of the Binomial (np) as the mean of the Poisson (λ): then with $\lambda = np$, proceed to calculate probabilities with the Poisson formula.

$$^{n}C_{x}\, p^{x} q^{(n-x)} \approx \frac{(np)^{x} e^{(-np)}}{x!}$$

(ii) If $np \geq 5$ and $nq \geq 5$ the Binomial may be approximated with the Normal.
The Normal approximations are carried out by using the mean and variance of the Binomial as the mean and variance of the Normal. Hence $\mu = np$, $\sigma^2 = npq$. This is best illustrated by a Worked Example.

(iii) The Poisson may also be approximated by Normal for large $\lambda(\lambda > 5)$. The mean and variance of the approximating Normal are $\mu = \lambda$, $\sigma^2 = \lambda$.

Worked Example 5.13: Approximating the Binomial with a Poisson

Organisers of a foreign property fair know from past records that in a certain location, 2 % of attendees will eventually go on to purchase.

Use (i) the Binomial formula; (ii) the Poisson approximation to the Binomial to calculate the probability that in a sample of 50:

(a) exactly one will go on to purchase
(b) at most two will go on to purchase.

Solution

For the Binomial, $p = \dfrac{2}{100}$, hence $q = \dfrac{98}{100}$; $n = 50$

(a) To calculate the exact probability, use **the Binomial formula**: $P(x) = {}^nC_x p^x q^{n-x}$

Substitute $x = 1; n = 50; p = \dfrac{2}{100}, q = \dfrac{98}{100}$ into the Binomial formula (5.1).

$$P(1) = {}^{50}C_1 p^1 q^{50-1} = 50 \times \left(\frac{2}{100}\right)^1 \times \left(\frac{98}{100}\right)^{49} = 0.3716$$

Use the Poisson formula $P(x) = \dfrac{\lambda^x e^{-\lambda}}{x!}$ to approximate the Binomial probability
Substitute the mean of the Binomial for the mean of the Poisson:

$$np = \lambda \rightarrow 50 \times \frac{2}{100} = 1 = \lambda$$

Substituting $\lambda = 1$ into the Poisson formula gives $P(x) = \dfrac{(1)^x e^{-(1)}}{x!}$

Then substitute $x = 1$ to evaluate $P(1) = \dfrac{(1)^1 e^{-(1)}}{1!} = e^{-1} = 0.3679$

(evaluate e^{-1} on the calculator: $e^{-1} = 0.3679$)

Hence the Binomial probability $P(1) = 0.3716$ is approximated by the Poisson
$P(1) = 0.3679$

Comment: the exact and approximate probabilities are close.

(b) 'At most two' means '0 OR 1 OR 2', hence
$P(\text{at most two}) = P(0) + P(1) + P(2)$
Use the Binomial (to calculate the exact probability):

> ... substitute $p = 0.02, q = 0.98$

$$P(0) + P(1) + P(2) = {}^{50}C_0 p^0 q^{50-0} + P(1) + {}^{50}C_2 p^2 q^{50-2} t$$

$$= 1(0.02)^0 (0.98)^{50} + 0.3716 + 1225(0.02)^2 (0.98)^{48}$$

$$= 0.3642 + 0.3716 + 0.1858 = 0.9216$$

Use the Poisson with $\lambda = 1$ (to calculate the approximate probability)

$$P(0) + P(1) + P(2) = \frac{(1)^0 e^{-(1)}}{0!} + P(1) + \frac{(1)^2 e^{-(1)}}{2!}$$

> Use $P(1)$ from part (a)

$$= \frac{(1)e^{-(1)}}{1} + 0.3679 + \frac{(1)e^{-(1)}}{2}$$

$$= 0.3679 + 0.3679 + 0.1839 = 0.9197$$

Comment: the exact and approximate probabilities are close, as expected since the conditions for approximation: $n \geq 20$ (in fact $n = 50$) and $p \leq 0.05$ (in fact $p = 0.02$) are satisfied.

Worked Example 5.14: Approximating the Binomial with a Normal

---

Organisers of a foreign property fair know from past records that in a different location, 12 % of attendees will eventually go on to purchase a property.

  Calculate the probability that in a sample of 50 attendees that:

(a) Exactly 3 will go on to purchase, using (i) the Binomial; (ii) the Normal approximation to the Binomial.

(b) Between 4 and 8 will go on to purchase using (i) the Binomial; (ii) the Normal approximation to the Binomial.

Solution

(a) (i) **Calculate the exact probability by the Binomial formula:** $P(x) = {^nC_x} p^x q^{n-x}$

  Hence substitute $x = 3; n = 50; p = \dfrac{12}{100}, q = \dfrac{88}{100}$ into the Binomial formula (5.1)

$$P(x = 3) = {^{50}C_3}\, p^3 q^{50-3} = 19600 \times \left(\frac{12}{100}\right)^3 \times \left(\frac{88}{100}\right)^{47} = 0.0833.$$

(ii) **Calculate the Normal approximation to the Binomial** by using the mean and variance of the Binomial as the mean and variance of the Normal

  Use the Binomial mean $= np$ as the mean for the Normal, hence

  $\mu = np = 50(0.12) = 6$

  Use the Binomial variance $= npq$ as the variance for the Normal, hence

  $\sigma^2 = npq = 50(0.12)(0.88) = 5.28$

  Hence $\sigma = \sqrt{npq} = 2.2978$

  Proceed with the Normal probability calculations with $\mu = 6$ and $\sigma = 2.2978$

  Step 1: In Figure 5.22, the approximation to $x = 3$ is the area under the Normal curve between $x = 2.5$ and $x = 3.5$.

  Steps 2 and 3: calculate Z: look up area in tables

$x = 2.5$	$x = 3.5$
$Z = \dfrac{x - \mu}{\sigma} = \dfrac{2.5 - 6}{2.2978} = -1.52$	$Z = \dfrac{x - \mu}{\sigma} = \dfrac{3.5 - 6}{2.2978} = -1.09$
From the tables, tail area for $Z = 1.52$ is 0.0643	From the tables, tail area for $Z = 1.09$ is 0.1379

  Step 4: The required area (shaded) under the curve is calculated by subtracting the smaller area from the larger area:

  Hence, $P(2.5 < x < 3.5) = 0.1379 - 0.0643 = 0.0736$

(b) (i) **Use the Binomial formula:** $P(x) = {^nC_x} p^x q^{n-x}$

  $n = 50; p = \dfrac{12}{100}, q = \dfrac{88}{100}$

  Using the calculator (or Excel):

$$P(4 \le x \le 8) = P(4) + P(5) + P(6) + P(7) + P(8)$$

$$= 0.1334 + 0.1674 + 0.1712 + 0.1467 + 0.0175 = 0.7263$$

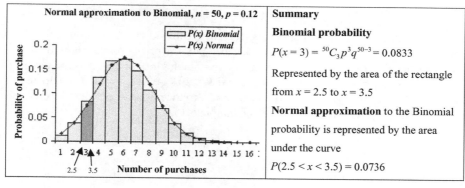

Figure 5.22   Normal approximation to $P(x = 3)$.

(ii) **To approximation the Binomial** with the Normal, use $\mu = 6$ and $\sigma = 2.2978$ as in part (a)(ii) above.

> **Step 1:** In Figure 5.23, the required area is the shaded under the Normal curve between $x = 3.5$ and $x = 8.5$

Steps 2 and 3: calculate Z: look up area in tables

$x = 3.5$	$x = 8.5$
$Z = \dfrac{x - \mu}{\sigma} = \dfrac{3.5 - 6}{2.2978} = -1.09$	$Z = \dfrac{x - \mu}{\sigma} = \dfrac{8.5 - 6}{2.2978} = 1.09$
From the tables, tail area for	From the tables, tail area for
$Z = 1.09$ is 0.1379.	$Z = 1.09$ is 0.1379
For $Z = -1.09$, this is the area in the lower tail	For $Z = 1.09$, this is the area in the upper tail

> **Step 4:** The required area (shaded) under the curve is calculated by subtracting the sum of the tail areas from one:
> Hence, $P(3.5 < x < 8.5) = 1 - 0.1379 - 01379 = 0.7242$

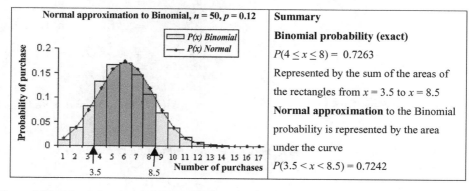

Figure 5.23   Normal approximation to $P(4 \le x \le 8)$.

5.5.3 Mean and variance of linear functions of random variables

Occasionally it is necessary to convert statistics (such as mean and variance) from one set of units to another. For example, if the mean and standard deviation for the price of a wine is given as € 8.50 and € 1.80 per bottle another retailer may require the same statistics in £ or $. If X is the price in € then price in pounds sterling, $Y = 1.5625X$ (assume £ 1 = € 1.5625). So, we say that Y is a linear function of the random variable, X. The formulae for the mean and variance linear functions of random variables are derived by using Expected values in Worked Example 5.15.

Worked Example 5.15: Mean and variance of the random variable *aX*

(a) Show that the mean and variance of the linear function of the random variable $Y = aX$ is $E(aX) = aE(X)$ and $V(aX) = a^2V(X)$ respectively.
(b) If the mean and variance of the price of wine (per bottle) is wine is given as € 8.50 and € 1.80 per bottle. Write down the mean and variance for the price in pounds sterling if € 1 = £ 0.64.

Solution

(a) From equations (5.8) and (5.11), the mean and variance for the random variable X is $E(X) = \sum xP(x)$. $V(X) = E(X^2) - [E(X)]^2$ respectively.

   If $Y = aX$, then the mean and variance for the random variable Y is derived from formulae (5.8) and (5.11) as follows:

(i) The mean value of $Y = aX$ The variance of $Y = aX$

If $Y = aX$, then $E(Y) = E(aX)$ If $Y = aX$, then $V(Y) = V(aX)$
Hence, by (5.8)   Hence

$$E(aX) = \sum (ax) \times P(x)$$
$$= \sum axP(x)$$
$$= a \sum xP(x)$$
$$= aE(x)$$

$$V(aX) = E\{(aX)^2\} - [E(aX)]^2$$
$$= E(a^2X^2) - [aE(X)]^2$$
$$= a^2E(X^2) - a^2[E(X)]^2$$
$$= a^2\{E(X^2) - [E(X)]^2\}$$
$$= a^2V(X)$$

$$E(Y) = E(aX) = aE(X) \qquad \textbf{(5.14)}$$
$$V(Y) = V(aX) = a^2V(X) \qquad \textbf{(5.15)}$$

(b) If € 1 = £ 0.64 then € X = £ 0.64X, hence the price in sterling is $Y = 0.64X$.
   $E(X) = $ € 8.5 and $V(X) = $ € 1.8.

The mean price (£) is $E(Y) = E(0.64X) = 0.64E(X) = 0.64(8.5) = £ 5.44$.
The variance of prices (£) is $V(Y) = V(0.64X) = (0.64)^2V(X) = (0.64)^2(1.8) = £ 0.7373$.

The mean and variance of the linear function $Y = aX + b$ of a random variable X.

It will be left as an exercise for the reader to use the expected value technique above to prove the following.

$$E(aX + b) = aE(X) + b \qquad (5.16)$$
$$V(aX + b) = a^2 V(X) \qquad (5.17)$$

Progress Exercises 5.5: Expected values and approximating distributions

1. Derive the formula for the expected value of a random variable from the formula for the mean value of grouped data.
2. See Worked Example 5.2. Calculate the expected number of red cards when three cards are selected, with replacement from a box containing six red and four white cards.
3. Derive the formula for the variance of a random variable from the formula for the mean value of grouped data.
4. See Progress Exercises 5.1, question six: calculate the expected value and variance for the rejects in a sample of 100 (assume a value of 7 for '6 or more') using formulae (5.8) and (5.11). Compare the answers with the means and variance calculated by formulae (2.5) and (2.18) in Chapter 2.
5. See Progress Exercises 5.1, question seven: calculate the expected value and variance for the number of part-time hours worked by students. Give a verbal explanation of your results.
6. (a) Under what conditions may Binomial probabilities be approximated by (i) Poisson; (ii) Normal?
   (b) A batch of key rings is known to contain 4 % defectives.
   Use (a) the Binomial; (b) the Poisson approximation to the Binomial to calculate the probability that a sample of (i) ten contains one defective; (ii) 50 contains two defectives; (iii) 200 contains six defectives; (iv) 200 contains at most 12 defectives.
7. 12 % of the population are known to be carriers of a common virus. Calculate the probability that between 20 and 50 individuals in a group of 400 are carriers. Use the Normal approximation to the Binomial.
8. Research on behalf of insurance companies has established 5 % of cars have 'worn' tyres.
   (a) State the mean and variance for the number of cars that have 'worn' tyres if (i) 20 cars; (ii) 250 cars are randomly checked?
   (b) Calculate the probability that (i) more than two cars in a random sample of 20 (ii) at least 15 cars in a sample of 250 have 'worn' tyres
9. According to the national roads authority, there are two pot holes per mile on tertiary roads.
   (a) On a stretch of five miles, what is the probability that there are (i) less than five (ii) between five and ten (iii) more than ten pot holes?
   (b) On a stretch of 25 miles, what is the probability that there are (i) less than 30 (ii) between 40 and 50 (iii) more than 50 pot holes? (iv) What is the variance of the number of pot holes?
10. (a) Use the definition of expected value and variance to prove the following:
    (i) $E(aX) = aE(X)$; (ii) $E(b) = b$; (iii) $E(aX + b) = aE(X) + b$.

(a) Weights of cabin baggage is Normally distributed with a mean weight of 7.5 kg and standard deviation of 1.85 kg. (i) Write down the mean and variance of weights in lbs (1 kg = 2.24 lbs) (ii) Calculate the percentage of bags that weight more than 10 kg (iii) Calculate the percentage of bags that weigh between 10 and 20 lbs.

11. (a) Use the definition of expected value and variance to prove the following:
    (i) $V(aX) = a^2 V(X)$; (ii) $V(b) = 0$; (iii) $V(aX + b) = a^2 V(X)$.
    (b) A medication is supplied in bottles from three manufacturing plants. Quality assurance from each of the plants report the variance in bottle weights as (i) 9.25 gm for plant A; (ii) 0.0145 ounces for plant B; and (iii) 1242.34 mg for plant C.
    If 1 ounce = 28 gm and 1 gm = 1000 mg, compare the variation in bottle weights.

12. A holiday resort claims that the temperature during July and August is normally distributed with mean, $\mu = 28°C$ with a standard deviation $\sigma = 1.8°C$.
    (a) Calculate the mean and variance of temperatures in Fahrenheit, if the temperature in Fahrenheit (Y) by the equation $Y = 32 + \dfrac{9}{5}X$, where X is the temperature in °C.
    (b) Calculate the probability that the temperature in July and August will be (i) less than 27°C; (ii) more than 80°F (iii) between 27°C and 30°C (iv) between 75°F and 85°F.

13. The variable cost (VC) of a good is Normally distributed with a mean cost of 25 per unit and standard cost of 2.4 per unit. If the fixed cost (FC) is 800, then total cost (TC) is given by the equation $TC = FC + VC$.
    (a) write down the probability distribution for total cost.
    (b) calculate the probability that total cost will be (i) between 820 and 825 (ii) greater than 825.

Summary and overview of probability distributions

Probability distributions give us rules or formulae for calculating probabilities in certain well defined circumstances.

In Chapter 4, **an empirical probability distribution** was defined as a list of every outcome of a random experiment with the corresponding probability. Examples were given on the calculations of joint, marginal and conditional probability distributions.

In Chapter 5 it was shown that all probability distributions have the following properties:

1. The sum of all the probabilities is one.
2. The total area under the probability histogram or curve is one.

5.1 Introduction: probability distributions and random variables

This section introduces the terminology and concepts that are used throughout the study of probability distributions.

5.2 Discrete probability distributions: Binomial

The formula for calculating Binomial probabilities (Binomial formula) is

$$P(x) = {}^nC_x p^x q^{n-x}$$

(5.1)

The probability of x successes in a sample size n. Probabilities calculated by (5.1) are valid provided the following conditions (**Binomial assumptions**) are satisfied:

1. There are only two mutually exclusive outcomes.
2. The probability of success is the same for each trial.
3. The trials are independent.
4. The number of trials (or sample size) is finite.

Binomial probabilities may be read from (i) various tables,; usually cumulative tables; (ii) Excel.

5.3   Discrete probability distributions: Poisson

The Poisson formula is used to calculate probabilities for events that occur randomly, but at a uniform rate. **The Poisson probability distribution** function is given by the formula

$$P(x) = \frac{\lambda^x e^{-\lambda}}{x!} \tag{5.2}$$

where λ is the mean rate at which X occurs in a given interval. The Poisson is a probability distribution in its own right, but may be used to approximate Binomial probabilities when $n > 30$ and $p < 0.10$. Poisson probabilities may be read from (i) tables, usually cumulative tables (ii) Excel

5.4   Continuous probability distribution: Normal

The Normal probability density function is given by the formula

$$f(x) = \frac{1}{\sigma\sqrt{2\pi}} e^{-\frac{1}{2}\left(\frac{x-\mu}{\sigma}\right)^2} \tag{5.3}$$

for values of a continuous random variable, $X: -\infty < x < \infty$.

Standard Normal Distribution

The formula $Z = \dfrac{x - \mu}{\sigma}$ transforms (or rescales) any Normal distribution that has a mean μ and standard deviation σ to the **standard Normal distribution**, that has a mean $\mu = 0$ and standard deviation $\sigma = 1$. Hence one set of tables may be used to look up areas (probabilities) for any Normal distribution. **Another application of the Normal** is the calculation of the limits that contain a given percentage of all values under the Normal curve. In general, if α represents the area in both tails of a Normal distribution (so $\alpha/2$ is the area in each), then $(1 - \alpha)$ represents that area between the tails. Hence $(1 - \alpha)100\%$ of all values under the normal curve will fall between

$$A = \mu - Z_{\alpha/2} \quad \text{and} \quad B = \mu + Z_{\alpha/2} \tag{5.5}$$

For example 95 % of all values fall between $(\mu - 1.96\sigma)$ and $(\mu - 1.96\sigma)$.

Sums or differences of Normal independent random variables (NIRV)

Then the sum of the random variables $X_1 + X_2$ is a random variable that is Normally distributed with mean: $\mu = (\mu_1 + \mu_2)$ and variance $\sigma^2 = \sigma_1^2 + \sigma_2^2$ $\tag{5.6}$

The difference between the random variables $X_1 - X_2$ is a random variable that is Normally distributed with mean: $\mu = (\mu_1 - \mu_2)$ and variance $\sigma^2 = \sigma_1^2 + \sigma_2^2$ (5.7) The differences between NIRVs will be feature again in chapter 7, 8 and 9 in the study of confidence intervals and hypothesis for differences between means and proportion.

5.5 Expected values

Expected values are an important mathematical tool used in many aspects of statistics. **The Expected value is the average value of a random variable, X** and is defined as

$$\mu = E(X) = \sum x P(x) \tag{5.8}$$

The variance of a random variable may also be expressed in terms of expected values

$$V(X) = \sigma^2 = E(X - \mu)^2 \tag{5.10}$$

Or, an alternative formula is

$$V(X) = E(X^2) - [E(X)]^2 \text{ or } V(X) = E(X^2) - [\mu]^2 \tag{5.11}$$

These definitions are used to derive the formulae for the mean value and variance of the parametric distributions such as the **Binomial, Poisson and Normal:**

Random Variable	Mean	Variance
Binomial (n,p)	np	npq
Poisson (λ)	λ	λ
Normal (μ, σ)	μ	σ^2

The approximation of the Binomial by the Poisson (when $n > 30$, $p < 0.1$) and Normal (when $np > 5$ and $nq > 5$) is made possible by using the mean and variance of the Binomial as the mean and variance of the approximating distribution. The Poisson may also be approximated by the Normal for large λ ($\lambda > 5$).

SAMPLING DISTRIBUTIONS FOR MEANS AND PROPORTIONS

6

6.1 Statistical inference and the sampling distribution of the mean
6.2 Sampling distribution of proportions
6.3 Some desirable properties of estimators

Chapter Objectives

Having carefully studied this chapter and completed the exercises you should be able to do the following

- Know the symbols for population parameters and the corresponding sample statistics
- Calculate and plot the sampling distribution of the mean for small populations
- State and apply the Central Limit Theorem
- Calculate probabilities for $\bar{x}$ for large samples
- Calculate the two limits that contain a given percentage of sample means.
- Calculate a means control chart and comment on the stability of the process mean
- Calculate and plot the sampling distribution of proportions for large n
- Calculate the two limits that contain a given percentage of sample proportions
- Plot a p-chart and comment on the stability of the process proportion of successes

---

6.1 Statistical inference and the sampling distribution of the mean

Inference about population parameters

Recall that the definition of statistics given in Collins Concise Dictionary was given in the introduction: *. . . a science concerned with the collection, classification and interpretation of quantitative data **and with the application of probability theory to the analysis and estimation of population parameters.***

Now that you have studied the 'collection, classification and interpretation of quantitative data' and 'probability theory' in Chapters 1 to 5, you have a solid foundation for the study of the really interesting part of statistics—statistical inference which is based on the **analysis and estimation of population parameters'**.

Parameters such as the population mean will be estimated by a sample mean, calculated from a *random* sample. At the outset, one might expect the sample mean to be close to the population mean but, in some cases, there will be a very large difference between them. In this chapter we will examine the variability in the values of the sample means (and proportions). We will show that under certain conditions, sample means are, in fact, Normally distributed and that the mean of all the sample means is the same as the population mean. Since the distribution of sample means in Normally distributed, we can calculate probabilities about the sample means and calculate intervals that contain a given percentage of all sample means. The confidence intervals and test hypothesis in later chapters are based on the probability distributions of sample statistics (such as sample means).

6.1.1   Sampling distributions of means and proportions

For the remainder of the text we will study statistical inference. We will use sample statistics to make inference about population parameters. Therefore it is necessary to use different symbols to represent sample characteristics and the corresponding population characteristics. In general the population characteristics will be represented by letters from the Greek alphabet while sample characteristics will be represented by the corresponding letters from the English alphabet. The symbols for mean, variance and proportions are listed below in Table 6.1. Some reminders are also given on definitions and other essential background and terminology.

Reminders

A statistic is a sample characteristic, such as the sample mean $\bar{x}$, the sample variance s^2 etc.
A parameter is a population characteristic such as the population mean and variance; μ and σ^2.

Symbols for sample and population characteristics

The symbols that represent population characteristics and the corresponding sample characteristics are listed in Table 6.1:

Table 6.1   Symbols for population and sample characteristics

	Population characteristic	Sample characteristic
Size	N = number in population	n = number in sample
Mean	μ: where $\mu = \dfrac{\sum x}{N}$	$\bar{x}$ where $\bar{x} = \dfrac{\sum x}{n}$
Standard deviation	σ where $\sigma = \sqrt{\dfrac{\sum (x - \mu)^2}{N}}$	s where $s = \sqrt{\dfrac{\sum (x - \bar{x})^2}{n - 1}}$
Proportion	π where $\pi = \dfrac{X}{N}$	p where $p = \dfrac{x}{n}$

Terminology

In statistical inference, the mean and variance calculated from sample data are used to estimate the population mean and variance, hence

$\bar{x}$ is called a **point estimator** of μ

s^2 is called a **point estimator** of σ^2

> Point estimators and point estimates

But individual values of $\bar{x}$ or s are called **a point estimate** of μ and σ respectively.

Symbols on calculators

The mean value is calculated by the same formula whether the data is a sample or a population. Hence calculators show only symbol for the mean, either $\bar{x}$ or μ.

If you recall in Chapter 2, it was stated that the formulae for standard deviation calculated from population and sample data were different, hence the symbols are different. The calculator gives standard deviations as follows:

1. $\sigma = \sqrt{\dfrac{\sum (x_i - \mu)^2}{N}}$ ... (2.12) is the standard deviation calculated from population data

On most calculators the symbol is σ, σ_x or $n\sigma_x$.

2. $s = \sqrt{\dfrac{\sum\limits_{i=1}^{n} (x_i - \bar{x})^2}{n - 1}}$ ... (2.13) is the standard deviation calculated from sample data

On the most calculators the symbol is s, s_x or $n(-1)\sigma_x$.

Note: Explanations for the different formulae for standard deviation is given in section 6.4.

6.1.2 Point estimates and the sampling distribution of the mean

One might easily assume that any sample mean would be close to the population mean. However, as mentioned above, this is not always the case. Consider a very small population of five numbers.

Population:	A	B	C	D	E
	3	1	5	6	2

> Check these answers on your calculator!

The population mean, $\mu = 3.4$, standard deviation, $\sigma = 1.8547$

From this population of five, list **every possible** sample of size 2 (column 1) and calculate the mean for each sample (column 2). Repeat this for samples of size 3 (columns 3 and 4). See Table 6.2. For samples size 2, calculate the mean value and the standard deviation of the 10 sample means. For samples size 3, calculate the mean value and the standard deviation of the 10 sample means (use your calculator for these calculations).

Note:

(i) the mean of **all** the sample means, written as $\mu_{\bar{x}}$

(ii) the standard deviation of **all** the sample means, written as $\sigma_{\bar{x}}$

> Note the symbols:
> $\mu_{\bar{x}}$ = mean; $\sigma_{\bar{x}}$ = standard deviation, of ALL sample means

Note: $\sigma_{\bar{x}}$ is the 'population' standard deviation calculated from the 10 samples' mean values:

$$\sigma_{\bar{x}} = \sqrt{\frac{\sum\limits_{i=1}^{10} (\bar{x}_1 - \mu_{\bar{x}})^2 + (\bar{x}_2 - \mu_{\bar{x}})^2 + \ldots (\bar{x}_{10} - \mu_{\bar{x}})^2}{10}}$$

The calculations are given towards the end of Table 6.2:

Table 6.2   A list of every possible sample (a) size 2 (b) size 3 from a population of 5

Sample of size 2	(a) Sample mean	Sample of size 3	(b) Sample mean
AB = 3, 1	2.0	ABC = 3, 1, 5	3.00
AC = 3, 5	4.0	ABD = 3, 5, 6	4.67
AD = 3, 6	4.5	ABE = 3, 1, 2	2.00
AE = 3, 2	2.5	ACD = 3, 5, 6	3.67
BC = 1, 5	3.0	ACE = 3, 5, 2	3.33
BD = 1, 6	3.5	ADE = 3, 6, 2	3.67
BE = 1, 2	1.5	BCD = 1, 5, 6	4.00
CD = 5, 6	5.5	BCE = 1, 5, 2	2.67
CE = 5, 2	3.5	BDE = 1, 6, 2	3.00
DE = 6, 2	4.0	CDE = 5, 6, 2	4.33

Mean of **all** sample means $\mu_{\bar{x}}$	$\mu_{\bar{x}} = 3.4$ for $n = 2$		$\mu_{\bar{x}} = 3.4$ for $n = 3$
Standard deviation of **all** sample means	$\sigma_{\bar{x}} = 1.1358$ for $n = 2$		$\sigma_{\bar{x}} = 0.7572$ for $n = 3$
Standard deviation of means by formula	$\sigma_{\bar{x}} = \dfrac{\sigma}{\sqrt{n}}\sqrt{\dfrac{N-n}{N-1}}$ $= \dfrac{1.8547}{\sqrt{2}}\sqrt{\dfrac{5-2}{5-1}} = 1.1358$		$\sigma_{\bar{x}} = \dfrac{1.8547}{\sqrt{3}}\sqrt{\dfrac{5-3}{5-1}}$ $= 0.7572$

From Table 6.2, note the following:

1. **The variability in the values of $\bar{x}$** (population mean, $\mu = 3.4$).
   For samples of size 2, one sample mean is as low as 1.5 and another as high as 5.5. For a sample of size 3, the situation is only slightly better, with sample means ranging from 2.00 to 4.33. The variability is reflected in the variance of all the means. For $n = 2$, $\sigma_{\bar{x}} = 1.1358$ while $\sigma_{\bar{x}} = 0.7572$ for $n = 3$.
   It is obvious that an individual sample mean (a point estimate) can be a very poor estimate of the population mean. This is the difficulty with point estimators: there is no way of knowing whether an individual sample mean is close to the population mean or not.
2. The mean of all the sample means is the same as the population mean: $\mu_{\bar{x}} = \mu$.
3. The standard deviation of all the sample means, $\sigma_{\bar{x}}$ decreases as n increases.

In this example, the standard deviation of all the sample means, $\sigma_{\bar{x}}$ was calculated from the 10 sample means directly. It is also possible to calculate $\sigma_{\bar{x}}$ by the formula (6.1) from the population standard deviation σ and population size (N) and sample size (n) as in the last row of Table 6.3.

$$\sigma_{\bar{x}} = \frac{\sigma}{\sqrt{n}}\sqrt{\frac{N-n}{N-1}} \qquad (6.1)$$

$$\sqrt{\frac{N-n}{N-1}} \text{ is called the finite population correction factor} \qquad (6.2)$$

The factor $\sqrt{\frac{N-n}{N-1}}$ must be used in the calculation of $\sigma_{\bar{x}}$ by formula (6.1) when sampling from finite populations. However, if N is large then $\sqrt{\frac{N-n}{N-1}} \cong 1$ hence $\sigma_{\bar{x}} = \frac{\sigma}{\sqrt{n}}$.

As a rule of thumb, when the sample size is less than 5 % of the population size ($n \leq 0.05N$), it is not necessary to use the correction factor.

Terminology

The standard deviation of all sample means is called the 'standard error of the mean'.

$$\sigma_{\bar{x}} = \frac{\sigma}{\sqrt{n}} \text{ is called the \textbf{standard error of the mean}} \qquad (6.3)$$

The difference between μ and its point estimate, $\bar{x}$ is called **Sampling error**.
Note: if all point estimates were exactly the same as the population mean, there would be no sampling error and the standard error would be zero!

Sampling distribution of the mean

Recall the definition of a probability distribution: the list of every possible outcome with the corresponding probability. The probability distribution for sample means is the list of every possible sample mean with the corresponding probability. The list of every possible sample mean is given in Table 6.2. The probability of each sample being selected is 0.10.
To plot the probability distribution of sample means for samples of size 2 in Table 6.2, proceed as follows.

Step 1. Summarise the data above in a frequency distribution table, Table 6.3. Calculate the relative frequencies: these are the probabilities that a sample mean value will be within that interval.

Table 6.3   Frequency distribution table for samples of size 2 from a population, $N = 5$

Intervals	$1.5 \leq \bar{x} < 3.0$	$3.0 \leq \bar{x} < 4.5$	$4.5 \leq \bar{x} < 6$
Frequency	3	5	2
Relative frequency	0.3	0.5	0.2

Step 2. Plot a relative frequency (or probability histogram) as in Figure 6.1.

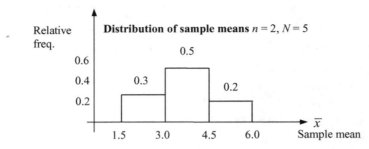

Figure 6.1   The distribution of sample means for samples size 2 from a population, $N = 5$.

The probability histogram in Figure 6.1 reveals that even for this very small population and sample size, the probability distribution is tending towards the Normal bell shape. It can be proved (mathematically) that the probability distribution of the sample means for samples of size $n \geq 30$, selected from any population (whose mean, μ and variance σ^2 are known), approaches a Normal distribution, with mean μ and standard deviation, $\sigma_{\overline{x}} = \dfrac{\sigma}{\sqrt{n}}$.

This is called the Central Limit Theorem.

$$\text{The distribution of sample means } \overline{x} \sim N\left(\mu, \frac{\sigma}{\sqrt{n}}\right) \text{ for sample sizes } n \geq 30 \qquad (6.4)$$

(where '$\sim$' means 'is distributed as')

In addition, it can also be proved (mathematically) that the Central Limit Theorem applies for small samples selected from Normal populations when the population variance, σ^2 is known.

$$\overline{x} \sim N\left(\mu, \frac{\sigma}{\sqrt{n}}\right) \text{ for samples of } any \text{ size from a Normal distribution, known variance } \sigma^2$$

A direct application of the Central Limit Theorem, when μ and σ^2 are known is the

 (i) calculation of probabilities regarding sample means;
(ii) calculation of the limits that contain various percentages of sample means.

See Worked Example 6.1.

In later chapters, when μ is not known, the Central Limit Theorem lays the foundation for constructing interval estimates for μ (confidence intervals) and testing hypothesis about μ.

Worked Example 6.1: Calculate probabilities for $\overline{x}$

An importer of Herbs and Spices claims that the average weight of packets of Saffron is 20 gms. However, packets are actually filled to an average weight, $\mu = 19.5$ gm, standard deviation, $\sigma = 1.8$ gm. A random sample of 36 packets is selected, calculate:

(a) the probability that the average weight is 20 gm or more;
(b) the two limits within which 95 % of all packets weight;

(c) the two limits within which 95 % of all average weights fall ($n = 36$).

(d) If the size of the random sample was 16 instead of 36 how would this affect the results in (a), (b) and (c)? State any assumptions made.

Solution

(a) The question asks for the probability: $P(\bar{x} \geq 20)$. To calculate probabilities about $\bar{x}$, it is necessary to use the distribution of $\bar{x}$. The population parameters, $\mu = 19.5$ gm and $\sigma = 1.8$ are given. Since sample size is greater then 30, the distribution of $\bar{x}$ is Normal.

Hence, the mean and standard error of the distribution of means is

$$\mu_{\bar{x}} = \mu = 19.5 \text{ and } \sigma_{\bar{x}} = \frac{\sigma}{\sqrt{n}} = \frac{1.8}{\sqrt{36}} = 0.3$$

Proceed to calculate Normal probabilities as outlined in Chapter 5, section 5.4.

Step 1. Draw the curve, mark in information given; see Figure 6.2.

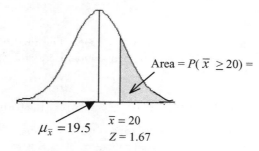

Figure 6.2   Calculating probabilities for sample means.

Step 2. Calculate Z

$$\text{For } \bar{x} = 20, Z = \frac{\bar{x} - \mu_{\bar{x}}}{\sigma_{\bar{x}}} = \frac{20 - 19.5}{0.3} = 1.67$$

Step 3. From Tables, area in tail $= 0.0457$.

Step 4. The required area (probability) is the area in the tail.
Hence $P(\bar{x} \geq 20\,\text{gm}) = 0.0475$.

(b) *Read this question carefully:* part (b) of the question asks about individual weights NOT average weights. The method for calculating the two limits within which 95 % of weights fall is given in Worked Example 5.8(b).

Step 1. Sketch a Normal curve. Label the area in each tail as 0.025. Hence the area between the mean and each tail is 0.475.

Step 2. Look up the Tables of Z-values for tail area is 0.025, hence $Z = 1.96$.

Step 3. The number of units (gm) from the mean $= Z \times \sigma = 1.96\,(1.8) = 3.528$ gm; see Figure 6.3(a).

Step 4. Hence the upper limit is $x_U = \mu + Z\sigma = 19.5 + 3.528 = 23.028$ gm.
The lower limit is $x_L = 19.5 - 3.528 = 15.672$ gm.
So, 95 % of packets weight between 15.672 and 23.028 gm.

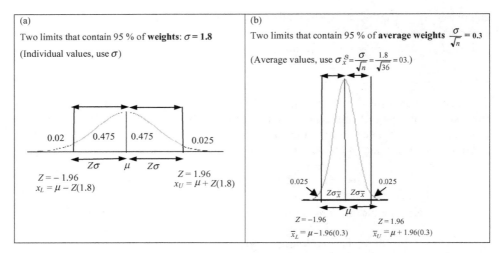

Figure 6.3   Calculating the two limits for a Normal distribution that contain 95 % of (a) values (b) mean values.

(c) This part of the question asks about sample averages. The method is the same as in part (b), except that the standard deviation (called standard error of the mean) for averages is $\sigma_{\bar{x}} = \dfrac{\sigma}{\sqrt{n}} = \dfrac{1.8}{\sqrt{36}} = 0.3.$

Step 1. The sketch in Figure 6.3(b) is similar to Figure 6.3(a), but with σ replaced by $\sigma_{\bar{x}} = \dfrac{\sigma}{\sqrt{n}}$.

Step 2. Look up tables for a tail area = 0.025, to find $Z = 1.96$.
Step 3. Calculate the number of units from the mean: $d = Z\sigma_{\bar{x}} = 1.96 \times 0.03 = 0.588$.
Step 4. Calculate the upper and lower limits for $\bar{x}$: $\mu \pm Z\sigma_{\bar{x}}$.

Hence $\bar{x}_U = 19.5 + 0.588 = 20.088$ and $\bar{x}_L = 19.5 - 0.588 = 18.912$.
Therefore 95 % of samples average weights ($n = 36$), fall between 18.912 and 20.088 gm.

(d) If the sample size is 16, the Central Limit Theorem applies if the population is Normally distributed, since the population standard deviation is known: $\sigma = 1.8$ gm.
If this is the case, then the standard error of the mean is $\dfrac{\sigma}{\sqrt{n}} = \dfrac{1.8}{\sqrt{16}} = 0.45$

In (a) the larger standard error will result in a smaller Z: $Z\dfrac{\bar{x} - \mu_{\bar{x}}}{\sigma_{\bar{x}}} = \dfrac{20 - 19.5}{0.45} = 1.11.$
Hence $P(\bar{x} \geq 20\,\mathrm{gm}) = P(Z \geq 1.11) = 0.1113.$
In (b) the question does not involve samples; therefore the answer is the same.
In (c) the larger standard error will result in a wider interval
$Z \times \sigma = 1.96(0.45) = 0.882$ gm.
Hence $\bar{x}_U = 19.5 + 0.882 = 20.382$ and $\bar{x}_L = 19.5 - 0.882 = 18.618.$
    If the population is NOT Normally distributed, then these methods cannot be used.

Control chart for means

In statistical process control, control charts are set up to monitor various aspects process stability. For example, a control chart for sample means consists of a **centre line** equal to μ, the population mean: the upper and lower warning lines (UWL and LWL) are the two limits within which 95 % of all sample means fall; the upper and lower control (or action) lines (UCL and LCL) are the two limits within

99.8 % of all sample means fall. If a process is stable then its sample means (sample size n) should be randomly and normally distributed within these limits. See Figure 6.4.

Skill Development Exercise SKD6_1: Control chart for means

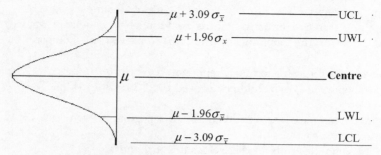

Figure 6.4   Explanation for centre line, warning lines and control lines for a means chart.

(a) Explain that the formulae for the centre, warning and control lines are those in given in Figure 6.4.

(b) In a call centre, the time taken to process calls is known to be normally distributed with a mean time of 12 minutes and standard deviation of 4.8 minutes. Samples of size 36 are to be selected to monitor stability. (i) Use the formulae in Figure 6.4 to calculate and plot the centre line, warning lines and control lines for a means chart for samples of 36; (ii) The mean call times for 10 samples of 36 were: 12.2; 11.5; 9.5; 13.2; 16.4; 14.2; 10.2; 8.4; 9.8; 15.2. Plot the sample means on the control chart. Would you consider the time taken to process calls to be stable?

Answer

(a) The two limits that contain $(1 - \alpha)\,100\,\%$ of all sample means are

$$\mu \pm Z_{\alpha/2}\sigma_{\overline{x}}$$

For a tail area of 2.5 %, $Z_{\alpha/2} = 1.96$ and a tail area of 0.5 %, $Z_{\alpha/2} = 3.09$
 (i) Centre = 12   UWL = 13.568   LWL = 10.432   UCL = 14.472   LCL = 9.528
(ii) The means chart with the sample means is given below. The process is not stable: four sample means are outside the control lines.

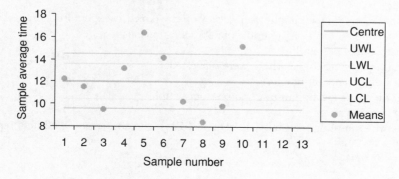

Means chart for time to process calls (sample size = 36)

Progress Exercises 6.1: Sampling distribution of the mean

1. A population consists of the numbers: 10; 8; 12; 7.
   (a) Calculate the population mean and standard deviation.
   (b) Write out every possible sample of size 2 that can be selected from this population. Calculate the means and standard deviation (σ) from the list of all sample means.
   (c) Calculate the standard error for means by formula (6.1). Hence state the mean and standard error for the sampling distribution of the mean. Compare the answers (b) and (c).
2. (a) Explain what the following formulae represent. μ; $\bar{x}$; $\mu_{\bar{x}}$; σ; $\sigma_{\bar{x}}$; s.
   (b) A group (a population) of 6 children are aged 10; 9; 14; 13; 11; 12.
      (i) Calculate: μ, σ.
      (ii) A random sample of three children from this population were aged 14, 12 and 10 years. Calculate $\bar{x}$ and s.
   Write down the sampling errors for these estimates of μ and σ.
   Why is it not possible to calculate $\mu_{\bar{x}}$ and $\sigma_{\bar{x}}$ from this sample?
3. Past studies established that the weight for adults in an isolated village was Normally distributed with $\mu = 69$ kg and standard deviation $\sigma = 8.5$ kg.
   (a) A random sample of 36 adults is required for further study. Define your target population. Describe how you would select the sample if there are 800 adults in the village.
   (b) Calculate the probability that
      (i) the weight of a randomly selected adult is greater than 72 kg.
      (ii) the average weight for a sample of 36 adults is greater than 72kg.
   Explain why the answers to parts (i) and (ii) are different.
4. A college accommodation office calculated (from its records) that the average monthly rent for a room in student accommodation was $\mu = €\,320$ with a standard deviation, $\sigma = €\,84$ during an academic year. Assume rents are Normally distributed.
   If random samples of size (i) $n = 4$; (ii) $n = 64$; (iii) $n = 100$, calculate, **for each sample size**
   (a) standard error of sample means.
   (b) the two limits within which 95 % of sample means will fall.
   (c) percentage of rents that fall between €\,300 and €\,340 per month.
   Describe verbally how the sample size affected the results (a), (b) and (c) above
5. Statistics collected by Eurostat for all 25 member countries in August 2004 reveal that then mean cost of a 10 minute local call is 36.36 cent with a standard deviation 10.03 cent.
   (a) Assuming costs are normally distributed calculate the percentage of calls that cost (i) less than 30 cent (ii) between 30 and 40 cent (iii) more than 40 cent.
   (b) Calculate the two limits that contain (i) 95 % of local call costs (ii) 95 % of average call cost for samples of 64.
6. See question 5. Suppose, for a follow-up survey on the monthly expenditure on mobile calls and cost of broadband connection, a sample of 2000 consumers is required.
   (a) Suggest a target population.
   (b) The following sampling methods were suggested.
      (i) a simple random sample.
      (ii) a simple proportional stratified random sample.
      (iii) a quota sample.
   Discuss the sampling methods (i), (ii) and (iii) in the context of yielding a representative sample.

(c) Two Methods of contact were proposed:
   (i) A questionnaire sent by email.
   (ii) A phone call.
   Discuss these methods of contact explaining the possible sampling errors that could arise and the likely response rate.

7. See question 5. Calculate a control chart for means for samples of size 36. The mean cost of a local call calculated for six independent samples were calculated as 36: 32.5; 42.3; 33.6; 29.4; 32.5; 35.3. Plot these sample means on the control chart. Would you consider the mean cost of a local call to be stable?

8. State, with explanations whether the following are true or false:
   (a) A simple random sample is more representative of the population than any other type of sample.
   (b) Sampling error is the same as standard error.
   (c) A sampling distribution of means is a probability distribution.
   (d) The standard error of the mean is less than the standard deviation of all sample means.

9. A hotel group has determined that wedding guests who stay overnight in the hotel spend an average of £175 with a standard deviation of £46.
   (a) Calculate the probability that the amount spent by a guest is between £100 and £200.
   (b) Calculate the probability that the average amount spent is between £150 and £200 per guest for a party (a sample) of (i) 40 guests (ii) 16 guests.
   (c) For a party of 64 guests, calculate the minimum amount spent by 90% of guests.

10. The mean age of emigrants from the UK in 2002 was 33, the standard deviation of 12 years. Assume ages are Normally distributed.
   (a) (i) Calculate the percentage of emigrants that are aged between 25 and 40; (ii) If there were 359,400 emigrants in 2003, how many were aged between 25 and 40?
   (b) If a random sample of 400 emigrants is selected. What is the probability that the sample average age is (i) less than 32; (ii) between 32 and 35; (iii) older than 35?

11. An official from the office of public works is given the task of estimating the mean daily wage for site workers in the construction industry.
   (a) If a random sample of 144 workers is selected, what is the probability that the sample mean differs from the true mean for the industry by at most €1 per day when the standard deviation for the industry is known to be €8 per day.
   (b) What sample size should be selected if the official wants to be 99% certain that the difference between the true mean and the sample mean is at most €1 per day (assume $\sigma = €8$, as above).

12. State, with explanations whether the following are true or false:
   (a) A quota controlled sample is used for statistical inference when there is no population list available.
   (b) Random samples are always unbiased.
   (c) The Central Limit Theorem applies for any sample size.
   (d) The standard error decreases as the sample size is increases.

6.2 Sampling distribution of proportions

A proportion is the number of elements with a given characteristic divided by the total number of elements in the group. For example, the proportion who votes in an election is the number who actually vote divided by the number eligible to vote. In every day situations proportions, when multiplied by 100, are quoted as percentages, for example 'the percentage unemployed'; 'the percentage who smoke'; 'the percentage of home owners', etc. Since the population proportion is estimated by a sample proportion each is represented by different symbols. See Table 6.4:

Table 6.4   Symbols for population proportions and sample proportions

	Population characteristic	Sample characteristic
Size	N = number in population	n = number in sample
Proportion	$\pi = \dfrac{X}{N}$ (6.5)	$p = \dfrac{x}{n}$ (6.6)
	X is the number of elements with a given characteristic in a population size N	x is the number of elements with a given characteristic in a sample size n

The sample proportion, p is a point estimate of the population proportion, π. To gain some insight into the variation in values among sample proportions, return to the small population of five numbers used earlier for sample means, but this time consider the proportion of even numbers:

A	B	C	D	E
3	1	5	6	2

The population proportion of even numbers, $\pi = \dfrac{2}{5} = 0.4$.

From this population take every possible sample of (a) size 2 (b) size 3 and calculate the sample proportion of even numbers. See Table 6.5.

From the sample proportions, calculate the mean and standard deviation of the 10 sample proportions in (a) and (b). The results are summarised in Table 6.5.

The standard error for proportions may be calculated by formula (6.7):

$$\sigma_p = \sqrt{\frac{\pi(1-\pi)}{n}}\sqrt{\frac{N-n}{N-1}} \qquad (6.7)$$

where $\sqrt{\dfrac{N-n}{N-1}}$ is the finite population correction factor, given in section 6.1.

From Table 6.5, note the following

1. The mean of all the sample proportions is equal to the population proportion.
2. The standard deviation of all the sample means decreases as n increases.
3. However, if N is large then $\sqrt{\dfrac{N-n}{N-1}} \cong 1$ hence $\sigma_p = \sqrt{\dfrac{\pi(1-\pi)}{n}}$.

Table 6.5   Distribution of sample proportions for (i) sample size 2 (ii) sample size 3

Sample of size 2	(a) Sample proportion	Sample of size 3	(b) Sample proportion
AB = 3, 1	0	ABC = 3, 1, 5	0
AC = 3, 5	0	ABD = 3, 5, 6	1/3
AD = 3, 6	0.5	ABE = 3, 1, 2	1/3
AE = 3, 2	0.5	ACD = 3, 5, 6	1/3
BC = 1, 5	0	ACE = 3, 5, 2	1/3
BD = 1, 6	0.5	ADE = 3, 6, 2	2/3
BE = 1, 2	0.5	BCD = 1, 5, 6	1/3
CD = 5, 6	0.5	BCE = 1, 5, 2	1/3
CE = 5, 2	0.5	BDE = 1, 6, 2	2/3
DE = 6, 2	1	CDE = 5, 6, 2	2/3
Mean of **all** sample proportions μ_p	0.4		0.4
Standard deviation of **all** sample proportions σ_p	0.3		0.2

Standard deviation of proportions by formula

$$\sigma_p = \sqrt{\frac{\pi(1-\pi)}{n}}\sqrt{\frac{N-n}{N-1}}$$

$$\sigma_p = \sqrt{\frac{(.4)(.6)}{2}}\sqrt{\frac{5-2}{5-1}} = 0.3 \qquad \sigma_p = \sqrt{\frac{(.4)(.6)}{3}}\sqrt{\frac{5-3}{5-1}} = 0.2$$

> Note the symbol:
> σ_p = standard deviation,
> of ALL sample proportions

As a rule of thumb, it is not necessary to use the correction factor when the sample size $n \geq 30$ and less than 5 % of the population size ($n \leq 0.05N$). Hence the standard error for proportions is given by equation (6.8).

$$\sigma_p = \sqrt{\frac{\pi(1-\pi)}{n}} \quad \text{is called the standard error for proportions} \qquad (6.8)$$

The sampling distribution of proportions

The list of every possible sample proportion, with its probability is called the sampling distribution of proportions.

To calculate the probability distribution of sample proportions for samples of size $n = 2$ summarise the data above in a frequency distribution table, Table 6.6. Plot a relative frequency (or probability histogram), Figure 6.5.

Table 6.6   Relative frequency (probability) distribution for sample proportions

p	0	0.5	1
Frequency	3	6	1
Relative frequency	0.3	0.6	0.1

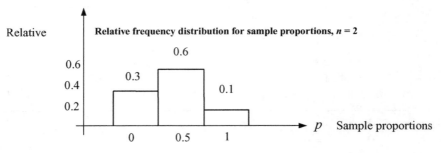

Figure 6.5   Relative frequency (probability) distribution for sample proportions.

The probability histogram in Figure 6.5 is reasonably Normal, even for this very small population. For large samples ($n \geq 30$), the distribution of sample proportions is approximately **Normal** with a mean, $\pi = \mu_p$, standard deviation, $\sigma_p = \sqrt{\dfrac{\pi(1-\pi)}{n}}$.

$$\text{For large samples } (n \geq 30), \qquad p \sim N\left(\pi, \sqrt{\dfrac{\pi(1-\pi)}{n}}\right) \qquad (6.9)$$

Since the sample proportions are Normally distributed, probabilities may be calculated regarding sample proportions.

See Appendix F for proof of the formulae for the mean and variance of the sampling for proportions.

Worked Example 6.2: probabilities for sample proportions

In a certain neighbourhood, it is known that 12 % of youths aged 16 to 24 years are unemployed. If a random sample of 150 youths are selected what is the probability that the sample contains

(a) at most 10 % unemployed;
(b) at most 15 unemployed.

Solution

(a) If 12 % are unemployed, then $\pi = 0.12$ and $\sigma_p = \sqrt{\dfrac{\pi(1-\pi)}{n}} = \sqrt{\dfrac{0.12(1-0.12)}{150}} = 0.0265$

So Normal probability calculations follow with $\mu_p = \pi = 0.12$ and $\sigma_p = 0.0265$
Calculate $P(p \leq 0.10)$ since 10 % means $p = 0.10$.
Step 1. Plot the curve and mark in information given; see Figure 6.6.

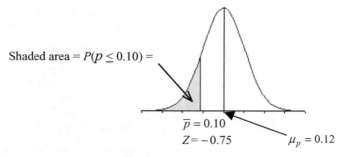

Figure 6.6   Calculating probabilities for sample proportions.

Step 2. Calculate Z

For $p = 0.10$ $Z = \dfrac{p - \pi}{\sigma_p} = \dfrac{0.10 - 0.12}{0.0265} = -0.75$

Step 3. From Tables, area in tail $= 0.2266$.

Step 4. Calculate the required area (probability).

Because of symmetry, the area in the lower tail is same as the area in the upper tail. Hence $P(p \leq 0.10) = 0.2266$.

The probability that at most 10 % of the sample is unemployed is 0.2266.

(b) First convert numbers to percentages and then proportions.

15 individuals is $\dfrac{15}{150} \times 100 = 10\%$

Hence the probability that the sample contains at most 15 unemployed is the same as the probability that it contains at most 10 %. This is the probability that was calculated in part (a).

Mean and standard errors for the distribution of proportions when π is unknown

The calculation of the mean and standard error of sample proportions in (6.9) depends on knowing the value of the population proportion, π, in the first place. Since π is seldom known, the mean and standard error for proportions are approximated by substituting p for π in (6.9). Hence $\bar{p} = p$, $s_p = \sqrt{\dfrac{p(1 - p)}{n}} = \sqrt{\dfrac{pq}{n}}$.

Control Charts for sample proportions

A quality control chart for sample proportions is similar to that for sample means given in Figure 6.7. The control chart for sample proportions is called a p-chart. The centre line is the population proportion or its estimate(p); the two limits that contain 95 % of sample proportions are called the warning lines; the two limits that contain 99.8 % of sample proportions are called the control or action lines, as given in Figure 6.7. When sample proportions are plotted on this control chart they should be normally distributed and randomly scattered about the centre line if the process is statistically stable.

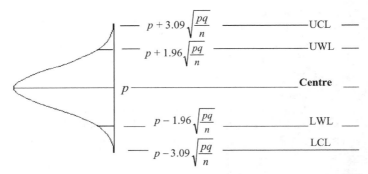

Figure 6.7   A *p*-chart with the formulae for centre, warning and control lines.

Skill Development Exercise 6.2 SKD6_2

Calculate the *p*-chart for the following situation.

Over a period of time it has been established that 66 % the amounts dispensed by a college ATM machine were £ 75 or less. Samples of 36 accounts are recorded at regular intervals from Monday to Friday. (a) Calculate and plot a control chart for proportions; (b) plot the following sample proportions on the chart: 0.58; 0.62; 0.57; 0.61; 0.64; 0.69; 0.64; 0.73; 0.71; 0.75; (c) do you consider the proportion of amounts withdrawn to be stable?

Answer

(a) and (b) Centre line, $p = 0.66$; UWL/LWL $= p \pm 19.6\sqrt{\dfrac{p(1-p)}{n}} = 0.8147, 05053$.

UCL/LCL $= p \pm 3.09\sqrt{\dfrac{p(1-p)}{n}} = 0.9040, 0.4160$.

With the values of the centre, warning and control lines, plot the *p*-chart. Plot the sample proportions on the chart.

(c)

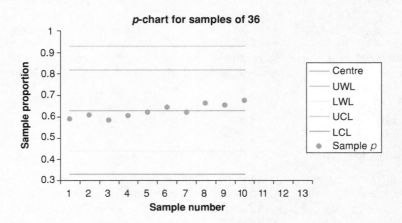

Sample proportions are within control lines, but are not Normally and randomly distributed – an upward trend is developing, therefore the process is not stable.

Progress Exercises 6.2: Sampling distribution of proportions

In the following exercises.
1. Round all calculations correct to four decimal places. 2. Assume large populations. 3. The continuity correction is not used.

1. A population consists of the numbers 15, 21, 17, 24, 28, 33.
   (a) Calculate the population proportion of even numbers.
   (b) Write out every possible sample of size five and calculate the sample proportion of even numbers.
       Calculate the mean and standard deviation from all the sample proportions.
   (c) State the mean and variance for the sampling distribution of proportions for samples of size 5. Calculate the standard error by formula (6.7). Compare the answers in (b) and (c).

2. Explain the following, illustrating your answer with an appropriate example and giving any relevant formulae: sampling error; sampling distribution of proportions; the standard error of proportions; the standard deviation of sample proportions.

3. (a) Explain the effect of sample size on the standard error for proportions.
   (b) In a certain manufacturing process for camera cases, it is known that 5 % will be defective. Calculate the probability that a sample proportion will be greater than 0.06 if the sample size selected is (i) 35 (ii) 100 (iii) 625 from a large population, $\pi = 0.05$.

4. According to tam ratings 45 % of viewers watch a news and politics show on Monday evenings.
   (a) If a poll is taken on a Monday evening, what is the probability that (i) more than 50 % of viewers in a sample of 40 watched the programme; (ii) less than 40 % in a sample of 100 watched the programme.
   (b) Calculate the two limits within which 90 % of all sample proportions fall if the sample size is 40. Describe this interval verbally.

5. Following an audit, the ABC bank disclosed that 28 % of customers with less than £ 1000 in their deposit accounts were overdrawn on their current accounts?
   (a) If a random sample of 50 such customers is selected, what is the probability that (i) 10 (ii) at least 8 are overdrawn on their current accounts?
   (b) If a random sample of 100 such customers is selected, what is the probability that (i) 30 (ii) less than 40 are overdrawn on their current accounts?

6. State, with explanations whether the following are true or false.
   (a) The value of the standard error for proportions is always between 0 and 1.
   (b) Increasing the sample size increases the value of the standard error.
   (c) A sample proportion is equal to the population proportion.
   (d) Opinion polls are more likely to give accurate predictions if larger samples are used.

7. In many mountainous rural areas tourism is an important industry. In the past, only 40 % of visitors stayed in hotels, the remainder preferring to stay in guest house or B&B accommodation. However, those working in the industry suspect that in recent years more and more visitors are choosing to stay in hotels. In a survey 96 out of 200 visitors indicate that their preference is for hotels.
   (a) According to the survey, what (i) proportion; (ii) percentage of visitors prefer to stay in hotels?
   (b) If it is known that 40 % of all visitors prefer hotel accommodation what is the probability of selecting a sample in which at least 96 out of 200 prefer hotel accommodation?
   Does this result confirm the suspicions expressed by the industry? Explain and justify your answer.

8. Candidate A was elected in a recent by-election with 52 % of the votes. In an opinion poll before the election 47 out of 100 voters indicated that they intended to vote for A. Hence the sampling error was five percentage points.

(a) Calculate the probability that the error was within five percentage points when 100 voters were polled.
(b) Calculate the minimum number of voters that should be polled if there is a 95 % probability that the error is within (i) five percentage points; (ii) one percentage point.
9. Three percent of the bovine population test positive for a certain disease. Calculate the probability that
(a) none test positive in a random sample of 30.
(b) at most two test positive in a random sample of 50.
(c) at least three test positive in a random sample of 100.
10. A UK labour force survey (spring 2000) found that 28 % of women and 6 % of men in the labour force work 20 hours per week or less.
(a) What is the probability that fewer than 100 women work 20 hours per week or less in a randomly selected sample of 400 (from the labour force)?
(b) What is the probability that a randomly selected sample of 400 men (from the labour force) will contain at least 350 that work more than 20 hours per week?
(c) Calculate the probability that the sample proportion of those who work 20 hours per week or less differs from the population proportion by 0.05 (five percentage points) or more when samples of (i) 400 females; (ii) 400 males are selected.
(d) Calculate the sample size of (i) females (ii) males that should be selected from the labour force for if the difference between the population and sample proportions is within 0.02 (two percentage points) with a maximum probability of 5 %.
11. In the holiday insurance business, 25 % of queries related to lost property. Samples of 40 calls are collected to monitor whether the percentage of queries regarding lost property is stable.
(a) Calculate and plot a p-chart.
(b) In successive samples of 40, the numbers of queries about lost property were 12; 14; 17; 12; 09; 11; 15; 10; 8; 12. Calculate and plot the sample proportions on the control chart.
(c) Do you consider the proportion of calls dealing with lost property to be stable?

6.3 Some desirable properties of estimators

Two of the most important properties of point estimators are that they should be

1. **Unbiased**
2. **Minimum variance**

1. Unbiased estimators
An estimator is said to 'unbiased' if the average value of all the point estimates is equal to the population parameter being estimated. Unbiased estimators are sometimes described as 'accurate': the average value of the estimates are equal to the population parameter (on average, the estimates are on Target). For example, in Table 6.2, the mean $\bar{x} = \dfrac{\sum x}{n}$ is equal to μ, hence is an unbiased estimator of μ. Similarly in Table 6.5, p is an unbiased estimator of π.

To prove that $\bar{x}$ is an unbiased estimator of μ, it is necessary to show that the expected value of the sample mean is the population mean: $E(\bar{x}) = \mu$. See Worked Example 6.3.

However, if the formulae, $\dfrac{\sum (x - \overline{x})^2}{n}$ is used to estimate the population variance, σ^2, it will underestimate σ^2 in the long run. Hence $\dfrac{\sum (x - \overline{x})^2}{n}$ is described as a **biased estimator** (one-sided) of the population variance.

$s^2 = \dfrac{\sum (x - \overline{x})^2}{n - 1}$ is an unbiased estimator of σ^2.

To prove that s^2 is an unbiased estimator of σ^2 it is necessary to show that

$$E(s^2) = E\left(\dfrac{\sum (x - \overline{x})^2}{n - 1}\right) = \sigma^2 \text{ (see Worked Example 6.3).}$$

2. Minimum variance

In the preceding sections, in Tables 6.2 and 6.5, the values of sample statistics (sample means and proportions) varied greatly about the population parameter being estimated. It is obviously desirable to keep this variation (as measured by standard error) as small as possible (minimum variance). Increasing sample size, as demonstrated in these examples, reduced the standard error.

An estimator is described as precise when the values of the estimates are close. The standard deviation of all the estimates (called standard error) is a measure of the precision of an estimator. Ideally an estimator should be unbiased and have minimum variance (be both accurate and precise). These ideas are illustrated in Figure 6.8, where μ is the population parameter to be estimated.

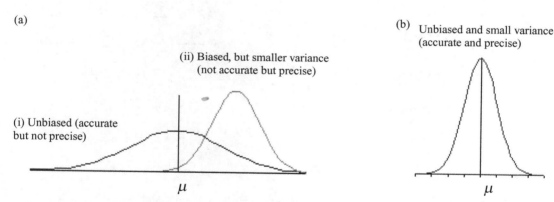

Figure 6.8   Distribution of (a)(i) an unbiased (ii) a biased estimator (b) an unbiased estimator.

Worked Example 6.3: Prove that $\overline{X}$ and s^2 are unbiased estimators of μ and σ^2.

If the population mean and variance for a random variable X are μ and σ^2, prove that $\overline{X}$ and s^2 are unbiased estimators of μ and σ^2

Solution

The population mean and variance are $E(X) = \mu$; $V(X) = \sigma^2$ respectively.

To prove that $\overline{X}$ is an unbiased estimator of μ, it is necessary to show that the expected value of the sample mean is the population mean: $E(\overline{X}) = \mu$.

$$E(\overline{X}) = E\,\frac{\sum\limits_{i=1}^{i=n} X_i}{n}$$

$$= \frac{1}{n}E\left(\sum X\right)$$

$$= \frac{1}{n}E(X_1 + X_2 + X_2 + \ldots X_n) \quad \ldots \text{ since } E(X) = \mu \text{ for every observation}$$

$$= \frac{1}{n}(\mu + \mu + \mu + \cdots + \mu) \quad \ldots n \text{ times}$$

$$= \frac{1}{n}(n\mu) = \mu$$

To prove that s^2 are unbiased estimator of σ^2.

The expected value of s^2 is

$$E(s^2) = E\left(\frac{\sum(X - \overline{X})^2}{n-1}\right)$$

$$= \frac{1}{n-1}E\left(\sum X^2 - n\left(\overline{X}\right)^2\right) \quad \ldots(1)$$

$$= \frac{1}{n-1}\left\{E\left(\sum X^2\right) - E\left(n\left(\overline{X}\right)^2\right)\right\}$$

$$= \frac{1}{n-1}\left\{\sum E(X^2) - nE\left((\overline{X})^2\right)\right\}$$

$$= \frac{1}{n-1}\left\{\sum(\mu^2 + \sigma^2) - n\left(\frac{\sigma^2}{n} + \mu^2\right)\right\}$$

see (2) and (3) for explanations

$$= \frac{1}{n-1}\{n\mu^2 + n\sigma^2 - \sigma^2 - n\mu^2\}$$

$$= \frac{1}{n-1}\{n\sigma^2 - \sigma^2\}$$

$$= \frac{1}{n-1}\{n-1\}\sigma^2 = \sigma^2$$

Explanation for points (1), (2) and (3)

(1)

$$\frac{\sum(X - \overline{X})^2}{n-1} = \frac{1}{n-1}\left(\sum X^2 - 2\overline{X}X + (\overline{X})^2\right)$$

$$= \frac{1}{n-1}\left(\sum X^2 - 2\overline{X}\sum x + \sum()^2\right)$$

$$= \frac{1}{n-1}\left(\sum X^2 - 2\overline{X}(n\overline{X}) + n(\overline{X})^2\right)$$

$$= \frac{1}{n-1}\left(\sum X^2 - n(\overline{X})^2\right)$$

(2)

$$\left.\begin{array}{l}\sigma^2 = E(X^2) - (E(X))^2\\ \sigma^2 = E(X^2) - \mu^2\end{array}\right\} \ldots \text{ Hence } E(X^2) = \mu^2 + \sigma^2$$

(3)

$$\left.\begin{array}{l}\dfrac{\sigma^2}{x} = E\{(\overline{X})^2\} - (E(\overline{X}))^2\\ \dfrac{\sigma^2}{n} = E\{(\overline{X})^2\} - (\mu)^2\end{array}\right\} \ldots E\{(\overline{X})^2\} = \dfrac{\sigma^2}{n} + \mu^2$$

Summary and overview of sampling distributions for means and proportions

Means and proportions calculated from random samples are used to estimate the corresponding population mean/proportion. These sample means/proportions can be very poor estimates of the corresponding population parameters. However, according to the Central Limit Theorem, for samples of size 30 or more, the probability distribution of sample means/proportions is Normal.

The probability distribution of sample means/proportions is called the sampling distribution of means/proportions. Hence, as with any Normal distribution, it is possible to calculate probabilities regarding sample means/proportions and to calculate intervals that contain a given percentage of all means/proportions.

In statistical inference, it is necessary to become familiar with the symbols used to represent sample statistics and the corresponding population parameters

6.1 The distribution of sample means

A parameter is a population characteristic, such as the population mean.

A statistic is a sample characteristic, such as the sample mean.

A point estimate of the population mean is the sample mean, calculated from the random sample data.

Sampling error is the difference between the population parameter and the sample statistic, for example, the difference between μ and $\overline{x}$.

The Sampling distribution of the mean is the distribution of all sample means.

Central Limit Theorem states that for samples of size, $n > 30$, the distribution of sample means from a population with mean μ and standard deviation σ is approximately normal:

$$N \left(\mu_{\overline{x}} = \mu, \sigma_{\overline{x}} = \frac{\sigma}{\sqrt{n}} \right)$$

Standard error of the mean is the standard deviation of all the sample means $\sigma_{\overline{x}} = \dfrac{\sigma}{\sqrt{n}}$.

As a consequence of the Central Limit Theorem, it is possible to calculate probabilities about $\overline{x}$ and calculate the limits that contain a given percentage of all sample mean. The control chart for means is a simple application based on the sampling distribution of means. In the remainder of the text, the sampling distribution of the mean will be used to calculate interval estimates, and in testing hypothesis about population mean(s).

6.2 The distribution of sample proportions

The population proportion is $\pi = \dfrac{X}{N}$, the number of elements with a given characteristic in a population size N.

The sample proportion is $p = \dfrac{x}{n}$

The Sampling distribution of proportions for samples of size, $n > 30$, is approximately Normal with mean π and **Standard error of proportions,** $\sigma_p = \sqrt{\dfrac{\pi (1 - \pi)}{n}}$.

As a consequence of the Central Limit Theorem, it is possible to calculate probabilities about p and calculate the limits that contain a given percentage of all sample proportions. The control chart for proportions is a simple application based on the sampling distribution of proportions.

In the remainder of the text, the sampling distribution of the proportions will be used to calculate interval estimates, and in testing hypothesis about population proportion(s).

6.3   Point estimators

It is desirable that point estimators **should be** (i) unbiased and (ii) have minimum variance. **Unbiased estimator** is an estimator whose mean value is equal to the population parameter. For example, $s^2 = \dfrac{\sum(x - \bar{x})^2}{n - 1}$ is an unbiased estimator of σ^2 since it can be proved that

$$E(s^2) = E\left(\frac{\sum(x - \bar{x})^2}{n - 1}\right) = \sigma^2.$$

CONFIDENCE INTERVALS FOR MEANS AND PROPORTIONS

7

Chapter Objectives

Having carefully studied this chapter and completed the exercises you should be able to do the following

- Give a verbal explanation of what is meant by a confidence interval
- Calculate a confidence interval for a population mean
- Calculate a confidence interval for population proportions
- Calculate confidence intervals for the differences between means, difference between proportions
- Describe any confidence verbally
- State the factors that determine the precision and accuracy of a confidence interval
- Calculate the sample size required for an interval estimate of a population mean or proportion for a given level of confidence and precision: (large n)
- Interpret and explain verbally Minitab printouts for confidence intervals and tests of hypothesis

7.1   Confidence intervals for the mean

Introduction

In Chapter 6, Worked Example 6.1 demonstrated that the sample mean $\bar{x}$ can vary greatly about the population mean μ. For example, for samples of size $n = 2$, the sample means range from $\bar{x} = 1.5$ to $\bar{x} = 5.5$. See Figure 7.1.

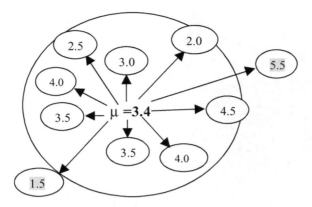

Figure 7.1   Sample means are point estimates of the population mean μ.

It was also noted that when the sample size was increased from $n = 2$ to $n = 3$ that the variation in point estimates was reduced. When the population mean is unknown there is no way of determining whether an individual sample mean is one of those that is close to the population mean or not. This situation is unsatisfactory, particularly in situations where accuracy is paramount: for example, estimates for (i) the mean breaking strength of suspension cables or (ii) the mean diameter of a precision lens. An interval estimate for μ of the form '$\overline{x}\pm$ a margin of error' would provide the user with a measure of the uncertainty associated with the point estimate $\overline{x}$. One would expect that the formula for the 'margin of error' should take into consideration the factors that determine the variation in values of $\overline{x}$, such as sample size, n and σ, the population standard deviation.

> Interval estimate for μ
> $\overline{x}\pm$ 'a margin of error'

7.1.1   Confidence interval for the population mean

In Worked Example 6.1, we calculated the interval that contained 95 % of sample means ($n \geq 30$). This is illustrated graphically in Figure 7.2:

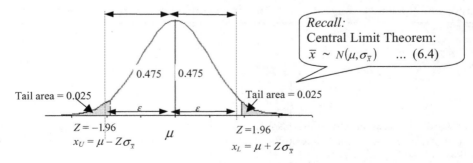

Figure 7.2   The interval $\mu \pm 1.96\sigma_{\overline{x}}$ that contains 95 % of all sample means.

To reiterate, the interval that contains 95 % of sample means is

$$\mu \pm 1.96\sigma_{\overline{x}}$$

(7.1)

or

$$\mu \pm \varepsilon \quad \ldots \text{where } \varepsilon = 1.96\sigma_{\overline{x}}$$

The interval '$\mu \pm \varepsilon$' is described as the interval with a fixed centre, μ and total width, $w = 2 \times \varepsilon$, that contains 95 % of all sample means.

In estimation, μ is unknown. Therefore replace μ in (7.1) by a point estimate, $\overline{x}$. Substitution of $\overline{x}$ for μ in equation (7.1) gives the interval

$$\overline{x} \pm 1.96\sigma_{\overline{x}}$$

or

$$\overline{x} \pm \varepsilon$$

(7.2)

The essential difference between (7.1) and (7.2) is that the centre of the interval is fixed at μ in (7.1), but in (7.2) the centre of the interval is no longer fixed: the centre moves according to the value of each new point estimate $\overline{x}$. Five such intervals centred on five different sample means are illustrated graphically in Figure 7.3:

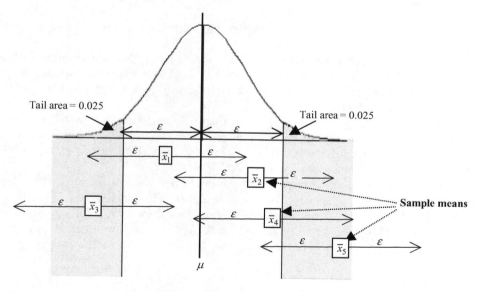

Figure 7.3   Five different interval estimate for μ based on sample means $\overline{x}$.

It is clear from Figure 7.3 that

- An interval estimate $\overline{x} \pm \varepsilon$ will contain μ if the sample mean, $\overline{x}$ is one of the 95 % of $\overline{x}$'s within the interval $\mu \pm \varepsilon$, such as $\overline{x}_1$, $\overline{x}_2$ and $\overline{x}_4$ in Figure 7.3.
- An interval estimate will NOT contain μ if the sample mean $\overline{x}$ is outside the interval $\mu \pm \varepsilon$. That is, if $\overline{x}$ is so far away from μ that it falls in either tail, such as $\overline{x}_3$ and $\overline{x}_5$, Figure 7.3.

- Each one of the 95 % of sample means that fall within a distance of '$1.96\sigma_{\bar{x}}$ from μ' will result in an interval $\bar{x} \pm 1.96\sigma_{\bar{x}}$ that contains the population mean **somewhere within the interval**.

The objective at the beginning of this section was to calculate an interval estimate of the form $\bar{x} \pm$ 'margin of error'. We now have such an interval in $\bar{x} \pm 1.96\sigma_{\bar{x}}$. The 'margin of error', $\varepsilon = 1.96\sigma_{\bar{x}}$ as illustrated in Figure 7.4. In addition, since 95 % such interval will contain μ, we can state that we are **95 % confident** that the population mean, μ, is in the interval.

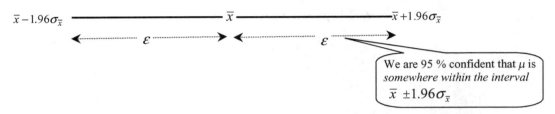

Figure 7.4   Construction of a 95 % confidence interval for means.

To construct a confidence interval for any level of confidence

The formula for an interval estimate for μ with **any level of confidence** may be deduced as a generalisation of the 95 % confidence interval above.

In the 95 % confidence interval for μ, the 'margin of error', $\varepsilon = 1.96\sigma_{\bar{x}}$ was calculated from the limits that contained 95 % of all sample means in Figure 7.2. The 5 % of sample means that fell outside these limits were equally divided between the two tails of the distribution, giving tail areas of 0.025, hence $Z_{0.025} = 1.96$ and $\varepsilon = 1.96\sigma_{\bar{x}}$.

Similarly, if a 99 % confidence interval is required, then the margin of error, would be calculated from the two limits that contain 99 % of sample means. The 1 % of sample means outside these limits is equally divided between the tails giving each tail area of 0.005, hence $Z_{0.005} = 2.5758$ and $\varepsilon = 2.5758\sigma_{\bar{x}}$. The 99 % confidence interval is $\bar{x} \pm 2.5758\sigma_{\bar{x}}$.

In general (see Figure 7.5), if we let the area in each tail be $\alpha/2$, then the corresponding Z-value will be referred to as $Z_{\alpha/2}$: hence, the margin of error, $\varepsilon = Z_{\alpha/2}\sigma_{\bar{x}}$. The area between the 2 tails is $(1 - \alpha)$. Then, $(1 - \alpha) \times 100\%$ is called the 'level of confidence' that the interval: $\bar{x} \pm Z_{\alpha/2}\sigma_{\bar{x}}$, contains μ somewhere within it. This is the definition for a $(1 - \alpha) \times 100\%$ confidence interval for μ given in (7.3).

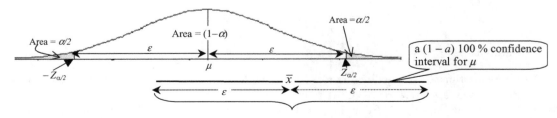

Figure 7.5   An example of a $(1 - \alpha) \times 100\%$ confidence for μ.

> Use 'CI' as an abbreviation for **Confidence Interval**

The $(1 - \alpha)\,100\,\%$ confidence interval (CI) is given by the formula

$$\bar{x} \pm Z_{\alpha/2}\sigma_{\bar{x}} \qquad\qquad (7.3)$$

$\bar{x} - Z_{\alpha/2}\sigma_{\bar{x}}$ is called the '**Lower confidence limit**' and
$\bar{x} + Z_{\alpha/2}\sigma_{\bar{x}}$ is called the '**Upper confidence limit**'.
$Z_{\alpha/2}$ is called the '**Critical Z value**' or the '**percentage point**'.

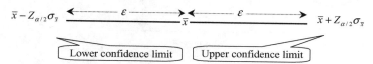

Figure 7.6 Terminology associated with confidence intervals.

Stated informally, in very general terms, a confidence interval is

Point estimate $\pm$ (Percentage point or Critical Z value) $\times$ (Standard error)

Confidence intervals when σ is unknown

In some applications the population standard deviation σ will be known. For example, the variation in the diameter of discs cut by a certain machine may have been established over a period of time. In other application σ will not be known. In such cases (provided $n \geq 30$), σ is estimated by s, the point estimate calculated from the sample data. Hence, when σ is unknown, the confidence interval for μ is $\bar{x} \pm Z_{\alpha/2}\, s_{\bar{x}}$.

Note: $s_{\bar{x}} = \dfrac{s}{\sqrt{n}}$ is called the **sample standard error of the mean**.

The $(1 - \alpha)\,100\,\%$ CI for μ when σ is not known, but $n \geq 30$ is

$$\bar{x} \pm Z_{\alpha/2} s_{\bar{x}} \qquad \text{where} \quad s_{\bar{x}} = \frac{s}{\sqrt{n}} \qquad\qquad (7.4)$$

The steps in calculating a confidence interval for means are summarised in Chart 7.1.

    Reminder on ASSUMPTIONS: The above discussion *assumes* that sample means are Normally distributed. This is true for samples from any population when $n \geq 30$ (Central Limit Theorem) or for samples of any size (<30) selected randomly from a Normal population with a known standard deviation, σ.

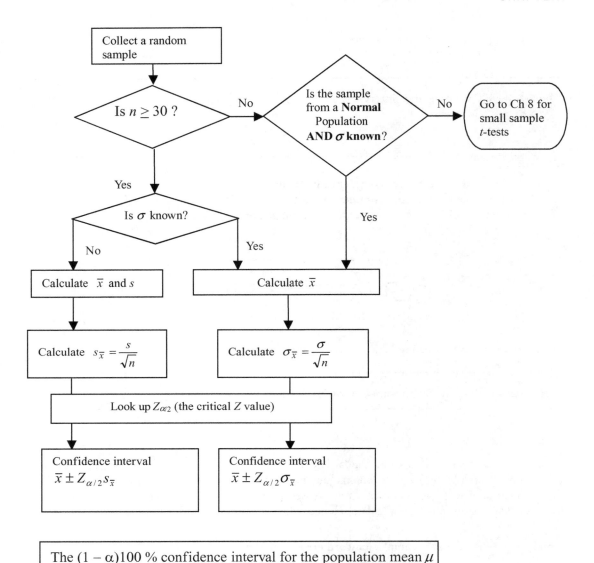

Chart 7.1   Construction of confidence interval for the mean.

Worked Example 7.1: Calculate the confidence interval for the mean σ known

An importer of Herbs and Spices claims that the average weight of packets of Saffron is 20gms. A random sample of 36 packets of saffron was collected. From the sample, the average weight was calculated as 19.35 gm. The population standard deviation is of weights is known to be 1.8 gm (see Worked Example 6.1).

Calculate

(a) 95 % confidence interval for the population average weight, μ.
(b) 99 % confidence interval for the population average weight, μ.
   In each case describe the interval estimate verbally.
   Do either of these results (a) or (b) support the claim that for the population average weight is 20 gm?
   NB *the population average (or mean) weight is sometimes called the true average (or mean) weight.*
(c) Estimate the range for total weight of saffron is 50 packets with 95 % confidence

Solution

Sample size is greater than 30 and σ is known. Hence, the confidence interval is $\bar{x} \pm Z_{\alpha/2}\sigma_{\bar{x}}$.

To construct the confidence interval you need $\bar{x}$, $\sigma_{\bar{x}} = \dfrac{\sigma}{\sqrt{n}}$ **and** $Z_{\alpha/2}$

$\bar{x} = 19.35$ was calculated from sample and given above.

$\sigma = 1.8$ and sample size $n = 30$ were given. Hence $\sigma_{\bar{x}} = \dfrac{\sigma}{\sqrt{n}} = \dfrac{1.8}{\sqrt{36}} = 0.30$

$Z_{\alpha/2} = 1.96$ when $\alpha/2 = 2.5\%$ as required in (a)
$Z_{\alpha/2} = 2.5758$ when $\alpha/2 = 0.5\%$ as required in (b)

(a) The 95 % confidence interval for μ is
   $\bar{x} \pm Z_{\alpha/2}\sigma_{\bar{x}}$
   $19.35 \pm 1.96(0.30)$
   19.35 ± 0.5880
   $19.35 - 0.5880$ to $19.35 + 0.5880$
   18.762 to 19.938

(b) The 99 % confidence interval for μ is
   $\bar{x} \pm Z_{\alpha/2}\sigma_{\bar{x}}$
   $19.35 \pm 2.5758\,(0.30)$
   18.5773 to 20.1227

   In (a) we can be 95 % confident that the population mean is somewhere between 18.762 to 19.938 gm.

   In (b) we can be 99 % confident that the population mean is somewhere between 18.5772 to 20.1227.

   Interval (a) does not support the claim that $\mu = 20$ since 20 outside the interval (a).
   Interval (b) does support the claim $\mu = 20$ since 20 is within the interval (b).

(c) Since, the mean value, $\bar{x} = \dfrac{\sum x}{n} = \dfrac{\text{Total weight of packets}}{\text{Number of packets}}$.

   Then $\bar{x} \times$ Number of packets = Total weight of packets.

   In (a), the mean weight per packet is between 18.762 to 19.938 gm, with 95 % confidence. Hence the total weight of 50 packets is between 50(18.762) and 50(19.938) gm: between 938.1 gm and 996.9 gm with 95 % confidence.

One-sided confidence intervals

One sided confidence intervals are constructed to give either

- the lower limit, above which we are $(1 - \alpha)\, 100\,\%$ confident the population mean lies:
  $\bar{x} - Z_\alpha \dfrac{\sigma}{\sqrt{n}}$ or $\bar{x} - Z_\alpha \dfrac{s}{\sqrt{n}}$ (if σ is unknown).
- the upper limit below which we are $(1 - \alpha)\, 100\,\%$ confident the population mean lies:
  $\bar{x} + Z_\alpha \dfrac{\sigma}{\sqrt{n}}$ or $\bar{x} + Z_\alpha \dfrac{s}{\sqrt{n}}$ (if σ is unknown)

Lower CI $\quad \bar{x} - Z_\alpha \sigma_{\bar{x}} \quad$ or $\quad \bar{x} - Z_\alpha s_{\bar{x}} \quad$ (if σ is unknown)   (7.5a)

Upper CI $\quad \bar{x} + Z_\alpha \sigma_{\bar{x}} \quad$ or $\quad \bar{x} + Z_\alpha s_{\bar{x}} \quad$ (if σ is unknown)   (7.5b)

Worked Example 7.2: Confidence interval for the mean σ unknown

A property investor claims that the average rental income per room in student accommodation is at most £ 5000 per year. The mean rent paid by a random sample of 36 students was calculated as £ 5200. The sample standard deviation was calculated as £ 735.

(a) Calculate a 90 % confidence interval for the true mean annual rental income. Do the sample results support the investor's claim?

(b) Calculate a (i) the lower limit; (ii) the upper limit for one-sided 95 % confidence intervals for μ. Describe each of these intervals verbally.

Solution

Sample size is greater than 30 and σ is unknown known but its estimate $s = 735$ was calculated from the sample.

Hence, the confidence interval is $\bar{x} \pm Z_{\alpha/2} s_{\bar{x}}$

To construct the confidence interval you need $\bar{x}$, $s_{\bar{x}} = \dfrac{s}{\sqrt{n}}$ and $Z_{\alpha/2}$

$\bar{x} = 5200$ was calculated from sample and given above.

$s = 735$ and sample size $n = 36$ were given. Hence $s_{\bar{x}} = \dfrac{s}{\sqrt{n}} = \dfrac{735}{\sqrt{36}} = 122.5$

$Z_{\alpha/2} = 1.6449$ when $\alpha/2 = 5\,\%$

(a) Hence, the 90 % confidence interval is
$\bar{x} \pm Z_{\alpha/2} s_{\bar{x}}$
$5200 \pm 1.6449(122.5)$
5200 ± 201.5
$5200 - 201.5$ to $5200 + 201.5$
4998.5 to 5401.5
This confidence interval contains 5000 and hence supports the investor's claim.

(b) You will need $Z_\alpha = 1.6449$ when $\alpha = 5\%$
  $\bar{x} = 5200$, $s_{\bar{x}} = 122.5$ and sample size $n = 36$ were given.
  (i) The 95 % lower confidence limit is $\bar{x} - Z_\alpha s_{\bar{x}} = 5200 - 1.6449(122.5) = \mathbf{4998.5}$.
     Hence we are 95 % confident that the mean rental income per room in student accommodation is equal to or greater than 4998.5.
     Similarly for (ii):
     The 95 % upper confidence limit is:
     $\bar{x} + Z_\alpha s_{\bar{x}} = 5200 + 1.6449(122.5) = \mathbf{5401.5}$.
     Hence we are 95 % confident that the mean rental income per room in student accommodation is equal to or less than 5401.5.

> 95 % confident that the mean is at least **4998.5**

> 95 % confident that the mean is at most **5401.5**

Progress Exercises 7.1: Confidence intervals for the mean

1. Explain the following terms in your own words with an appropriate example:
   (a) a parameter;
   (b) an estimator;
   (c) a point estimate;
   (d) a confidence interval.
2. (a) Give the formula used for the 'margin of error' for a (i) 95 % confidence interval; (ii) a 90 % confidence interval. Explain how these formulae are derived.
   (b) The air travel industry is concerned about the increase in the numbers of adults in the population who are overweight or obese. The mean weight of a random sample of 36 adults was calculated as 78 kg. The standard deviation for adult weights is known to be 36 kg.
      (i) Calculate a 95 % confidence interval for the mean weight of all adults. Describe this interval as you would to a non-statistician. Will this interval provide useful information?
      (ii) If the sample mean weight had been calculated from a sample of 84 adults how would this affect the standard error and the precision of a confidence interval?
3. State, with explanations, whether the following statements are true or false:
   (a) The population mean is always at the centre of the confidence interval.
   (b) If the sample size increases the width of confidence interval is reduced.
   (c) The greater the value of the population variance, σ the greater the width of the confidence interval for the population mean.
   (d) A 90 % confidence interval is more likely to contain the population mean than a 99 % confidence interval.
   (e) The formula $\bar{x} \pm Z_{\alpha/2}\, \sigma_{\bar{x}}$ may only be used if the sample is selected from a Normal population.
4. In the horticulture industry predicting the demand for products is of paramount importance, since most products are perishable. This is particularly true in the case of flowers. For example, growers are concerned about the demand for daffodils in January.
   (a) State possible target populations for a survey on the amount (money!) that consumers are willing to spend on daffodils. Briefly describe a suitable sampling method and method of contact for such a survey.

(b) Fifty households in a particular neighbourhood were surveyed on the amount they intend to spend on daffodils in January. The sample mean and standard deviation were calculated as £8.55 and £3.14, respectively.
Construct a 95 % confidence for the mean amount these households are willing to spend.

(c) If there are 940 households in the neighbourhood in (b), what is the minimum and maximum revenue that growers could expect from sales of daffodils in January with 95 % confidence?

5. The time taken to travel to and from work is one of the factors considered by new companies setting up in a neighbourhood. From a sample of 84 potential employees, the sample mean and standard deviation were calculated as 2.44 and 2.12 hours travelling time per day.

(a) Calculate a (i) 95 % (ii) 99.8 % confidence for the mean daily travelling time.

(b) Is it likely that the true mean travelling time is greater than 3.00 hours?

(c) Explain why it would not be unusual for an individual's daily travelling time to be outside the confidence limits.

6. Explain the difference between one-sided and two-sided confidence intervals. Online booking, particularly for low fares airlines or short haul flights is just one of the changes in the travel industry in recent years. A travel agent has estimated that his business is still healthy as the majority of current customers use the agency to arrange long-haul flights, business flights or package holidays and expenditure per booking is higher than in the past. Data for 104 randomly selected customers is given in Table PE7.1 Q6:

Table PE7.1 Q6   Data collected from 104 customers from the travel agent | File: Online flights |

	(i) Long-haul flights	(ii) Package holidays	(iii) Business
Numbers of customers	32	54	18
Average amount spent (€)	483.5	854	58.5
Sample standard deviation	35.5	28.4	12.2

For each group of customers (i), (ii) and (iii)

(a) Calculate a 95 % confidence interval for the mean amount spent. Describe each interval verbally, stating any assumptions made.

(b) From the confidence intervals in (a), can you conclude that the mean amount spent by the three categories of customers is different?

(c) Construct a one-sided confidence interval for the minimum amount spent by 90 % of customers. Describe each interval verbally.

7. **Excel file 'Savings 5'.**
(See question 15 PE5.4).
From all the data calculate a 90 % confidence interval for the mean amount saved. Is it likely that the average amount saved is £30 during the school year?

8. **Excel file 'Hours worked UK 2000'.**
(See question 12 PE2.4).
From the data given in this file calculate 95 % confidence intervals for the average hours worked per week for (a) males (b) females. Give a verbal description of these intervals and explain the high precision of the intervals.

7.2 Confidence intervals for proportions

In Chapter 6, the Central Limit Theorem (CLT) stated that sample means were Normally distributed

$$\bar{x} \sim N\left(\mu, \frac{\sigma}{\sqrt{n}}\right) \text{ for } n \geq 30.$$

Then based on the CLT, we derived the formula for the confidence interval for the mean as 'point estimate' $\pm$ (Z percentage point/critical point) $\times$ (standard error for the mean)

$\bar{x} \pm Z_{\alpha/2}\, \sigma_{\bar{x}}$ ...(7.3) or, if σ is unknown, $\bar{x} \pm Z_{\alpha/2}\, s_{\bar{x}}$ ...(7.4).

Similarly

In Chapter 6, the CLT stated that sample proportions were Normally distributed,

$$p \sim N\left(\pi, \sqrt{\frac{\pi(1-\pi)}{n}}\right) \text{ for } n \geq 30.$$

Based on the CLT, the formula for the confidence interval for the population proportion is given as 'point estimate' $\pm$ (Z percentage point/critical point) $\times$ (standard error for proportions)

$$p \pm Z_{\alpha/2} \sqrt{\frac{p(1-p)}{n}}$$

where the population proportion, π is estimated by the sample proportion p.

The $(1-\alpha)\,100\,\%$ confidence for the population proportion is

$$p \pm Z_{\alpha/2} \sqrt{\frac{p(1-p)}{n}} \tag{7.6}$$

Graphically, the $(1-\alpha)\,100\,\%$ confidence interval is represented in Figure 7.7. The interval estimate is centred on a sample proportion p with a margin of error, $\varepsilon = Z_{\alpha/2}\sqrt{\dfrac{p(1-p)}{n}}$.

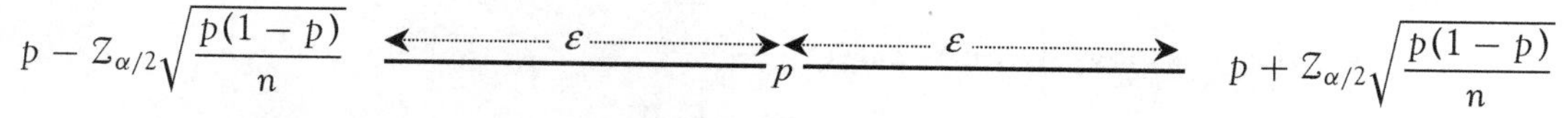

Figure 7.7   Construction of confidence interval for proportions.

The method for constructing confidence intervals for proportions is summarised in Chart 7.2:

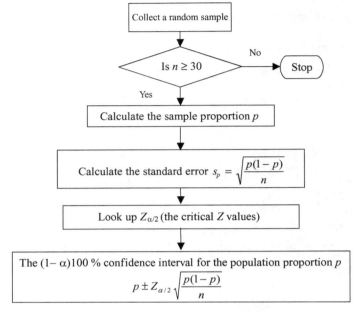

Chart 7.2   The method for constructing confidence intervals for proportions.

Worked Example 7.3: Confidence interval for population proportion

In a poll of 200 voters 88 stated that they will vote for the Green party candidate. Construct a 95 % confidence interval for the true proportion of support for the Green party candidate. Comment on the precision of the interval. Is the result useful to the Green party election team?

Solution

The confidence interval is given by (7.5), $p \pm Z_{\alpha/2} \sqrt{\dfrac{p(1-p)}{n}}$

You will need sample size, n, $Z_{\alpha/2}$ and the point estimate p. Then use p to calculate the standard error for proportions.

$n = 200$.

$Z_{\alpha/2} = 1.96$ when $\alpha/2 = 2.5\,\%$

$p = \dfrac{x}{n} = \dfrac{88}{200} = 0.44$.

Hence the standard error $= \sqrt{\dfrac{p(1-p)}{n}} = \sqrt{\dfrac{0.44(1-0.44)}{200}} = 0.0351$

Hence the confidence interval $= p \pm Z_{\alpha/2} \sqrt{\dfrac{p(1-p)}{n}}$

$$= 0.44 \pm (1.96)(0.0351)$$

$$= 0.44 \pm 0.0688$$

The population proportion is between 0.3712 and 0.5088 with 95 % confidence. Stated another way, we are 95 % confident that between 37.12 % and 50.88 % will vote for the Green party. This interval is too wide to be of any practical use. There is a very wide margin of error, 0.0688 or 6.88 percentage points.

Progress Exercises 7.2: Confidence intervals for proportions

1. 1000 commuters on a suburban bus route are asked to indicate whether they are very satisfied, satisfied or dissatisfied with the bus service. The results are summarised in Table PE7.2 Q1.

Table PE7.2 Q1   Survey on satisfaction with bus commuter service   File: Bus

	Age	Very satisfied	Satisfied	Dissatisfied	Row totals
Male	20–25 years	64	48	68	180
	26–45 years	30	82	28	140
	Over 46 years	30	40	10	80
Female	20–25 years	24	86	140	250
	26–45 years	48	48	184	280
	Over 46 years	24	16	30	70
	Column totals	220	320	460	1000

(a) From the sample of 1000 commuters, calculate (i) the proportion of males (ii) the proportion of females (iii) the proportion males aged between 20 and 25 years (iv) the proportion of females aged between 20 and 25 years (v) the proportion of very satisfied customers.

(b) Calculate the 90 % confidence for the proportion of all male commuters.

(c) Calculate the 90 % confidence for the proportion of commuters not satisfied.
    Give a verbal description of each interval (b) and (c). Do the intervals support the claim that (i) half of all commuters are male (ii) half of all commuters are dissatisfied with the bus service?

2. **See question 1:** Consider the female respondents in the survey in Table PE7.2 Q1. Calculate the 90 % confidence for the proportion of female commuters that are dissatisfied with the service. Does this interval support the claim that more than half of the female commuters are dissatisfied with the service?

3. In a random sample of 220 customers who called into a hardware shop, 88 made no purchase, 33 made purchases € 100 or less, the reminder made purchases exceeding € 100.

(a) Calculate a 95 % confidence interval for the proportion of customers that
    (i) made no purchases; (ii) spend € 100 or less; (iii) spend less than € 100.

(b) Do the above intervals support the claim that 50 % of those who come into the shop (i) make no purchases; (ii) make purchases exceeding € 100?

4. State, with explanations, whether the following statements are true or false:

(a) The population proportion is always at the centre of the confidence interval.

(b) If the sample size increases the width of confidence interval is reduced.

(c) The greater the level of confidence, the greater the width of the confidence interval.

(d) The width of a confidence interval for a population proportion is smallest when the sample proportion is 0.5 (assuming the same sample sizes).

5. A Health and Safety survey of 200 industrial accidents claims revealed that 53 were due to untidy working conditions.
   (a) Calculate a 90 % confidence interval for the proportion of accidents that are due to untidy working conditions.
   (b) Calculate the 'margin of error' at 99 % confidence.

7.3   The precision of confidence intervals for the mean

It has already been noted that very wide interval estimates are of little practical use. For example, it would be much more useful to Green party activists to know that the percentage of votes for the party is within 44 % ± 2 % rather than the 44 % ± 7 % or between 37 % and 51 % as in Worked Example 7.2. It has been noted several times that increasing sample size results in a reduction in the width or precision of a confidence interval. To calculate the exact sample size required to give an interval estimate of a specified precision, return to the formulae used to calculate confidence intervals for (i) means; (ii) proportions.

7.3.1   The precision of confidence intervals for the mean

The precision of the confidence interval $\bar{x} \pm Z_{\alpha/2}\sigma_{\bar{x}}$ may be stated in terms of the width (w) of the confidence interval or in terms of the margin of error (ε) where $w = 2 \times \varepsilon$ as shown in Figure 7.6 (reproduced here to aid discussion).

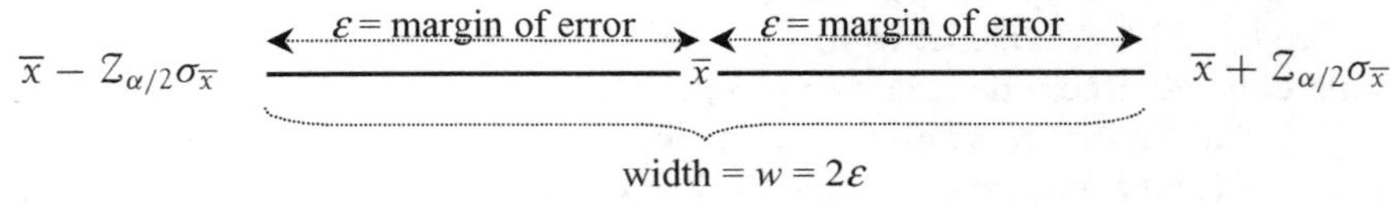

$$\bar{x} - Z_{\alpha/2}\sigma_{\bar{x}} \qquad \underset{\varepsilon\,=\,\text{margin of error}}{\longleftrightarrow} \quad \bar{x} \quad \underset{\varepsilon\,=\,\text{margin of error}}{\longleftrightarrow} \qquad \bar{x} + Z_{\alpha/2}\sigma_{\bar{x}}$$

$$\text{width} = w = 2\varepsilon$$

Figure 7.6   The width of the interval is twice the 'margin of error'.

The size of the margin of error, $\varepsilon = Z_{\alpha/2}\dfrac{\sigma}{\sqrt{n}}$ is determined by values of σ, $Z_{\alpha/2}$ and n.

The effect of σ, $Z_{\alpha/2}$ and n on the precision of the interval $(2 \times \varepsilon)$ is discussed below.

The population standard deviation σ is fixed for a given population; therefore we cannot adjust σ. But it is worth noting that the margin of error (and hence the width) will be wider when the population standard deviation is large.

The level of confidence: the greater the level of confidence, the greater the value of the confidence coefficient, $Z_{\alpha/2}$ and hence the wider the interval. For example, $Z_{\alpha/2} = 1.96$ for 95 % confidence but 2.5758 for 99 % confidence. Therefore a more precise interval estimate could be achieved at the expense of lower confidence. This is not usually an option: a lower level of confidence reduces the credibility of the estimate.

The sample size, n. The sample size is the divisor in the formula for the margin of error: $\varepsilon = \dfrac{Z_{\alpha/2}\sigma}{\sqrt{n}}$. Therefore when n is large, the margin of error will be small and hence the interval will be narrow and 'more precise'.

For example, consider a $(1 - \alpha)100\,\%$ confidence interval calculated from a random sample from a population whose standard deviation σ is known then

(i) for $n = 100$, the maximum margin of error, $\varepsilon = Z_{\alpha/2}\dfrac{\sigma}{\sqrt{100}} = \dfrac{Z_{\alpha/2}\sigma}{10} = \dfrac{1}{10}\left(Z_{\alpha/2}\sigma\right)$.

(ii) for $n = 4$, the maximum margin of error, $\varepsilon = Z_{\alpha/2}\dfrac{\sigma}{\sqrt{4}} = \dfrac{Z_{\alpha/2}\sigma}{2} = \dfrac{1}{2}\left(Z_{\alpha/2}\sigma\right)$.

The interval for $n = 4$ in (ii) is 5 times greater, $5 \times \left(\dfrac{Z_{\alpha/2}\sigma}{10}\right)$, than the interval for $n = 100$ in (i).

Hence, sample size has a major effect on the width of a confidence interval. If time and cost permit then collecting a larger sample will dramatically improve the precision of interval estimates.

7.3.2   Sample size for a confidence interval of a given precision

To calculate the sample size that will result in a confidence interval of a specified precision ($\pm\,\varepsilon$), substitute the required value for ε into equation (7.7a), then solve for n (ε is sometimes called 'tolerable error' or 'error bound').

$$\varepsilon = \frac{Z_{\alpha/2}\sigma}{\sqrt{n}} \text{ is the precision, } \varepsilon, \text{ of a } (1 - \alpha)100\,\% \text{ CI for means} \tag{7.7a}$$

Alternatively use (7.7b) which gives the solution for (7.7a).

$$n = \left(\frac{Z_{\alpha/2}\sigma}{\varepsilon}\right)^2 \text{ is the sample size for } (1 - \alpha)100\,\% \text{ CI for } \mu, \text{ precision} \pm \varepsilon \tag{7.7b}$$

The sample size for a confidence interval for proportions with precision $\pm\varepsilon$.

Similarly, to calculate the sample size that will give a confidence interval for proportions with a specified precision ($\pm\varepsilon$), substitute the required value for ε into equation (7.8a) and solve for n.

$$\varepsilon = Z_{\alpha/2}\sqrt{\frac{p(1 - p)}{n}} \text{ is the precision, } \varepsilon, \text{ for } (1 - \alpha)100\,\% \text{ CI for proportions} \tag{7.8a}$$

Alternatively n may be calculated directly from the solution to (7.8a) given in (7.8b).

$$n = \left(\frac{z_{\alpha/2}}{\varepsilon}\right)^2 \times p(1-p) \text{ is the sample size for } (1-\alpha)100\% \text{ CI for } p, \text{ precision } \pm\varepsilon \qquad (7.8b)$$

BUT, in estimation, π is usually unknown. It is estimated by the sample proportion p. It can shown (with some maths) that the largest value of n (and hence the narrowest or most precise interval) is found when $p = 0.5$. See Appendix G.

For maximum precision, substitute $p = 0.5$ for p in (7.8b).

Hence $\varepsilon = z_{\alpha/2}\sqrt{\dfrac{0.5(1-0.5)}{n}} = z_{\alpha/2}\sqrt{\dfrac{(0.5)^2}{n}} = z_{\alpha/2}\dfrac{0.5}{\sqrt{n}}$

Then solve for n. The solution is given in (7.8c).

$$n = \left(\frac{z_{\alpha/2}}{2\varepsilon}\right)^2 \text{ is the sample size for } (1-\alpha)100\% \text{ CI for } p \text{ for precision } \leq\varepsilon \qquad (7.8c)$$

See worked Example 7.4.

Worked Example 7.4: Sample size for confidence intervals of given precision

(a) (i) For the data in Worked Example 7.1, calculate the sample size that will give a 99 % confidence interval for the population mean with a margin of error ±0.5 when $\sigma = 1.8$.
   (ii) Confirm your answer by recalculating the margin of error for the 99 % confidence interval.
(b) (i) For the data in Worked Example 7.3, calculate the sample size that will give a 95 % confidence interval with a margin of error ±0.01, for the population proportion when p is unknown.
   (ii) Confirm your answer by recalculating the margin of error for the 95 % confidence interval.

Solution

(a) (i) Calculate n from either $\varepsilon = \dfrac{z_{\alpha/2}\sigma}{\sqrt{n}}$ ...(7.7a) or $n = \left(\dfrac{z_{\alpha/2}\sigma}{\varepsilon}\right)^2$ ...(7.7b)

$$\bar{x} - z_{\alpha/2}\frac{\sigma}{\sqrt{n}} \quad \longleftarrow \quad \varepsilon \quad \longrightarrow \!\!\longleftarrow \quad \varepsilon \quad \longrightarrow \quad \bar{x} + z_{\alpha/2}\frac{\sigma}{\sqrt{n}}$$
$$\bar{x}$$

Since $\varepsilon = 0.500$. $\sigma = 1.8$ (given) and $z_{0.005} = 2.5758$ for 99 % confidence. Substitute the values for ε, σ and $Z_{\alpha/2}$ into (7.7a).

$$\varepsilon = \frac{Z_{\alpha/2}\sigma}{\sqrt{n}}$$

$$0.50 = \frac{(2.5758)(1.8)}{\sqrt{n}}$$

$$\sqrt{n} = \frac{(2.5758)(1.8)}{0.50} = 9.27288$$

$$n = (9.27288)^2 = 85.986$$

Or use (7.7b) $n = \left(\frac{Z_{\alpha/2}\sigma}{\varepsilon}\right)^2 = \left(\frac{2.5758 \times 1.8}{0.50}\right)^2 = 85.986$

Therefore a sample size of $n = 86$ will give a 99 % confidence interval estimate for μ within 0.50 of the point estimate ($\bar{x} \pm 0.500$).

(ii) Recalculate the margin of error for the 99 % confidence interval for μ

$$\varepsilon = Z_{\alpha/2}\frac{\sigma}{\sqrt{n}} = 2.5758\frac{1.8}{\sqrt{86}} = 0.49996.$$

Hence margin of error is within the required specification. ±0.50.

(b) (i) Since π is unknown, use either (7.8b) or (7.9) with $p = 0.5$ to calculate the sample size to give the 95 % confidence interval of width $2 \times \varepsilon = 0.2$.

Substitute $\varepsilon = 0.01$, $Z_{0.025} = 1.96$ and $p = 0.5$ into (7.8b)

$$0.01 = 1.96\sqrt{\frac{(0.5)^2}{n}}$$

$$\frac{0.01}{1.96} = \sqrt{\frac{(0.5)^2}{n}} \dots \text{dividing both sides by 1.96}$$

$$\left(\frac{0.01}{1.96}\right)^2 = \frac{(0.5)^2}{n} \quad \dots \text{squaring both sides}$$

$$n = \frac{(0.5)^2}{\left(\frac{0.01}{1.96}\right)^2} = \frac{0.25}{0.0000026} = 9604 \dots \text{solving for } n$$

Or, take the easier option and use equation (7.9): $n = \left(\frac{Z_{\alpha/2}}{2\varepsilon}\right)^2 = \left(\frac{1.96}{2(0.01)}\right)^2 = 9604.$

(ii) $\varepsilon = Z_{\alpha/2}\sqrt{\frac{\bar{p}(1 - \bar{p})}{n}} = 1.96\sqrt{\frac{0.5(1 - 0.5)}{9604}} = 0.01$

The margin of error is exactly equal to the specified maximum error required.

Progress Exercises 7.3: Confidence intervals: precision and sample size

---

1. (a) Define the 'precision' of a confidence interval.
   (b) State the factors that determine the precision of confidence intervals for means.
2. The following are three 95 % confidence intervals for a population mean calculated from three independent samples: [22.00, 23.96], [22.87, 26.13], [27.31, 28.69].
   (a) State which of the intervals is the most 'precise'.
   (b) Calculate the value of the sample mean in each interval.
   (c) Calculate the sample sizes for each of the confidence intervals, given that the population standard deviation is 5.
3. See question 4, progress exercises 7.1. The horticulture industry requires an estimate of the mean amount consumers are willing to spend within $\pm £\, 0.5$ with 98 % confidence. Calculate the sample size that should be selected, using the sample standard deviation $s = £\, 3.14$.
4. (a) State the factors that determine the precision of a confidence interval for the population proportion.
   (b) The 95 % confidence intervals for p: [0.28, 0.52], [0.32, 0.48] were calculated from two independent random samples.
   (i) State which of the confidence intervals for p is the most 'precise'.
   (ii) Show that the sample proportion was $p = 0.40$ for each interval.
   (iii) Calculate the sample size used in the calculation of each interval.
5. Calculate the value of product pq for values of $p = 0, 0.1, 0.2 \ldots 0.9, 1$.

   Do the calculated values of pq suggest that largest value of $\sqrt{\dfrac{pq}{n}}$ could be at ($p = q = 0.50$)?

6. In question 5, PE7.2, calculate the sample size that the Health and Safety researchers should select if the proportion of industrial accidents due to unsafe working conditions is to be estimated within two percentage points with 90 % confidence.
7. An opinion poll of 410 voters taken a week before an election showed that 138 intended to vote for the current party in power (referred to as Party A), 102 for the largest opposition party (Party B), 67 for others, 82 don't know, 21 won't vote at all.
   (a) Calculate a 95 % confidence interval for the proportion that will vote for the current party in power. What is the margin of error for this estimate? Is this party likely to gain more than 50 % of the vote?
   (b) Calculate a 95 % confidence interval for the proportion that will vote for the main opposition party.
   (c) Calculate a 95 % confidence interval for the proportion that will not vote for the current party in power.
   (d) Would the inclusion of don't knows with the groups in (a), (b) or (c) make a difference?
   (e) What sample size should be selected of the maximum error in the estimates is to be within one percentage point?

7.4 Confidence intervals for differences between means and proportions

While the estimation of a single population mean or proportion is important, there are situations where we may be more interested in estimating the difference between two means or proportions. For example, we may be interested in whether the percentage that intend to vote for party B is higher that for party A or whether commuting time is faster by train than by car etc.

7.4.1 Difference between means

In Chapter 5 equation (5.7), it was stated that the distribution for the differences between two Normal independent random variables was also Normal, with a mean equal to the difference between the two means and the variance equal to the sum of the two variances.

If $X_1 \sim N(\mu_1, \sigma_1^2)$ and $X_2 \sim N(\mu_2, \sigma_2^2)$ then $(X_1 - X_2) \sim N(\mu_1 - \mu_2, \sigma_1^2 + \sigma_2^2)$.

Similarly, since the distribution of sample means is Normally distributed $(n \geq 30)$

$\overline{X}_1 \sim N\left(\mu_1, \dfrac{\sigma_1^2}{n_1}\right)$ and $\overline{X}_2 \sim N\left(\mu_2, \dfrac{\sigma_2^2}{n_2}\right)$ is given by (7.9), then the distribution of the differences between every possible pair of sample means is given by (7.9).

$$\text{Distribution for } \overline{X}_1 - \overline{X}_2 \sim N\left(\mu_1 - \mu_2, \frac{\sigma_1^2}{n_1} + \frac{\sigma_2^2}{n_2}\right) n_1 \text{ and } n_2 \geq 30 \qquad (7.9)$$

The distribution of the differences between every possible pair of sample means is illustrated graphically in Figure 7.8a:

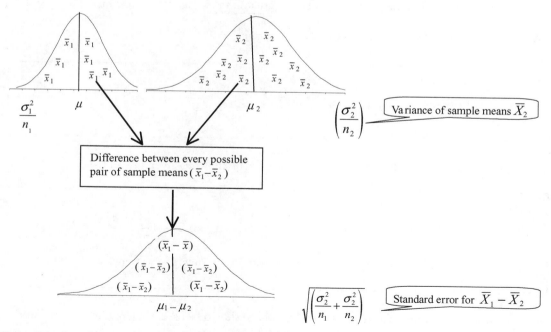

Figure 7.8   Distribution for difference between means: n_1 and $n_2 \geq 30$.

Confidence intervals for differences between population means

Since the difference between sample means is Normally distributed the confidence interval is described, in words, in equation (7.4) as

Point estimate $\pm$ (Percentage point or critical value) $\times$ (Standard error)

The point estimate for the difference between two population means $(\mu_1 - \mu_2)$ is the difference between the corresponding sample means $(\bar{x}_1 - \bar{x}_2)$: the standard error for the difference between sample means is given in (7.10a).

$$\text{The standard error for } \overline{X}_1 - \overline{X}_2 \text{ is } \sigma_{\bar{x}_1 - \bar{x}_2} = \sqrt{\frac{\sigma_1^2}{n_1} + \frac{\sigma_2^2}{n_2}} \qquad (7.10a)$$

Hence, (7.11a) is the confidence interval for $(\mu_1 - \mu_2)$ the difference between population means.

$$\text{The } (1 - \alpha)\,100\,\%\,\text{CI for } (\mu_1 - \mu_2)\colon\ (\bar{x}_1 - \bar{x}_2) \pm Z_{\alpha/2} \sqrt{\frac{\sigma_1^2}{n_1} + \frac{\sigma_2^2}{n_2}} \qquad (7.11a)$$

If the sample sizes are 30 or more and σ_1 and/or σ_2 are unknown they may be estimated by s_1 and s_2. The sample standard error is estimated as $s_{\bar{x}_1 - \bar{x}_2} = \sqrt{\dfrac{s_1^2}{n_1} + \dfrac{s_2^2}{n_2}}$.

$$\text{The sample standard error for } \overline{X}_1 - \overline{X}_2 \text{ is } \sigma_{\bar{x}_1 - \bar{x}_2} = \sqrt{\frac{s_1^2}{n_1} + \frac{s_2^2}{n_2}} \qquad (7.10b)$$

The confidence interval for $(\mu_1 - \mu_2)$, σ_1 and/or σ_2 are unknown is given in (7.11b):

$$\text{Confidence interval for } (\mu_1 - \mu_2)\colon\ (\bar{x}_1 - \bar{x}_2) \pm Z_{\alpha/2} \sqrt{\frac{s_1^2}{n_1} + \frac{s_2^2}{n_2}} \qquad (7.11b)$$

$$\text{where } n_1 \text{ and } n_2 \geq 30, \text{ unknown variance estimated by } s_1^2 \text{ and } s_2^2$$

Strictly speaking, the t-percentage point (see Chapter 8) should be used when σ is unknown, but the Z percentage point is a good approximation for large n.

7.4.2 Difference between proportions

The sample proportions are Normally distributed for n_1 and $n_2 \geq 30$ according to the Central Limit Theorem, Chapter 6, formula (6.9) as follows

$$p_1 \sim N\left(\mu_{p_1} = \pi_1, \sigma^2_{p_1} = \frac{\pi_1(1 - \pi_1)}{n_1}\right) \text{ and } p_2 \sim N\left(\mu_{p_2} = \pi_2, \sigma^2_{p_2} = \frac{\pi_2(1 - \pi_2)}{n_2}\right)$$

Hence the difference between sample proportions is also Normally distributed

$$p_1 - p_2 \sim N\left(\mu_{p_1-p_2} = \pi_1 - \pi_2, \quad \sigma^2_{p_1-p_1} = \frac{\pi_1(1 - \pi_1)}{n_1} + \frac{\pi_2(1 - \pi_2)}{n_2}\right) \qquad (7.12)$$

The probability distribution for the difference between sample proportions is illustrated graphically in Figure 7.7b:

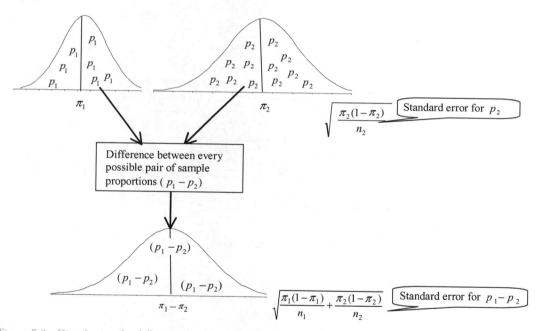

Figure 7.9   Distribution for difference between proportions: n_1 and $n_2 \geq 30$.

Confidence intervals for differences between population proportions

Since the difference between sample proportions is Normally distributed the confidence interval is described, in words, in equation (7.4) as

Point estimate ± (Percentage point or critical value) × (Standard error)

The point estimate for the difference between two populations is $(p_1 - p_2)$, the standard error for the difference between sample proportions is given in (7.13).

$$\text{Sample standard error for } p_1 - p_2: \ S_{p_1-p_2} = \sqrt{\frac{p_1(1 - p_1)}{n_1} + \frac{p_2(1 - p_2)}{n_2}} \qquad (7.13a)$$

Note: p_1 and p_2 are the point estimates of the unknown population proportions π_1 and π_2.
Hence the confidence interval for the difference between population proportions $(\pi_1 - \pi_2)$ is:

$$\text{Confidence interval for } \pi_1 - \pi_2: \ \ p_1 - p_2 \pm Z_{\alpha/2}\sqrt{\frac{p_1(1 - p_1)}{n_1} + \frac{p_2(1 - p_2)}{n_2}} \qquad (7.13b)$$

These methods are summarised in Chart 7.3:

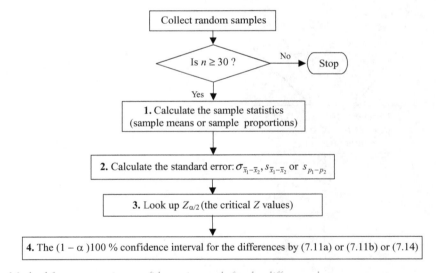

Chart 7.3   Method for constructing confidence intervals for the difference between means or proportions.

Interpreting confidence intervals for differences between means or proportions

The objective in calculating confidence intervals for difference between means (or proportions) is to determine whether a difference does exist between the population parameters and if so to estimate the size of the difference. Guidelines for the interpretation of the intervals for difference between means (or proportions) are as follows:

Let the confidence interval for the difference between means be $A < \mu_1 - \mu_2 < B$.

Case 1: If A and B are both positive, then we can be $(1 - \alpha)100\%$ confident that μ_1 is greater than μ_2 by an amount ranging from A to B.

Case 2: If A and B are both negative, then we can be $(1 - \alpha)100\%$ confident that μ_1 is less than μ_2 by an amount ranging from A to B.

Case 3: If A and B have different signs then the interval contains $\mu_1 - \mu_2 = 0$. Then we can be $(1 - \alpha)100\%$ confident that A is equal to B.

These guidelines refer to population means and proportions. Care must be taken about making inference about difference between individual values. See Worked Example 7.5(b)(ii).

Worked Example 7.5: Confidence intervals or the difference between means

Designers of rowing equipment investigate the difference between the mean weights (kg) of male and female rowing teams. Random samples of male and female rowers are selected: the sample sizes and average weights and sample standard deviations are given in Table 7.1:

Table 7.1 Sample mean and standard deviations for random samples of male and female rowers

	Male rowers	Female rowers
Sample size	42	30
Sample mean	60.5	52.6
Sample standard deviation	6.8	4.5

(a) Calculate the 95 % confidence interval for the difference in mean weights between male and female rowers. Describe your results verbally.

(b) What inference can be drawn from your results about (i) the difference between the population means; (ii) the difference between the individuals in each population?

Solution

To calculate the confidence interval for difference between population means when variances are unknown is $(\bar{x}_1 - \bar{x}_2) \pm Z_{\alpha/2}\sqrt{\dfrac{s_1^2}{n_1} + \dfrac{s_2^2}{n_2}}$ ...(7.11b).

You will need to calculate $(\bar{x}_1 - \bar{x}_2)$; the standard error and $Z_{\alpha/2}$

1. $(\bar{x}_1 - \bar{x}_2) = 60.5 - 52.6 = 7.9$

2. The standard error is $\sqrt{\dfrac{s_1^2}{n_1} + \dfrac{s_2^2}{n_2}} = \sqrt{\dfrac{(6.8)^2}{42} + \dfrac{(4.5)^2}{30}} = \sqrt{1.7760} = 1.3326$

3. $Z_{\alpha/2} = 1.96$, for 95 % confidence

(a) Substitution of the above into (7.12b) gives

$$(\bar{x}_1 - \bar{x}_2) \pm Z_{\alpha/2}\sqrt{\frac{s_1^2}{n_1} + \frac{s_2^2}{n_2}} = 7.9 \pm 1.96(1.3326) = 7.9 \pm 2.6119$$

Hence the confidence interval for the difference between population means is $5.2881 \le (\mu_1 - \mu_2) \le 10.5119$. **In words**, we are 95 % confident that the difference between the mean weights of male and female rowers range from 5.2881 to 10.5991.

(b) (i) Both limits of the interval for $(\mu_1 - \mu_2)$ are positive, thus indicating that μ_1 (average weight for male rowers) is greater than μ_2 (average weight for female rowers with 95 % confidence).

The lower confidence limit: $(\mu_1 - \mu_2) = 5.2881$. Hence μ_1 is greater than μ_2 by 5.2881. **The upper confidence limit:** $(\mu_1 - \mu_2) = 10.5991$. Hence μ_1 is greater than μ_2 by 10.5991.

We are 95 % confident that the mean weights of male rowers exceeds the mean weight for female rowers by 5.2881 to 10.5991 kg.

(ii) While we are very confident that the mean weight of male rowers is greater than the mean weight of female rowers we cannot assume that any individual male rower will be heavier than any individual female rower. This is because the variance for means is $\sigma_{\bar{x}}^2 = \dfrac{\sigma^2}{n}$ while the variance for individual values is σ^2, which is n times greater than the variance for means.

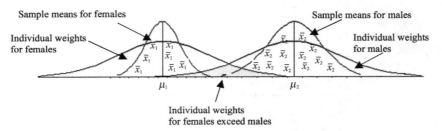

Figure 7.10 Sample means for males exceed sample means for females but some individual female weights exceed individual male weights (shaded area).

Worked Example 7.6: Confidence intervals: difference between proportions

Table 7.4 gives the results for polls taken in 2 localities.

Table 7.2 Opinion polls in A and B

	Area A (Data from Worked Example 7.2)	Area B
Sample size	200	160
Vote for Green party	88	54

(a) Calculate the 90 % confidence interval for the difference in the proportions who will vote for the Green party in areas A and B.

(b) Describe your results verbally. What inference can be drawn from your results?

Solution

To calculate the confidence interval for difference between population proportions is

$$p_A - p_B \pm Z_{\alpha/2}\sqrt{\frac{p_A(1 - p_A)}{n_A} + \frac{p_B(1 - p_B)}{n_B}} \tag{7.14}$$

you will need to calculate $(p_A - p_B)$; the standard error and $Z_{\alpha/2}$

1. $(p_A - p_B) = \dfrac{88}{200} - \dfrac{54}{160} = 0.44 - 0.3375 = 0.1025$
2. The standard error is

$$s_{P_A - P_B} = \sqrt{\frac{p_A(1 - p_A)}{n_A} + \frac{p_B(1 - p_B)}{n_B}} = \sqrt{\frac{0.44(1 - 0.44)}{200} + \frac{0.3375(1 - 0.3375)}{160}}$$

$$= \sqrt{0.001232 + 0.0013975} = 0.0513$$

3. $Z_{\alpha/2} = 1.6449$, for 90 % confidence
   (a) Substitution of the above into (7.13) gives $-0.0225 \pm 1.6449(0.0513) = 0.1025 \pm 0.0843$
   Hence the 90 % confidence interval for the difference between population proportions is $0.0182 \le (\pi_A - \pi_B) \le 0.1868$.
   (b) **In words**, we are 90 % confident that the difference between the two population proportions is between 0.0182 and 0.1868.
   The lower confidence limit: $= 0.0182$
   Hence π_A is greater than π_B by 0.0182
   The upper confidence limit: $= 0.1868$
   Hence π_A is greater than π_B by 0.1868
   Hence we are 90 % that π_A is greater than π_B by anything from 0.0182 to 0.1868. Since the interval does not contain zero we are 90 % confident that π_A is greater than π_B.

Progress Exercises 7.4: Confidence intervals for differences between means and proportions

1. Explain the derivation of the formulae for confidence intervals for the difference between population means. Based on a confidence interval, how would you decide whether (i) there is no difference; (ii) there is a difference between population means at a given level of confidence?
2. State whether there is a difference between means in the following 99 % confidence intervals for $(\mu_1 - \mu_2)$. If there is a difference, state which mean is greatest.
   (a) 4.56 ± 5.24; (b) 110.4 ± 25.35; (c) -1.4375 ± 1.4214.
3. The county council manager claims that the mean time taken to process online motor vehicle tax renewals is the same for all district offices. The mean time for random samples of 100 renewals in two district council offices (The Hill and Park Lane) was calculated as 6.72 and 7.24 minutes respectively. Calculate the 95 % confidence intervals for the difference between the population mean times between the two offices if standard deviation for each office is 1.50 minutes.
   Does your result support the manager's claim?
   Would the conclusion be the same if a 99 % instead of a 95 % confidence was calculated?

4. **Refer to question 3**. Suppose the claim that the standard deviation is 1.50 minutes the same for all district offices is in doubt. Hence the sample variance was calculated from the data for each of the samples:

Table PE7.4 Q4   Difference between means   File: Motor tax 1

| Motor Tax Offices | Calculations from sample data (time in minutes) | | |
	Size	Mean	Variance
The Hill	100	6.72	2.88
Park Lane	100	7.24	1.48

(a) Calculate the standard error for the difference between the population means.
(b) Calculate (i) the 95 % and (ii) the 99 % confidence intervals for the difference between the population means. Does the result support them managers claim that the mean processing time is the same for all offices?
(c) Explain why the confidence intervals in (a) and (b) are different.

5. A survey of regular customers in (i) restaurants and (ii) pubs were asked whether they approve of non-smoking in their restaurant/pub. Their responses are given below:

Table PE7.4 Q5   Difference between proportions

	Restaurant customers	Pub customers
Approve	64	36
Total surveyed	112	72

(a) Write down the proportion of (i) restaurant and (ii) pub customers who approve of smoking.
(b) Construct a 95 % confidence interval for the difference in the proportions that approve of smoking in restaurant and pubs. What do you conclude?

6. A pharmaceutical company has developed a new medication for the relief of tension headaches. In an initial test on the effectiveness of the new medication a random sample of 40 patients who use the existing treatment are asked to record their recovery times for each medication (on different occasions!). The sample means and variances are given in Table PE7.4 Q6:

Table PE7.4 Q6   Difference between means   File: Headache

| | Sample calculations (time in minutes) | | |
	Size	Mean	Variance
Existing medication	40	50	8.43
New medication	40	43	6.45

(a) Calculate a 90 % confidence interval for the difference between the mean recovery times. Does this interval support the claim that recovery time with the new medication is 'faster'?
(b) Could the results given above be biased? Describe how a (i) blind; (ii) double blind experiment would reduce bias in the results.

7. What are the assumptions necessary for the use of formula (7.13b) to calculate the confidence interval for the difference between proportion?

In a dental practice the number of patients who do not show up for appointments is thought to be particularly high on Mondays. From past records, the number of 'no shows' on randomly selected Mondays and Tuesdays are as follows:

Table PE7.4 Q7   Difference between proportions   File: No shows dentist

	Monday	Tuesday
Number of appointments	248	240
Number of 'no shows'	52	31

Calculate a 98 % confidence interval for the difference in the proportions of 'no shows' between Mondays and Tuesdays. Interpret your results.

8. The manufacturing company CoolSnap are concerned about the high proportion of domestic fridges that are returned within the guaranteed time of one year. Records show that the thermostat failed in the majority of fridges returned. The design team subsequently developed a new type of thermostat (thermo B). 198 of the new thermostats and 250 of the existing thermostats (thermo A) were tested on an accelerated life test and the numbers failing with the 'year' recorded.

The results for the existing and new thermostats are given as follows:

Table PE7.4 Q8   Difference between proportions (thermostats)   File: Thermo

	Thermo A	Thermo B
Sample size	250	198
Number failed	20	10

(a) Calculate the 95 % confidence interval for proportion of failed thermostats for (i) thermo A; (ii) thermo B. Could you conclude that there is a difference between the new and existing types of thermostat?

(b) Calculate the 95 % confidence interval for the difference between the proportion of failed thermostats for thermo A and thermo B. Could you conclude that there is a difference between the new and existing types of thermostat? Does this conclusion agree with that in (a)?

(c) On the basis of these results, should the company replace the existing thermostat with the new type?

9. **See question 8.**

(a) Explain why the interval estimate for the existing thermostat is more precise than that for the new thermostat.

(b) Would the results in question 8 be different if the all the figures in the table above are multiplied by four (i.e., 80 existing types failed out of 1000 within the year while 40 new types failed out of 792)? Give explanations for the answers.

10. A mobile phone provider is alarmed about the numbers of subscribers that appeared to be moving to a competing provider during the previous month. The company tracks a sample of customers who have left. The results were sorted into those who had left within the last month and those who left earlier and whether or not they had moved to the competitor.

Table PE7.4 Q10   Difference between proportions   File: Mobile move

	Left within the month	Left more than one month ago	Totals
Moved to competitor	422	326	748
Moved to other provider	154	298	452
Totals	576	624	1200

(a) Calculate the proportion who moved to the competitor out of those (i) who left within the last month; (ii) who left more than one month ago
(b) Calculate the 90 % confidence interval for the difference between the proportions who moved to the competitor in the last month and more than one month ago. Is the provider's alarm justified?

11. An article in a daily newspaper headed 'Signs of a growing gun culture' quoted the following figure: possession of guns increased by 16 % in 2005 to 424: discharging of firearms increased by 7 % to 313.
(a) Calculate (i) the number of guns in possession before 2005; (ii) the number of 'discharging of firearms' before 2005.
(b) Calculate the 'discharging of firearms' as a proportion of 'guns in possession' in (i) 2005 and (ii) pre-2005. Hence calculate the 95 % confidence interval for the difference between these proportions. Give a verbal interpretation of the confidence interval.

12. Health advisory services state that 50 % of heart attack patients die before reaching hospital: the chances of surviving a heart attack are dramatically increased by receiving medical help as quickly as possible. A survey on 420 heart attack victims carried out by a school of nursing discovered men sought help more urgently than women.

Table PE7.4 Q12   Difference between proportions   File: Heart attack

	Number	Mean time to reach hospital after the initial attack (hours)	Referred by GP	Time in hospital before being treated (hours)
Men	318	$\bar{x} = 3.40: s = 1.42$	65 %	$\bar{x} = 1.24: s = 0.25$
Women	102	$\bar{x} = 8.24: s = 4.24$	72 %	$\bar{x} = 1.43: s = 0.38$

(a) Calculate the proportion of (i) men and (ii) women who were referred by their GP. Hence calculate the 95 % confidence interval for the difference between the proportion of men and women who were referred by their GP.
(b) Calculate the 95 % confidence interval for the difference between the mean times in hospital before receiving treatment.

(c) Is it possible to infer that there is no difference between the mean times taken to reach hospital from the data given above?

(d) Based on the results in (a) and (b) above describe the difference between the behaviour and treatment of male and female heart attack victims.

13. **Excel file** 'Savings 5'

This file gives the amount saved by five groups of children.

(a) How many different pairs can be selected from these five groups?

(b) Calculate a 90 % confidence interval for the difference in the mean amount saved by the group with the lowest and the group with the highest average. Can you conclude that the mean amount saved is different for all groups?

14. **Excel file** 'Hours worked UK 2000'

From the data given in this file calculate 95 % confidence interval for the difference between the average hours worked per week by males and females. What can you conclude? (see PE2.4, question 12)

Summary and overview of confidence intervals for means and proportions

See also summary charts: Chart 9.2 and Chart 9.3.

7.1 Confidence intervals for means

It was shown in Chapter 6 that it is difficult to determine the accuracy of **a point estimate** for an unknown **population parameter**.

Based on the sampling distribution of means for samples of size, $n \geq 30$, an interval estimate for μ is calculated by the formula

Point estimate $\pm$ 'margin of error'

Common confidence interval (CI) for μ	**$(1 - \alpha)100\,\%$ confidence interval for μ**
The 95 % CI for μ is $\bar{x} \pm 1.96\sigma_{\bar{x}}$ The 99 % CI for μ is $\bar{x} \pm 2.5758\sigma_{\bar{x}}$ The 90 % CI for μ is $\bar{x} \pm 1.6449\sigma_{\bar{x}}$	Area in tail $= \alpha/2$ Area in the middle is $(1 - \alpha)$. See Figure 7.5 Hence the $(1 - \alpha)\,100\,\%$ confidence interval is $\bar{x} \pm Z_{\alpha/2}\,\sigma_{\bar{x}}$

Generalising from a 95 % confidence interval to a $(1 - \alpha)\,100\,\%$ confidence interval.

When σ is unknown use the sample standard error: $s_{\bar{x}} = \dfrac{s}{\sqrt{n}}$, hence the approximate CI is $\bar{x} \pm Z_{\alpha/2}\,s_{\bar{x}}$.

The 'margin of error' takes into account (i) σ, the variation in the population; (ii) the level of confidence I $Z_{\alpha/2}$ and (iii) n, the sample size.

Terminology

The upper and lower limits of a confidence interval are referred to as the '**Upper and lower confidence limit**' $(\overline{x} + Z_{\alpha/2}\sigma_{\overline{x}})$ and $(\overline{x} - Z_{\alpha/2}\sigma_{\overline{x}})$ respectively and $Z_{\alpha/2}$ is called the '**critical value**' or the '**percentage point**'.

7.2   Confidence intervals for proportions

Since the sampling distribution of proportions is approximately Normal for $n \geq 30$ the confidence interval for the population proportion is

$$\overline{p} \pm Z_{\alpha/2}\sqrt{\frac{\overline{p}(1-\overline{p})}{n}}$$

One-sided confidence intervals may also be calculated: $\overline{x} + Z_{\alpha}\sigma_{\overline{x}}$ and $\overline{x} - Z_{\alpha}\sigma_{\overline{x}}$ are referred to as upper and lower confidence intervals respectively.

7.3   The precision of a confidence interval

A confidence interval is described as 'precise' when the 'width' of the interval is narrow: The width (precision) of the interval is given by $Z_{\alpha/2}\sigma_{\overline{x}}$. Hence the precision is greater when $Z_{\alpha/2}$ and σ are small and the sample size n is large.

The sample size that will give a confidence interval of precision, $\pm\varepsilon$ at a stated level of confidence may be calculated as

$$n = \left(\frac{Z_{\alpha/2}\sigma}{\varepsilon}\right)^2 \text{ for confidence intervals for means.}$$

$$n = \left(\frac{Z_{\alpha/2}}{2\varepsilon}\right)^2 \text{ for confidence intervals for proportions.}$$

Describing a confidence interval verbally: If A $< \mu <$ B is, say, a 95 % a confidence interval for μ then we can claim that we are 95 % confident that the population mean is 'somewhere' within the interval. It follows, that if a proposed value of μ is within the interval, then this claim is supported with 95 % confidence (similarly for proportions).

7.4   Confidence intervals for the difference between means (and proportions) for independent samples

These are deduced from the sampling distribution for differences between means for NIRVs:

$$(\overline{X}_1 - \overline{X}_2) \sim N\left(\mu_1 - \mu_2, \frac{\sigma_1^2}{n_1} + \frac{\sigma_2^2}{n_2}\right)$$

Hence $\overline{x}_1 - \overline{x}_2 \pm Z_{\alpha/2}\sqrt{\dfrac{\sigma_1^2}{n_1} + \dfrac{\sigma_2^2}{n_2}}$ is the $(1-\alpha)100$ % confidence interval for $\mu_1 - \mu_2$.

Similarly, $\overline{p}_1 - \overline{p}_2 \pm z_{\alpha/2} \sqrt{\dfrac{\overline{p}_1(1 - \overline{p}_1)}{n_1} + \dfrac{\overline{p}_2(1 - \overline{p}_2)}{n_2}}$ is the $(1 - \alpha)$ 100 % confidence interval for $p_1 - p_2$.

Describing a confidence interval verbally: If [A, B] is, say, a 95 % a confidence interval for $\mu_1 - \mu_2$ then we can claim that we are 95 % confident that the difference between the population means is 'somewhere' between A and B. It follows that if a proposed value of $\mu_1 - \mu_2$ is within the interval, then this claim is supported with 95 % confidence. If A and B have different signs, the interval contains zero and so the interval supports the claim that $\mu_1 - \mu_2 = 0$: there is no difference between the means. Also, if both A and B are positive then we can state that $\mu_1 > \mu_2$ with 95 % confidence: similarly, if both A and B are negative then $\mu_1 < \mu_2$ with 95 % confidence.

TESTS OF HYPOTHESIS FOR MEANS AND PROPORTIONS

Chapter Objectives

Having carefully studied this chapter and completed the exercises you should be able to do the following

- Define a null and alternative hypothesis
- Calculate the p-value: the probability of obtaining the given sample or a more extreme sample assuming the null hypothesis is true
- Explain the decision rule at a given level of significance for one-sided and two-sided tests
- Test hypothesis for population means, proportions and difference between two means and two proportions
- Explain the terms Type I error, Type II error, the power of a test
- Explain the effect of $Z_{\alpha/2}$, σ and n on the power of a test
- Calculate the sample size required for an interval estimate of a population mean or proportion for a given level of confidence and precision: (large n)
- Use Excel and Minitab to calculate confidence intervals and test hypothesis
- Interpret and explain verbally Minitab printouts for confidence intervals and tests of hypothesis

8.1 Hypothesis tests for means

8.1.1   Null and alternative hypotheses

Up to this point we have used sample means (and proportions) to estimate population means (and proportions). Some inference about the population mean can be made from a confidence interval; for example in Worked Example 7.1 it was concluded that the true mean weight of packets of saffron was not 20 since 20 was not within the 95 % confidence interval. Similarly inference about differences between two population means or proportions can be made as outlined in Worked Examples 7.4 and 7.5.

Testing a statistical hypothesis is a different approach to making inference about a population parameter on the basis of a random sample. Instead of estimating the population parameter, a claim is made regarding its value. Such a claim is called a **Null Hypothesis**.

For example, a property developer makes a claim that the average annual rental income per room in student accommodation is at most £ 5000. That is, mean income is equal to or less than 5000: $\mu \leq 5000$. This statement is the Null Hypothesis.

The **Null Hypothesis** (H_0) is a claim or conjecture made about a population characteristic. For example $H_0: \mu \leq 5000$ (8.1)

To test the null hypothesis a random sample is selected from the population under investigation: the sample mean, $\bar{x}$ is calculated if the sample average, $\bar{x}$ is *sufficiently close* to the claimed population mean then we may conclude that there is evidence to support the Null Hypothesis.

> The test assumes H_0 is true: $\mu = 5000$

To determine what we mean by '*sufficiently close*' we will use the most extreme value of μ in the statement of the null hypothesis, i.e. $\mu = \mu_{H_0} = 5000$, to test the null hypothesis.

Recall that even if the population mean is 5000, the values of sample means will vary above and below it because of sampling error. See Figure 8.1a. In this chapter we will consider two cases:

Case 1: sample sizes are 30 or more. In this case sample means are Normally distributed, according to the Central Limit Theorem.

Case 2: sample sizes are less than 30. Provided the variable of interest is Normally distributed in the population and σ is known, then sample means are Normally distributed.

In each case: $\bar{x} \sim N\left(\mu_{\bar{x}} = \mu, \sigma_{\bar{x}} = \dfrac{\sigma}{\sqrt{n}}\right)$.

Note: for large samples, $(n \geq 30)$, if σ is unknown it may estimated by s: the sampling distribution $\bar{x}$ is approximately Normal; $\bar{x} \sim N\left(\mu_{\bar{x}} = \mu, s_{\bar{x}} = \dfrac{s}{\sqrt{n}}\right)$.

Assume the null hypothesis, $\mu = \mu_{H_0}$ **is true** until there is sufficient evidence to reject it

---

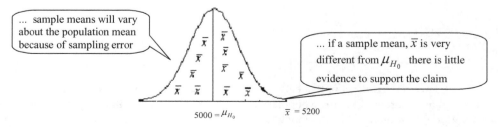

Figure 8.1a   Sampling distribution of means for testing $H_0: \mu = 5000$.

Since sample means are Normally distributed, we can calculate the probability that a sample mean differs from a population mean by a value equal to or greater than the difference between the hypothesised population mean μ_{H_0} and the present sample mean $\bar{x}$. These calculations are demonstrated in WE 8.1. If the probability is very small we will conclude that there is insufficient evidence to support the null hypothesis and it is rejected. If the probability is large (usually 0.05 or higher) then we say there is evidence to support the null hypothesis and it is accepted. See Figure 8.1b.

The rationale for hypothesis is summarised graphically as follows:

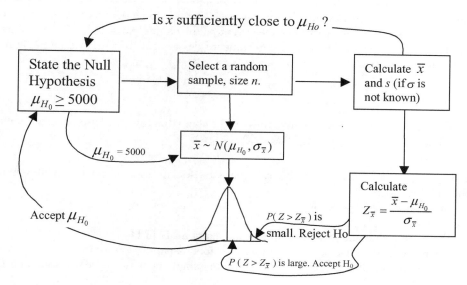

Figure 8.1b   Sketch illustrating the rationale for hypothesis tests for means. $\bar{x} \sim N(\mu_{H_0}, \sigma_{\bar{x}})$.

The alternative hypothesis

If the null hypothesis is rejected then the situation that prevails is defined by the **alternative hypothesis**, H_1. Hence the alternative hypothesis is the complement of the null hypothesis.

An alternative hypothesis (H_1) is the complement of the null hypothesis: for example if
$H_0: \mu \leq 5000$ then $H_1 \mu > 5000$

(8.2)

The null hypothesis is usually a statement of the status quo and its equation always contains the 'equality'.

The alternative hypothesis is sometimes called the 'research hypotheses' when the status quo is being tested. The statement of the alternative does not contain the 'equality'. Hence the alternative may be any one of the following statements:

$H_1: \mu > 5000$: the mean is greater than 5000.
$H_1: \mu < 5000$: the mean is less than 5000.
$H_1: \mu \# 5000$: the mean is not equal to 5000.

Note: the notation H_a is also used for the alternative hypothesis.

In the example above the null and alternative hypotheses are stated as follows:

$H_0: \mu_{H_0} \leq 5000$.
$H_1: \mu > 5000$.

This hypothesis is tested in Worked Example 8.1.

Worked Example 8.1: Testing a statistical hypothesis: the p-value method

The same property developer claims that the average rental income per room in student accommodation is at most £5000 for year. The mean rent paid by a random sample of 36 students was calculated as £5200. Assume the population standard deviation is £735. Do the sample results support the investor's claim?

Solution

1. **State the null hypothesis**: the null hypothesis is the claim that the population average is at most 5000. That is, up to and including 5000 (5000 or less).

   $H_0: \mu \leq 5000$.

   We use $\mu = \mu_{H_0} = 5000$ to set up the test. Some statisticians use the equality in all statements of the null hypothesis.
2. **State the alternative hypothesis:**
   $H_1: \mu > 5000$
3. **State the sample characteristic used to test the claim made in H_0.**
   A sample mean is the obvious way to test a claim about a population mean.
4. **Calculate the probability of selecting the present sample or more extreme, assuming H_0 is true.**
   Since $n = 36$, according to the central limit theorem, the distribution of sample means, $\bar{x} \sim$
   $N\left(\mu, \sigma_{\bar{x}} = \dfrac{\sigma}{\sqrt{n}}\right)$, where $\mu = \mu_{H_0} = 5000$ and the standard error is calculated $\sigma_{\bar{x}} = \dfrac{\sigma}{\sqrt{n}} = \dfrac{735}{\sqrt{36}} = 122.5$. In this example $\bar{x} = 5200$. A more extreme sample is a sample that lends less credibility to Ho: a sample whose mean is even greater that 5200.
   Hence calculate the probability that $\bar{x} \geq 5200$ for a random sample from a population with $\mu_{H_0} = 5000$, see Figure 8.2(b). If $P(\bar{x} \geq 5200)$ is very small, then the sample does not provide evidence to support H_0.

(a) $\bar{x} \sim N(\mu = 5000, \sigma_{\bar{x}} = 122.5)$ **(b)** p-value $= P(\bar{x} \geq 5200) = 0.0515$ **(c)** Classical method, $\alpha = 0.05$

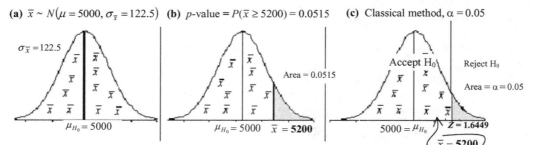

Figure 8.2   How likely is it that this sample came from the population with $\mu = 5000$?

The probability, $P(\bar{x} \geq 5200)$, is calculated by the usual Normal probability method. When $\bar{x} = 5200$, $Z_{\bar{x}} = \dfrac{\bar{x} - \mu_{H_0}}{\sigma_{\bar{x}}} = \dfrac{5200 - 5000}{122.5} = 1.63$. See Figure 8.2(b).

From the Normal tables, the area to the right of $Z = 1.63$ is $P(\bar{x} \geq 5200) = 0.0515$. So, there is a 5.15 % chance of selecting a sample whose mean, $\bar{x} \geq 5200$ when the population mean is $\mu_{H_0} = 5000$. This probability is called the p-value of the test.

5. **Conclusion**

   The p-value is an indication of the level of support for the null hypothesis provided by the sample. Therefore if the population mean is 5000, there is a 5.15 % chance of selecting a sample with mean, $\bar{x} = 5200$ or higher. If the developer is satisfied that a 5.15 % chance is sufficiently high to support H_0, then H_0 is accepted. The mean income per room is at most £ 5000.

> The p-value indicates the level of
> support for H_0 provided by the sample

The probability of selecting the given sample or a more extreme sample assuming H_0 is true is called the p-value for the test   (8.3)

8.1.2 The classical method testing hypothesis. (Significance testing)

The classical method of testing hypotheses involves declaring *in advance* that the null hypothesis will be rejected if the probability of selecting the present sample is less than a given value, usually 0.05 or 0.01. In other words the null hypothesis will be rejected if the p-value is less than a given probability (usually 0.05 or 0.01). The given probability is referred to as α. The probability, α is called the level of significance of the test. Graphically, α is the tail area(s) under the Normal curve as illustrated in Figure 8.2(c). If the probability that the difference between the sample mean and the hypothesised population mean is less than α, then H_0 is rejected and we say that the difference is significant.

The classical method is used in most literature and in Statistical software packages, in addition to the p-value method.

Worked Example 8.2: The classical method for testing hypothesis

Test the hypothesis in Worked Example 8.1 for $\alpha = 0.05$.

Solution

The first two steps of the method have already been dealt with above: these are

1. **H_0** $\mu \leq 5000$.
2. **H_1** $\mu > 5000$.
3. **$\alpha = 0.05$**: reject H_0 if the probability of obtaining the given sample is less than 0.05.

4. State the test statistic that will be used to test H_0.

The **test statistic** is the sample mean, $\bar{x}$ or the Z-value calculated from $\bar{x}$.

$$Z_{\bar{x}} = \frac{\bar{x} - \mu_{H_0}}{\sigma_{\bar{x}}}$$

5. State the decision rule and define the critical region.

The **critical region** is the region under the Normal curve of the sampling distribution, mean $\mu_{H_0} = 5000$, where H_0 will be rejected. For $\alpha = 0.05$, H_0 will be rejected if $P(\bar{x} > \bar{x}_{\text{critical}}) < 0.05$. i.e. $P(Z > Z_{\text{critical}}) < 0.05$.

If the area in the upper tail is 0.05 (5 %) then $Z_{0.05} = 1.6449$

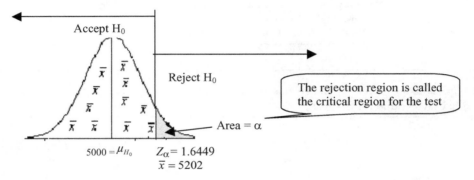

Figure 8.2(c)   Define the critical region for the test.

Hence, the decision rule may be stated in terms of Z or in terms of $\bar{x}$ as follows:

> $Z_{0.05}$ is the value of Z for which the tail area = 0.05

Accept H_0 if the sample Z falls in the region below $Z_{0.05} = 1.6449$
Reject H_0 if a sample Z falls in the region above $Z_{0.05} = 1.6449$ } In terms of Z

OR

Accept H_0 if $\bar{x} \leq \mu_{H_0} + 1.6449\sigma_{\bar{x}}$, that is $\bar{x} \leq 5000 + 1.6449(122.5) = 5202$
Reject H_0 if $\bar{x} > 5202$ } In terms $\bar{x}$

In most cases the decision rule is stated in terms of Z as this involves fewer calculations

6. Calculate the Test Statistic from sample data.

When $\bar{x} = 5200$, Z-value is 1.63, previously calculated. | The **Test Statistic** is $Z = 1.63$ |

Test Statistic: $Z_{\bar{x}} = \dfrac{\bar{x} - \mu_{H_0}}{\sigma_{\bar{x}}} = \dfrac{5200 - 5000}{122.5} = 1.63$ \hfill (8.4a)

is described as the sample characteristic, calculated from the random sample to test H_0
The test statistic is sometimes referred to as T

Informally, the test statistic may be described verbally as

$$\textbf{Test statistic: } T = \frac{\text{Point estimate} - (\text{H}_0 \text{ claim})}{\text{Standard error}} \qquad \text{(8.4b)}$$

7. **Compare the test statistic with critical region.**
   The sample Z-value is 1.63 it falls in the acceptance region, see Figure 8.2(c): H_0 is accepted. We accept the average rental income per room in student accommodation is at most £ 5000.
8. **Calculate and interpret the p-value.**
   In step 7 H_0 was accepted, but only just! The p-value will reflect whether H_0 was only just accepted or whether it was accepted very comfortably.

Recall that the p-value $= P$(selecting the present sample or a more extreme sample given H_0 is true).
   In this example the p-value $= 0.0515$, illustrated in Figure 8.2(b).
   So the chance of selecting the present sample is only slightly greater than α. In conclusion, while H_0 is accepted–the mean rental income is at most £ 5000–it is not very conclusive.

8.1.3 Type I and Type II errors and related topics

Statistical significance and Type I error

Having stated the decision rule for the test (based on a given value of α) if a sample mean falls within the acceptance region then we say there is *no significant difference* between the sample mean $\bar{x}$ and the hypothesised population mean, μ_{H_0}. Hence the difference between the present sample mean and the hypothesised population mean μ_{H_0} is attributed to sampling error due to the chance random variation of sample means about the population mean. See Figure 8.3. On the other hand, **if a sample mean falls in the rejection region then we say that there is a *significant difference* between $\bar{x}$ and μ_{H_0}.** A significant difference between $\bar{x}$ and μ_{H_0} indicates the difference is too great to attribute to chance. (In Worked Example 8.1, a sample mean that is greater than 5240 would be significantly different from $\mu = 5000$ at the 5 % level of significance).

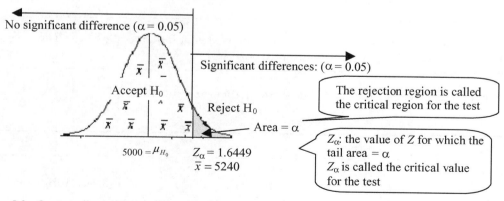

Figure 8.3   Statistically significant differences fall in the rejection region.

Type I error

In Figure 8.3, if H_0 is true, hence $\mu_{H_0} = 5000$, then 95 % of sample means will be less than 5202. The remaining 5 % of sample means will fall in the upper tail of the distribution. But, according to the decision rule, when a sample mean falls in the critical region H_0 will be rejected. Since 5 % of the sample means fall in the critical region then H_0 will be rejected when it is true, 5 % of the time. Rejecting a true null hypothesis is called Type I error.

> **Rejecting H_0 when it is true is called Type 1 error.** (8.5)

In classical tests of hypothesis, such as Worked Example 8.2, the level of significance is α. But, since α = the area of the rejection region, then α = the probability of rejecting H_0 when it is true. Hence α is the probability of making Type I error.

> **The level of significance of the test (α)** is the maximum probability of making Type 1 error:
> $\alpha = P\,(\text{Reject } H_0/H_0 \text{ is true})$. (8.6)

The size of the level of significance of the test is decided by the researcher who takes into account the consequences of rejecting a true null hypothesis. For example, if a new production method is to be tested against an existing method (by comparing the proportion of defective components for each method, for instance) then the rejection of the existing method may involve huge costs in terms of equipment, retraining of staff, possible redundancies etc. Hence the level of significance will be set at a low value, for example 0.01 or 0.0001, or even less.

Terminology

If H_0 is rejected at $\alpha = 0.10$ (10 %), we say 'there is moderate evidence that H_0 is not true'
If H_0 is rejected at $\alpha = 0.05$ (5 %), we say 'there is strong evidence that H_0 is not true'
If H_0 is rejected at $\alpha = 0.01$ (1 %), we say 'there is very strong evidence that H_0 is not true'

Type II error

> The error made by rejecting the alternative hypothesis when it is true is called Type II error
> The probability of making Type II error denoted by β (8.7)

The relationship between Type I and Type II error is summarised in Table 8.1.

Table 8.1   The relationship between Type I and Type II errors

	H_0 is True	H_1 is True
Accept H_0	Correct $1 - \alpha = P\,(\text{Accept } H_0/H_0 \text{ is true})$.	Type II error $\beta = P\,(\text{Reject } H_1/H_1 \text{ is true})$.
Accept H_1	Type I error $\alpha = P\,(\text{Reject } H_0/H_0 \text{ is true})$.	Correct $1 - \beta = P\,(\text{Accept } H_1/H_1 \text{ is true})$

To illustrate the relationship between Type I and Type II error consider the upper tail test $H_0 \: \mu \leq \mu_{H_0}$ vs. $H_1 \: \mu > \mu_{H_0}$. Assume the alternative hypothesis is true and the true population mean is μ_1, Figure 8.4(a). According to the decision rule for this test, if a sample mean falls into the acceptance region for the null hypothesis then H_0 will be accepted and consequently, the true alternative H_1 will be rejected in error – Type II error. See Figure 8.4(a). The probability that H_1 is rejected in error (β) is represented by the lower tail of the Normal curve $(\mu_1, \sigma_{\bar{x}}^2)$ representing the alternative hypothesis that falls inside the acceptance region for H_0. Figures 8.4(a) and (b) illustrate graphically that a reduction in the size of Type I error results in an increase in the size of Type II error.

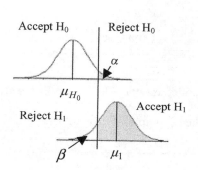

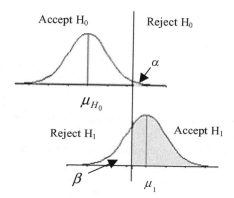

Figure 8.4(a)   When H_1 is true it is rejected if $\bar{x}$ falls in the acceptance region.

Figure 8.4(b) β increases when α decreases.

The Power of a test is the probability of accepting the true alternative hypothesis. This power is given in Table 8.1 as $1 - \beta = P(\text{Accept } H_1/H_1 \text{ is true})$. Alternatively, power may be defined as the probability of rejecting a false null hypothesis. Hence $1 - \beta = P(\text{Reject } H_0/H_0 \text{ is false})$. The power of the test is represented by the shaded area under the alternative hypothesis in Figure 8.4(a) and (b). These diagrams illustrate the rule that the Power of a test increases when α increases.

(Type II error and the power of the test are calculated for specific values of μ_1).

8.1.4   One- and two-sided tests of hypothesis

One-sided tests

The null hypothesis in Worked Example 8.1 is rejected if a sample mean falls in the upper tail (critical region) of the sampling distribution; hence the test was called a right tailed or upper tailed one-sided test. If the researcher is investigating whether the population mean is *at least 5000* (in other words, $\mu \geq 5000$) then H_0 is rejected if the sample mean falls in the critical lower tail of the distribution of means. These one-sided tests are illustrated in Figure 8.5(b) and (c).

Two-sided tests

If the researcher is investigating whether the population is equal to 5000 or not then H_0 will be rejected when the sample means is either too high or too low. Hence the test will have 2 critical regions. Such a test is called a **two-tailed test or a two-sided test**. In two-sided tests the level of significance, α is divided equally between the upper and lower tail: see Figure 8.5(a).

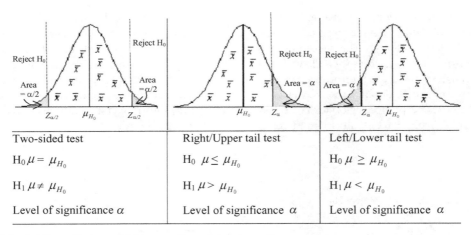

Two-sided test	Right/Upper tail test	Left/Lower tail test
$H_0 \, \mu = \mu_{H_0}$	$H_0 \, \mu \leq \mu_{H_0}$	$H_0 \, \mu \geq \mu_{H_0}$
$H_1 \, \mu \neq \mu_{H_0}$	$H_1 \, \mu > \mu_{H_0}$	$H_1 \, \mu < \mu_{H_0}$
Level of significance α	Level of significance α	Level of significance α

Figure 8.5   (a) Two-sided test (b) Upper tail test (c) Lower tail test.

The critical Z-values are referred to as Z_α or $-Z_\alpha$ for upper and lower one sided tests respectively and $\pm Z_{\alpha/2}$ for two-sided tests.

A step-by-step method for testing hypothesis used in Worked Example 7.8 may be summarised into the following steps:

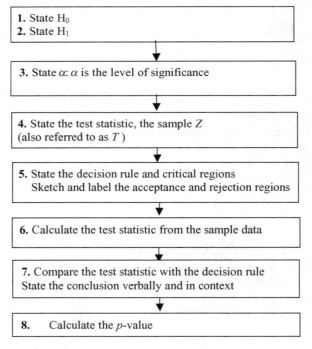

1. State H_0
2. State H_1

3. State α: α is the level of significance

4. State the test statistic, the sample Z (also referred to as T)

5. State the decision rule and critical regions
    Sketch and label the acceptance and rejection regions

6. Calculate the test statistic from the sample data

7. Compare the test statistic with the decision rule
    State the conclusion verbally and in context

8.   Calculate the p-value

Chart 8.1   Method for testing hypotheses.

These steps are illustrated in outline in Worked Example 8.3 for a two-sided test. In Worked Example 8.4 right and left tailed tests are compared. All tests are based on the example of the rental income (again!) for student accommodation to simplify comparisions between one-sided and two-sided tests.

Worked Example 8.3: Method for testing a two-sided statistical hypothesis

A property investor claims that the average rental income per room in student accommodation is £ 5000 per year. The mean rent paid by a random sample of 36 students was calculated as £ 5200. Assume the population standard deviation is £ 735. Do the sample results support the investor's claim?

Solution

This is a two-sided test since the claim is that μ is £ 5000.

1. **H_0** $\mu_{H_0} = 5000$
2. **H_1** $\mu \neq 5000$
3. **Level of significance** $\alpha = 0.05$
4. **Test statistic:** Since $n \geq 30$, sample means will be normally distributed. Hence the test statistic is the sample Z where $Z_{\bar{x}} = \dfrac{\bar{x} - \mu_{H_0}}{\sigma_{\bar{x}}}$.
5. **Decision rule** and critical regions:
   This is a two-sided test, so the size of each rejection region is $\alpha/2 = 0.025$, hence $Z_{\alpha/2} = 1.96$ (from the normal probability tables).
   Accept H_0 if the sample Z is equal to or between -1.96 and 1.96. Reject H_0 if the sample Z is less than -1.96 or greater than $+1.96$ as in Figure 8.6.

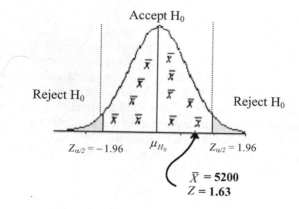

Figure 8.6   Critical regions for two-sided test.

6. **Calculate the test statistic from sample data:** $n = 36$; $\sigma = 735$; $\bar{x} = 5200$

$$Z_{\bar{x}} = \frac{\bar{x} - \mu_{H_0}}{\sigma_{\bar{x}}} = \frac{5200 - 5000}{122.5} = 1.63$$

where the standard error $\sigma_{\bar{x}} = \dfrac{\sigma}{\sqrt{n}} = \dfrac{735}{\sqrt{36}} = 122.5$.

7. **Compare the test statistic with the critical region.**
   The sample $Z = 1.63$. This is in the acceptance region, therefore accept the null hypothesis.
8. **Calculate and interpret the p-value.**
   Note: in two-sided tests the p value is the probability that a sample mean will be differ from μ_{H_0} by more than 200 units *in either direction*.

   Hence, the p-value $= P(\overline{x} \geq 5200) + P(\overline{x} \leq 4800)$
   $$= P(Z \geq 1.63) + P(Z \leq -1.63)$$
   $$= 2(0.0515)$$
   $$= 0.1030$$

In conclusion, we are reasonably confident that the claim that the mean rental income is £ 5000 is true. There is a 10.26 % chance of getting the present sample or a more extreme sample when $\mu = 5000$.

Worked Example 8.4: Comparison of the one-sided test for means

For the data in Worked Example 8.1, test that the hypothesis that the mean rental income is (a) at least £ 5000; (b) at most £ 5000.
   Compare the results of (a) and (b).

Solution

(a) 'at least 5000' means 5000 or more. Hence the null hypothesis is, H_0 $\mu_{H_0} \geq 5000$. If sample means are greater than 5000 then the null hypothesis is supported. But if a sample mean is much smaller than 5000 then the null hypothesis is rejected and the alternative hypothesis, H_1 $\mu < 5000$ is accepted. This test is carried out in column 1 of Table 8.2.
   The p-value $= 0.9485$: (the probability of selecting a sample whose mean $\overline{x} \leq 5200$): this probability is represented by the area under the curve to the left of $\overline{x} = 5200$.
(b) The calculations for (b) have already been completed in Worked Example 8.2 and are presented again in column 2 of Table 8.2 for comparison. The p-value $= 0.0515$ is the probability of selecting a sample whose mean $\overline{x} \geq 5200$.

Conclusion

The fact that two opposing null hypotheses are accepted demonstrates that the sample evidence was not sufficiently strong to reject either null hypothesis conclusively. However the large p-value: $p = 0.9485$, means that there is a 94.85 % chance of selecting the given sample or a more extreme sample when the mean rental income is at 5000 for (a) H_0 $\mu_{H_0} \geq 5000$ indicates this hypothesis is more plausible.
   Recall: a null hypothesis is assumed true until there is sufficient evidence to reject it. With insufficient evidence a null hypothesis will be accepted no matter how it is declared.

Table 8.2   Comparison of left and right tailed tests

1. $H_0\ \mu_{H_0} \geq 5000$
2. $H_1\ \mu < 5000$
3. Level of significance $\alpha = 0.05$
4. Test statistic: $Z_{\bar{x}} = \dfrac{\bar{x} - \mu_{H_0}}{\sigma_{\bar{x}}}$
5. **Decision rule**: Reject H_0 if the sample Z is less than -1.6449

1. $H_0\ \mu_{H_0} \leq 5000$
2. $H_1\ \mu > 5000$
3. Level of significance $\alpha = 0.05$
4. Test statistic: $Z_{\bar{x}} = \dfrac{\bar{x} - \mu_{H_0}}{\sigma_{\bar{x}}}$
5. **Decision rule**: Reject H_0 if the sample Z is greater than 1.6449

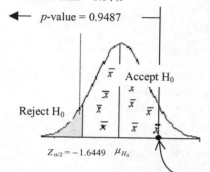

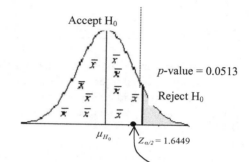

6. $n = 36;\ s = 735;\ \bar{x} = 5200$

$$Z_{\bar{x}} = \frac{\bar{x} - \mu}{\sigma_{\bar{x}}} = \frac{5200 - 5000}{122.5} = 1.63$$

since $s_{\bar{x}} = \dfrac{s}{\sqrt{n}} = \dfrac{735}{\sqrt{36}} = 122.5$

7. The sample $Z = 1.63$ is in the acceptance region.
8. The p-value $= 0.9485$

6. $n = 36;\ s = 735;\ \bar{x} = 5200$

$$Z_{\bar{x}} = \frac{\bar{x} - \mu}{\sigma_{\bar{x}}} = \frac{5200 - 5000}{122.5} = 1.63$$

since $s_{\bar{x}} = \dfrac{s}{\sqrt{n}} = \dfrac{735}{\sqrt{36}} = 122.5$

7. The sample $Z = 1.63$ is in the acceptance region.
8. The p-value $= 0.0515$

So, one should always enquire about how the null hypothesis is declared when statements such as the following are made: 'the results were inconclusive' or 'the results were not statistically significant'.

8.1.5   Relationship between confidence intervals and tests of hypothesis

The random sample of size, $n = 36$, the sample mean and standard deviation were calculated as 5200 and 735 respectively. The sample standard error was calculated as $s_{\bar{x}} = 122.5$. This sample was used to test hypotheses and calculate confidence intervals.

In Worked Example 8.3, in the two-sided test, H_0: $\mu = 5000$ vs. H_1: $\mu \neq 5000$ the null hypothesis was accepted at $\alpha = 5\%$.

The 95 % confidence interval is calculated from (7.4).

$$\bar{x} \pm Z_{\alpha/2}\, s_{\bar{x}} \rightarrow 5200 \pm 1.96(122.5) \rightarrow 4959.9 \text{ to } 5440.1$$

Hence we are 95 % confident that the population mean is somewhere between 4959.9 and 5440.1. Since $\mu = 5000$ is within the interval then there is sufficient evidence to accept that this is true. However, the null hypothesis would have been accepted if the claim had been any value between $\mu = 4959.9$ and $\mu = 5440.1$.

This example demonstrates that the two-sided confidence interval is a set of acceptable two-sided null hypotheses at the same level of significance (the same argument applies to one-sided tests and confidence intervals).

Progress Exercises 8.1: Tests of hypothesis for means

1. Define and give an example (with diagrams and equations when appropriate) the following: A Null hypothesis: an alternative hypothesis: a one-sided test: a two-sided test.
2. Formulate the null and alternative hypothesis from the following statements:
   (a) The mean daily temperature is at least 16°C in July.
   (b) The average taxi fare is £ 5.
   (c) The average journey time between Paris to Versailles takes longer than 45 minutes.
   (d) The average amount spent on flowers is at most £ 8 during the month of January.
3. Define the level of significance of a test.
   At the final stage in the manufacture of cappuccino, the mix is filled into sachets by a machine set to dispense 16 g into each sachet. A random sample of 36 packets was weighed and the mean weight was calculated as 15.42 g. The standard deviation of the amount of fill is known to be 1.25 g. State the null and alternative hypotheses, the test statistic and decision rule.
4. Use the data in question 3 to test the claim that the mean amount of cappuccino per sachet is 16 g at the 5 % level of significance. Calculate the p-value of the test. Is there sufficient evidence to state that the mean amount per sachet is 16 g?
5. A taxi company claims that the mean taxi fare for the route from the railway station to Maple Drive is less than £ 7.50, but customers claim that the fare is over £ 8. A random sample of 30 taxi fares for the route from the railway station to Maple Drive was collected over a period of one week. The sample mean and sample standard deviation were calculated as £ 7.75 and £ 1.50, respectively.
   At the 5 % level of significance,
   (a) test the claim made by customers that the mean fare exceeds £ 8.00.
   (b) test the taxi company's claim that the mean fare is less than £ 7.50.
   Is there sufficient evidence to reject either (or both) claims in (a) and (b)?
6. Fill in the missing sections (??) in Table PE8.1 Q6:

Table PE8.1 Q6   Test hypothesis for means

H_0	??
H_1	$\mu > 25$
The Decision rule ($\alpha = 0.05$) is	??
Sample data	Sample size, $n = 32$: $\bar{x} = 26.54$: $s = 2.45$
The test statistic is	??
Conclusion	??

7. See PE 7.1 question 4. Fifty households from an apartment block were surveyed. From this sample, the mean and standard deviation of the amount they are willing to spend on daffodils in January was calculated as £ 8.55 and £ 3.14.

   (a) Test the claim that mean amount households are willing to spend is different from £ 9. How does your answer compare with the 95 % confidence for the mean?

8. **Refer to Worked Example 6.1.** A random sample of 36 packets of saffron was collected. The sample average weight was calculated as 19.35 gm. The population standard deviation of weights is known to be 1.8 gm.

   (a) Test the claim that mean weight is 20 g at the (i) 5 % and (ii) 1 % levels of significance.

   (b) How do these conclusions compare with the 95 % and 99 % confidence intervals in Worked Example 6.1?

9. State, with explanations, whether the following statements are true or false:

   (a) Type I error is rejecting the null hypothesis while Type II error is rejecting the alternative hypothesis.

   (b) The p value is always greater than the level of significance when H_0 is accepted.

   (c) The null hypothesis may be tested with a confidence interval.

   (d) The size of the p value indicates the level of support for the null hypothesis given by the sample.

10. Fill in the missing sections (??) in the following table:

Table PE8.1 Q10   Test hypothesis for means

H_0	??
H_1	$\mu \neq 120$
The Decision rule ($\alpha = 0.01$) is	??
The test statistic is	$Z = -2.65.$
Conclusion	??
p-value =	??

11. The Minitab output for daily hours of sunshine in a resort is given in Table PE8.1 Q11:

Table PE8.1 Q11   Minitab printout where the variable being tested is daily hours of sunshine

One-Sample Z

```
Test of mu = 10 vs not = 10
The assumed standard deviation = 2.2
```

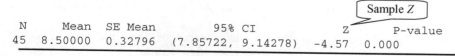

```
 N     Mean   SE Mean          95% CI              Z      P-value
45   8.50000  0.32796   (7.85722, 9.14278)      -4.57   0.000
```

... where N = sample size, Mean = sample mean, SE Mean = standard error of the mean

(a) Explain the Minitab printout.

(b) What do you conclude about the daily hours of sunshine in this resort?

Note: instructions on using Minitab are given in section 8.4.

12. The Minitab output for the annual sick days for a sample of employees is given in Table PE8.1 Q12:

Table PE8.1 Q12   Minitab printout for the annual sick days for a sample of employees

One-Sample Z: Company A

```
Test of mu = 5 vs not = 5
The assumed standard deviation = 5

Variable    N    Mean    StDev   SE Mean        95% CI              Z      P
Company A   30   5.60000 6.26209 0.91287  (3.81081, 7.38919)  0.66   0.511
```

Interpret this output. What is the most likely average annual sick days taken by employees?
Note: instructions on using Minitab are given in section 8.4.

8.2   Hypothesis tests for proportions

In section 6.2 the distribution of sample proportions, p ($n \geq 30$) was stated to be approximately Normal with a mean, $\mu_p = \pi$, standard error $\sigma_p = \sqrt{\dfrac{\pi(1-\pi)}{n}}$.
Hence the test statistic for proportions is

$$\frac{\text{Point estimate} - (\text{H}_0 \text{ claim})}{\text{Standard error}} = \frac{p - \pi_{H_0}}{\sigma_p} \text{ given in (8.8a).}$$

Since the population proportion is unknown, it is estimated by the sample proportion, p. It follows that the standard error must also be estimated: the formula for the sample standard error, s_p is given in (8.8b).

The test statistic:	$Z_p = \dfrac{p - \pi_{H_0}}{s_p}$	(8.8a)
The sample standard error $s_p = \sqrt{\dfrac{\pi_{H_0}(1 - \pi_{H_0})}{n}}$		(8.8b)

The method for testing hypotheses for proportions is similar to that for means and given in Chart 8.1.

Worked Example 8.5: Test the hypothesis for a population proportion

A budget airline claims that 96 % of its flights depart on time. A researcher working for a competitor records departure information for 80 randomly selected flights and discovers that five departed late. Test the airline's claim at the 5 % and 1 % level of significance.

Solution

The sample proportion of late departures is $p = \dfrac{x}{n} = \dfrac{5}{80} = 0.0625$.

The airline's claim is that the 4 % of flights are late or the proportion of late flights is $\pi_{H_0} = 0.04$. So set out the test following the steps in Chart 8.1:

1. $H_0 \; \pi_{H_0} = 0.04$
2. $H_1 \; \pi_{H_0} \neq 0.04$
3. Level of significance (i) $\alpha = 0.05$; (ii) $\alpha = 0.01$
4. Test statistic: Since $n \geq 30$, sample proportions will be Normally distributed. Hence the test statistic is $Z_p = \dfrac{p - \pi_{H_0}}{s_p}$.
5. **Decision rule**: This is a 2-sided test,
   For $\alpha = 0.05$, $\alpha/2 = 0.025$ hence $Z_{\alpha/2} = 1.96$.
   Accept H_0 if the sample Z is equal to or between -1.96 and 1.96. Reject H_0 if the sample Z is less than -1.96 or greater than 1.96. See Figure 8.7.
   For $\alpha = 0.01$, $\alpha/2 = 0.005$ hence $Z_{\alpha/2} = 2.5758$.
   Reject H_0 if the sample Z is less than -2.5758 or greater than 2.5758 (these points are not marked on Figure 8.7).

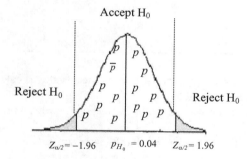

Figure 8.7   Critical regions for two-sided test.

6. Calculate the test statistic from sample data $n = 80$. $p = 0.0625$;
   Calculate the sample standard error by (8.8b)

$$s_p = \sqrt{\frac{\pi_{H_0}(1 - \pi_{H_0})}{n}} = \sqrt{\frac{0.04(1 - 0.04)}{80}} = \sqrt{\frac{0.0384}{80}} = \sqrt{0.00048} = 0.0219$$

Hence the test statistic (8.8a)

$$Z_p = \frac{p - \pi_{H_0}}{s_p} = \frac{0.0625 - 0.04}{0.0219} = \frac{0.0225}{0.0219} = 1.0274$$

7. Compare the test statistic with the decision rule.
   The sample $Z = 1.0274$. This is in the acceptance region, therefore accept the null hypothesis: accept that 96 % of the airline's departures are on time.

8. The difference between the sample proportion and the hypothesised proportion $|\pi_{H_0} - p| = |0.04 - 0.0625| = 0.0225$. The p value may be described as the probability that the difference between π_{H_0} and p will be equal to or greater than 0.0225 in either direction. As illustrated below:

$$p = 0.0175 \xleftarrow{\quad p_{H_0} - p = 0.0225 \quad} \pi_{H_0} = 0.04 \xleftarrow{\quad p_{H_0} - p = 0.0225 \quad} p = 0.0625$$

The p-value $= P(p \geq 0.0625) + P(p \leq 0.0175)$

$$= P(Z \geq 1.03) + P(Z \leq -1.03)$$

$$= 2(0.1515)$$

$$= 0.3030$$

In conclusion, the null hypothesis that 4 % of the airline's departures are late is accepted. The p-value $= 0.3030$: states that there is a 30.30 % chance of selecting the given sample or a more extreme sample (that is five or more late departures in 80 or 1.4 or less late departures in 80) when the true percentage of late departures is 4.

Progress Exercises 8.2: Tests of hypothesis for proportions

1. What is the p-value of a test of hypothesis? How is the p-value is calculated for (i) a one-sided; (ii) a two-sided test for proportions.
2. (a) Explain the relationship between a two-sided test of hypothesis and a confidence interval for proportions.
   (b) An opinion poll of 200 voters revealed that 85 were satisfied with government performance.
      (i) Calculate a 95 % confidence interval for the true proportion satisfied with the government performance.
      (ii) Test the hypothesis that at most 50 % are satisfied with government performance at the 5 % level of significance. Calculate the p-value for the test.
   Write brief comments describing the level of satisfaction with government performance.
3. (a) What is a test statistic? How is the test statistic calculated for a (i) one-sided (ii) two-sided test for proportions?
   (b) In a survey consumers were asked if they had ever used online shopping. The numbers of Y/N responses were YES = 72: NO = 93.
      (i) Formulate the null and alternative hypothesis to test the claim that at least 50 % of consumers used online shopping.
      (ii) Calculate the test statistic and test the hypothesis in (i).
      State your conclusion verbally, using the p-value.
   (c) Would your conclusion be different if the claim was that at most 50 % of consumers used online shopping?
4. See PE 7.2, question 5. A Health and Safety survey of 200 industrial accidents claims revealed that 53 were due to untidy working conditions.

(a) A factory manager claims that the above the sample is just a 'one off', that less than 20 % of accidents are a result of untidy working conditions. How would you explain that hypothesis tests allow for such sampling errors?

(b) Use the sample (53 out of 200 accidents) to test the manager's claim that the proportion of accidents due to untidy working conditions is less than 0.20 at 5 % and 0.5 % levels of significance. Calculate the p value. State your conclusion verbally.

5. Fill in the missing entries indicated by (??) in the following table:

Table PE8.2 Q5   Test hypothesis for proportions

H_0	??
H_1	$p < 0.5$
The Decision rule ($\alpha = 0.05$) is	??
The test statistic is	$Z = 1.99$.
Conclusion	??
p-value =	??

6. **See PE 7.2, question 1.** The survey of 1000 commuters was summarised in Table 7.2 (reproduced below). From this data, test the following claims using the 5 % and 1 % levels of significance and calculate the p values.

(a) At least 50 % of commuters are satisfied/very satisfied.

(b) 54 % of females aged 45 and younger are not satisfied.

(c) Over 70 % of males are satisfied.

Write a brief report on commuters' views on the bus service based on the results in (a), (b) and (c) above.

Table PE7.2 Q1   Survey on satisfaction with bus commuter service   File: Bus

	Age	Very satisfied	Satisfied	Not satisfied	Row totals
Male	20–25 years	64	48	68	180
	26–45 years	30	82	28	140
	Over 46 years	30	40	10	80
Female	20–25 years	24	86	140	250
	26–45 years	48	48	184	280
	Over 46 years	24	16	30	70
	Column totals	220	320	460	1000

8.3   Hypothesis tests for the difference between means and proportions

The sampling distributions for the difference between sample means (for independent samples, $n \geq 30$) was given in section 7.4, formula (7.9) as

$$(\overline{X}_1 - \overline{X}_2) \sim N\left(\mu_{\overline{X}_1 - \overline{X}_2} = \mu_1 - \mu_2, \sigma^2_{\overline{X}_1 - \overline{X}_2} = \frac{\sigma_1^2}{n_1} + \frac{\sigma_2^2}{n_2}\right)$$

Since the sampling distribution is Normal, the test statistic used to test the difference between two population means is

$$Z_{\overline{X}_A - \overline{X}_B} = \frac{\text{Point estimate} - (H_0 \text{ claim})}{\text{Standard error}}$$

$$Z_{\overline{X}_A - \overline{X}_B} = \frac{(\overline{x}_1 - \overline{x}_2) - (H_0 \text{ claim})}{\sqrt{\dfrac{\sigma_1^2}{n_1} + \dfrac{\sigma_2^2}{n_2}}} = \frac{(\overline{x}_1 - \overline{x}_2) - (H_0 \text{ claim})}{\sigma_{\overline{x}_1 - \overline{x}_2}} \qquad (8.9a)$$

Note: the sample variances may be used when the population variances unknown for $n_1, n_1 \geq 30$.

$$\text{Hence, the test statistic } Z_{\overline{X}_A - \overline{X}_B} = \frac{(\overline{x}_A - \overline{x}_B) - (\mu_A - \mu_B)}{s_{\overline{x}_A - \overline{x}_B}} \qquad (8.9b)$$

Also, in section 7.4, the sampling distributions for the difference between two sample proportions (for independent samples $n \geq 30$) was

$$p_1 - p_2 \sim N \left(\mu_{p_1 - p_2} = \pi_1 - \pi_2, \quad \sigma_{p_1 - p_1}^2 = \frac{\pi_1(1 - \pi_1)}{n_1} + \frac{\pi_2(1 - \pi_2)}{n_2} \right) \qquad (7.12a)$$

Hence the test statistic for the difference between two population proportions

$$Z_{p_A - p_B} = \frac{\text{Point estimate} - (H_0 \text{ claim})}{\text{Standard error}}$$

$$Z_{p_A - p_B} = \frac{(p_1 - p_2) - (H_0 \text{ claim})}{\sqrt{\dfrac{\pi_1(1 - \pi_1)}{n_1} + \dfrac{\pi_2(1 - \pi_2)}{n_2}}} = \frac{(p_1 - p_2) - (H_0 \text{ claim})}{\sigma_{p_1 - p_2}} \qquad (8.10)$$

Since the population proportions are unknown, it follows that the standard error must also be estimated, formula given by (7.13).

$$\text{Hence the test statistic is } \frac{(p_1 - p_2) - (H_0 \text{ claim})}{s_{p_1 - p_2}} \qquad (8.10b)$$

The procedure for testing the hypothesis for differences between population means and proportions is the same as for single means and proportions and is demonstrated in the following Worked Examples.

Worked Example 8.6: Test for difference between means

In Worked Example 7.2, the sample mean and standard deviation for rental expenditures was £5200 and £735 respectively for a sample of 36 students. A sample of 45 young professionals was surveyed. The sample mean and standard deviation for annual expenditure on rent and calculated as £4920 and £225 respectively. Is there a difference in the average rental expenditures between the two groups?

Solution

The sample data may be summarised as follows in Table 8.3:

Table 8.3   Annual expenditure on rent by students and professionals

Annual expenditure on rent	Students (Data in Worked Example 7.2)	Young professionals
Sample size	36	45
Sample mean	5200	4920
Sample standard deviation	735	225

The hypothesis is tested as follows:

1. H₀ $\mu_A = \mu_B$ or $\mu_A - \mu_B = 0$
2. H₁ $\mu_A \neq \mu_B$
3. Level of significance $\alpha = 0.05$ and $\alpha = 0.01$
4. Test statistic: The test statistic is calculated from (8.9b):

$$Z_{\overline{x}_A - \overline{x}_B} = \frac{(\overline{x}_A - \overline{x}_B) - (\mu_A - \mu_B)}{s_{\overline{x}_A - \overline{x}_B}} = \frac{\text{Point estimate} - (\text{H}_0 \text{ claim})}{\text{Standard error}}$$

5. Decision rule and critical regions: This is a two-sided test. Use two levels of significance as guides ($\alpha = 5\%$ and $\alpha = 1\%$).
For $\alpha = 0.05$, accept H₀ if the sample Z is equal to or between -1.96 and 1.96. See Figure 7.16.
For $\alpha = 0.01$, accept H₀ if the sample Z is equal to or between -2.5758 and 2.5758 (this is not marked on Figure 7.16).
6. Calculate the test statistic from sample data summarised in Table 8.3.

$$\text{The test statistic is } Z_{\overline{x}_A - \overline{x}_B} = \frac{(\overline{x}_A - \overline{x}_B) - (\mu_A - \mu_B)}{s_{\overline{x}_A - \overline{x}_B}} \tag{8.9b}$$

You will need to calculate (i) $s_{\overline{x}_A} - s_{\overline{x}_B}$, (ii) $\overline{x}_A - \overline{x}_B$ and (iii) $\mu_A - \mu_B$, then substitute into (8.9b)

(i) The standard error is $s_{\overline{x}_A - \overline{x}_B} = \sqrt{\dfrac{s_A^2}{n_A} + \dfrac{s_B^2}{n_B}} = \sqrt{\dfrac{(735)^2}{36} + \dfrac{(225)^2}{45}}$

$$= \sqrt{15006.25 + 1125} = \sqrt{16131.25} = 127.0089$$

The standard error $s_{\overline{x}_A - \overline{x}_B} = 127.0089$.
(ii) $\overline{x}_A - \overline{x}_B = 5200 - 4920 = 280$
(iii) The claim made in the null hypothesis $\mu_A - \mu_B = 0$ (from H₀)
Hence $Z_{\overline{x}_A - \overline{x}_B} = \dfrac{(5200 - 4920) - (0)}{127.0089} = 2.2046$

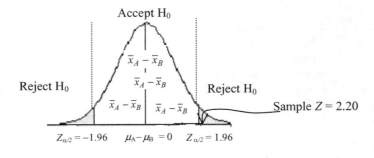

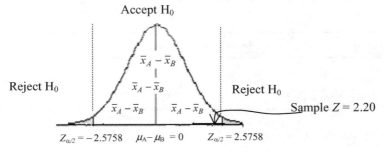

Figure 8.8   Critical regions for two-sided test for difference between means.

7. Compare the test statistic with the critical region.
   The sample $Z = 2.20$ is in the rejection region when $\alpha = 0.05$.
   The sample $Z = 2.20$ is in the acceptance region when $\alpha = 0.01$. The result is not conclusive; H_0 is accepted at one level of significance and rejected at another level of significance.
8. Calculate and interpret the p-value.

$$\begin{aligned} \text{The } p\text{-value} &= P(Z \geq 2.20) + P(Z \leq -2.20) \\ &= 2(0.0139) \\ &= 0.0278 \end{aligned}$$

The p-value is 0.0278, indicating that there is a 2.78 % chance of that 2 sample means will differ by 280 or more when the population means are equal. If the researcher is satisfied with this then H_0 is accepted: if not, then H_0 is rejected.
Note: the p-value is between the levels of significance 1 % and 5 %. If H_0 is rejected then the p-value is the level of significance.

Pooled estimate of variance used in testing difference between proportions

In testing the difference between proportions, the null hypothesis will state that the population proportions are equal: H_0: $\pi_A = \pi_B$. Hence the sample proportions are two independent point estimates of the common population proportion (let the π_c = common population proportion). A better estimate of π_c is obtained by combining the two estimates p_A and p_B. The combined or pooled estimate of π_c calculated from all the data is given by the following equations:

Combined estimate of population proportion

$$p_c = \frac{\text{Number of items with given characteristic from both samples}}{\text{Sum of all items in both samples}} \qquad (8.10)$$

Substituting the combined estimate of the population proportion into the formula for sample standard error for the difference between proportions gives the combined or 'pooled' standard error in (8.11).

Combined estimate of standard error for proportions

$$s_{p_A - p_B} \rightarrow s_{combined} = \sqrt{\frac{p_c(1 - p_c)}{n_A} + \frac{p_c(1 - p_c)}{n_B}} = \sqrt{p_c(1 - p_c)\left(\frac{1}{n_A} + \frac{1}{n_B}\right)} \qquad (8.11)$$

In Worked Example 8.7, the pooled proportion of voters for the Green party will be used.

Worked Example 8.7: Test for difference between proportions

In Worked Example 7.6 the results of polls on the voting intentions of voters in two constituencies were recorded as follows:

(Data from Worked Example 7.6)	Constituency A	Constituency B
Sample size (numbers polled)	200	160
Vote for Green party	88	54

(a) Calculate a combined estimate of the proportion of votes for the Green party from the sample data from constituencies A and B. Hence calculate the sample standard error for the difference between proportions.

(b) Test the hypothesis that
  (i) the support for the Green party is the same in both constituencies (use 10 % and 1 % levels of significance).
  (ii) the support for the Green party in constituency A is at least 5 % higher (use 10 % level of significance).

(c) Compare the inference from the 90 % confidence interval calculated in Worked Example 7.5: $0.0182 \le (\pi_A - \pi_B) \le 0.1868$ to the inference from the tests of hypothesis in part (a) above.

Solution

(a) $p_c = \dfrac{\text{Total voters for Green}}{\text{Total voters polled}} = \dfrac{88 + 54}{200 + 160} = \dfrac{142}{360} = 0.3944$

Hence the standard error based on the pooled estimate of p in (8.11) is

$$s_{combined} = \sqrt{\frac{p_c(1 - p_c)}{n_A} + \frac{p_c(1 - p_c)}{n_B}} = \sqrt{p_c(1 - p_c)\left(\frac{1}{n_A} + \frac{1}{n_B}\right)}$$

$$= \sqrt{0.3944(0.6056)\left(\frac{1}{200} + \frac{1}{160}\right)} = 0.0518$$

(b) (i)

> Note: In H_0 no values were specified for π_A or π_B-simply that they are equal. Hence $s_{p_A - p_B}$ is calculated from p_c the pooled estimate

1. H_0 $\pi_A = \pi_B$ or $\pi_A - \pi_B = 0$
2. H_1 $\pi_A \neq \pi_B$
3. **Levels of significance** $\alpha = 0.10$ and $\alpha = 0.01$
4. **Test statistic:**

$$Z_{\overline{X}_A - \overline{X}_B} = \frac{\text{Point estimate} - (H_0 \text{ claim})}{\text{Standard error}}$$

$$Z_{p_A - p_B} = \frac{(p_A - p_B) - (\pi_A - \pi_B)}{s_{p_A - p_B}}$$

5. **Decision rule** and critical regions.
   This is a 2-sided test
   For $\alpha = 0.10$.
   Accept H_0 if the sample Z is equal to, or between -1.6449 and 1.6449.
   For $\alpha = 0.05$.
   Accept H_0 if the sample Z is equal to or between -2.5758 and 2.5758.
   See Figure 8.9.
6. Calculate the test statistic $Z_{p_A - p_B} = \dfrac{(p_A - p_B) - (\pi_A - \pi_B)}{s_{p_A - p_B}} \dots (8.10b)$

To evaluate the test statistic you will need to calculate
(i) *the standard error:* $s_{combined}$ (ii) $p_A - p_B$ and (iii) $\pi_A - \pi_B$, then substitute the results into formula (8.10b).
  (i) *The standard error:* $s_{combined} = 0.0518$, calculated in (a)
  (ii) *The difference between sample proportions:*

$$(p_A - p_B) = \frac{88}{200} - \frac{54}{160} = 0.44 - 0.3375 = 0.1025$$

(iii) *The claim made in the null hypothesis is* $\pi_A - \pi_B = 0$
Hence, substituting the values for (i), (ii) and (iii) into the formula for the test statistic into (8.10b), gives

$$Z_{p_A - p_B} = \frac{(0.1025) - (0)}{0.0518} = 1.9981 = 2.00$$

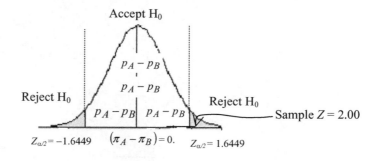

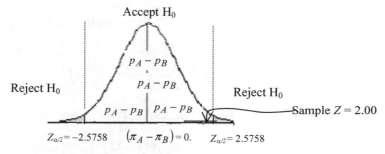

Figure 8.9   Critical regions for two-sided test for difference between proportions.

7. Compare the test statistic with the critical region.
   The sample $Z = 2.00$ is in the rejection region when $\alpha = 0.10$.
   The sample $Z = 2.00$ is in the acceptance region when $\alpha = 0.01$. The result is not conclusive; H_0 is accepted at one level of significance and rejected at another.
8. Calculate and interpret the p-value
   $P(Z > 2.00) = 0.02275$.
   But this is a two-sided test, hence the p-value $= 2 \times 0.02275 = 0.0455$.
   If there is no difference between the support for the Green party in constituencies A and B there is a 4.55 % chance that the difference between two sample proportions will be 0.1025 or more (sample sizes 200 and 160).

(b) (ii)   If π_A is at least 5 % greater than π_B then $\pi_A - \pi_B \geq 0.05$. This is the null hypothesis. The calculation of the test statistic is given below.

1. **H_0** $\pi_A - \pi_B \geq 0.05$
2. **H_1** $\pi_A - \pi_B < 0.05$
3. **Level of significance** $\alpha = 0.10$ and $\alpha = 0.01$
4. **Test statistic**: Hence the test statistic is the sample Z
   where $Z_{p_A - p_B} = \dfrac{(p_A - p_B) - (\pi_A - \pi_B)}{s_{p_A - p_B}}$
5. **Decision rule** and critical regions:
   This is a one-sided test. See Figure 8.10.
   For $\alpha = 0.10$, reject H_0 if the sample Z is less than or equal to 1.28.

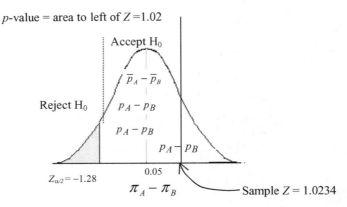

Figure 8.10   Critical regions for one-sided test for difference between proportions.

6. Calculate the test statistic $Z_{p_A - p_B} = \dfrac{(p_A - p_B) - (\pi_A - \pi_B)}{s_{p_A - p_B}} \ldots (8.10b)$

You will need to calculate
(i) *the standard error:* $s_{p_1 - p_2}$; (ii) $p_A - p_B$; and (iii) $\pi_A - \pi_B$, then substitute the results into (8.10b).
Note: the pooled estimate of variance, $s_{combined}$ cannot be used here because the null hypothesis does not claim $\pi_A - \pi_B = 0$, it claims that $\pi_A - \pi_B \geq 0.05$
Hence the standard error is for the difference between proportions,

$$s_{p_A - p_B} = \sqrt{\frac{p_A(1 - p_A)}{n_A} + \frac{p_B(1 - p_B)}{n_B}}$$

$$= \sqrt{\frac{0.44(1 - 0.44)}{200} + \frac{0.3375(1 - 0.3375)}{160}} = \sqrt{0.001232 + 0.0013975} = 0.0513$$

(i) $s_{p_A - p_B} = 0.0513$.
(ii) $p_A - p_B = 0.1025$
(iii) $\pi_A - \pi_B = 0.05$ (from H_0)
Hence the test statistic is $Z_{p_A - p_B} = \dfrac{(0.1025) - (0.05)}{0.0513} = 1.0233$

7. Compare the test statistic with the critical region.
The sample $Z = 1.0233$ is in the acceptance region when $\alpha = 0.10$, H_0 is accepted. The support for the Green party in constituency A exceeds that in B by at least 5 %.

8. Calculate and interpret the p-value.
The p-value $= P(Z < 1.02) = 1 - 0.1539 = 0.8461$. If the support for the Green party in constituency A exceeds that in B by at least 5 % then there is a 84.61 % chance that the difference between 2 sample proportions will be 0.1025 or less. (Sample sizes 200 and 160).
There is strong evidence to conclude that the support for the Green party is at least 5 % higher in constituency A.

(c) The 90 % confidence interval was: $0.0182 \leq (\pi_A - \pi_B) \leq 0.1868$. Since the difference is positive, ranging from 0.0182 to 0.1868, it was concluded that. $\pi_A > \pi_B$.

In (b)(i), the null hypothesis $H_0\, \pi_A = \pi_B$ was rejected at the 10% level of significance, inferring that the proportion were not equal.

However, the p value $= 0.0474$, indicating less than 5% support for H_0. In (b)(ii), the null hypothesis: $H_0\, \pi_A - \pi_B \geq 0.05$ was accepted with a p-value $= 0.8463$. The sample provides strong evidence to support H_0: that the Green party will get at least 5% more votes in constituency A.

So, while the confidence interval indicated that Green party support in A is greater then in B, the hypotheses tests gave more information. The large p-value $= 0.8463$ in (b)(ii) indicates that we can be very confident that the support in constituency A is at least 5% higher than in B (based on the evidence from two independent random samples $n = 200$ and 160).

Progress Exercises 8.3: Tests of hypothesis for differences

In the following questions, test hypothesis using at least two levels of significance and calculate the p-value for each test.

1. The prices (in £) of two different brands of T-shirts (brand X and brand Y) in the various retail outlets are normally distributed, with means and standard deviations: $\mu_X = 9.10$, $\sigma_X = 1.87$ and $\mu_Y = 8.70$, $\sigma_Y = 2.24$, respectively. Random samples of each brand of T-shirt of size $n_X = 30$ and $n_Y = 45$ are selected.
   (a) Write down the sampling distributions of the mean for each brand.
   (b) Write down the sampling distribution for the difference between means. Hence calculate the probability that the difference between the two sample means is greater than zero.
   (c) The mean price of a sample of 30 brand X T-shirts and a sample of 45 brand Y T-shirts were calculated as $\bar{x} = 10.15$ and $\bar{y} = 9.25$ respectively. Test the hypothesis that (i) the mean price of brand X is greater than the mean price of brand Y; (ii) the mean price of brand X is £1 more than the mean price of brand Y.
   State your conclusions carefully.
2. Fill in the missing sections (??) in the following table:

Table PE8.3 Q2   Test hypothesis for two means

H_0	??
H_1	$\mu_1 \neq \mu_2$
The Decision rule ($\alpha = 0.10$) is	??
The test statistic is	$Z = 1.85$.
Conclusion	??
p-value $=$	??

Would the conclusion be different if a 1% level of significance is used?
3. (see question 3 PE 7.4). The mean time to process online motor tax renewals for random samples of 100 renewals in two district offices (The Hill and Park Lane) was calculated as 6.72 and 7.24 minutes respectively. If standard deviation for each office is 1.50 minutes test the hypothesis that there is no difference between the mean process times for the two offices. State the p-value for the test.

4. Refer to question 3. The standard deviation of 1.50 minutes for all district offices was in doubt, hence the sample variance were calculated, in addition to the sample means given in question 3.

Table PE7.4 Q4   Difference between means   File: Motor tax 1

Motor Tax Offices	Size	Sample calculations (time in minutes)		
			Mean	Variance
The Hill	100		6.72	2.88
Park Lane	100		7.24	1.48

(a) Use the sample variances to calculate the standard error for the difference between the population means.
(b) Test the claim that the mean process time is the same for both offices.
(c) At the 1 % levels of significance can you conclude that every individual in The Hill processes the online tax faster than those in Park Lane?

5. (see question 6 PE 7.4) From the recovery times for the new and existing medications.

Table PE7.4 Q6   Difference between proportions   File: Headache

Customers	Size	Mean	Variance
Existing medication	40	50	8.43
New medication	40	43	6.45

Test the claim that the recovery time for the new medication is faster than that for the existing medication.

6. Airline booking policy allows for the number of 'no shows'. From records, the proportion of 'no shows' on internal flights is (i) 0.15 on morning flights (earlier than 9.00 am) and (ii) 0.25 on evening flights (later than 4.30 pm). Planes used for morning flights have 80 seats while those used for evening flights have 120 seats (hence sample sizes are 80 and 120 for morning and evening flights, respectively).

(a) Write down the sampling distribution for proportions for
(i) morning (ii) evening flights. Hence calculate the probability that the number of 'no shows' is at least 10 in (i) and (ii).
(b) Write down the sampling distribution for difference between proportions for 'no shows' on evening and morning flights. Calculate the probability that the percentage of 'no shows' for an evening flight is at least 15 % higher than for morning flight.
(c) On randomly selected flights on a given day there were 12 'no shows' on both the morning and evening flight. Test the hypothesis that (i) the percentage of 'no shows' for an evening flight is at least 15 % higher than for morning flight (ii) the percentage of 'no shows' is the same morning and evening. Use 1 % and 5 % levels of significance. Write a brief report on your conclusions.

7. **See question 7 PE 7.4.** The number of 'no shows' on randomly selected Mondays and Tuesdays are as follows:

Table PE7.4 Q7   Difference between proportions   File: No shows dentist

	Monday	Tuesday
Number of appointments	248	240
Number of 'no shows'	52	31

Test the claim that the proportion of no shows is greater on Mondays.

8. **See question 8 PE 7.4.** The numbers of failed thermostats for Thermo A (existing model) and Thermo B(new design) are given as follows:

Table PE7.4 Q8   Difference between proportions (thermostats)   File: Therom

	Thermo A	Thermo B
Sample size	250	198
Number failed	20	10

(a) Test the hypothesis that there is no difference between the proportion that fail for Thermo A and Thermo B.

(b) On the basis of these results, should the company replace the existing model with the new design?

9. Fill in the missing entries indicated by (??) in the following table:

H_0	??
H_1	$\pi_1 < \pi_2$
The Decision rule ($\alpha = 0.05$) is	??
Sample data	$n_1 = 82, x_1 = 25: n_2 = 52, x_2 = 12$
p_c : pooled estimate of p	??
The test statistic is	??.
Conclusion	??

10. **See question 6, PE.7.4** From the data in Table PE7.1 Q6:

Table PE7.1 Q6   Data collected from 104 customers from the travel agent   File: No shows

	(i) Long-haul flights	(ii) Package holidays	(iii) Business
Numbers of customers	32	54	18
Average amount spent (€)	483.5	854	58.5
Sample standard deviation	35.5	28.4	12.2

(a) Test the hypothesis that the mean amount spent by package holiday customers is at least €350 more than the mean amount spent by long-haul customers.
(b) Why would we not test hypothesis for the difference between the mean amount spent by long-haul customers and schedule flight customer by the present methods?

11. See question 10, PE 7.4

Table PE7.4 Q10   Difference between proportions

File: Mobile move

	Left within the month	Left more than one month ago	Totals
Moved to competitor	422	326	748
Moved to other provider	154	298	452
Totals	576	624	1200

Test the hypothesis that the proportion customers who moved to the competitor in greater for those who left in the last month than for those who left more than one month ago.

12. See question 12PE 7.4.

Table PE7.4 Q12   Difference between proportions

File: Heart attack

	Number	Mean time to reach hospital after the initial attack (hours)	Referred by GP	Time in hospital before being treated (hours)
Men	318	$\bar{x} = 3.40: s = 1.42$	65 %	$\bar{x} = 1.24: s = 0.25$
Women	102	$\bar{x} = 8.24: s = 4.24$	72 %	$\bar{x} = 1.43: s = 0.38$

(a) Test the hypothesis that the proportion of women referred by their GP is greater than the proportion of men.
(b) Test the hypothesis that the mean time before being treated is greater for women than for men.
(c) Based on the results in (a) and (b) write a brief report on the difference between the treatment of males and female heart attack victims.

13. A survey is carried out on the numbers of motorists who pay their car tax in advance of the 'due month' in two different council areas. The Minitab output for the analysis of the data collected is given in Table PE8.3 Q13.
Describe this output. What do you conclude about the difference in the proportions who pay in advance in these two council areas?
Note: instructions on using Minitab are given in section 8.4.

14. Excel file 'Savings 5'
Calculate the mean and variance of the amounts saved by each of the five groups of children. Test the hypothesis that the mean amount saved is the same for the groups with the highest and lowest sample means. Can you conclude that the mean amount saved is the different for all groups?

Table PE8.3 Q13   The Minitab output for motorists who pay car tax in advance of the due month

Test and CI for Two Proportions

```
Sample    X     N   Sample p
1        126   400   0.315000
2         85   400   0.212500


Difference = p (1) - p (2)
Estimate for difference: 0.1025
95% CI for difference: (0.0418425, 0.163158)
Test for difference = 0 (vs not = 0): Z = 3.29 P-Value = 0.001
```

15. **Excel file 'Hours worked UK 2000'**
    Test the hypothesis that, in the labour force, men work more than 10 hours longer than women per week on average. What can you conclude?

8.4   Minitab and Excel for confidence intervals and tests of hypothesis

8.4.1   Excel

Excel was introduced in Chapter 2, Section 2.7. Excel offers two options for testing hypothesis and confidence intervals for means proportions and their differences:

1. Use the 'formula' facility to do basic calculations of sample means, variances, proportions. The formulae function in Excel may then be used to calculate standard errors, etc.
2. Excel has built in functions under 'data analysis' that calculate confidence intervals and produce the essential calculations for standard errors, point estimates, test statistics etc. for testing hypothesis for difference between means from raw data with known variances. From the menu bar, select Tools → Data Analysis as illustrated in Figure 2.12a, Chapter 2. From the dialogue box select 'Z-test two sample for means'.

The next dialogue box requests the location of the data, the population variances, etc. necessary for the analysis. As an example of the output, see question 4, PE 8.4. The user must set out the test of hypothesis (see Chart 8.1) or calculate confidence intervals and state their conclusion.

Excel's Data Analysis also covers topics in the remainder of the text: Regression, Correlation, ANOVA and t-tests.

8.4.2   Minitab

Minitab is a statistical package: it consists of a 'project manager', a 'worksheet' window (similar to Excel) and a 'session window' where the results of analysis are displayed. Data is entered into Minitab worksheets in the same way as into Excel. Data may also be copied from Excel to Minitab worksheets. Graphs, descriptive and statistical analysis of data in the worksheet is carried out by selecting the appropriate options from the menu bar. Minitab may be used to calculate confidence intervals and test hypothesis from summarised data or from raw data. The Minitab analysis for means, proportions and their differences gives confidence intervals, test statistics and descriptive statistics on a single printout. For example, the summarised data in Worked Example 7.1 is the following: sample size $n = 36$ $\bar{x} = 19.35$ and $\sigma = 1.8$. To calculate the 95 % confidence interval in Minitab and test the hypothesis that the true mean weight is 20 make the following selection from the menu bar. Click on 1. **Stat** → 2. **Basic Statistics** → 3. **1-Sample Z** as demonstrated in Figure 8.11.

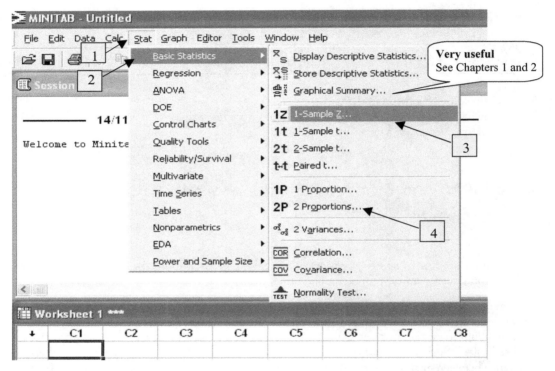

Figure 8.11   Menus for calculating confidence intervals and hypothesis testing in Minitab.

The next dialogue box allows the user to enter the summarised data: as shown in Figure 8.12. Clicking on **Options** opens the 'Options' dialogue box into which the level of confidence and alternative hypothesis is entered.

The Minitab printout is given in Figure 8.13.

Confirm that the confidence interval given in the printout is calculated by the usual formula

$$\bar{x} \pm Z_{\alpha/2}\sigma_{\bar{x}}$$

$$19.35 \pm 1.96(0.30)$$

$$19.35 \pm 0.5880$$

$$19.35 - 0.5880 \text{ to } 19.35 + 0.5880$$

$$\underline{18.762 \text{ to } 19.938 \text{ as given in the printout}}$$

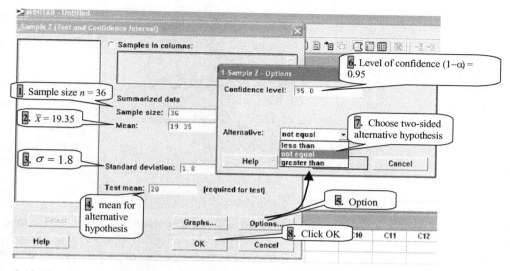

Figure 8.12 Entering the summarised data: selecting the level of significance level in Minitab.

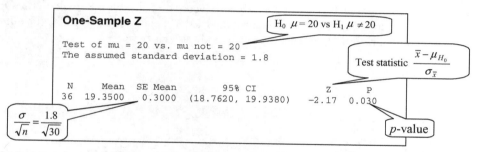

Figure 8.13 Minitab printout for confidence interval and test of hypothesis for data in Worked Example 7.1.

Compare the test of hypothesis with Minitab: Minitab statistics are underlined:

1. H_0 $\mu_{H_0} = 20$
2. H_1 $\mu \neq 20$
3. **Level of significance** $\alpha = 0.05$
4. **Test statistic:** $Z_{\bar{x}} = \dfrac{\bar{x} - \mu_{H_0}}{\sigma_{\bar{c}}}$
5. **Decision rule** Reject H_0 if the sample Z is less than -1.96 or greater than $+1.96$

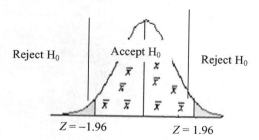

6. **Calculate the test statistic** from sample data: $n = 36$; $s = 1.8$; $\bar{x} = 19.35$

$$Z_{\bar{x}} = \frac{\bar{x} - \mu_{H_0}}{\sigma_{\bar{x}}} = \frac{19.35 - 20}{0.30} = -2.17 \text{ where } \sigma_{\bar{x}} = \frac{1.8}{\sqrt{36}} = 0.30 = \text{SE MEAN}$$

7. **Compare the test statistic with the critical region.**
   The sample $Z = -2.17$ is in the rejection region. Conclude that the mean weight is not 20.
8. **Calculate and interpret the p-value**
   The p-value $= P(\bar{x} \geq -2.17) + P(\bar{x} \leq 2.17) = 2(0.0150) = \underline{0.0300}$.

One-sided tests of hypothesis: for an alternative hypothesis, H_0 $\mu < 20$, select 'less that' at point 7 in Figure 8.12. Similarly select 'greater than' for $H_0 > 20$.

8.4.3 Minitab difference between means and proportions

The confidence interval and test of hypothesis for the difference between proportions were explained in Worked Examples 7.6 and 8.7(b)(i), respectively. These examples will be repeated in Minitab. To explain the Minitab output, compare original worked examples with the statistics given in Minitab.
The instructions for Minitab are as follows:
From the menu bar in the Minitab session window click on 1. **Stat** → 2. **Basic Statistics** → 4. **2P 2 Proportions** see Figure 8.11.
The summarised data is entered into the dialogue box in Figure 8.15. In this box click on the **Options** button to open the 'Options dialogue box' and enter the the level of confidence (0.90): the difference between proportions claimed in H_0 and alternative hypothesis ($H_1 \pi_A \neq \pi_B$).

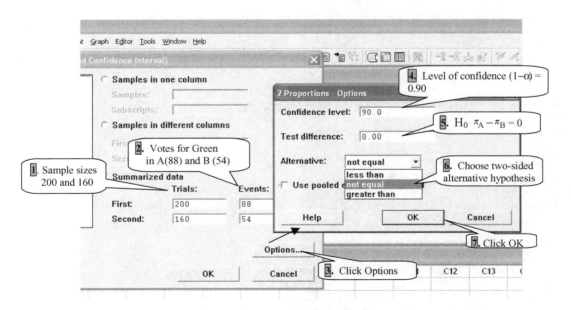

Figure 8.14 Entering the summarised data: selecting the level of significance level in Minitab for differences between proportions.

The Minitab printout is given in Figure 8.15.

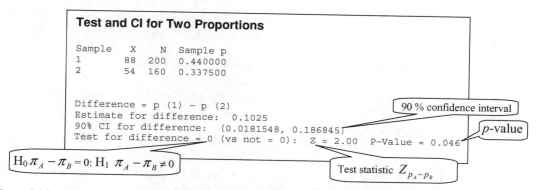

Figure 8.15 Minitab printout for confidence interval and test of hypothesis for data in Worked Example 7.6 and 8.7(b)(i).

Worked Example 8.8: Difference between proportions in Minitab

The null and alternative hypotheses were:

$$H_0\ \pi_A - \pi_B = 0.50\ (\text{or}\ \pi_A - \pi_B \geq 0.05): H_1\ \pi_A - \pi_B < 0.05\ \text{and}\ \alpha = 0.10.$$

These details are entered in the Options dialogue box (see Figure 8.14) (i) at point 5. type 0.05 ($\pi_A - \pi_B = 0.05$) and (ii) at point 6. select 'less than' Also select 'use pooled estimate of p for test' The Minitab printout is given in Figure 8.16.

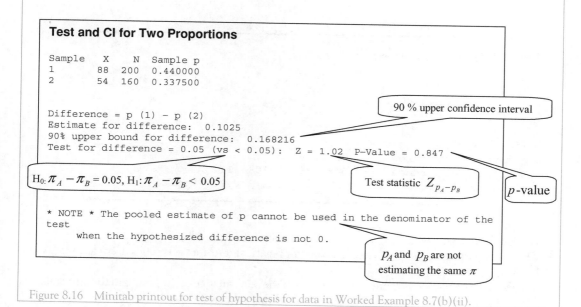

Figure 8.16 Minitab printout for test of hypothesis for data in Worked Example 8.7(b)(ii).

Progress Exercises 8.4: Excel and Minitab

1. Review the exercises in this chapter, using Minitab and/or Excel when appropriate. Compare the Minitab and /or Excel printout with the calculated results; hence give full explanations for the printouts.
2. **Excel files (a) cholesterol; (b) Hours worked UK 2000; and (c) Savings 5**
   For the data in each of these files, do a graphical summary in Minitab.
   Review and describe the printout for each set of data. In particular, can you conclude that each set of data is normally distributed?
   (a) **For the data given in Cholesterol**
       (i) construct a 95 % confidence interval for the difference in the mean level of LDL and HDL cholesterol;
       (ii) test claim that the difference between the average LDL and HDL cholesterol is 4.
   (b) **For the data given in the file Hours worked UK 2000**
       (i) construct a 95 % confidence interval for the difference in the mean hours worked by males and females;
       (ii) test the claim that males work 10 hours more per week than females;
       (iii) Give the 90 % confidence interval and test the hypothesis that the mean weekly hours worked by males exceeds that worked by females by at least 12 for those who work between 20 and 30 hours.
   (c) **For the data given in the file Savings 5**
       (i) state the number of paired comparisons that can be made from the five data sets given;
       (ii) construct a 95 % confidence interval for the difference in the mean weekly saving between (i) groups 2 and 5 and (ii) groups 3 and 4;
       (iii) test the claim that group 5 saves an average of € 10 more per week than group 2.
3. Explain the following Minitab output:

Table PE8.4 Q3   Minitab output

Text and CI for Two Proportions

```
Sample   X       N   Sample p
1       24     140   0.171429
2       36     250   0.144000

Difference = p (1) - p (2)
Estimate for difference: 0.0274286
95% CI for difference: (-0.0486734, 0.103531)
Test for fifference = 0 (vs not = 0): z = 0.72 P-value = 0.471
```

Do you think that this data would be easier to interpret if it had been given in context?
4. The Excel output for the difference in scores for practical tests for two groups of students is given below (File: Practical scores):
   (a) Explain the output. Is there any difference in the performance of the groups in the assessment?
   (b) If the assumed variances were reduced would this affect your conclusions in (a)?

Table PE8.4 Q4   Excel output for end of course assessments

z-Test: Two Sample for Means

	B	A
Mean	95.61290323	83.38709677
Known Variance	1000	1500
Observations	31	31
Hypothesised Mean Difference	0	
z	1.361408189	
P(Z<=z) one-tail	0.086692423	
z Critical one-tail	1.644853476	
P(Z<=z) two-tail	0.173384847	
z Critical two-tail	1.959962787	

Summary and overview of testing hypothesis for means and proportions

8.1   Testing Hypothesis for a population mean

A Null hypothesis is a claim or conjecture made about a population characteristic, such as $H_0: \mu = 5000$. The alternative hypothesis is the complement of the Null hypothesis, such as $H_0: \mu \neq 5000$.

The test involves selecting a random sample from the population and calculating a point estimate for the claimed population characteristic. Then, based on sampling theory, we can decide whether the point estimate is sufficiently close to the claimed population parameter provides sufficient evidence to support the claim made in the Null hypothesis.

The steps for testing hypotheses are given in Chart 8.1:

1. State H_0
2. State H_1
3. State α: α is the level of significance
4. State the test statistic: Hence state the sampling distribution
5. State the decision rule and critical regions.
   Sketch and label the acceptance and rejection regions
6. Calculate the test statistic from the sample data
7. Compare the test statistic with the decision rule
   State the conclusion verbally and in context
8. Calculate the p-value

The **p-value** of a test is the probability of selecting this present sample or a more extreme sample assuming the null hypothesis is true. Hence the p-value gives a measure of the supporting evidence for the null hypothesis provided by the sample.

Important concepts and terminology include: Test statistic: Type I error: Level of significance (α): One-sided tests: Two-sided tests Type II error:

Tests for population means: summary:

$$\text{The test statistic, } T \text{ (sample } Z) = \frac{(\text{estimate}) - (H_0 \text{ claim})}{\text{standard error}} = \frac{\bar{x} - (\mu_{H_0})}{\sigma_{\bar{x}}}$$

where $\sigma_{\bar{x}} = \dfrac{\sigma}{\sqrt{n}}$. OR, if σ is unknown use s etc.

Hypothesis	Two-sided	Left tail	Right Tail
H_0:	$\mu = \mu_{H_0}$	$\mu \leq \mu_{H_0}$	$\mu \geq \mu_{H_0}$
H_1:	$\mu \neq \mu_{H_0}$	$\mu > \mu_{H_0}$	$\mu < \mu_{H_0}$
Decision rule based on the test statistic, T,	Reject H_0 if T is outside $\pm Z_{\alpha/2}$	Reject H_0 if T is greater than $Z_{\alpha/2}$	Reject H_0 if T is less than $Z_{\alpha/2}$

8.2 Tests of hypothesis for a population proportion: summary

$$\text{The test statistic, } T \text{ (or } Z_p) = \frac{(\text{estimate}) - (H_0 \text{ claim})}{\text{standard error}} = \frac{\pi - \pi_{H_0}}{s_p}$$

where $s_p = \sqrt{\dfrac{\pi_{H_0}(1 - \pi_{H_0})}{n}}$

Hypothesis	Two-sided	Left tail	Right Tail
H_0:	$\pi = \pi_{H_0}$	$\pi \leq \pi_{H_0}$	$\pi \geq \pi_{H_0}$
H_1:	$\pi \neq \pi_{H_0}$	$\pi > \pi_{H_0}$	$\pi < \pi_{H_0}$
Decision rule based on the test statistic, T,	Reject H_0 if T is outside $\pm Z_{\alpha/2}$	Reject H_0 if T is greater than $Z_{\alpha/2}$	Reject H_0 if T is less than $Z_{\alpha/2}$

8.3 Tests of hypothesis for difference between means: summary

To test claims about two population means μ_1 and μ_2

$$\text{The test statistic, } T = \frac{(\text{estimate}) - (H_0 \text{ claim})}{\text{standard error}} = \frac{(\bar{x}_1 - \bar{x}_2) - (\mu_1 - \mu_2)}{\sigma_{\bar{x}_1 - \bar{x}_2}} \text{ where } \sigma_{\bar{x}_1 - \bar{x}_2} = \sqrt{\frac{\sigma_1^2}{n_1} + \frac{\sigma_2^2}{n_2}}$$

OR, if σ is unknown use s, etc.

Hypothesis	Two-sided	Left tail	Right Tail
H_0:	$\mu_1 - \mu_2 = 0$	$\mu_1 - \mu_2 < 0$	$\mu_1 - \mu_2 > 0$
H_1:	$\mu_1 - \mu_2 \neq 0$	$\mu_1 - \mu_2 > 0$	$\mu_1 - \mu_2 <$
Decision rule based on the test statistic, T,	Reject H_0 if T is outside $\pm Z_{\alpha/2}$	Reject H_0 if T is greater than $Z_{\alpha/2}$	Reject H_0 if T is less than $Z_{\alpha/2}$

8.4 Tests of hypothesis for difference between proportions: summary

To test claims about two population proportions p_1 and p_2

The test statistic, $T = \dfrac{(p_1 - p_2) - (H_0 \; claim)}{s_{p_1-p_2}}$

where the sample standard error is $s_{\bar{p}_1 - \bar{p}_2} = \sqrt{\dfrac{p_1(1-p_1)}{n_1} + \dfrac{p_2(1-p_2)}{n_2}}$

If the test statistic is calculated using $\pi_A - \pi_B \pi_1 - \pi_2 = 0$ as the H_0 claim, then use the combined estimate the population proportion

$$\bar{P}_c = \frac{\text{Number of items with given characteristic from both samples}}{\text{Sum of all items in both samples}}$$

Hence the corresponding combined estimate of standard error for proportions

$$s_{p_2-p_2} \rightarrow s_{combined} = \sqrt{\frac{\bar{P}_c(1-\bar{P}_c)}{n_1} + \frac{\bar{P}_c(1-\bar{P}_c)}{n_2}} = \sqrt{\bar{P}_c(1-\bar{P}_c)\left(\frac{1}{n_2} + \frac{1}{n_2}\right)}$$

Hypothesis	Two-sided	Left tail	Right Tail
H_0: H_1: Decision rule based on the test statistic, T,	$\pi_A - \pi_B = 0$ $\pi_A - \pi_B \neq 0$ Reject H_0 if T is outside $\pm Z_{\alpha/2}$	$\pi_A - \pi_B \leq 0$ $\pi_A - \pi_B > 0$ Reject H_0 if T is greater than $Z_{\alpha/2}$	$\pi_A - \pi_B \geq 0$ $\pi_A - \pi_B < 0$ Reject H_0 if T is less than $Z_{\alpha/2}$

8.5 Excel and Minitab for confidence intervals and tests of hypothesis

Both Excel and Minitab provide the confidence limits and test statistic for raw data or summarised data. The user must be able to interpret the output and use the statistics given to explain hypothesis tests and confidence intervals.

INFERENCE FROM SMALL SAMPLES. CONFIDENCE INTERVALS AND TESTS OF HYPOTHESIS

Chapter Objectives

Having carefully studied this chapter and completed the exercises you should be able to do the following

- Calculate confidence intervals and test hypothesis for population mean from a small independent sample from Normal populations with σ known
- Describe the characteristics of Student's t distribution and use the t tables
- State the assumptions necessary for the use of the Student's t distribution
- Calculate a confidence interval for a population mean based on a small sample with σ unknown
- Calculate a confidence interval for the difference between means from small independent samples
- Calculate a confidence interval for the differences between means from paired samples
- Test hypothesis for means, difference between means for independent and paired samples
- State the assumptions underlying the confidence intervals and tests of hypothesis based on small samples
- Carry out an F-test to test for differences between population variances.
- Interpret and explain verbally Minitab printouts for confidence intervals and tests of hypothesis
- Use Minitab to check the assumptions for inference from small samples

9.1 Inference from small samples: Normal populations, σ known

In chapter 6 it was stated in (6.4a) that the sampling distribution for sample means for small ($n < 30$) is Normal, $\bar{x} \sim N\left(\mu, \frac{\sigma}{\sqrt{n}}\right)$ under the following conditions

1. samples are selected from a Normal population
2. the population standard deviation (σ) is known

Hence confidence intervals and test statistics for μ are calculated by the formulae given in Chapters 7 and 8 as

$$\bar{x} \pm Z_{\alpha/2}\sigma_{\bar{x}} \quad (7.3) \quad \text{and} \quad T = \frac{\bar{x} - \mu_{H_0}}{\sigma_{\bar{x}}} \quad (8.4a) \text{ respectively}$$

See Worked Example 9.1

Worked Example 9.1: Inference from a small sample, population Normal, σ known

The temperature of a cool storage unit (in °C) is taken on 8 consecutive days. 4.5; 4.8; 5.2; 4.7; 3.8; 3.7; 4.1; 3,9.
   Temperatures for this type of cold storage unit are known to be Normally distributed with a standard deviation $\sigma = 0.35$.

(a) Construct a 90 % confidence interval for the true mean temperature.
(b) Test the hypothesis that the mean temperature is 4°C.
   State your conclusions verbally.

Solution

Calculate the sample mean $\bar{x}$ since this is required for both the confidence interval and the test statistic. $\bar{x} = \dfrac{\sum x}{n} = \dfrac{34.7}{8} = 4.3375$

The standard error is $\sigma_{\bar{x}} = \dfrac{\sigma}{\sqrt{n}} = \dfrac{0.35}{\sqrt{8}} = 0.1237$ ($\sigma = 0.35$ was given)

(a) The 90 % confidence interval $= \bar{x} \pm Z_{\alpha/2}\sigma_{\bar{x}}$

$$= \bar{x} \pm 1.6449 \frac{\sigma}{\sqrt{n}}$$

$$= 4.3375 \pm 1.6449 \frac{0.35}{\sqrt{8}}$$

$$= 4.3375 \pm 0.2035$$

$$= 4.1340 \text{ to } 4.5410$$

(b) Setting out the test of hypothesis by the usual 8 steps
   1. H_0 $\mu_{H_0} = 4$
   2. H_1 $\mu \neq 4$
   3. Level of significance $\alpha = 0.10$
   4. Test statistic: $Z_{\bar{x}} = \dfrac{\bar{x} - \mu_{H_0}}{\sigma_{\bar{x}}}$
   5. **Decision rule**: Reject H_0 if the test statistic is outside the interval -1.6449 to 1.6449, see Figure 9.1:

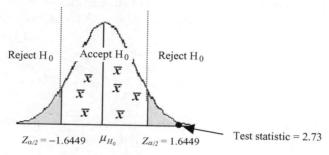

Figure 9.1   Accept H_0 if test statistics falls between $Z = -1.6449$ and $Z = 1.6449$.

   6. $n = 8; \sigma = 0.35; \bar{x} = 4.3375$
   $$Z_{\bar{x}} = \frac{\bar{x} - \mu}{\sigma_{\bar{x}}} = \frac{4.3375 - 4}{0.1237} = 2.7284 = 2.73 \text{ (correct to 2D)}$$
   7. The sample $Z = 2.73$ is in the reject region, therefore there is insufficient evidence to accept the null hypothesis.
   8. The p-value $= 2 \times P(Z > 2.73) = 2 \times (0.0032) = 0.0064$.

Conclusion

According to the confidence interval we are 90 % confident that the true mean temperature is within the range 4.1340 to 4.5410, hence greater than 4. This is supported by the hypothesis test which rejects the claim that the mean temperature is 4. However, the p-value $= 0.0064$, indicating that there is a 0.64 % chance of selecting the given sample (or more extreme) if the true mean temperature is 4°C. Hence the results are not highly significant.

The Minitab output for the test is given in Figure 9.2:

One-Sample Z: Temp

```
Test of mu = 4 vs not = 4
The assumed standard deviation = 0.35


Variable  N     Mean     StDev   SE Mean       90% CI           Z      P
Temp      8   4.33750  0.54232  0.12374  (4.13396, 4.54104)   2.73  0.006
```

Figure 9.2   Minitab printout Z test for small samples.

It is left as an exercise for the reader to use the results in (a) and (b) to explain the Minitab output.

9.2 The Student's *t* distribution

William Sealey Gosset

Much of the small sample analysis in this chapter is based on a probability distribution called the Student's *t* distribution. The *t* distribution was derived by William Sealey Gosset who was employed as a chemist by the Guinness brewery in Dublin between 1899 and 1934. He invented the *t* distribution in order to monitor quality in brewing using small samples from **Normal populations with σ unknown**. He corresponded with a large number of eminent statisticians: Fisher, Neyman and Pearson of the time. Gosset used the pen name 'Student' when publishing the results of his statistical research

The Student's t distribution

If random samples of size n are selected from a Normal population, mean μ, but standard deviation, σ **is unknown** then the distribution of sample means is a t-distribution

$$\overline{x} \sim t_{n-1}\left(\mu,\ s_{\overline{x}}\right) \tag{9.1}$$

where the unknown σ is estimated by the sample standard deviation s and $s_{\overline{x}} = \dfrac{s}{\sqrt{n}}$.

The subscript '$n - 1$' refers to a parameter of the t distribution: $\nu = n - 1$, the sample size less one. ν is called the '**degrees of freedom**'. (see Appendix H for an explanation for degrees of freedom).

The t-distribution is similar to the Normal in many of its properties:

- it is bell shaped,
- it is symmetrical about the mean;
- $t = \dfrac{\overline{x} - \mu}{s_{\overline{x}}}$ is the number standard errors between $\overline{x}$ and μ in a t-distribution.

> Remember $Z = \dfrac{\overline{x} - \mu}{\sigma_{\overline{x}}}$ is the number of standard errors between $\overline{x}$ and μ in a Normal distribution

However, t-distribution is wider than the Normal for small sample sizes, but as sample sizes increase its shape approaches that of the Normal. See Figure 9.3.

Table 9.1 gives an extract of the t percentage point tables (see Table 5, Appendix G) for selected tail areas (α) and degrees of freedom $\nu = (n - 1)$. Tail areas ranging from 0.1 to 0.0005 are given in the top margin of the table: degrees of freedom are given in the left margin (column) of the table: t-values are given in the body of the table. As an example to illustrate how to use the tables, we will

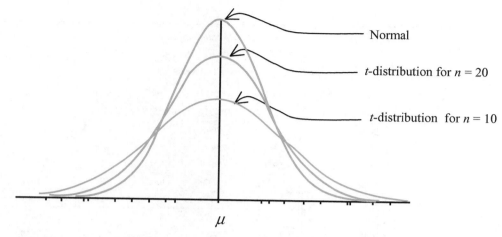

Figure 9.3   Comparison of the Normal and t-distribution for sample sizes $n = 20$ and $n = 10$.

find t-value that gives a tail area of 0.05 (5 %) when the sample size is 10. First calculate the degrees of freedom: $v = n - 1 = 10 - 1 = 9$. Then find $v = 9$ in the left margin and $\alpha = 0.05$ in the top margin of the tables. Follow the lines down from $\alpha = 0.05$ and across from $v = 9$. The required is at the point of intersection of these lines at t-value $= 1.8331$.

The notation $t_{n-1,\alpha}$ is the critical t-value (t percentage point) for which the upper tail area is α when the sample size is n. Hence $t_{9,0.05} = 1.8331$ for $\alpha = 0.05$, $n = 10$.

Compare the critical Z-values and the critical t-values for $\alpha = 0.05$ (or $\alpha = 5$ %)

Examples of t-values and the Z-values for a tail area $= 0.05$ (5 %) for samples of size 10, 20, 30 and 601 are given in Table 9.2. These examples demonstrate that the t-distribution approaches the Normal as n increases.

The 5 % points for the Normal distribution and for the t-distributions with $n = 10$ and $n = 20$ are illustrated graphically in Figure 9.4.

Using Z-percentage points instead of t-percentage points for large samples

It is common practise in statistics to substitute Z-percentage points for t-percentage points when calculating confidence intervals and testing hypotheses for 'large' samples ($n \geq 30$) when σ is NOT known (except in very accurate work). For samples of size 30 or more the Z- and t-percentage points are close: see Table 9.2. It was on this basis that the Z-values (percentage points) were used in confidence intervals (7.3b) and hypothesis tests in Chapters 7 and 8.

But, statistical software packages, such as Excel and Minitab, always use the t-percentage points, even for very large samples.

Table 9.1 Extract from the Student's t distribution

Percentage points of the Student's t-distribution

Tail areas 0.05

$\alpha =$ **Tail Area**

$\upsilon = n - 1$	0.1	0.05	0.025	0.01	0.005	0.001	0.0005
1	3.0777	6.3137	12.7062	31.8210	63.6559	318.2888	636.5776
2	1.8856	2.9200	4.3027	6.9645	9.9250	22.3285	31.5998
3	1.6377	2.3534	3.1824	4.5407	5.8408	10.2143	12.9244
4	1.5332	2.1318	2.7765	3.7469	4.6041	7.1729	8.6101
5	1.4759	2.0150	2.5706	3.3649	4.0321	5.8935	6.8685
6	1.4398	1.9432	2.4469	3.1427	3.7074	5.2075	5.9587
7	1.4149	1.8946	2.3646	2.9979	3.4995	4.7853	5.4081
8	1.3968	1.8595	2.3060	2.8965	3.3554	4.5008	5.0414
9	1.3830	1.8331	2.2622	2.8214	3.2498	4.2969	4.7809
10	1.3722	1.8125	2.2281	2.7638	3.1693	4.1437	4.5868
11	1.3634	1.7959	2.2010	2.7181	3.1058	4.0248	4.4369
12	1.3562	1.7823	2.1788	2.6810	3.0545	3.9296	4.3178
13	1.3502	1.7709	2.1604	2.6503	3.0123	3.8520	4.2209
14	1.3450	1.7613	2.1448	2.6245	2.9768	3.7874	4.1403
15	1.3406	1.7531	2.1315	2.6025	2.9467	3.7329	4.0728
16	1.3368	1.7459	2.1199	2.5835	2.9208	3.6861	4.0149
17	1.3334	1.7396	2.1098	2.5669	2.8982	3.6458	3.9651
18	1.3304	1.7341	2.1009	2.5524	2.8784	3.6105	3.9217
19	1.3277	1.7291	2.0930	2.5395	2.8609	3.5793	3.8833
20	1.3253	1.7247	2.0860	2.5280	2.8453	3.5518	3.8496
21	1.3232	1.7207	2.0796	2.5176	2.8314	3.5271	3.8193
~~	~~~~~	~~~~~	~~~~~	~~~~~	~~~~~	~~~~~	~~~~~
29	1.3114	1.6991	2.0452	2.4620	2.7564	3.3963	3.6595
30	1.3104	1.6973	2.0423	2.4573	2.7500	3.3852	3.6460
35	1.3062	1.6896	2.0301	2.4377	2.7238	3.3400	3.5911
40	1.3031	1.6839	2.0211	2.4233	2.7045	3.3069	3.5510
600	1.2830	1.6474	1.9639	2.3326	2.5841	3.1039	3.3068

9 degrees of freedom

Table 9.2 t-value and the Z- values for tail areas of 5 % in the Normal and t-distributions

Sample size n	Degrees of freedom $\upsilon = n - 1$	$t_{n-1,\alpha}$ distribution $\alpha = 5\%$	Normal distribution $\alpha = 5\%$
$n = 10$	$\upsilon = 10 - 1 = 9$	$t_{9,0.05} = 1.8330$	$Z = 1.6449$
$n = 20$	$\upsilon = 20 - 1 = 19$	$t_{19,0.05} = 1.7291$	$Z = 1.6449$
$n = 30$	$\upsilon = 30 - 1 = 29$	$t_{29,0.05} = 1.6991$	$Z = 1.6449$
$n = 601$	$\upsilon = 601 - 1 = 600$	$t_{600,0.05} = 1.6474$	$Z = 1.6449$

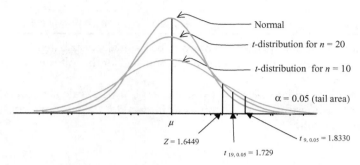

Figure 9.4   5 % points for the Normal and the t-distributions when $n = 10$ and $n = 20$.

9.3   Inference from small samples: Normal populations, σ NOT known

The methods for the construction of confidence intervals and testing hypothesis from small samples from a Normal population with σ unknown parallels those described in Chapters 7 and 8, except that the t-percentage or critical points are substituted for the Z-percentage or critical points.

9.3.1   Student's t-distribution: confidence intervals and tests of hypothesis

Confidence intervals, small samples, σ is NOT known

The formula for the confidence interval for the population μ calculated from small samples ($n < 30$) selected from a Normal population when σ is NOT known is given by formula (9.2).

The confidence interval for the population mean is

$$\bar{x} \pm t_{n-1,\,\alpha/2}\, s_{\bar{x}} \tag{9.2}$$

for random samples from a Normal population, $n < 30$ and σ is unknown

Tests of hypothesis, small samples, σ is NOT known

To test of hypothesis for the population mean based on small samples, when σ is NOT known, the formula for the test statistic is given by formula (9.2). BUT the decision rule is based on critical points from the t distribution as demonstrated in Worked Example 9.3.

The test statistic is

$$T = \frac{\bar{x} - \mu_{H_0}}{s_{\bar{x}}} \tag{9.3}$$

for random samples from a Normal population, $n < 30$ and σ is unknown

All the above assume that the random sample is selected from a NORMAL population.

Worked Example 9.2: Confidence interval based on the *t*-distribution

The queuing time at an airline check-in is known to be Normally distributed. A random sample of 5 passengers were interviewed and reported that they queued for the following times (in minutes): 15.5; 21.2; 12.6; 18.4; 22.9.

State the assumptions under which the *t*-distribution is used to calculate confidence intervals for population means. Construct a 90 % confidence interval for the mean queuing time. Interpret and comment on the result.

Solution

The calculations for the 90 % confidence interval are summarised in Chart 9.1.

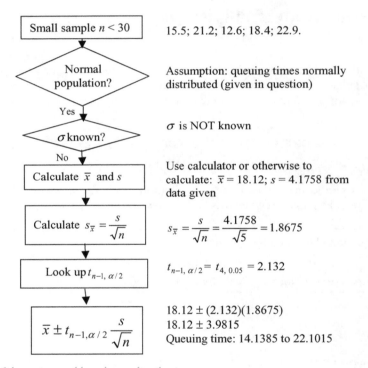

Chart 9.1   Confidence interval based on *t*-distribution.

We are 90 % confident that the mean queuing time is between 14.1385 and 22.1015 minutes.

Worked Example 9.3: tests of hypothesis using a small sample, σ unknown

Test the hypothesis the mean queuing time is at most 20 minutes based on the sample data supplied in Worked Example 9.2.

Solution

The step-by-step method given in Chart 7.4 applies, but since σ is estimated the critical values of the decision rule are based on the Student's t-distribution.

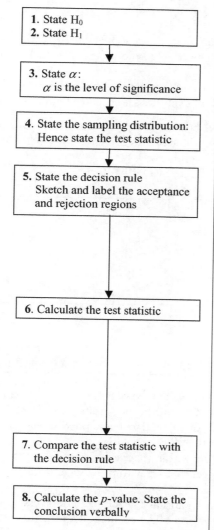

1. State H_0
2. State H_1

3. State α:
   α is the level of significance

4. State the sampling distribution:
   Hence state the test statistic

5. State the decision rule
   Sketch and label the acceptance and rejection regions

6. Calculate the test statistic

7. Compare the test statistic with the decision rule

8. Calculate the p-value. State the conclusion verbally

Solution

1. H_0 $\mu_{H_0} \leq 20$

2. H_1 $\mu_{H_0} > 20$

3. $\alpha = 0.05$

4. The sampling distribution is t_{n-1}.

   Hence the **test statistic** is : $t = \dfrac{\bar{x} - \mu}{s_{\bar{x}}}$.

5. **Decision rule**

   Critical t-value: $t_{n-1,\alpha} = t_{4,0.05} = 2.132$

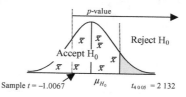

6. **Calculations**

   $\bar{x} = 18.12$; $s = 4.1758$ (calculator)

   $$s_{\bar{x}} = \frac{s}{\sqrt{n}} = \frac{4.1758}{\sqrt{5}} = 1.8675$$

   $$t = \frac{\bar{x} - \mu}{s_{\bar{x}}}. = \frac{18.12 - 20}{1.8675} = -1.0067.$$

7. -1.0067 is in the acceptance region, hence accept H_0. Conclude that queuing time is at most 20 minutes.

8. The p-value is greater than 0.50 (see diagram for the decision rule above).

Chart 9.4

Note: the calculation of the exact p-value requires the use of the t-distribution tables which are not covered in this text. But an approximate p-value is obtained from the t-percentage point tables or from the decision rule, such as above where the p-value is greater than 0.05.

Skill Development Exercise SKD_9.1: Small samples: *t* tests

In Worked Example 9.1, the variation of temperature in the cold storage unit is assumed to be Normally distributed. If the standard deviation is not known, calculate the 90 % confidence interval and test the hypothesis that the mean temperature is 4°C.

Answers

(a) The 90 % confidence interval for the true mean temperature is 3.9472 to 4.7008.
   A true mean temperature of 4°C is supported by the 90 % confidence, since 4 is within the interval.
(b) The test statistic is 1.76. This is within the acceptance region: -1.8946 to 1.8946.
   The p-value is between 0.10 and 0.20. The null hypothesis is accepted but not conclusively.

The results are different from Worked Example 9.1 because of the population standard deviation: In WE 9.1 σ was assumed to be 0.50. The estimated value of σ in SKD_9.1 $s = 0.5423$. Because of this larger value of s the results are not statistically significant.

   The answers are given in the Minitab output in Figure SKD_9.1 below:

```
One-Sample T: Temp

Test of mu = 4 vs not = 4


Variable   N    Mean     StDev   SE Mean        90% CI           T      P
Temp       8  4.33750  0.54232  0.19174  (3.97424, 4.70076)   1.76  0.122
```

Figure SKD_9.1   Minitab printout for SKD_9.1.

Progress Exercises 9.1: Inference for a population mean from small samples

1. An antibiotic powder is supplied in single sachets. The powder is filled to a mean weight of 50 mg and the net weight of the powder should be within the specification 50 ± 0.5 mg. The content of the sachets is known to be Normally distributed with a standard deviation of 0.21 mg (μ is assumed to be the 50 mg, the setting on the filling machine).
   (a) Calculate the percentage of sachets that are within the specification 50 ± 0.5 mg.
   (b) Calculate the percentage of sample averages that are within the specification 50 ± 0.5 mg for samples of size 20.
   (c) A sample of 20 sachets is selected from a particular machine. The average weight was calculated as 50.18 mg. (i) Calculate the 95% confidence interval for the true mean weight of packets filled by this machine. (ii) Test the hypothesis that the mean weight of all sachets is 50 mg. Do your results indicate that the average weight per sachet for the entire batch is 50 mg?
2. (a) Write down the formula for the 90 % confidence interval and the test statistic for a population mean calculated from a random sample of size 8, selected from a Normal population which has a known standard deviation, $\sigma = 0.15$.

(b) The weight (in kg) of eight randomly selected wooden CD racks is recorded: 2.35, 2.41, 2.62, 2.19, 2.39, 2.72, 2.52, 2.43. Assume the population standard deviation is known to be 0.15 kg.
   (i) Calculate a 90 % confidence interval for the population mean weight.
   (ii) Test the claim that the population mean weight is greater than 2.5 kg at the 10 % level of significance. What do you conclude?

3. The time taken to accurately type a technical report by each of five individuals was recorded as 15.5, 22.5, 18.6, 20.8, 24.1 minutes. If times are Normally distributed and the standard deviation, $\sigma = 2.2$ minutes, calculate a 95 % confidence interval for the true mean time. Does your result support the claim that the true average time taken to type the report is 17 minutes?

4. An athlete, training for the triathlon, records the times taken to cycle five miles on 10 occasions: 12.5, 10.5, 14.2, 8.8, 9.6, 10.7, 11.2, 10.5, 12.6, 9.8 minutes. Assuming times are Normally distributed with a known standard deviation of 1.5 minutes.
   (a) Calculate the 99 % confidence interval for the mean cycling time.
   (b) At the 1 % level of significance can you conclude that the mean time is 10 minutes?

5. Explain the conditions under which t-percentage points are used in calculating a confidence interval for a population mean?
   Write out the formula for calculating a confidence interval for a population mean when the only information available is the sample mean ($\bar{x}$) and sample standard deviation (s) calculated from a sample of 12 values. State any assumptions made.

6. In question 2 it was noted, in recent times, that the weights of CD racks had become more variable. Hence the assumed value, $\sigma = 0.15$kg may not be accurate.
   (a) Calculate the sample standard deviation from the data in question 2.
   (b) Calculate a 95 % confidence interval for the true mean weight using the sample standard deviation. State any assumptions used.
   (c) Test the claim that the population mean weight is 2.5kg at the 5 % level of significance. Explain the results to (a), (b) and (c) as you would to a non-statistician.

7. The refreshment shop in a large cinema complex records the amount spent (in £) by twelve randomly selected customers on one particular night: 3.85, 5.28, 6.74, 1.96, 4.85, 4.28, 6.73, 4.56, 5.45, 8.35, 10.50, 4.25. Does this sample data provide evidence that the mean amount spent by customers is at least £5?

8. In question 3, the assumed population standard deviation, $\sigma = 2.2$ minutes is in doubt.
   (a) Calculate the sample standard deviation from the data in question 3.
   (b) Calculate a 95 % confidence interval for the true mean time using the sample standard deviation. State any assumptions used.
   (c) Test the claim that the population mean time is 17 minutes.
   Explain why these results differ from those in question 3.

9. In the production of a certain beer, the specification for the pH is 5 ± 0.2. On two consecutive days, the pH for random samples of 10 beers were recorded:
   Day 1: 5.2, 5.3, 5.3, 5.4, 5.5, 4.8, 4.8, 5.0, 5.2, 5.1
   Day 2: 4.9, 5.1, 5.0, 5,0, 5.1, 5.2, 5.1, 5.1, 5.0, 5.2
   (a) Calculate a 90 % confidence interval for the true mean pH over the two days
   (b) Calculate a 90 % confidence interval for the true mean pH on (i) day 1 (ii) day 2.
   Is there a difference in the pH of beers produced on day 1 and day 2?

10. A hospital clinic deals with routine monitoring of blood pressure/blood tests on groups of patients suffering from diabetes. The time (in minutes) taken to complete the tests for each patient was recorded for three randomly selected clinics. The mean time and the standard deviation per patient were calculated for each clinic in Table PE9.1 Q10:

Table PE9.1 Q10   Summary statistics for routine blood pressure/blood tests for diabetic patients

	Clinic 1	Clinic 2	Clinic 3
Sample size, n	15	24	8
Sample mean $\bar{x}$	15.4	14.6	20.6
Sample standard deviation, s	3.2	2.5	6.7

(a) Calculate a 95 % confidence interval for the true mean time taken to perform the tests in each clinic. Can you conclude that the true mean times are different?
(b) For each of the clinics, test the claim that the average time to perform the tests is less than 15 minutes. State any assumptions made.
Describe the results and your conclusions in (a) and (b) verbally.

9.4   Difference between means. Small independent samples

In Chapter 7, an inference on difference between means from large independent samples was based on the fact that the sampling distribution for the difference between means was Normally distributed. See equation (7.9):

$$(\overline{X}_1 - \overline{X}_2) \sim N\left(\mu_1 - \mu_2, \frac{\sigma_1^2}{n_1} + \frac{\sigma_2^2}{n_2}\right) \tag{7.9}$$

In this chapter the differences between population means will be tested for independent and paired samples.
In section **9.4.1** inferences on difference between means is based on small independent samples.
In section **9.4.2** more powerful tests for inference on difference between means is based on 'paired' samples.

9.4.1   Difference between means for small independent samples

The methods and procedures for inference from small independence samples are similar to those used for large samples provided certain conditions are satisfied.
Two cases will be considered:

Case 1: variances σ_1^2 and σ_2^2 are known then the distribution for differences between sample means is given by (7.9) above.
Case 2: variances σ_1^2 and/or σ_2^2, are NOT known, but the following assumptions are satisfied
   (i) the samples were selected from Normal populations and
   (ii) the population variances are equal.

In Case 2, the distribution for differences between sample means is a Student's t-distribution with $(n_1 + n_2 - 2)$ *degrees of freedom* (df). See Figure 9.5.

$$(\overline{X}_1 - \overline{X}_2) \sim t_{(n_1+n_2-2)} \left(\mu_1 - \mu_2, \frac{s_p^2}{n_1} + \frac{s_p^2}{n_2} \right) \qquad (9.4)$$

where s_p^2 is called the pooled estimate of variance

The formula for s_p^2, the estimator of the common but unknown variance is given in (9.5).

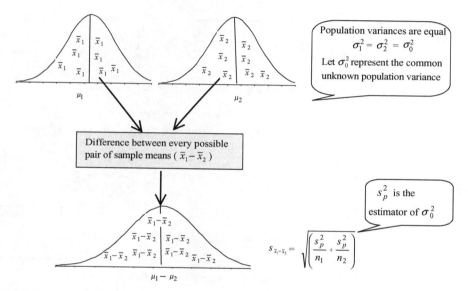

Figure 9.5   Distribution of $(\overline{x}_1 - \overline{x}_2)$ for n_1 and $n_2 < 30$ from normal populations with unknown, but equal variances.

The equality of population variances will be tested in section 9.5.

The formula for the pooled estimate of variance

Since the population variances are assumed equal: $\sigma_1^2 = \sigma_2^2$ (call the common variance σ_0^2) then the sample variances s_1^2 and s_2^2 are two independent estimates of σ_0^2. Hence a more accurate estimate of σ_0^2 will be obtained by 'pooling' the two independent estimates by calculating the **weighted average** of s_1^2 and s_2^2.

Recall (Chapter 2) a weighted average is calculated as $\dfrac{w_1 s_1^2 + w_2 s_2^2}{w_1 + w_2}$, where w_1 and w_2 are the weights. The weights for the 'pooled estimate' of σ_0^2 are $w_1 = (n_1 - 1)$ and $w_2 = (n_2 - 2)$, reflecting the fact that larger sample sizes give more accurate and reliable estimates.

Hence the 'pooled estimate' of σ_0^2 is $\dfrac{(n_1 - 1)s_1^2 + (n_2 - 1)s_2^2}{(n_1 - 1) + (n_2 - 1)}$.

Hence s_p^2 is given in (9.5):

$$s_p^2 = \frac{(n_1 - 1)s_1^2 + (n_2 - 1)s_2^2}{(n_1 - 1) + (n_2 - 1)} = \frac{(n_1 - 1)s_1^2 + (n_2 - 1)s_2^2}{(n_1 + n_2 - 2)} \tag{9.5}$$

To calculate the standard error for the difference between means return to formula (7.10a), but use the common variance σ_0^2 (since $\sigma_1^2 = \sigma_2^2 = \sigma_0^2$), then substitute s_p^2 for the unknown σ_0^2.

$$\sqrt{\left(\frac{\sigma_0^2}{n_1} + \frac{\sigma_0^2}{n_2}\right)} \rightarrow \sqrt{\frac{s_p^2}{n_1} + \frac{s_p^2}{n_2}} = \sqrt{s_p^2\left(\frac{1}{n_1} + \frac{1}{n_2}\right)} = s_p\sqrt{\left(\frac{1}{n_1} + \frac{1}{n_2}\right)} = s_{\bar{x}_1 - \bar{x}_2}$$

Note: if sample sizes are the same, then s_p^2 is simply the average value of s_1^2 and s_2^2.

The sample standard error for the difference between sample means, small independent samples is

$$s_{\bar{x}_1 - \bar{x}_2} = \sqrt{\frac{s_p^2}{n_1} + \frac{s_p^2}{n_2}} = \sqrt{s_p^2\left(\frac{1}{n_1} + \frac{1}{n_2}\right)} = s_p\sqrt{\left(\frac{1}{n_1} + \frac{1}{n_2}\right)} \tag{9.6}$$

The formulae for the confidence interval and test statistic follow in (9.7) and (9.8):

The confidence interval for the difference between population means, $\mu_1 - \mu_2$ is

$$(\bar{x}_1 - \bar{x}_2) \pm t_{(n_1+n_2-2), \alpha/2}\sqrt{s_p^2\left(\frac{1}{n_1} + \frac{1}{n_2}\right)} \tag{9.7}$$

Assume: small independent samples from Normal population with equal variances

where $t_{(n_1+n_2-2), \alpha/2}$ is the percentage point (critical t value)

The test statistic for the difference between means based on small independent samples from Normal population with equal variances: $df = n_1 + n_2 - 2$.

$$T = \frac{(\bar{x}_1 - \bar{x}_2) - (\mu_1 - \mu_2)}{\sqrt{s_p^2\left(\frac{1}{n_1} + \frac{1}{n_2}\right)}} \tag{9.8}$$

Worked Example 9.4: Difference between means, small independent samples

Promoters of e-learning software design a test for effectiveness of an online course based on typing tutor software. Two groups are randomly selected: Group 1 consists of 10 subjects who have completed a course that did not use any supporting software while Group 2 consists of eight subjects who have completed the test online.

The typing speeds (in words per minute) for the subjects in each group were recorded as follows:

Group 1: 23, 35, 37, 12, 26, 60, 13, 24, 27, 53

Group 2: 56, 30, 55, 48, 35, 40, 33, 23

(a) Construct a 90 % confidence interval for the difference in the mean typing speed between the two groups. Can you conclude that those who completed the online typing course can type faster than those who didn't.
(b) Test the hypothesis that the mean typing speed is faster for those who have completed the online course based on the typing tutor software.
State any assumptions made.

What inference can be drawn from the analysis in (a) and (b)

Solution

To answer parts (a) and (b) the sample means, sample variances must be calculated from the sample data. Use your calculator if possible. Hence, for group 1, $\bar{x}_1 = 31.00$ and $s_1 = 15.69147$. For group 2, $\bar{x}_2 = 40.00$ and $s_2 = 12.0000$.

The calculation the confidence interval is given below with the formulae required are given in the left column, data and calculations in the right.

Formulae	Data and Calculations
Sample means $\bar{x} = \dfrac{\sum x}{n}$	**Group 1** $\qquad$ **Group 2** $n_1 = 10 \qquad\qquad n_2 = 8$ $\bar{x}_1 = 31.00 \qquad\quad \bar{x}_2 = 40.00$ $s_1 = 15.69147 \qquad s_2 = 12.0000$
Sample standard deviation $s = \sqrt{\dfrac{\sum(x - \bar{x})^2}{n - 1}}$	
Pooled estimate of variance $s_p^2 = \dfrac{(n_1 - 1)s_1^2 + (n_2 - 1)s_2^2}{(n_1 + n_2 - 2)}$ **Degrees of freedom** $= n_1 + n_2 - 2$	Pooled estimate of variance $s_p^2 = \dfrac{(10 - 1)(15.69147)^2 + (8 - 1)(12)^2}{(10 + 8 - 2)} = 201.5$ Degrees of freedom $= 10 + 8 - 2 = 16$ Percentage point: $t_{16,\,0.05} = 1.746$
Standard error $s_{\bar{x}_1 - \bar{x}_2} = \sqrt{s_p^2\left(\dfrac{1}{n_1} + \dfrac{1}{n_2}\right)}$	Standard error $\sqrt{201.5\left(\dfrac{1}{10} + \dfrac{1}{8}\right)} = 6.7333$

(a) Calculation of 90 % confidence interval

90 % Confidence for $\mu_1 - \mu_2$

$$\bar{x}_1 - \bar{x}_2 \pm t_{n_1+n_2-2,\frac{\alpha}{2}} \sqrt{s_p^2 \left(\frac{1}{n_1} + \frac{1}{n_2}\right)}$$

90 % Confidence for $\mu_1 - \mu_2$

$$40 - 31 \pm 1.746(6.7333)$$
$$9 \pm 11.7563$$
$$-2.7563 \text{ to } 21.7563$$

The difference between the mean score for students from Group 1 and Group 2 lies between
-2.7563 to 21.7563 with 90 % confidence.
So this interval ranges from a negation lower confidence limit to a positive upper confidence
limit. Since the interval contains $\mu_1 - \mu_2 = 0$ there is sufficient evidence to conclude that
typing speeds are the same for both groups at the 90 % level of confidence.

(b) Set out the step-by-step procedure for testing hypothesis
   1. $\mathbf{H_0}$ $\mu_2 = \mu_1$ or $\mu_2 - \mu_1 = 0$
   2. $\mathbf{H_1}$ $\mu_2 > \mu_1$ or $\mu_2 - \mu_1 > 0$
   3. $\alpha = 0.05$
   4. Since we have small independent samples from Normal populations with equal but
      unknown variances, the test statistic is

$$T = \frac{(\bar{x}_1 - \bar{x}_2) - (\mu_1 - \mu_2)}{\sqrt{s_p^2 \left(\frac{1}{n_1} + \frac{1}{n_2}\right)}} \qquad (9.8)$$

   5. **Decision rule**
      The degrees of freedom are $n_1 + n_2 - 2 = 10 + 8 - 2$ and $t_{16, 0.05} = 1.746$
      This is a right tail test, hence accept H_0 if the test statistic is less than or equal to 1.746.
      Reject H_0 if the test statistic is greater than 1.746.

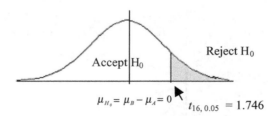

Reject H_0

Accept H_0

$\mu_{H_o} = \mu_B - \mu_A = 0$

$t_{16, 0.05} = 1.746$

   6. **Calculations:**
      The test statistic: $T = \dfrac{(\bar{x}_2 - \bar{x}_1) - (\mu_1 - \mu_2)}{\sqrt{s_p^2 \left(\frac{1}{n_2} + \frac{1}{n_1}\right)}}$

      You will need to calculate the following:

      (i) s_p^2 and $s_{\bar{x}_1-\bar{x}_2} = \sqrt{s_p^2 \left(\frac{1}{n_1} + \frac{1}{n_2}\right)}$, (ii) $\bar{x}_2 - \bar{x}_1$, (iii) $\mu_2 - \mu_1$

Most of these calculations are given in (a) above for confidence interval.
(i) The pooled estimate of variance, $s_p^2 = 201.5$, $n_1 = 10$; $n_2 = 8$

$$\text{Hence } s_{\bar{x}_1 - \bar{x}_2} = \sqrt{201.5 \left(\frac{1}{10} + \frac{1}{8} \right)} = 6.7333$$

(i) $\bar{x}_2 - \bar{x}_1 = 40 - 31 = 9$
(iii) $\mu_2 - \mu_1 = 0$ according to H_0
Hence

$$\text{The test statistic: } t = \frac{(9) - (0)}{\sqrt{201.5 \left(\frac{1}{10} + \frac{1}{8} \right)}} = \frac{9}{6.7333} = 1.3366$$

7. The test statistic is in the acceptance region, hence accept H_0: typing speeds are the same.
8. p-value is greater than 0.10

Assumptions:

1. Small independent samples
2. Samples are randomly selected from Normal populations
3. The variance is the same for both populations $\sigma_1^2 = \sigma_2^2 = \sigma_0^2$

The inference from the hypothesis test is the same as that from the confidence interval.

Progress Exercises 9.2: Difference between means, small independent samples

1. State the conditions under which (i) the Z (ii) the t critical values are used to calculate of confidence intervals for differences between means from small independent samples.
2. The following data was collected on prices in different branches of competing supermarket chains Hyper Store and Hipo Store as part of a consumer price (prices in €) survey.

Table PE9.2 Q2   Supermarket prices (NA not available)   File: Hyper Hipo

	Hipo	Hyper	Hipo	Hyper	Hipo	Hyper
	Strawberries (5 kg)	Strawberries (5 kg)	Litre of Chablis	Litre of Chablis	Litre of petrol	Litre of petrol
n	10	10	18	12	24	24
$\bar{x}$	18.5	19.9	12.45	10.25	1.089	1.149
σ	NA	NA	NA	NA	12.53	19.82
s	1.25	2.16	2.34	1.75	NA	NA

(a) Calculate the pooled estimate of variance for the difference between prices in Hyper Stores and Hipo Stores for (i) strawberries (ii) Chablis (iii) petrol prices.
(b) Calculate the sample standard error for the difference between mean prices Hyper Stores and Hipo Stores for (i) strawberries (ii) Chablis (iii) the mean petrol prices.

(c) Calculate 95 % confidence intervals for the difference between mean prices in Hyper Stores and Hipo Stores for (i) strawberries (ii) Chablis (iii) petrol.

Write a brief report, comparing the prices of these three goods in the two stores.

3. The sales and marketing team in motor business is investigating the quality of 'almost new' second-hand cars. One indicator of quality is the number of miles on the clock. The miles on the clock for samples for two different models are given in the following table.

Table PE9.2 Q3   Milometer miles on cars less than one year old

| Model 1.8 | 1 2125 | 8021 | 2876 | 9874 | 3156 | 7489 | 8436 | 6184 | 1 0546 | 9684 |
| Model 1.4 | 6574 | 5743 | 4865 | 6695 | 7364 | 5474 | 5476 | 7254 | | |

(a) Calculate the sample means and standard deviations.
(b) Calculate the 'pooled' estimate of variance.
(c) Construct a 98 % confidence interval for the difference between the mean miles for Model 1.8 and Model 1.4.
(d) Test the hypothesis that the average miles for Model 1.8 is greater that for Model 1.4 at the 1 % level of significance.

Write a brief report on your findings.

4. An experiment is set up to test the effect of using computer aided learning (CAL) software to support traditional teaching methods. At the beginning of a 10 week course, the students were randomly divided into two groups of 12 and were taught by the same person but at different scheduled times. Both groups were taught by traditional methods but students in the test group, group A, were also trained to use CAL software and encouraged to do the exercises. All students were given the same test before they started the 10 week course and again at the end of the course. Their results are given in Table PE9.2 Q4:

Table PE9.2 Q4   Test results before and after 10 week course

Before course		After	
Control before	A before	Control after	A after
82	12	102	64
86	40	100	150
92	78	172	160
52	70	118	104
72	18	136	90
0	46	20	80
60	64	116	134
36	92	62	172
82	80	118	84
40	46	108	112
126	59	136	88
52	10	100	56

Source: P Sharpe

(a) Test whether there is a difference in the mean scores between the control group and the test group (A) (i) before the course (i) after the course.

(b) Test whether there is a difference in the mean scores before and after the course for (i) the control group and (ii) the test group.

Is there any evidence to support the claim that CAL has improved the students' results?

5. Engineers recruited by an agency are assessed on the basis of a personal interview and a technical interview. There are concerns that the marking scheme for technical interviews gives consistently lower marks than for the personal interviews. A quick check on the marks (maximum mark is 100) for 8 random candidates is given below:

Table PE9.2 Q5   Marks for technical and personal interviews

| Technical | 48 | 72 | 58 | 63 | 71 | 49 | 28 | 74 |
| Personal | 78 | 74 | 52 | 71 | 77 | 50 | 66 | 75 |

(a) Calculate the pooled estimate of variance.

(b) Test the claim that the mean technical mark is lower than the mean personal mark at the 5 % level of significance.

Are the agency's concerns justified?

6. It has come to the notice of schedule managers that underground trains are delayed because of the practise of 'holding' the automatic doors open to allow late-comers board the train. These delays jeopardise the whole scheduling system. To investigate the extent of the delays the time taken to complete a 10 minute journey between 'trouble spot' stations was recorded at specific times: 6–8 am (pre-morning rush), 8–10 am (morning rush 'hour'), 10–4 pm (normal time), 4–7 pm (evening rush 'hour'). The results are given in Table PE9.2 Q6:

Table PE9.2 Q6   Times taken to complete a 10 minutes scheduled journey between two stations on an underground

Time	Sample size, n	Sample mean, $\bar{x}$	Sample standard deviation, s
A: 6–8 am	24	12.18	2.76
B: 8–10 am	25	15.26	3.35
C:10 am–4 pm	18	10.18	1.12
D: 4–7 pm	29	14.32	1.65

(a) How many different pairs of sample means are possible from the data in this table?

(b) Construct 95 % confidence intervals for the differences in the mean journey times between time slots (i) A and C (ii) B and C (iii) B and D.

Is there any evidence of a difference in journey times in (i), (ii) and (iii)?

(c) Suppose that a transcription error has been discovered in the table and that the sample size for D should be 9 and not 29. How would this affect the confidence interval between B and D?

7. Two quick random samples on the overnight cost of 3 star hotel accommodations were selected from hotels countries A and B. The Excel analysis for this data is given in Table PE9.2 Q7:

Table PE9.2 Q7   Excel analysis for 3 star hotel overnight cost in countries A and B

	t-Test: Two-Sample Assuming Equal Variances	
	A	B
Mean	82.66666667	59.125
Variance	226.2666667	216.125
Observations	6	8
Pooled Variance	220.3506944	
Hypothesised Mean Difference	0	
df	12	
t Stat	2.936544559	

Note: t-stat is the test statistic.

(a) From Table PE9.2 Q7 for each country write down the sample size, mean and standard deviation. Explain each of the figures given in Table PE9.2 Q7.

(b) What do you conclude about the cost of accommodation in the two countries?

(c) Do you think the assumptions for small independent t tests are satisfied?

9.5 F-test for equality of two variances

In section 9.2, one of the assumptions necessary for the validity of inference based on small independent samples is the equality of variances of the Normal populations from which they are selected. To test the equality of population variances it would seem reasonable to look at the ratio $\dfrac{s_1^2}{s_2^2}$ where s_1^2 and s_2^2 are independent estimates of σ_1^2 and σ_2^2, samples size n_1 and n_2. If this ratio is close to 1 then the hypothesis $H_0\ \sigma_1^2 = \sigma_2^2$ would seem plausible; if it is very different from 1 then there may be insufficient evidence to support H_0. To see how large (or how small) this ratio should be before we reject null hypothesis we require the sampling distribution for the test statistic $\dfrac{s_1^2}{s_2^2}$. The sampling distribution of $\dfrac{s_1^2}{s_2^2}$ is an F-distribution, when the samples are independent and selected from Normal populations with equal variances. The F-distribution is non-symmetrical and its shape is dependent on 2 parameters: $v_1 = n_1 - 1$, $v_2 = n_2 - 1$. The F-distributions for ($v_1 = 6$ and $v_2 = 20$) and for ($v_1 = 6$ and $v_2 = 12$) are sketched in Figure 9.6:

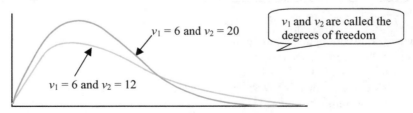

Figure 9.6   Sketches of F-distributions.

Tables for F percentage point for various combinations of the degrees of freedom, v_1 and v_2 that give an upper tail area of a are given in Appendix G, Table 6.

Table 9.3 is an extract of tables for critical F-values (percentage points) for selected combinations of the degrees of freedom, v_1 and v_2 when the right-tail area is 5 %.

Table 9.4 is an extract of tables for critical F-values for a right-tail area is 2.5 %.

As an example on reading the tables consider two samples sizes $n_1 = 10$ and $n_2 = 8$. To find the F-value that gives 5 % of the area in the upper tail, calculate the degrees of freedom: $v_1 = n_1 - 1 = 9$, $v_2 = n_2 - 1 = 7$. Follow the line down from $v_1 = 9$ in the top margin and the line across from $v_2 = 7$ in the left margin. These lines meet at the critical F-value $= 3.68$:

Table 9.3   5 % points for the F-distribution

5% points for the F-distribution

v_2	1	2	3	4	5	6	7	8	9	10
1	161.45	199.50	215.71	224.58	230.16	233.99	236.77	238.88	240.54	241.88
2	18.51	19.00	19.16	19.25	19.30	19.33	19.35	19.37	19.38	19.40
3	10.13	9.55	9.28	9.12	9.01	8.94	8.89	8.85	8.81	8.79
4	7.71	6.94	6.59	6.39	6.26	6.16	6.09	6.04	6.00	5.96
5	6.61	5.79	5.41	5.19	5.05	4.95	4.88	4.82	4.77	4.74
6	5.99	5.14	4.76	4.53	4.39	4.28	4.21	4.15	4.10	4.06
7	5.59	4.74	4.35	4.12	3.97	3.87	3.79	3.73	3.68	3.64
8	5.32	4.46	4.07	3.84	3.69	3.58	3.50	3.44	3.39	3.35
9	5.12	4.26	3.86	3.63	3.48	3.37	3.29	3.23	3.18	3.14
10	4.96	4.10	3.71	3.48	3.33	3.22	3.14	3.07	3.02	2.98
11	4.84	3.98	3.59	3.36	3.20	3.09	3.01	2.95	2.90	2.85

Table 9.4   2.5 % points for the F-distribution

2.5 % points of the F-distribution

$v1$

$v2$	1	2	3	4	5	6	7	8	9	10
1	647.79	799.48	864.15	899.60	921.83	937.11	948.20	956.64	963.28	968.63
2	38.51	39.00	39.17	39.25	39.30	39.33	39.36	39.37	39.39	39.40
3	17.44	16.04	15.44	15.10	14.88	14.73	14.62	14.54	14.47	14.42
4	12.22	10.65	9.98	9.60	9.36	9.20	9.07	8.98	8.90	8.84
5	10.01	8.43	7.76	7.39	7.15	6.98	6.85	6.76	6.68	6.62
6	8.81	7.26	6.60	6.23	5.99	5.82	5.70	5.60	5.52	5.46
7	8.07	6.54	5.89	5.52	5.29	5.12	4.99	4.90	4.82	4.76
8	7.57	6.06	5.42	5.05	4.82	4.65	4.53	4.43	4.36	4.30
9	7.21	5.71	5.08	4.72	4.48	4.32	4.20	4.10	4.03	3.96
10	6.94	5.46	4.83	4.47	4.24	4.07	3.95	3.85	3.78	3.72

Worked Example 9.5: Test the equality of two variances

Given the two independent samples from Normal populations in Worked Example 9.4:
Sample 1: $n_1 = 10$, $s_1^2 = 246.22$
Sample 2: $n_2 = 8$, $s_2^2 = 144$
Test the hypothesis that

(a) the variances are equal
(b) the variance of population 1 is greater than the variance of population 2

Solution

Set out the hypothesis test

Note: in (b) the statement 'the variance of population 1 is greater than the variance of population 2 is written as $\sigma_1^2 > \sigma_2^2$. This is the alternative hypothesis since there is no equality in this statement.

(a)

1. H_0: $\sigma_1^2 = \sigma_2^2$
2. H_1: $\sigma_1^2 \neq \sigma_2^2$
3. α: 5 %
4. **Test Statistic** F-ratio $= \dfrac{s_1^2}{s_2^2}$.

   Degrees of freedom: $v_1 = 9$. $v_2 = 7$
5. **Decision Rule:** If the test statistic exceeds the critical F value, reject H_0 The critical F-value is 4.82 for a 2.5 % tail area (see Table 9.4)

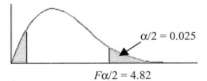

$F\alpha/2 = 4.82$

6. **Calculations:**

   Test statistic $= \dfrac{s_1^2}{s_2^2} = \dfrac{246.22}{144} = 1.7099$

7. Compare test statistic with decision rule
   Since the test statistic is in the acceptance region there is insufficient evidence to reject H_0
   Conclusion: Accept H_0 the population variances are equal

(b)

1. H_0: $\sigma_1^2 \leq \sigma_2^2$
2. H_1: $\sigma_1^2 > \sigma_2^2$
3. α: 5 %
4. **Test Statistic** F-ratio $= \dfrac{s_1^2}{s_2^2}$.

   Degrees of freedom: $v_1 = 9$. $v_2 = 7$
5. **Decision Rule:** If the test statistic exceeds the critical F-value, reject H_0 The critical F-value is 3.68 for a 5 % tail area (see Table 9.3)

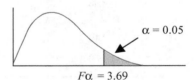

$F\alpha = 3.69$

6. **Calculations:**

   Test statistic: $\dfrac{s_1^2}{s_2^2} = 1.7099$

7. Since the test statistic is in the acceptance region there is insufficient evidence to reject H_0. Conclusion: Accept H_0: the variance of population 1 is less than or equal to the variance of population 2.

Note: in the two-sided test, the test statistic was calculated by putting the larger of the s^2's in the numerator (above the line) so that the test statistic will be in the right tail of the F distribution when there is a significant difference between the variances.

If required, the equation $F_{1-\alpha,v_1,v_2} = \dfrac{1}{F_{\alpha,v_2,v_1}}$ gives the critical value for the left-tail as the inverse of that for the right tail with the degrees of freedom reversed. Hence, in the Worked Example above, the left-tail critical value is $F_{0.95,9,7} = \dfrac{1}{F_{0.05,7,9}} = \dfrac{1}{4.20} = 0.2381$.

Progress Exercises 9.3: F-tests for equality of variances

1. Look up the following critical F-values (percentage points). (i) $F_{4,8,0.05}$ (ii) $F_{2,10,0.05}$ (iii) $F_{2,2,0.025}$ (iv) $F_{5,3,0.025}$.
2. See question 2, PE9.2. Test the equality of the variances for each product.
3. See question 3, PE9.2. Test the equality of the variances for both models of car.
4. See question 5, PE9.2. Test the equality of the variance of the control group and group A (a) before the course (b) after the course.
5. See question 6, PE9.2. Test the equality of the variances for technical and personal tests.
6. See question 7, PE9.2. Test the equality of the variances for the following pairs of time slots: (a) (i) A and C; (ii) B and C; (iii) B and D (b) B and D for sample size 9 for D.
7. The Excel data analysis gave the following printout for the difference between means for the average annual sick days for samples of employees from Companies A and B. File: Sick days.

Table PE9.3 Q7   Excel printout for differences in average sick days

	t-Test: Two-Sample Assuming Equal Variances	
	Company A	Company B
Mean	5.6	12.43333333
Variance	39.2137931	173.5643678
Observations	30	30
Pooled Variance	106.3890805	
Hypothesised Mean Difference	0	
df	58	
t Stat	−2.565840809	

(a) Interpret the Excel output.
(b) Test the equality of variances by the F-test (variances given in the printout are sample variances).
8. See question 7. The Minitab printout for the data in Table PE9.3 Q7 is given below with the box plots for the data for companies A and B.

```
              N     Mean    StDev    SE Mean
Company A    30     5.60     6.26      1.1
Company B    30    12.4     13.2       2.4


Difference = mu (Company A) - mu (Company B)
Estimate for difference: -6.83333
95% CI for difference: (-12.16430, -1.50237)
T-Test of difference = 0 (vs not =): T-Value = -2.57 P-Value = 0.013
DF = 58. Both use Pooled StDev = 10.3145
```

Figure PE9.3 Q7   Minitab printout for differences in average sick days.

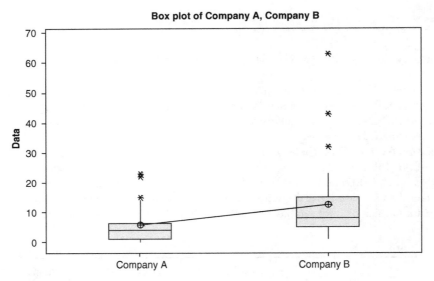

Figure PE9.3 Q8   Box plots for sick days in company A and B.

(a) Interpret the Minitab output.
(b) Describe the box plots and compare the distribution of data for A and B. Do you think the assumption of Normality is satisfied?

The data for the above analysis is given in File: Sick days.

9.6   Difference between means, paired samples

In Worked Exercise 9.4, the test for difference between means based on small independent samples, there was insufficient evidence to conclude that there was a difference in typing-speeds between the two groups. One factor that makes it difficult to detect a significant difference is the large variation in typing speeds within each group: this is reflected in the estimates of variance which were 246.22 and 144 for groups 1 and 2, respectively.

A more powerful test for the effectiveness of the online course is a test that considers group 2 only and records the typing speeds of each individual in group 2 before starting the course and again when the course is completed, as in Table 9.4. The difference in speed (after – before) for each subject in the group is calculated; these differences are called 'paired differences'.

The paired t-test

Paired t-tests are used when data consists of pairs of measurements on the same (or matched) subject(s), for example, the '**before** and **after**' situation above.

Let X = value 'before' and Y = value 'after' so pairs $(x_1, y_1), (x_2, y_2), \ldots\ldots, (x_n, y_n)$ is the sample data 'before and after' for subject 1, subject 2, ... subject n.

Then the differences $(x_i - y_i)$ for each subject are $d_1 = x_1 - y_1, d_2 = x_2 - y_2, \ldots\ldots\ldots\ldots\ldots d_n = x_n - y_n.$

Table 9.5   Paired (before and after) data related to individual subjects

Subject	After	Before	Difference
JM	46	32	14
AC	18	10	8
TB	58	65	−7
A MC F	50	39	11
A O'G	36	24	12
PD	24	10	14
FF	21	24	−3

The paired differences $d_1, d_2 ....d_n$ constitute a single sample.

If there is no difference in the average speed before and after the course then the average of the paired differences $(\bar{d})$ should be close to zero.

To test whether $\bar{d}$ is sufficiently close to zero to support the claim that there is no difference before and after the course we must use the sampling distribution of means, $\bar{d}$.

The population of d's is Normally distributed with mean μ_d and standard deviation σ_d

By the Central Limit Theorem, the sampling distribution of $\bar{d}$ is also Normal, mean $= \mu_d$; standard error $= \dfrac{\sigma_d}{\sqrt{n}}$. Hence $\bar{d} \sim N\left(\mu_{\bar{d}}, \dfrac{\sigma_d}{\sqrt{n}}\right)$. See Figure 9.7.

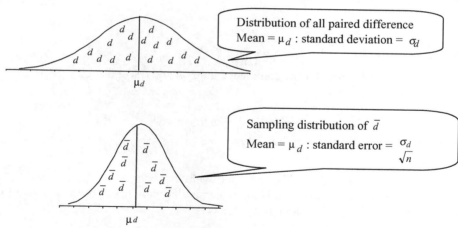

Figure 9.7   Sampling distribution for the mean of paired differences

Note: $\mu_d = \mu_{\bar{d}}$

Consider a sample of paired d's, such as the typing speeds given above.

The sample mean, $\bar{d}$ is an estimate of μ_d.

The sample standard deviation, s_d is an estimate of σ_d.

The sample standard error is $s_{\bar{d}} = \dfrac{s_d}{\sqrt{n}}$, since σ_d is almost always unknown.

Hence the sampling distribution of $\bar{d}$ is $\bar{d} \sim t_{n-1}\left(\mu_{\bar{d}}, \dfrac{s_d}{\sqrt{n}}\right)$ when σ_d is unknown.

The formulae for confidence intervals and the test statistic for paired samples are given in (9.9) and (9.10). See also Table 9.7 in the chapter summary for comparison with t-tests for means.

Confidence Interval for difference between means (paired samples) | Compare these with (9.2) and (9.3) for $\bar{x}$

$$\bar{d} \pm t_{n-1,\alpha/2} \frac{s_d}{\sqrt{n}}$$

(9.9)

Test Statistic

$$T = \frac{estimate - H_0\ claim}{standare\ error} = \frac{\bar{d} - H_0\ claim}{\left(\frac{s_d}{\sqrt{n}}\right)}$$

(9.10)

Worked Example 9.6: Paired *t*-test and confidence intervals

The typing speeds for seven individuals recorded before and after completing a course based on the typing tutor software are given in Table 9.5.

(a) Construct a 90 % confidence interval for the difference between the average typing speed before and after the completion of the typing course.
(b) Test the hypothesis that typing speeds have increased after completing the course.

Solution

(a) The confidence interval is $\bar{d} \pm t_{n-1,\alpha/2} \dfrac{s_d}{\sqrt{n}} \ldots$ (9.9).

You will need to calculate (for the confidence interval and test statistic)
  (i) the paired differences the d's
  (ii) $\bar{d}$, the mean value of the d's
  (iii) s_d, the standard deviation of the d's
  (iv) the sample standard error of the d's $s_{\bar{d}} = \dfrac{s_d}{\sqrt{n}}$

You will also need $t_{n-1,\alpha/2}$
  (i) The paired differences from Table 9.5 and again in 9.6.
     The calculation of the mean and sample standard deviation may be carried out by calculator or manually as illustrated in Table 9.6:

Table 9.6   Calculation of the mean and standard error for paired differences

Subject	After	Before	d_i (After – before)	$(d_i - \bar{d})^2$
JM	46	32	14	49.00
AC	18	10	8	1.00
TB	58	65	−7	196.00
A MC F	50	39	11	16.00
A O'G	36	24	12	25.00
PD	24	10	14	49.00
FF	21	24	−3	100.00
		Total	49	436.00

The d's are 14, 8, −7, 11, 12, 14, −3

(ii) The sample mean of the d's is $\bar{d} = \dfrac{\sum d_i}{n} = \dfrac{49}{7} = 7$

(iii) Sample standard deviation $s_d = \sqrt{\dfrac{\sum (d_i - \bar{d})^2}{n - 1}} = \sqrt{\dfrac{436}{6}} = \sqrt{72.67} = 8.5245$

(iv) $s_{\bar{d}} = \dfrac{s_d}{\sqrt{n}} = \dfrac{8.5245}{\sqrt{7}} = 3.2219$

(Use your calculator for this calculation: key in $d_1, d_2 ... d_n$. Then read out $\bar{d}$ and s_d)
The degrees of freedom $= (7 - 1) = 6$, hence $t_{6, 0.05} = 1.9432$

Hence the 90 % confidence interval from (8.5) is

$\bar{d} \pm t_{n-1, \alpha/2} \dfrac{s_d}{\sqrt{n}}$

$7 \pm 1.943 \dfrac{8.5245}{\sqrt{7}}$

7 ± 6.2602

0.7398 to 13.2602

(b) **Testing the Hypothesis**

If speeds have increases then $\mu_{\text{after}} > \mu_{\text{before}}$. Hence $\mu_{\text{after}} - \mu_{\text{before}} > 0$, i.e., $\mu_d > 0$. Since there is no equality in this statement it is the alternative hypothesis: $H_1 : \mu_d > 0$
1. $H_0 : \mu_d \leq 0$ (after − before)
2. $H_1 \quad \mu_d > 0$
3. $\alpha = 5 \%$ and 2.5%

4. Test Statistic $T = \dfrac{estimate - H_0\ claim}{standard\ error} = \dfrac{\bar{d} - H_0\ claim}{\left(\dfrac{s_d}{\sqrt{n}}\right)}$

5. Decision rule: This is an upper tail test for a t-distribution with 6 degrees of freedom. The critical t-value $= 1.9432$ at the 5 % level of significance:
The critical t-value $= 2.447$ at the 2.5 % level of significance

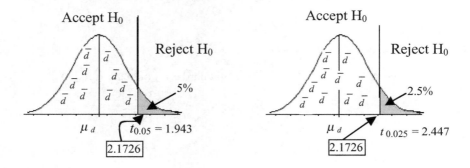

6. Calculation of test statistic: $T = \dfrac{estimate - H_0\ claim}{standard\ error} = \dfrac{\bar{d} - 0}{\left(\dfrac{s_d}{\sqrt{n}}\right)} = \dfrac{7}{3.2219} = 2.1726$

7. Compare the test statistic with the critical regions:
When $\alpha = 5\,\%$ the test statistic $t = 2.1726$ falls in the reject region
When $\alpha = 2.5\,\%$ the test statistic $t = 2.1726$ falls in the acceptance region
The result is marginal with a p value between 0.025 and 0.05

Skill Development Exercise SKD_9.2: Paired *t*-tests

Consider 10 individuals about to embark on CPD course.

Before starting each individual takes a test on (a) current affairs (b) legalisation on Health and safety.

The results of the tests for each person are summarised below:

Participants	A	B	C	D	E	F	G	H	I	J
Current affairs	25	36	45	25	52	12	28	23	44	30
H&S legalisation	42	34	72	39	44	30	37	44	45	23

Construct a 95 % confidence interval for the difference in scores between the two tests.
Test the claim that the average score on H&S legalisation is greater than on current affairs.

Solution

Formulae/Reminders

Formula for Confidence Interval

$$\bar{d} \pm t_{n-1,\,\alpha/2}\ \frac{s_d}{\sqrt{n}}$$

Sample mean $= \dfrac{\sum d_i}{n}$

Sample standard deviation (use your calculator)

$$s_d = \sqrt{\frac{\sum (d - \bar{d})^2}{n - 1}}$$

Standard error of mean

$$s_{\bar{d}} = \frac{s_d}{\sqrt{n}}$$

Data and Calculations

From the Table above, the sample d_i are

$17,\ -2,\ 27,\ 14,\ -8,\ 18,\ 9,\ 21,\ 1,\ -7$

Sample mean for d's : $\bar{d} = \dfrac{\sum d_i}{10} = \dfrac{90}{10} = 9$

Estimate of population standard deviation (for d's)

$s_d = 12.32883$

Standard error of mean (for d's)

$s_{\bar{d}} = \dfrac{s_d}{\sqrt{n}} = \dfrac{12.32883}{\sqrt{10}} = 3.898718$

Degrees of freedom: $n - 1$

$\alpha/2 = 0.025$

Formula for Confidence Interval

$\bar{d} \pm t_{n-1,\alpha/2} \dfrac{s_d}{\sqrt{n}}$

Assumptions:
1. Paired/dependent samples
2. d's are Normally distributed

Degrees of freedom $= 10 - 1 = 9$

$t_{9,\,0.025} = 2.262$

Construct the confidence interval

$9 \pm 2.262 \dfrac{12.32883}{\sqrt{10}}$

9 ± 8.8189

0.1811 to 17.8189
We are 95 % confident that there is a difference between means. The interval does not contain $\mu_1 - \mu_2 = \mu_d = 0$ etc.

Progress Exercises 9.4: Differences and paired differences between means

1. (a) What is a 'paired' difference?
   (b) The examination results in Marketing and Statistics for a group of six randomly selected students are given in Table PE9.4 Q1:

   Table PE9.4 Q1 Marks for six students in Marketing and Statistics | File: Marks Stats |

Student initials	M.N.M.	C.O'N.	M.H.	G.K.	H.H.J.	D.W.-J.
Marketing	44	32	72	75	35	45
Statistics	65	40	67	98	48	53

   Calculate the difference between the Statistics and Marketing marks for each student: $(d_i, i = 1$ to 6).
   Calculate the mean and sample standard deviation for the differences $(\bar{d}, s_d)$.
   Calculate the standard error of the mean for the differences $s_{\bar{d}} = \dfrac{s_d}{\sqrt{n}}$.
   Construct a 95 % confidence interval for the mean difference between Statistics and Marketing marks.
   Can you conclude that the Statistics marks are higher than Marketing marks on average?

2. Why are paired differences used? Would you expect different results from an independent t-test and a paired t-test on the same data?
   Suppose in question 1 that it was discovered that in the population of students taking the examinations there were several students with the initials M.N.M., M.H. and G.K. Hence the data cannot be considered 'paired'. Repeat question 1 treating the data as independent samples. What do you conclude?

3. The downtime for computer systems in six branches of a major bank in 2005 and 2007 is given in Table PE9.4 Q3:

Table PE9.4 Q3 Downtime for computer systems (hours)   File: Downtime 2

Branch	2005	2007
A	40	30
B	54	50
C	32	22
D	38	36
E	80	68
F	44	40

(a) Could this sample be treated as 'paired' data? Explain.
(b) Calculate a 95 % confidence interval for the difference between the mean downtimes in 2005 and 2007.
(c) The technical support team claim that downtime in 2007 is less than in 2005 because of new developments in computer systems. Test the claim at the 1 % level of significance.

Give a brief verbal description of your conclusions in (b) and (c).

4. State the assumptions for the 'paired t-test'

The average gross weekly earnings (not seasonally adjusted) for England, Wales, Scotland and Northern Ireland are given in Table PE9.4 Q4:

Table PE9.4 Q4 Gross weekly earnings (£)   File: Earnings

	England	Wales	Scotland	N. Ireland
Summer 2003	373	340	314	298
Spring 2004	375	327	346	328
Autumn 2004	387	339	345	344
Spring 2005	404	333	365	346

Source: Office for National Statistics. Crown copyright material is reproduced with the permission of the Controller of HMSO. Labour Force Survey: **File: Earnings**

Between Summer 2003 and Spring 2005,
(a) test the claim that weekly earnings are higher in England than in (i) Wales; (ii) Northern Ireland.
(b) Construct a 95% confidence interval for the difference between earnings in Wales and Scotland.

What can you conclude from the analysis in (a) and (b)?

5. The gross household disposable income (£, 000m) for four regions in the UK is given in Table PE9.4 Q5 for years 1991, 1995 and 1999:

Table PE9.4 Q5 Gross disposable household income   File: Disposable income

	1991	1995	1999
East	37.602	47.373	57.647
London	55.093	70.785	88.93
South East	55.978	72.84	89.299
South West	31.681	41.542	49.718

Source Office for National Statistics. Crown copyright material is reproduced with the permission of the Controller of HMSO.: File: Disposable income

(a) Test whether the average gross household disposable income (£, 000m) is same in London and the South East over the three years, 1991, 1995 and 1999.
(b) Test whether the average gross household disposable income (£, 000m) is greater in the South East than the South West over the three years, 1991, 1995 and 1999.
(c) Construct a 99 % confidence interval for the difference between the average gross household disposable income (£, 000m) in 1999 and 1995.

6. The data in the Table PE9.4 Q6 gives the daily revenue from parking fines (£, 000) in three different areas collected by five traffic wardens:

Table PE9.4 Q6   Revenue from parking fines (£, 000)   File: Parking fines

Wardens	Main street	Shop street	Old Market
T	2.5	3.4	5.3
R	1.2	1.0	3.4
A	1.1	1.8	2.0
F	1.5	1.8	2.9
C	1.3	1.5	0.5

(a) Construct a 90 % confidence interval for the difference in average parking fine revenue collected by wardens T and F. What do you conclude?
(b) Test the claim that the average parking fine revenue is highest from Old Market. State your conclusions verbally.
(c) The Minitab printout for Main street and Shop street are given below:

Paired T-Test and CI: Main street, Shop street

Table PE9.4 Q6   Minitab output for Main street and Shop street

```
Paired T for Main street - Shop street

               N      Mean        StDev      SE Mean
Main street    5    1.52000     0.56745     0.25377
Shop street    5    1.90000     0.90000     0.40249
Difference     5   -0.380000    0.432435    0.193391

95% CI for mean difference: (-0.916939, 0.156939)
T-Test of mean difference = 0 (vs not = 0): T-Value = -1.96 P-Value
= 0.121
```

Describe this output. Can you conclude that there is a difference between the average parking fines revenue from the two streets?

7. On reviewing the data recorded in the experiment to test whether CAL improved students' results in PE9.4 Q2 it was discovered that the data was 'paired'; the assessment score was in fact recorded for each student before and after the course (file: Experiment P S).

The Minitab printout for the paired t-test is given in Table PE9.4 Q7:

Table PE9.4 Q7   Minitab output or paired data on assessment results

Paired T-Test and CI: A after, A pre
```
Paired T for A after - A pre

                N       Mean    StDev   SE Mean
A after         12     107.833  38.133   11.008
A pre           12      51.250  27.479    7.933
Difference      12      56.5833 28.9716   8.3634

95% CI for mean difference: (38.1756, 74.9910)
T-Test of mean difference = 0 (vs not = 0): T-Value = 6.77 P-Value
= 0.00
```

Paired T-Test and CI: Control after, Control pre
```
Paired T for Control after - Control pre

                 N      Mean    StDev   SE Mean
Control after    12    107.333  38.075   10.991
Control pre      12     65.000  32.435    9.363
Difference       12     42.3333 24.2687   7.0058

95% CI for mean difference: (26.9137, 57.7529)
T-Test of mean difference = 0 (vs not = 0): T-Value = 6.04 P-Value
= 0.000
```

Describe the output for the control group and the test group (A). Is there any evidence that the test group performed better than the control group?

Summary and overview inference from small samples

A summary comparing inference from large and small samples:
   A summary of confidence intervals and test statistics for difference between means from large and small samples:
   Note: if $n \geq 30$, σ unknown then $Z_{\alpha/2}$ may be used as an approximation to $t_{(n_1+n_2-2), \alpha/2}$.

Chart summary of confidence intervals for a single mean or proportions and paired differences

(i) **means** (large and small samples)
(ii) **proportions** (large samples)
(iii) **Difference between means** (paired samples)

Table 9.7   Summary: Confidence intervals and test statistic for the population mean based on large and small independent samples and paired samples (paired *t*-test)

	$n \geq 30$ for any population OR $n < 30$, σ **known** and Normal population	$n < 30$: σ **unknown** and Normal population	$n < 30$ **Paired differences**
Sampling distribution	$\bar{x} \sim Z(\mu_{\bar{x}}, \sigma_{\bar{x}})$	$\bar{x} \sim t_{n-1}(\mu_{\bar{x}}, s_{\bar{x}})$	$\bar{d} \sim t_{n-1}\left(\mu_{\bar{d}}, \dfrac{s_d}{\sqrt{n}}\right)$
Confidence Interval	$\bar{x} \pm Z_{\alpha/2}\,\sigma_{\bar{x}}$	$\bar{x} \pm t_{n-1,\,\alpha/2}\, s_{\bar{x}}$	$\bar{d} \pm t_{n-1,\,\alpha/2}\, \dfrac{s_d}{\sqrt{n}}$
Test statistic	$Z = \dfrac{\bar{x} - \mu_{H_0}}{\sigma_{\bar{x}}}$	$T = \dfrac{\bar{x} - \mu_{H_0}}{s_{\bar{x}}}$	$T = \dfrac{\bar{d} - H_0 \text{ claim}}{\left(\frac{s_d}{\sqrt{n}}\right)}$

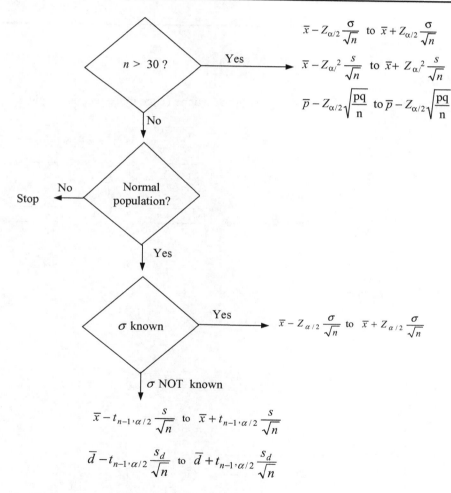

$$\bar{x} - Z_{\alpha/2}\frac{\sigma}{\sqrt{n}} \ \text{ to } \ \bar{x} + Z_{\alpha/2}\frac{\sigma}{\sqrt{n}}$$

$n > 30$? — **Yes** →

$$\bar{x} - Z_{\alpha/}2\frac{s}{\sqrt{n}} \ \text{ to } \ \bar{x} + Z_{\alpha/}2\frac{s}{\sqrt{n}}$$

$$\bar{p} - Z_{\alpha/2}\sqrt{\frac{pq}{n}} \ \text{ to } \ \bar{p} - Z_{\alpha/2}\sqrt{\frac{pq}{n}}$$

No ↓

No ← **Stop**   Normal population?

Yes ↓

σ known — **Yes** →

$$\bar{x} - Z_{\alpha/2}\frac{\sigma}{\sqrt{n}} \ \text{ to } \ \bar{x} + Z_{\alpha/2}\frac{\sigma}{\sqrt{n}}$$

σ NOT known ↓

$$\bar{x} - t_{n-1,\,\alpha/2}\frac{s}{\sqrt{n}} \ \text{ to } \ \bar{x} + t_{n-1,\,\alpha/2}\frac{s}{\sqrt{n}}$$

$$\bar{d} - t_{n-1,\,\alpha/2}\frac{s_d}{\sqrt{n}} \ \text{ to } \ \bar{d} + t_{n-1,\,\alpha/2}\frac{s_d}{\sqrt{n}}$$

Difference between means for paired samples: (where each $d_i = x_i - y_i$ etc.)

Chart 9.2   Confidence intervals for a single population mean or proportion or paired differences

Table 9.8   Summary: Confidence intervals and test statistic for the difference between population means based on large and small independent samples

	Independent samples $n \geq 30$ for any population OR small samples $n < 30$, σ **known** from Normal population	Independent samples ($n < 30$): from normal populations equal but unknown variances
Standard error	$\sqrt{\dfrac{\sigma_1^2}{n_1} + \dfrac{\sigma_2^2}{n_2}}$	$\sqrt{s_p^2 \left(\dfrac{1}{n_1} + \dfrac{1}{n_2} \right)}$
Confidence Interval	$(\overline{x}_1 - \overline{x}_2) \pm Z_{\alpha/2} \sqrt{\dfrac{\sigma_1^2}{n_1} + \dfrac{\sigma_2^2}{n_2}}$	$(\overline{x}_1 - \overline{x}_2) \pm t_{(n_1 + n_2 - 2),\alpha/2} \sqrt{s_p^2 \left(\dfrac{1}{n_1} + \dfrac{1}{n_2} \right)}$
Test statistic	$Z = \dfrac{(\overline{x}_1 - \overline{x}_1) - (\mu_1 - \mu_2)}{\sigma_{\overline{X}_A - \overline{X}_2}}$	$T = \dfrac{(\overline{x}_1 - \overline{x}_2) - (\mu_1 - \mu_2)}{\sqrt{s_p^2 \left(\frac{1}{n_1} + \frac{1}{n_2} \right)}}$

Chart Summary of confidence intervals for differences between means and proportions

(a) **means** (large and small independent samples); (b) **proportions** (large independent samples)

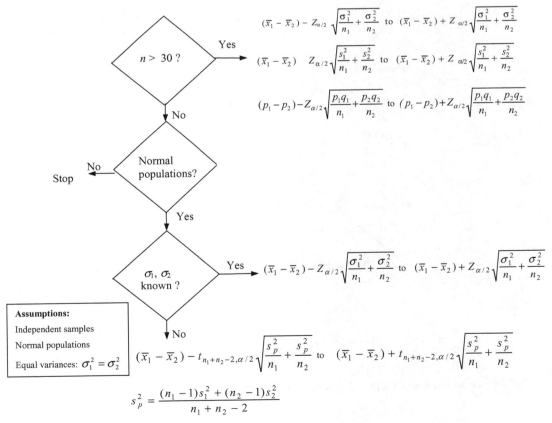

Chart 9.3   Confidence intervals for differences between population means and proportions

ANALYSIS OF VARIANCE

10

Chapter Objectives

Having carefully studied this chapter and completed the exercises you should be able to do the following

- Explain and calculate variance 'within' samples and 'between' samples
- Carry out the F-test for equality of variances and hence test the equality of several means
- Calculate a one-way ANOVA table
- Use the one-way ANOVA table to test the equality of several means
- Explain the relationship between one-way and two-way ANOVA
- Calculate a two-way ANOVA table
- Use the two-way ANOVA table to test the equality of several means
- Explain the difference between designed and observational experiments
- Calculate simultaneous confidence intervals to identify which means are different
- Carry out ANOVA in Excel and Minitab and interpret the outputs
- Use Minitab to check the assumptions of ANOVA

10.1   The rationale behind one-way analysis of variance

In Chapters 8 and 9 confidence intervals and tests of hypothesis were used to compare two population means. But, if the problem involves comparing several population means, the number of paired comparisons increases rapidly – for example, in a comparison of the average weekly savings for five groups of students (File: Savings 5) there are 15 different pairings, hence 15 tests or confidence intervals for differences between means.

$^5C_2 = 15$, see Chapter 4

Analysis of variance (ANOVA) is a test for the equality of several population means. If no significant difference between the means is found, then the test is concluded that at least one mean is different from the others. Further tests may be used to determine which mean or means are different.

The null hypothesis is H_0: $\mu_1 = \mu_2 = \mu_3 = \ldots = \mu_c$
The alternative hypothesis is H_1: at least one mean is different.

The null hypothesis for ANOVA is stated as follows:
The method for testing the null hypothesis is derived on the assumption that the following conditions are true:

ANOVA assumptions
The test for the equality of several means is set up on the assumption that

1. The samples are independent.
2. The samples are from Normal populations.
3. All populations have equal variances $\sigma_1^2 = \sigma_2^2 = \sigma_3^2 = \ldots = \sigma^2$ (let $\sigma^2 =$ common variance).

Note:

(i) These assumptions are identical to those for the t-test for differences between two population means based on small independent samples in Chapter 9.
(ii) The assumptions may be tested. There are numerous tests for Normality (assumption 2). The equality of variances (assumptions 3), may be tested by the F-test as explained in Chapter 9.

The F-test for the equality of several population means

If H_0 and the **assumptions are true** then there is only one population, mean value $= \mu$ and variance $= \sigma^2$. See Figure 10.1.

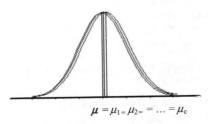

$$\mu = \mu_1 = \mu_2 = \ldots = \mu_c$$

Figure 10.1 H_0 and the assumptions state that all populations are $N(\mu, \sigma^2)$.

Assume we have c samples, each of size n (unequal samples sizes will be covered later). Then the sample means are independent estimates of the common population mean μ and the sample variances are independent estimates of the population variance σ^2.

The population variance, σ^2 is estimated by two different methods as follows:

Method A: the population variance is estimated by averaging the c independent sample variances:

$$s^2 = \frac{s_1^2 + s_2^2 + s_3^2 + \cdots + s_c^2}{c}.$$

> s^2, (hence s_w^2), is an estimator of σ^2

Since each of the sample variances: $s_1^2, s_2^2, s_3^2 \ldots s_c^2$ is the variance '*within*' a sample, this estimate of σ^2, calculated by averaging $s_1^2, s_2^2, s_3^2 \text{ K } s_c^2$ will be referred to as s_w^2.

An estimate of the common variance σ^2 from within c independent samples

$$s_W^2 = \frac{s_1^2 + s_2^2 + s_3^2 + \cdots + s_c^2}{c} \qquad (10.1)$$

Method B: The c independent sample means, $\overline{x}_1, \overline{x}_2 \ldots \overline{x}_c$, are used to estimate $s_{\overline{x}}^2$, the variance of the distribution of means. Since $s_{\overline{x}}^2 = \dfrac{s^2}{n}$, then $s^2 = n \times s_{\overline{x}}^2$. Hence s^2 is an estimate of σ^2 and will be referred to as s_B^2 because it is based on the variance *between* sample means $\overline{x}_1, \overline{x}_2 \ldots \ldots \overline{x}_c$. Hence $s_B^2 = n \times s_{\overline{x}}^2$.

> s^2, (refer to as s_B^2), is an estimator of σ^2

An estimate of the common variance, σ^2 based on the c independent samples means, each of size n.

$$s_B^2 = n \times s_{\overline{x}}^2 = n \times \frac{\sum\limits_{j=1}^{c} (\overline{x}_j - \overline{\overline{x}})^2}{c - 1} \qquad (10.2)$$

where $\overline{x}_j$ is the mean value of sample j and $j = 1, 2, 3 \ldots c$

It can be proved (proof beyond the scope of this text) that s_B^2 **will be greater than** s_W^2 **when at least one population mean is different**. However, an informal explanation illustrating how $s_B^2 > s_W^2$ when at least one population mean is different from the others is given in case 1 and case 2 as follows.

> When Ho is true s_w^2 and s_B^2 will not be significantly different

Case 1 when H_0 is true

If the samples are from identical populations (hence H_0 is true) then the estimates of variance, σ^2 by methods A and B: s_W^2 and s_B^2 will not be significantly different. See Figure 10.2(a).

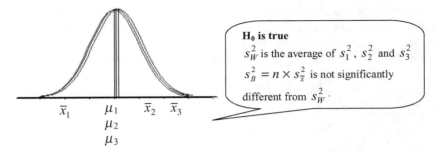

Figure 10.2(a) H_0 is true: three identical populations.

Case 2, when H_0 is NOT true

If at least one population mean is different from the others then the sample means, $\bar{x}_1$, $\bar{x}_2$ and $\bar{x}_3$ will more dispersed than those in case 1. Hence $s^2_{\bar{x}}$, which is calculated from $\bar{x}_1$, $\bar{x}_2$ and $\bar{x}_3$ will be larger than in case 1. Consequently s^2_B (since $s^2_B = n \times s^2_{\bar{x}}$) will be also be greater. Compare Figure 10.2(a) and Figure 10.2(b).

s^2_B is larger

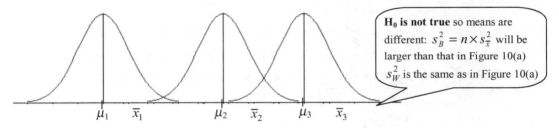

Figure 10.2(b) H_0 is not true: three populations with equal variance but different means.

But, on the other hand, s^2_W is the same in case 1 and case 2. This is because s^2_W is the average of the variance within the samples and the variance within each sample is the same in case 1 and case 2.

s^2_W is not affected

Therefore if s^2_B significantly greater than s^2_W the inference is that at least one population mean is different from the others.

This is expressed in the alternative hypothesis: $H_1 : \sigma^2_B > \sigma^2_W$.

So the test for the equality of several means culminates in an F-test for $\sigma^2_B > \sigma^2_W$.

Hence the name 'Analysis of Variance' or ANOVA.

(The above explanation for ANOVA is summarised at the end of the chapter.)

To formulate the test for the equality of several means

To test the equality of c population means from independent random samples, each of size n state the null and alternative hypotheses and test statistic as follows.

Recall: The equality of variances is tested by the F-test as described in Chapter 9.

1. **H_0:** $\mu_1 = \mu_2 = \mu_3 \ldots \mu_c$ (hence $\sigma^2_B = \sigma^2_W$)
2. **H_1:** at least one mean is different from the others (hence $\sigma^2_B > \sigma^2_W$)
    (*Hence, ANOVA is always a one-sided test.*)

3. α: 5%

4. Test Statistic F-ratio where $F = \dfrac{s_B^2}{s_W^2}$

5. Decision Rule: If the sample-F exceeds the critical-F, reject H_0.
Look up the F-tables with v_1 and v_2 degrees of freedom
$v_1 = (c - 1)$ is the degrees of freedom associated with s_B^2
since s_B^2 is calculated from c values $(\overline{x}_1, \overline{x}_2 \ldots \ldots \overline{x}_c)$
$v_2 = c(n - 1)$ is the degrees of freedom associated with s_W^2
since s_W^2 is the average of c sample variances, $s_1^2, s_2^2, s_3^2 \ldots s_c^2$ each of which was $(n - 1)$ df

6. Calculation of s_W^2 and s_B^2 and the test statistic, $F = \dfrac{s_B^2}{s_W^2}$ from the sample data.

7. Compare the test statistic with the critical regions given in the decision rule. If the test statistic exceeds the critical F, then reject H_0.

The test is set out formally in Worked Example 10.1.

10.2 One-way analysis of variance

The difference between several means is initially tested by the F-test as described above and demonstrated in Worked Examples 10.1 and 10.2. This F-test is then adapted for use in the ANOVA table in section 10.2.2.

10.2.1 Test the equality of several means by the F-test

The following worked examples illustrate the F test for the equality of several means.

Worked Example 10.1: ANOVA test for the equality of three means

A company is concerned about that excessive overtime hours worked by staff in various departments could lead to stress related illnesses. To investigate, an HR assistant randomly selected five employee records from the following departments (i) Marketing, (ii) Accounts and (iii) Personnel. The overtime hours worked by the employees in the last month are given in Table 10.1, with the sample means and variances.

Table 10.1   Overtime hours for five employees from departments A, B and C

File: Overtime 1

	Marketing	Accounts	Personnel
	14	8	13
	9	13	9
	11	15	6
	12	10	10
	7	13	9
Mean $\overline{x}$	10.6	11.8	9.4
Variance s^2	7.300	7.700	6.300

(a) Explain why the different values of the sample means could be due to chance.
(b) Estimate the population variance by (i) Method A (s_W^2) (ii) Method B (s_B^2).
(c) Test the hypothesis that the population means are equal against the alternative that at least one mean is different. $H_1: \sigma_B^2 > \sigma_W^2$.
(d) State the assumptions for the test.

Solution

(a) According to the Central Limit Theorem, sample means are Normally distributed, mean $= \mu$ variance $= \sigma^2$. Hence $\overline{x}_1, \overline{x}_2$ and $\overline{x}_3$ may be different due to sampling error. See Figure 10.3.

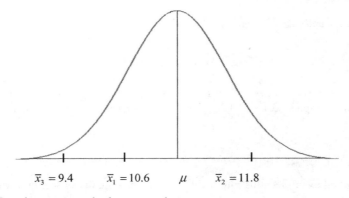

$$\overline{x}_3 = 9.4 \qquad \overline{x}_1 = 10.6 \qquad \mu \qquad \overline{x}_2 = 11.8$$

Figure 10.3   Sample means may be from a population, mean μ.

(b) The population variance is estimated as follows:

Method A: This method estimates σ^2 as the average of the three sample variances by (10.1)

$$s_w^2 = \frac{s_1^2 + s_2^2 + s_3^2}{3} = \frac{7.3 + 7.7 + 6.3}{3} = 7.1$$

or the average of the variances 'within' the individual samples

Method B: From $\overline{x}_1, \overline{x}_2 \ldots \ldots \overline{x}_c$ calculate $s_{\overline{x}}^2$. Since $s_{\overline{x}}^2 = \frac{s^2}{n}$ then $s^2 = n \times s_{\overline{x}}^2$

Recall: s^2 is referred to as s_B^2, since it is derived from the difference **between** the sample means: hence $s_B^2 = n \times s_{\overline{x}}^2$ by (10.2).
To calculate $s_{\overline{x}}^2$ manually, first calculate the mean of the sample means:

$$\overline{\overline{x}} = \frac{10.6 + 11.8 + 9.4}{3} = 10.6$$

$$s_{\bar{x}}^2 = \frac{\sum\limits_{j=1}^{c} (\bar{x}_j - \bar{\bar{x}})^2}{c - 1} = \frac{(10.6 - 10.6)^2 + (11.8 - 10.6)^2 + (9.4 - 10.6)^2}{3 - 1} = \frac{2.88}{2} = 1.44$$

Hence, $s_B^2 = n \times s_{\bar{x}}^2 = 5 \times 1.44 = 7.2$, for samples of size, $n = 5$.

(c) Set out the procedure for testing the hypothesis in the seven steps as follows:

1. **H$_0$:** $\mu_1 = \mu_2 = \mu_3$ (hence $\sigma_B^2 = \sigma_W^2$)
2. **H$_1$:** at least one mean is different from the others (hence $\sigma_B^2 > \sigma_W^2$)
3. **α:** 5 %

4. **Test Statistic** F-ratio where $F_{v_1, v_2} = \dfrac{s_B^2}{s_W^2}$

5. **Decision Rule**: If the sample-F exceeds the critical-F, reject H$_0$
   The degrees of freedom are $v_1 = c - 1 = 2$, $v_2 = c(n - 1) = 3(4) = 12$ since all four samples are of size 5. The critical F-value is 3.89 at $\alpha = 0.05$.

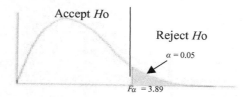

6. **Calculations:**
   Some calculations are given above, but the steps are as follows:

   1. Calculate the sample mean and variance from each sample
   2. Estimate the population variance by methods A and B
   3. Calculate the test statistic or sample $F = \dfrac{s_B^2}{s_W^2}$
   4. Sample $F = \dfrac{s_B^2}{s_W^2} = \dfrac{7.2}{7.1} = 1.014$

7. **Compare the test statistic with the decision rule.** The test statistic is in the acceptance region; therefore H$_0$ is accepted at the 5 % level of significance. Conclude that there is no difference between population means.

(d) The assumptions are:

1. The samples are independent
2. The samples are from Normal populations
3. All populations have equal variances. $\sigma_1^2 = \sigma_2^2 = \sigma_3^2 = \cdots = \sigma_c^2$

Worked Example 10.2: ANOVA test for the equality of four means

In addition to the data in Worked Example 10.1, the overtime for five randomly selected individuals from the Production department is included in Table 10.2:

Table 10.2   Overtime hours for five employees from four departments   File: Overtime 2

	Marketing	Accounts	Personnel	Production
	14	8	13	19
	9	13	9	16
	11	15	6	20
	12	10	10	15
	7	13	9	21
Mean $\bar{x}$	10.6	11.8	9.4	18.2
Variance s^2	7.300	7.700	6.300	6.7

The sample average for department D (Production department) is much greater than for other departments, indicating that this population mean may be different, as illustrated in Figure 10.4. If this is the case then s_B^2 will be significantly greater than s_W^2.

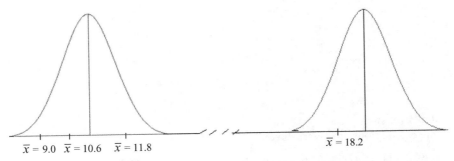

$\bar{x} = 9.0$ $\bar{x} = 10.6$ $\bar{x} = 11.8$ $\bar{x} = 18.2$

Figure 10.4   One population mean is different from the others.

(a) Estimate the population variance by (i) Method A (s_W^2) (ii) Method B (s_B^2).
(b) Test the hypothesis that all population means are equal against the alternative that at least one mean is different.
(c) State the assumptions for the test.

Solution

(a) The estimates of the common population variance are again calculated by methods A and B:

Method A: estimate σ^2 as the average the four sample variances

$$s_w^2 = \frac{s_1^2 + s_2^2 + s_3^2 + s_4^2}{4} = \frac{7.3 + 7.7 + 6.3 + 6.7}{4} = 7.0$$

Method B: From $\bar{x}_1, \bar{x}_2 \ldots \ldots \bar{x}_c$ calculate $s_{\bar{x}}^2 = \frac{\sum_{j=1}^{c} (\bar{x}_j - \bar{\bar{x}})^2}{c - 1}$. Then $s_B^2 = n \times s_{\bar{x}}^2$

First calculate $\bar{\bar{x}}$.

$$\bar{\bar{x}} = \frac{10.6 + 11.8 + 9.4 + 18.2}{4} = 12.5 \ldots \text{the mean of sample means}$$

$$s_{\bar{x}}^2 = \frac{\sum\limits_{j=1}^{c} (\bar{x}_j - \bar{\bar{x}})^2}{c - 1} = \frac{(\bar{x}_1 - \bar{\bar{x}})^2 + (\bar{x}_2 - \bar{\bar{x}})^2 + \cdots + (\bar{x}_c - \bar{\bar{x}})^2}{c - 1}$$

$$= \frac{(10.6 - 12.5)^2 + (11.8 - 12.5)^2 + (9.4 - 12.5)^2 + (27.8 - 12.5)^2}{4 - 1} = \frac{46.2}{3} = 15.4$$

(or, use the calculator to calculate $s_{\bar{x}}^2$ from $\bar{x}_1, \bar{x}_2 \ldots \ldots \bar{x}_c$)
Then $s_B^2 = n \times \sigma_{\bar{x}}^2 = 5 \times 15.4 = 77$
In this example the two estimates of σ^2 are very different: $s_B^2 = 77$ is much greater than $s_w^2 = 7$. To determine whether this is significant, the F-test is formulated to test whether $\sigma_B^2 > \sigma_W^2$ in part (b) below.

(b) Set out the hypothesis test as follows
   1. H_0: $\mu_1 = \mu_2 = \mu_3 = \mu_4$ (hence $\sigma_B^2 = \sigma_W^2$)
   2. H_1: at least one mean is different from the others (hence $\sigma_B^2 > \sigma_W^2$)
   3. α: 5 %
   4. **Test Statistic** F-ratio where $F_{v_1, v_2} = \dfrac{s_B^2}{s_W^2}$.
   5. **Decision Rule**: If the sample-F exceeds the critical-F, reject H_0.

   The degrees of freedom are $v_1 = c - 1 = 3$, $v_2 = c(n - 1) = 4(4) = 16$ since all four samples are of size $n = 5$. The critical F-value is 3.24 at $\alpha = 0.05$

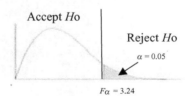

Accept Ho

Reject Ho

$\alpha = 0.05$

$F\alpha = 3.24$

6. **Calculations**:
   These calculations are given above in (a), but the steps are as follows:

   1. Calculate the sample mean and variance for samples A, B, C and D
   2. Estimate the population variance by methods A and B
   3. Calculate the test statistic or sample $F = \dfrac{s_B^2}{s_W^2}$

$$\text{Sample } F = \frac{s_B^2}{s_W^2} = \frac{77}{7} = 11$$

7. **Compare the test statistic with the decision rule**. The test statistic is in the reject region, therefore reject H_0 at the 5 % level of significance and conclude that at least one population mean is different.

(c) The assumptions are stated in (10.2)

Skill Development Exercise SK10_1: *F*-test for the equality of several means

Data on food prices collected by researchers is collected at regular intervals and used in the calculation of the Consumer Price Index. Prices for bread in three districts are given in Table SKD_10.1:

Table SK_10.1 Price of the standard loaf of bread in three districts

File: Bread

District A	District B	District C
1.7	1.9	1.8
1.8	1.8	1.6
1.6	2.2	1.6
1.9	2.1	1.7
1.8	1.9	1.7

Test whether the mean price of bread is the same in districts A, B and C.
State the null and alternative hypothesis of the test and state the assumptions.

Answer

H_0 is not accepted for levels of significance greater than 1 %. The test statistic is

$$F = \frac{s_B^2}{s_W^2} = \frac{0.1207}{0.0157} = 7.7$$

where degrees of freedom are $v_1 = c - 1 = 2$, $v_2 = c(n - 1) = 3(4) = 12$.
For $\alpha = 5\%$, the critical $F = 3.885$

10.2.2 The ANOVA table

The Analysis of Variance table or **ANOVA table** is a table in which the calculation of the test statistic or sample $F = \frac{s_B^2}{s_W^2}$ is set broken down into step-by-step stages.

To set up the ANOVA table three estimates σ^2 are required: s_B^2, s_W^2 and s_T^2.
s_W^2 and s_B^2 are the estimates of variance from 'within' samples and 'between' samples given in formulae (10.1) and (10.2).
s_T^2 is the estimate of variance from *all* c samples treated as a single sample, size $k = n \times c$.

$$\text{Hence, } s_T^2 = \frac{\sum_{k=1}^{k=nc} (x_k - \bar{\bar{x}})^2}{nc - 1} \text{ or simply } s_T^2 = \frac{\sum_{All} (x_k - \bar{\bar{x}})^2}{nc - 1}$$

Each estimate of population variance σ^2 is separated into numerator and denominator: the numerator is called the Sums of Squares: the denominator is called the degrees of freedom

$$s^2 = \frac{numerator}{denomintor} = \frac{sums\ of\ squares}{degrees\ of\ freedom}$$

The Sums of Squares (SS) and degrees of freedom (df) for s_T^2, s_B^2 and s_W^2 are described and calculated as follows:

$$s_T^2 = \frac{\displaystyle\sum_{k=1}^{k=nc} (x_k - \bar{\bar{x}})^2}{nc - 1} = \frac{TSS}{df} = \frac{TSS}{nc - 1} \quad \begin{array}{l} \leftarrow \text{ TSS: Sum of Squares from all the data} \\ \leftarrow \text{ Degrees of freedom for TSS} \end{array}$$

$$\boxed{\textbf{TSS} = \sum_{k=1}^{k=nc} (x_k - \bar{\bar{x}})^2 \textbf{ and degrees of freedom are } nc - 1} \qquad (10.3)$$

s_B^2 **is the estimate of variance from 'between' sample means (Method B):**

$$s_B^2 = n \times s_{\bar{x}}^2 = \frac{n \times \displaystyle\sum_{j=1}^{c} (\bar{x}_j - \bar{\bar{x}})^2}{c-1} = \frac{SSB_{col}}{c-1} \quad \begin{array}{l} \leftarrow \text{ SSB: Sum of Squares Between sample (column) means} \\ \leftarrow \text{ Degrees of freedom for SSB}_{col} \end{array}$$

$$\boxed{\textbf{SSB}_{col} = n \times \sum_{j=1}^{c} (\bar{x}_j - \bar{\bar{x}})^2 \textbf{ and degrees of freedom is } c - 1} \qquad (10.4)$$

s_W^2 **is the estimate of variance from within samples (Method A):**

$$s_W^2 = \frac{s_1^2 + s_2^2 + \cdots + s_c^2}{c}$$

Write out the formula for the calculation of each sample variance

$$s_W^2 = \frac{1}{c} \left(\frac{\displaystyle\sum_{i=1}^{n} (x_{i,1} - \bar{x}_1)^2}{n-1} + \frac{\displaystyle\sum_{i=1}^{n} (x_{i,2} - \bar{x}_2)^2}{n-1} + \ldots + \frac{\displaystyle\sum_{i=1}^{n} (x_{i,c} - \bar{x}_c)^2}{n-1} \right)$$

$$\boxed{s_1^2} \quad + \quad \boxed{s_2^2} \quad \ldots \ldots \quad + \quad \boxed{s_c^2}$$

Simplify to give s_W^2 in its simplest form is

$$s_W^2 = \left(\frac{\sum\limits_{i=1}^{n} (x_{i,1} - \overline{x}_1)^2 + \sum\limits_{i=1}^{n} (x_{i,2} - \overline{x}_2)^2 + \ldots + \sum\limits_{i=1}^{n} (x_{i,c} - \overline{x}_c)^2}{c(n-1)} \right) = \frac{SSW_1}{df}$$

$$s_W^2 = \frac{SSW_1}{c(n-1)} \quad \leftarrow \text{ SSW: Sum of Squares Within samples}$$
$$\quad\quad\quad\quad \leftarrow \text{ Degrees of freedom for SSW}$$

> The subscript, 1, is used to distinguish the SSW for one-way ANOVA from SSW for two-way ANOVA later

---

$$SSW_1 = \sum_{i=1}^{n} (x_{i,1} - \overline{x}_1)^2 + \sum_{i=1}^{n} (x_{i,2} - \overline{x}_2)^2 + \ldots + \sum_{i=1}^{n} (x_{i,c} - \overline{x}_c)^2 \qquad (10.5)$$

Degrees of freedom is $c(n-1)$

A final requirement to complete the ANOVA table is the 'Sum of Square identity for one-way ANOVA

It can be proved (mathematically, see Appendix K) that numerators in s_T^2, s_B^2 and s_w^2 satisfy the 'Sums of Squares' identity in (10.6):

The Sums of Squares identity for one-way ANOVA

$$\sum_{k=1}^{k=nc} (x_k - \overline{\overline{x}})^2 = n \times \sum_{j=1}^{c} (\overline{x}_j - \overline{\overline{x}})^2 + \sum_{i=1}^{n} (x_{i,1} - \overline{x}_1)^2 + \ldots + \sum_{i=1}^{n} (x_{i,c} - \overline{x}_c)^2$$

$$\text{TSS} = \text{SSB}_{\text{column means}} + \text{SSW}_1 \qquad (10.6)$$

SSW_1 is the Sum of Squares 'Within' for one-way ANOVA.
In addition, the corresponding denominators (degrees of freedom) satisfy (10.7):

$$nc - 1 = (c - 1) \qquad\qquad + c \times (n - 1)$$

$$df(\text{Total}) = df(\text{Between column means}) \quad + df(\text{Within columns}) \qquad (10.7)$$

The Sum of Squares identity (10.6) and (10.7) form the basis for the ANOVA table. Hence in the ANOVA table:
The first column states the source of data from which the Sums of Squares are calculated.
The second column gives the Sums of Squares which satisfies the equation: $TSS = SSB_{col} + SSW_1$.
The third column gives the degrees of freedom which satisfy the equation: $nc - 1 = (c - 1) + c(n - 1)$.

In the fourth column, the Sums of Squares are divided by the corresponding degrees of freedom, thus reconstituting s_B^2 and s_W^2. These are referred to as the 'Mean Sum of Squares'.

The test statistic $F = \dfrac{s_B^2}{s_W^2}$ (or sample F or F-ratio) is given in *the fifth column*.

The Null hypothesis is then tested with the F-test as in Worked Examples 10.1 and 10.2.
The ANOVA table is set out as follows:

Table 10.3 ANOVA Table for one-way analysis of variance

Source of Variation	Sum of Squares	Degrees of freedom	Mean Sum of Squares	Sample-F (Test Statistic)
Between Samples (or between Columns)	SSB_{col}	$v_1 = c - 1$	$MSSB_{col} = \dfrac{SSB_{col}}{c-1} = s_B^2$	$\dfrac{MSSB_{\text{column means}}}{MSSW}$
Within Samples (or within Columns)	SSW_1	$v_2 = c \times (n-1)$	$MSSW_1 = \dfrac{SSW_1}{c(n-1)} = s_W^2$	where $v_1 = c - 1$ $v_2 = c(n-1)$
Total	TSS	$nc - 1$		

Note: in calculating the Sums of Squares for the ANOVA table, it is only necessary to calculate any two of the Sums of Squares. As the Worked Examples will demonstrate, the TSS and SSB_{col} are the easier to calculate than SSW_1.

$SSW_1 = TSS - SSB_{col}$, by the Sums of Squares identity (10.6).

Worked Example 10.3: One way ANOVA table by direct calculations

Calculate the one–way ANOVA table to test the equality of the mean overtime hours for the four departments in Table 10.2.

Solution

The following steps outline a method for testing the equality of several means by one-way ANOVA:

Step 1. State c and n: the number of samples and the sample sizes.
Step 2. Calculate *any 2* of the Sums of Squares and degrees of freedom. Calculate the remaining SS and *df* from (10.6) and (10.7) by subtraction.
It is usually easier to calculate TSS and SSB_{col}. Hence $SSW_1 = TSS - SSB_{col}$.
Step 3. Calculate the ANOVA Table.
Step 4. Set out the test of hypothesis that there is no difference between the mean overtime hours worked by the usual 7-step method.

Hence

Step 1. There are $c = 4$ samples: each sample is of size, $n = 5$.
Step 2. Calculate *any 2* of the Sums of Squares:

$$TSS = \sum_{k=1}^{k=nc} (x_k - \bar{\bar{x}})^2 \tag{10.3}$$

$$SSB_{\text{col means}} = n \times \sum_{j=1}^{c} (\bar{x}_j - \bar{\bar{x}})^2 \text{ for equal sample sizes} \tag{10.4}$$

$$SSW_1 = \sum_{i=1}^{n} (x_{i,1} - \bar{x}_1)^2 + \sum_{i=1}^{n} (x_{i,2} - \bar{x}_2)^2 + \ldots + \sum_{i=1}^{n} (x_{i,c} - \bar{x}_c)^2 \tag{10.5}$$

The remaining Sum of Squares is then calculated from (10.6):
In this example we will calculate (i) SSB_{col} and (ii) TSS. Then (iii) $SSW_1 = TSS - SSB_{\text{col}}$

Abbreviation for $SSB_{\text{col means}}$

(i) Calculate **SSB_{col}** $= n \times \sum_{j=1}^{c} (\bar{x}_j - \bar{\bar{x}})^2$

Start by calculating the mean value for each sample (each column) and $\bar{\bar{x}}$, the mean of all the means as set out in Table 10.4, where the overall mean is

$$\bar{\bar{x}} = \frac{10.6 + 11.8 + 9.4 + 18.2}{4} = \frac{50}{4} = 12.5$$

Table 10.4 Calculation of Sums of Squares

	$j = 1$ Marketing	$j = 2$ Accounts	$j = 3$ Personnel	$j = 4$ Production
	14	8	13	19
	9	13	9	16
	11	15	6	20
	12	10	10	15
	7	13	9	21
Sample Means	10.6	11.8	9.4	18.2 $\bar{\bar{x}} = 12.5$
$n(x_j - \bar{\bar{x}})^2$	$5(10.6 - 12.5)^2$	$5(11.8 - 12.5)^2$	$5(9.4 - 12.5)^2$	$5(18.2 - 12.5)^2$
$n(x_j - \bar{\bar{x}})^2$	18.05	2.45	48.05	162.45

SSB_{col} Calculate the squared difference between each column mean from the overall mean. Add the squared deviations: then multiply the answer by n, the common sample size.

$$SSB_{\text{col}} = n \times \sum_{j=1}^{c} (\bar{x}_j - \bar{\bar{x}})^2 = 18.05 + 2.45 + 48.05 + 162.45 = 231, \text{ see Table 10.4.}$$

Degrees of freedom for SSB_{col} is $4 - 1 = 3$.

(ii) **TSS**: Calculate the squared deviation of each datum from the overall mean ($\bar{\bar{x}} = 12.5$): then add the squared differences.

$$TSS = (14 - 12.5)^2 + (9 - 12.5)^2 + 11 - 12.5)^2 + \cdots + (15 - 12.5)^2 + (21 - 12.5)^2 = 343$$
Degrees of freedom for TSS $= 20 - 1 = 19$.

(iii) From (10.6) $SSW_1 = TSS - SSB_{col} = 343 - 231 = 112$.
Degrees of freedom for SSW $= 19 - 3 = 16$.

Hence the ANOVA table

Table 10.5   ANOVA table for data in Table 10.2

Source of Variation	Sum of Squares	Degrees of freedom	Mean Sum of Squares	Sample-F
Between departments (or between Columns) SSB_{col}	231	$4 - 1 = 3(v_1)$	$MSSB = \dfrac{231}{3} = 77$	$\dfrac{77}{7} = 11$
Within departments (or within Columns) SSW_1	112	$19 - 3 = 16(v_2)$	$MSSW = \dfrac{112}{16} = 7$	where
Total **TSS**	343	$20 - 1 = 19$		$v_1 = 3$ $v_2 = 16$

Note: the SSW_1 and MSSW are frequently called the Error Sum of Squares (SSE) and the Mean Sum of Squares Error (MSE) respectively.

Remember, the ANOVA table gives the calculations leading to the test statistics (Sample-F). You must then set out the formal test of hypothesis, step by step, as in Worked Example 10.1 and 10.2 (b).

10.2.3   Calculation of Sums of Squares on the calculator

The calculation of the Sums of Squares in the ANOVA table may be carried out manually, as in Worked Example 10.3 or use the built in function for variance on the calculator as follows.

When data $x_1, x_2, x_3 \ldots . x_R$ are keyed into the calculator, the built in function for variance calculates

σ^2 as $\dfrac{\sum\limits_{k=1}^{k=R}(x_k - \bar{x})^2}{R}$ (R is the number of data entered).

Hence the Sums of Squares $\sum\limits_{k=1}^{k=R}(x_k - \bar{x})^2 = R\sigma^2$ 　　　　　　　(10.8)

OR use the built in function for sample variance $s^2 = \dfrac{\sum\limits_{k=1}^{k=R}(x_k - \bar{x})^2}{R-1}$

Hence the Sum of Squares is $\sum\limits_{k=1}^{k=R}(x_k - \bar{x})^2 = (R-1)s^2$ 　　　　　　　(10.9)

Note: the sum of the squared deviation of a set of data from its mean value, $\sum_{k=1}^{k=R} (x_k - \bar{x})^2$ is called the Sums of Squares.

TSS on the calculator

Key in all the data (that is $n \times c$ values): read out the built in function σ^2 and refer to it as σ^2_{All}.

$$TSS = \sigma^2_{All} \times n \times c$$

$$df\,(\text{Total}) = nc - 1$$

> Using σ^2 may look strange- but remember we are calculating Sums of Squares (SS) at this point
> If we use σ^2, the same multiplier, $n \times c$ is used in
> (i) $TSS = (n \times c)\,\sigma^2_{All}$ ...(10.10)
>
> (ii) $SSB_{col.} = (n \times c)\,\sigma^2_{col}$ ...(10.11)

(10.10)

SSB_{col} on the calculator

Key in the column means; $\bar{x}_1, \bar{x}_2 \ldots \ldots \bar{x}_c$, read out the built in function σ^2 and refer to as σ^2_{col}. This

variance is calculated as $\sigma^2_{col} = \dfrac{\sum\limits_{j=1}^{c} (\bar{x}_j - \bar{\bar{x}})^2}{c} \rightarrow \sum\limits_{j=1}^{c} (\bar{x}_j - \bar{\bar{x}})^2 = c \times \sigma^2_{col}$

Hence, $SSB_{\text{col means}} = n \times \sum\limits_{j=1}^{c} (\bar{x}_j - \bar{\bar{x}})^2 = n \times (c \times \sigma^2_{col})$ (10.11)

$df\,(\text{between column means}) = c - 1$

SSW_1 from the Sum of Squares identity

From the Sum of Squares identity (10.6), $SSW = TSS - SSB_{col}$
$df\,(\text{Within}) = c(n - 1)$
OR from (10.7), $df\,(\text{Within}) = df\,(\text{Total}) - df\,(\text{Between})$

Note: a simple check that data is entered correctly on the calculator!
Since the mean of all the data is identical to the mean of all the column means, use the built in function, $\bar{x}$, on the calculator to read the overall mean when calculating TSS and SSB on the calculator. The values of $\bar{x}$ must be identical.

Worked Example 10.4: Calculations of ANOVA Sums of Square by calculator

Calculate TSS, SSB_{col} and SSW_1 for the data given in Table 10.2.

Solution

TSS
Key in each datum in the table i.e. all four samples ($c = 4, n = 5$).

The built in function for the variance and the mean gives $\sigma^2_{All} = 17.15$: $\overline{\overline{x}} = 12.5$:
TSS $= \sigma^2_{All} \times n \times c = 17.15 \times 5 \times 4 = 343$

SSB_{col}
Key in the values of the four sample means $\overline{x}_1, \overline{x}_2, \overline{x}_3, \overline{x}_4$
The built in function for the variance and the mean gives $\sigma^2_{col} = 11.55$: $\overline{\overline{x}} = 12.5$
$SSB_{col} = \sigma^2_{col} \times n \times c = 11.55 \times 5 \times 4 = 231$
df (Between means) $= 4 - 1 = 3$

Now use the Sums of Squares identity (10.6)
$SSW_1 = TSS - SSB_{col} = 343 - 231 = 112$
df (Within) $= df$ (Total) $- df$ (Between means) $= 19 - 3 = 16$

These are the essential calculations for the ANOVA in Table10.5, Worked Example 10.3 above.

Remember!
Following the calculation of the ANOVA table is imperative to state the Null and alternative hypothesis, look up the critical F-values for the test and state your conclusion very carefully. You cannot state that you accept or reject a null hypothesis without stating the hypothesis in the first place!

Skill Development Exercise SKD_10.2: One-way ANOVA Table

Calculate the ANOVA table for the data on the price of a standard loaf of bread in districts A, B and C given in table SKD_10.1.
Test whether the mean price of bread is the same in districts A, B and C.

Answer

Table SKD_ 10.2   ANOVA table for data in Table SKD_10.1

ANOVA Table

Source of Variation	Sum of Squares	Degrees of freedom	Mean Sum of Squares	Sample-F
Between Areas SSB_{col}	0.2418	$3 - 1 = 2\ (v_1)$	$MSSB = \dfrac{0.2413}{2} = 0.1207$	$\dfrac{0.1207}{0.0157} = 7.7021$
Within Areas (or within Columns) SSW_1	0.1880	$14 - 2 = 12\ (v_2)$	$MSSW = \dfrac{0.1880}{12} = 0.0157$	where $v_1 = 3$ $v_2 = 16$
Total **TSS**	0.4293	$25 - 1 = 14$		

10.2.4   One-Way ANOVA with unequal sample sizes

When sample sizes are unequal, the calculation of the TSS for one-way ANOVA is the same as when sample sizes are equal. However the calculation of SSB_{col} must be carried out directly:

$$\sum_{j=1}^{j=c} n_j \times (\bar{x}_j - \bar{\bar{x}})^2.$$

The calculation of an ANOVA table for unequal sample sizes is illustrated in Worked Example 10.5.

Worked Example 10.5: ANOVA table for unequal sample sizes

The bi-annual expenditure on electricity for randomly selected households in Areas A, B and C is given in Table 10.6.
Test the claim that expenditure is the same in all areas.

Table 10.6   Bi-annual expenditure on electricity in Areas A, B, C   File: Electricity 1

	Area A	Area B	Area C
	780	650	640
	690	689	689
	735	700	673
	750	655	702
	766	680	681
	689		
Sample means	735.00	674.80	677.00

Solution

Step 1. $c = 3$ samples of sizes 5, 5 and 6
Step 2. Calculation of Sums of Squares

$$SSB_{col} = \sum_{j=1}^{j=c} n_j \times (\bar{x}_j - \bar{\bar{x}})^2$$

$$= 6(735.00 - 698.06)^2 + 5(674.80 - 698.06)^2 + 5(677.00 - 698.06)^2$$

$$= 13\,110.14$$

$$TSS = 24\,502.9 \text{ (from the calculator: } \sigma_{All}^2 \times \text{ number of data)}$$

$$SSW_1 = 24\,502.9 - 13\,110.14 = 11\,392.76$$

Step 3. Degrees of freedom

$$df \text{ for TSS} = \text{total number of data in table} - 1 = (6 + 5 + 5) - 1 = 15$$

$$df \text{ for } SSB_{col} = \text{number of samples} - 1 = 3 - 1 = 2$$

$$df \text{ for } SSW_1 = df \text{ for TSS} - df \text{ for SSB} = 15 - 2 = 13$$

Step 4. Set up the One-way **ANOVA Table**

Table 10.7   One-way ANOVA table for data in Table 10.6

Source of Variation	Sum of Squares	Degrees of freedom	Mean Sum of Squares	Sample-F
Between Samples SSB_{col}	13 110.14	$3 - 1 = 2$	$MSSB = \dfrac{13\,110.14}{2} = 6555.07$	$\dfrac{6555.07}{876.34} = 7.48$
Within Samples SSW_1	11 392.76	$15 - 2 = 13$	$MSSW = \dfrac{11\,392.76}{13} = 876.34$	where $v_1 = 2,$ $v_2 = 13$
Total **TSS**	24 502.9	$16 - 1 = 15$		

Step 5. Set out the steps for the hypothesis test.

1. H_0: $\mu_1 = \mu_2 = \mu_3$: (that is $\sigma_B^2 = \sigma_W^2$)
2. H_1: At least one mean is different: ($\sigma_B^2 > \sigma_W^2$)
3. α: 5 %, 1 %

4. **Test Statistic:** F-ratio where $F_{v_1, v_2} = \dfrac{s_B^2}{s_w^2}$

5. **Decision Rule:**

If the sample-F exceeds the critical-F, ($df : v_1 = 2, v_2 = 13$), reject H_0.

Tables $F\alpha = 6.701$ for $\alpha = 0.01$
Tables $F\alpha = 3.806$ for $\alpha = 0.05$

6. **Calculations:** From ANOVA table, $F_{v_1, v_2} = \dfrac{\sigma_B^2}{\sigma_w^2} = 7.48$

7. **Conclusion:** Since the sample-F is greater than the tables-F, reject H_0. Conclude that the mean expenditure is different in at least one area.

10.2.5   Simultaneous confidence intervals: which means are different?
Confidence intervals for difference between means for individual pairs

When the null hypothesis is rejected there may be a difference between two or more pairs of population means. Constructing confidence intervals (or test hypothesis) for the difference between every possible pair of means would be the obvious way to identify which of the means are different. If the equality of three means is tested $H_0: \mu_A = \mu_B = \mu_C$ then three pair-wise confidence intervals are required for $(\mu_A - \mu_B)$, $(\mu_A - \mu_C)$ and $(\mu_B - \mu_C)$.

For example, for the data in Table 10.2 the 95 % confidence interval for the difference $\mu_4 - \mu_1$ is

$$\bar{x}_4 - \bar{x}_1 \pm t_{\alpha/2, df \text{ for SSW}} \sqrt{MSSW\left(\frac{1}{n_1} + \frac{1}{n_2}\right)} = 10.6 - 18.2 \pm 2.120\sqrt{7\left(\frac{1}{5} + \frac{1}{5}\right)} = -7.6 \pm 3.547$$

. . . where $t_{\alpha/2, df \text{ for SSW}} = t_{0.025, 16} = 2.120$, $MSSW = 7$ from ANOVA, Table 10.5, $n_1 = n_2 = 5$.
In words, we are 95 % confident that the difference between the mean overtime between those in the Marketing and Production departments is $[-11.147 \le \mu_1 - \mu_4 \le -4.053]$. This interval does not contain zero. Hence there is evidence that $\mu_1 - \mu_4 \ne 0$. Furthermore, since the confidence limits are both negative the evidence is that $\mu_4 > \mu_1$.
Note: the MSSW is an estimate of the common variance based on three independent estimates. This is a better estimate than the pooled variance s_p^2 in the independent t-test which was based on two estimates of the common variance

Probability of the correct conclusion for all confidence intervals

Two problems arise when testing the difference between every possible pair of means following the rejection of the null hypothesis in ANOVA:

- If the equality of c means is tested in H_0 then the number of confidence intervals required is
  $*M = \dfrac{c(c-1)}{2}$. The number of confidence intervals increases rapidly as the number of means
  increases. For example, a test for the equality of four means requires $\dfrac{4(4-1)}{2} = 6$ confidence
  intervals!
  *Recall Chapter 4 – the number of different samples of size 2 from a population of c is given by the
  formula: $^cC_2 = \dfrac{c!}{(c-2)!2!} = \dfrac{c(c-1)}{2}$
- Another more serious problem that arises as the number of multiple comparisons increases is that the size of Type 1 error also increases and hence the probability that the true null hypothesis is accepted decreases.

For example, if $\alpha = 0.05$ then the probability of accepting H_0 when it is true is 0.95 for a single comparison either in the test $H_0: \mu_A = \mu_B$ or the confidence interval for $(\mu_A - \mu_B)$. For six comparisons the probability of accepting H_0 when it is true in all pair-wise comparison is P (all six confidence intervals contain the true difference).

$$= P(H_0 \text{ is accepted in all 6 tests})$$

$$= (0.95)^6 = 0.7351.$$

Therefore to maintain an overall probability of $(1 - \alpha)$ of accepting all true null hypothesis, the level of significance for each interval must be reduced.

Bonferroni's method calculates the reduced level of significance for each interval as $\frac{\alpha}{M}$ for M comparisons. For six comparisons, each has a level of significance of $\frac{\alpha}{M} = \frac{0.05}{6} = 0.0083$.

Hence P (all six intervals contain the true difference) $= \left(1 - \frac{0.05}{6}\right)^6 = 0.95103$.

Hence, simultaneous confidence intervals for M comparisons with ovarall type I error $= \alpha$ is

$$\overline{x}_A - \overline{x}_B \pm t_{df\,(SSW),\,\alpha/(2M)} \sqrt{MSSW_1\left(\frac{1}{n_A} + \frac{1}{n_B}\right)} \tag{10.12}$$

$MSSW_1$ is the pooled **estimate** of variance from the ANOVA table; the degrees of freedom is $df(SSW_1)$ from the ANOVA table.

The Scheffé method

There are numerous formulae for simultaneous confidence intervals. The Scheffé method uses the standard F-tables and is given in (10.13) for an ANOVA test for the equality of c population means. Where $F_{\alpha,c-1,c(n-1)}$ is the critical F value for the testing the equality of all c means at the α % level of significance.

The **Scheffé's** simultaneous confidence intervals for $\mu_A - \mu_B$ are calculated by the formula

$$\overline{x}_A - \overline{x}_B \pm \sqrt{MSSW_1(c-1)F_{\alpha,c-1,c(n-1)}\left(\frac{1}{n_A} + \frac{1}{n_B}\right)} \tag{10.13}$$

Worked Example 10.6: Simultaneous confidence intervals

Calculate confidence intervals for the difference between pairs of means for the data in Table 10.2 (File: Overtime 2) with an overall level of significance of 5 % by
(a) Bonferroni's method; (b) Scheffé's method.

Solution

The sample means (see Table 10.2) are $\overline{x}_1 = 10.6$; $\overline{x}_2 = 11.8$; $\overline{x}_3 = 9.4$. $\overline{x}_4 = 18.2$.
From Table 10.5, $MSSW_1 = 7.2$ is the pooled variance, $df = 16$.

(a) **Bonferroni's method**

The level of significance for each individual interval is $\frac{0.05}{6} = 0.0083$.

The t percentage point for a tail area of $\dfrac{0.0083}{2} = 0.00415$ and 16 df is 3.3405 (use Excel or interpolate between $\alpha = 0.005$ and $\alpha = 0.001$ at 16 df in the t percentage Tables)

$$\overline{x}_A - \overline{x}_B \pm t_{df\,(SSW),\,\alpha/(2M)} \sqrt{MSSW_1 \left(\frac{1}{n_A} + \frac{1}{n_B} \right)}$$

The margin of error for the difference between any two means is

$$\pm t_{df\,(SSW),\,\alpha/(2M)} \sqrt{MSSW_1 \left(\frac{1}{n_A} + \frac{1}{n_B} \right)} = \pm 3.0083 \sqrt{7.0 \left(\frac{1}{5} + \frac{1}{5} \right)}$$

$$= \pm 5.0338 \text{ or } \pm 5.03$$

correct to two decimal places.

Hence the confidence intervals for the three pairs of means is $\overline{x}_A - \overline{x}_B \pm 5.59$.

Table 10.8   Simultaneous confidence intervals

Sample means		Difference between means	Margin of error	Lower end of interval	Upper end of interval	Significant difference?
$\overline{x}_1 = 10.6$	$\overline{x}_3 = 9.4$	1.2	5.03	−3.83	6.23	No
$\overline{x}_1 = 10.6$	$\overline{x}_3 = 11.8$	−1.2	5.03	−6.23	3.83	No
$\overline{x}_1 = 10.6$	$\overline{x}_4 = 18.2$	−7.6	5.03	−12.63	−2.57	Yes
$\overline{x}_3 = 9.4$	$\overline{x}_3 = 11.8$	−2.4	5.03	−7.43	2.63	No
$\overline{x}_3 = 9.4$	$\overline{x}_4 = 18.2$	−8.8	5.03	−13.83	−3.77	Yes
$\overline{x}_3 = 11.8$	$\overline{x}_4 = 18.2$	−6.4	5.03	−11.43	−1.37	Yes

If the magnitude of the difference between two sample means is less than the margin of error then the interval will contain zero ($\mu_i - \mu_j = 0$) hence no difference between means is detected. For example, the difference between sample 1 and sample 3 is 1.2: hence the simultaneous confidence interval for the difference is $1.2 \pm 5.03 \rightarrow -3.83$ to 6.23.

Therefore, scan the differences between means and pick out those that are greater than the margin of error. The conclusion in this example is that the sample mean, $\overline{x} = 18.2$ is different from the other three samples. There is no evidence of differences between the first three samples.

(b) **Scheffé's method**

$$\overline{x}_A - \overline{x}_B \pm \sqrt{MSSW_1 (c - 1) F_{\alpha,\,c-1,\,c(n-1)} \left(\frac{1}{n_A} + \frac{1}{n_B} \right)}$$

Hence, the margin of error is $\pm \sqrt{(7)(3)(3.24) \left(\dfrac{1}{5} + \dfrac{1}{5} \right)} = 5.22$

From Table 10.8, three differences exceed the Scheffé's margin of error, indicating that the mean overtime hours for the Production department is greater than that for any of the other three departments.

Comments

1. In the above example, the simultaneous confidence intervals must be less precise than individual (pair-wise) confidence intervals to maintain the same probability of Type I error for all intervals.

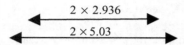

Width of interval for the single comparison for $\mu_4 - \mu_1$
Width of interval for each of the six comparisons

Individual confidence intervals are more precise than simultaneous confidence intervals for the same level of confidence.

2. There are several other methods for constructing confidence intervals for multiple comparisons, such as Tukey's method or Duncan's method.

Skill Development Exercise SK10_3: Simultaneous confidence intervals

Calculate the 95 % simultaneous confidence intervals for the differences in the price of the standard loaf of bread data in three districts given in Table SKD_10.1, File: Bread.
Use (a) Bonferroni's method; (b) Scheffé's method.

Answer

(a) $\text{Difference} \pm t_{df(SSW),\alpha/(2M)} \sqrt{MSSW_1 \left(\dfrac{1}{n_A} + \dfrac{1}{n_B} \right)}$

$\text{Difference} \pm t_{12,0.0083} \sqrt{(0.0157) \left(\dfrac{1}{5} + \dfrac{1}{5} \right)} = \text{Difference} \pm 0.2203$

(b) $\bar{x}_A - \bar{x}_B \pm \sqrt{MSSW_1(c-1)F_{\alpha,c-1,c(n-1)} \left(\dfrac{1}{n_A} + \dfrac{1}{n_B} \right)}$

$\text{Difference} \pm \sqrt{0.0157(3-1)F_{0.05,2,12} \left(\dfrac{1}{5} + \dfrac{1}{5} \right)} = \text{Difference} \pm 0.2209$

Progress Exercises 10.1: One-way ANOVA

1. A large banking organisation carried out an analysis on social circumstance of joint account holders. The annual amount in savings accounts for families of two adults and two children was shown to be Normally distributed with a mean and standard deviation of $\mu = €\ 6554$ and $\sigma = €\ 378$.
   (a) If a sample of 64 accounts is selected what is the probability that the sample average is
      (i) less that € 6500; (ii) greater than € 6650; (iii) between € 6500 and € 6650.

(b) The sample mean and standard deviation for a random sample of 64 accounts was calculated as $\bar{x} = 6,385$ $s = 453$. Use this data to calculate the 99 % confidence interval for the true population mean. Does your result support the bank's claim that the average account is € 6554?

2. In the banking sector profits are affected when computer systems are down in local branches.

(a) The number of hours down-time in six branches during 2000 were recorded as

| 30 | 50 | 22 | 36 | 68 | 40 |

Calculate a 95 % confidence interval for the mean down-time, stating any assumptions made.

(b) Data for the down-time for three years is given in Table PE10.1 Q2b:

Table PE10.1b Q2   Down time in hours for years 2003, 2004, 2005   File: Downtime 1

	2003	2004	2005
	35	40	30
	45	54	50
	44	32	22
	36	38	36
	30	80	68
	38	44	40
Mean	38	48	41
s^2	32.4	299.2	263.6

(i) State the assumptions for a one-way ANOVA.
(ii) State the null and alternative hypothesis.
(iii) Calculate s_W^2 and s_B^2, the estimates of variance from within samples and between sample means respectively. Hence test the claim that these is no difference between the mean down-times in 2003, 2004 and 2005.

(c) The partially completed ANOVA Table for the data is in Table PE10.1 Q2c:

Table PE10.1 Q2c   ANOVA table for down times for years 2003, 2004, 2005

ANOVA

Source of variation	SS	df	MSS	F	F critical (5 %)
Between Groups	??	2	158	0.80	??
Within Groups	2976	??	??	??	
Total	??	17	??		

Fill in the missing entries (??). Show how the ANOVA table confirms the values of s_W^2, s_B^2 and test statistic in (b)

3. An experiment was conducted to record the time taken, in seconds, for three brands of mobile phone to power up. See Table PE10.1 Q3.

Table PE10.1 Q3   Power-up time for three brands of mobile phone   File: Battery

Brand A	Brand B	Brand C
1.9	1.4	4.2
1.4	1.0	4.9
2.1	1.8	3.0
2.5	1.8	4.9
2.3	2.5	4.8

(a) Test the claim that the time is the same for brands A, B and C at the 1 % level of significance.
(b) Calculate simultaneous confidence intervals by; (i) Bonferroni's method; (ii) Scheffé's methods. State and explain your conclusions.

4. A recruitment agency claims that by the age of 35, an individual's annual salary is largely determined by qualification. A random sample of individuals whose annual salary (£ 000) and highest qualification is primary degree, diploma, a trade or a professional qualification is given in Table PE10.1 Q4:

Table PE10.1 Q4   Annual salaries in £ ,000   File: Salary vs. qualification

Degree	Diploma	Trade	Professional
42	25	44	45
37	32	25	56
26	36	47	76
35	28	42	54
31	29	36	65

(a) Test the claim that all salaries are the same for all qualifications.
(b) Calculate the 95 % individual confidence intervals for the annual salary for each qualification.
(c) Calculate simultaneous confidence intervals for the pair-wise differences between salaries.
(d) Based on the analysis in (a), (b) and (c) write brief comments on the agency's claim.

5. The bonding strength of resins is of major importance in the production of a range of components. The bonding strength was measured for random sample of the three most commonly used resins. The results are given in Table PE10.1 Q5:

Table PE10.1 Q5   Bonding strength for three brands of resin   File: Resins

Resins	EPXY	ERA	ERB
	1.26	0.75	1.16
	1.65	0.98	1.45
	1.98	1.23	1.84
	0.85	1.05	0.92
	1.21	1.72	1.14
Sample mean	1.39	1.146	1.302

Test whether the bonding strengths are different (use $\alpha = 0.05$).
Explain why it is not necessary to calculate confidence intervals in this case.

6. Anecdotal evidence suggests that an individual's performance in mathematics is an indicator of academic performance in general. Table PE10.1 Q6 gives the leaving certificate points (total points for six subjects) and grades in mathematics for a random sample of students registered for an IT course.

Table PE10.1 Q6   Total leaving certificate points for students with the following maths grades

OD	OC	OB	OA	HD
300	330	325	310	365
195	335	425	415	300
310	335	385	350	335
325	350	300	300	315
310	305	360	440	325
		395	355	395
		340		335
		320		
		400		

Source: LIT student survey 2000

File: Maths indicator

Where **OD**: grade D in Ordinary level mathematics **OA**: grade A in Ordinary level mathematics **HD**: grade D in Higher level mathematics etc.
(a) Calculate the ANOVA table for this data. Test the claim that there is no difference in the average leaving certificate points for the different grades in mathematics.
(b) Calculate the 95 % individual confidence intervals the difference in average points between (i) OA and OD; (ii) OB and OD. What do you conclude?

7. In waste management, the effects of landfill on surface soil characteristics is recorded each month. Particle size is of particular interest since it is indicative of the water holding capacity of the soil; large particles are well separated giving free draining but easily eroded soil. At the other extreme, clay consisting of very fine particles is easily waterlogged.
Table PE10.1 Q7 gives the results for particle size (μm) analysis for samples taken at five sites.
(a) Test the claim that the particle size is the same at all sites. State any assumptions.
(b) Write a brief report comparing the particle size for soil from the five sites.

Table PE10.1 Q7   Particle size at five sites

File: Soil 1

Site 1	Site 2	Site 3	Site 4	Site 5
965	1092	1292	1325	970
1700	1130	1900	1150	1150
1725	1400	1700	1300	1100
1500	1400	1600	1400	1010

10.3 Two-way ANOVA

When testing the difference between two population means with independent samples in Chapter 9, it was noted that large variations within samples acted like a 'clouding effect' and made it difficult to detect a difference between means. If it were possible to use paired data (hence the paired t-test) then the effect of the variation within the individual samples was reduced, making the detection of a difference much more likely. Two-way ANOVA is, in a sense, an extension of the idea of 'pairing', but 'pairing' across several samples (called 'blocking' rather than 'pairing').

10.3.1 Blocking to reduce the unexplained variation within samples

In the previous section, the identity $TSS = SSB + SSW_1$ formed the basis for the one-way ANOVA table.

The **TSS** reflects the total variation of all the data in the entire table from the overall mean. The $\mathbf{SSB_{col}}$ the variation between the different column means reflects the variation that is 'explained' by the column factor. In Tables 10.1 and 10.2 the column factor is the different departments.

The $\mathbf{SSW_1}$ is the unexplained variation within each sample in the one-way ANOVA. When the unexplained variation within samples is large relative to the difference between sample means is difficult to detect. For example, the plots of the overtime for the individuals in the Marketing and Accounting departments (data in Table 10.1) given in Figure 10.5 illustrate the large variation within samples relative to the difference between the means.

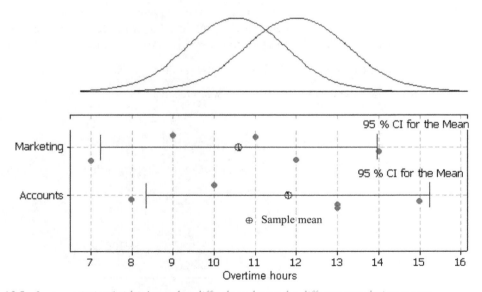

Figure 10.5   Large variation 'within' samples: difficult to detect the different population means.

On the other hand, when the unexplained variation within samples is small relative to the difference between sample means, it is highly likely that the difference between means will be detected. This situation is illustrated in Figure 10.6 in the plots of the overtime for the individuals in the Marketing and Production departments.

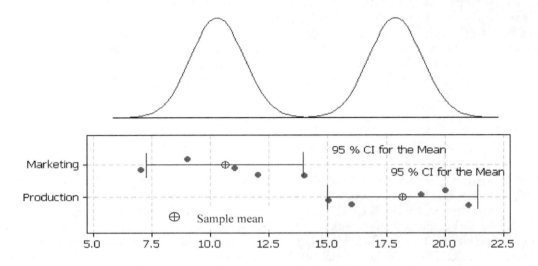

Figure 10.6   Small variation 'within' samples relative to between samples: easier to detect the different population means.

Hence, if the variation within samples in the one-way ANOVA can be reduced by explaining or attributing some of it to another factor, then s_W^2 becomes smaller. As a consequence, the test statistic $F = \dfrac{s_B^2}{s_W^2}$ becomes larger, so that it is more likely that a false null hypothesis (all means are equal) will be rejected, thus providing evidence of a difference between means.

10.3.2   Blocking to reduce the unexplained variation within samples

In ANOVA, the idea of 'pairing' is called 'blocking'. For example, suppose individuals, in Table 10.1, are 'blocked according to age in Table 10.9:

Table 10.9   Overtime hours for individuals by department and age (File: Overtime 3)

File: Overtime 3

Age	Marketing	Accounting	Personnel
20 < 25	14	15	13
25 < 30	12	13	10
30 < 40	11	13	9
40 < 50	9	10	9
50 < 60	7	8	6

If the 'blocking' factor (age) explains a significant amount of the variation within samples then the unexplained variation within samples in the one-way ANOVA (SSW_1) is expressed as

$$SSW_1 = SSB_{bl} + SSW_2$$

... where SSW_1 is the Sum of Squares for the one-way ANOVA

SSB_{bl} is the Sum of Squares between 'blocks' or 'rows'

SSW_2 is the remaining unexplained variation within samples

Hence the expression for the Sum of Squares for a two-way analysis of variance is related to that for the one-way as illustrated in equation (10.14).

Mathematical proof is given in Appendix L.

Sum of Squares for one-way ANOVA: $TSS = SSB_{col} + SSW_1$

Sum of Squares for two-way ANOVA: $TSS = SSB_{col} + SSB_{bl} + SSW_2$ **(10.14)**

The two-way model based on (10.14) assumes no interaction* between row and column variables and hence is called an 'additive model'.

*An interaction is when any combination of row and column variables produces an unusually high or low response: for example, if the individual from the Accounting department aged $20 < 25$ works 50 hours overtime this would be unusually high and not typical. Hence this age/department combination is called an interaction.

Assumptions for two-way ANOVA

1. The probability distributions for all column-block combinations are Normal.
2. The variances of all distributions are equal.
3. There is no interaction between the row and column variables.

10.3.3   Calculation of Sums of Squares for a two-way ANOVA table

The various formulas for the Sums of Squares for the two-way ANOVA necessitates the careful use of subscripts in order to refer to table entries by their row and column location. To illustrate the use of subscripts the data in Table 10.9 is presented again in Table 10.10 where row (block) means, column means, and the overall mean are given.

Each datum is referred to as $x_{i,j}$ or $x_{row, column}$ where i refers to the row and j refers to the column that contains $x_{i,j}$ OR i gives the position of $x_{i,j}$ in the sample list ($i = 1$ to n) and j gives the sample number ($j = 1$ to c).

The notation for the mean of row i (block i) is $\overline{x}_{i,\bullet}$

For example, the mean value for block 2 is $\overline{x}_{2,\bullet} = \dfrac{x_{2,1} + x_{2,2} + x_{2,3}}{3}$ where the first subscript references the block and the second subscript '$\bullet$' indicates that the calculation included all columns (column 1, 2 and 3) in block 2.

Similarly the notation for the column means is $\overline{x}_{\bullet, j}$ for $j = 1, 2, 3$.

Table 10.10   Notations and calculations of block and column means in a two-way table

	Age	Column factor (Department)			Row/Block means
		Marketing	Accounting	Personnel	
Block Factor (Age)	**Block 1** 20 < 25	$x_{1,1} = 14$	$x_{1,2} = 15$	$x_{1,3} = 13$	$\overline{x}_{1\bullet} = 14$ mean of block 1
	Block 2 25 < 30	$x_{2,1} = 12$	$x_{2,2} = 13$	$x_{2,3} = 10$	$\overline{x}_{2\bullet} = 11.67$ mean of block 2
	Block 3 30 < 40	$x_{3,1} = 11$	$x_{3,2} = 13$	$x_{3,3} = 9$	$\overline{x}_{3\bullet} = 11$ mean of block 3
	Block 4 40 < 50	$x_{4,1} = 9$	$x_{4,2} = 10$	$x_{4,3} = 9$	$\overline{x}_{4\bullet} = 9.33$ mean of block 4
	Block 5 50 < 60	$x_{5,1} = 7$	$x_{5,2} = 8$	$x_{5,3} = 6$	$\overline{x}_{5\bullet} = 7$ mean of block 4
	Column means	$\overline{x}_{\bullet 1} = 10.6$ mean of column 1	$\overline{x}_{\bullet 2} = 11.8$ mean of column 2	$\overline{x}_{\bullet 3} = 9.4$ mean of column 3	$\overline{\overline{x}}_{\bullet\bullet} = 10.6$ Overall mean

Double summations

When discussing the one-way ANOVA, the SSW_1 was given as follows:

$$\sum_{i=1}^{n}(x_{i,1} - \overline{x}_1)^2 + \sum_{i=1}^{n}(x_{i,2} - \overline{x}_2)^2 + \cdots\cdots + \sum_{i=1}^{n}(x_{i,c} - \overline{x}_c)^2$$

With the notation given above, the column means should be $\overline{x}_{\bullet,j}$, hence

$$SSW_1 = \sum_{i=1}^{n}(x_{i,1} - \overline{x}_{\bullet,1})^2 + \sum_{i=1}^{n}(x_{i,2} - \overline{x}_{\bullet,2})^2 + \cdots\cdots + \sum_{i=1}^{n}(x_{i,c} - \overline{x}_{\bullet,c})^2$$

This expression is unnecessarily cumbersome: each of the c summation is carried out for $i = 1$ to n. Therefore use a second subscript, j where $j = 1$ to c to refer to each summation. The sum of the c summations may then be written as $SSW_1 = \sum_{j=1}^{j=c}\sum_{i=1}^{n}(x_{i,j} - \overline{x}_{\bullet,j})^2$

Similarly, TSS may be expressed as a double summation. In the description of one-way ANOVA, the TSS was given informally as $TSS = \sum_{k=1}^{k=nc}(x_k - \overline{\overline{x}})^2$.

A more rigorous expression for TSS calculated from a table of n rows and c columns is

$$TSS = \sum_{i=1}^{c}\sum_{j=1}^{n}(x_{i,j} - \overline{\overline{x}})^2.$$

In the remainder of the chapter we will describe subscripts in summations verbally in addition to the 'i, j' notation. See appendices I and J for further explanations on summations.

Calculation of Sums of Squares in a two-way table

You will need (i) TSS, (ii) SSB_{col}, (iii) SSB_{row} and (iv) SSW_2

The calculation of TSS and SSB_{col} are the same as in the one-way ANOVA, given in equations (10.3a) and (10.4a). The formulae for TSS and SSB_{col} are given in double summation notation in equations (10.3b) and (10.4b).

$$TSS = \sum_{i=1}^{c} \sum_{j=1}^{n} (x_{i,j} - \overline{\overline{x}})^2 \qquad (10.3b)$$

$$SSB_{col} = n \times \sum_{j=1}^{c} (\overline{x}_{\bullet j} - \overline{\overline{x}})^2 \qquad (10.4b)$$

(i) The calculation of the Sum of Squares between the row (block) means in (10.15) below is similar to (10.14) for SSB_{row} BUT the column means and column sizes are replaced by row means and row sizes: $SSB_{row} = c \times \sum_{all\ row\ means} (\overline{x}_{row\ mean} - \overline{\overline{x}})^2$

Note: the 'size' of a row is c, the number of data column.
Row means $\overline{x}_{row\ mean}$ are written as $\overline{x}_{i,\bullet}$, $i = 1$ to n.

$$\mathbf{SSB_{row}} = c \times \sum_{i=1}^{n} (\overline{x}_{i\bullet} - \overline{\overline{x}})^2 \qquad (10.15)$$

(ii) Finally, SSW_2 is calculated from (10.14) by subtraction, when TSS, SSB_{col} and SSB_{row} are known

$$SSW_2 = TSS - SSB_{col} - SSB_{row}$$

The degrees of freedom for two-way ANOVA associated with the TSS and SSB are the same as for the one-way table:

TSS: $df = nc - 1$ ... number of values in the entire table, less 1 ⎫ **Note** the degrees of
SSB_{col}: $df = v_1 = (c - 1)$ ... number of columns, less 1 ⎪ freedom for SSB_{col}, SSB_{row}
SSB_{row}: $df = v_3 = (n - 1)$ ... number of blocks, less 1 ⎬ and SSW_2 total to $(nc - 1)$,
SSW_2: $df = v_2 = (n - 1)(c - 1)$ ⎪ the degrees of freedom for
⎭ the TSS ..(10.7)

Testing hypothesis for difference between (i) blocks (ii) columns

First, test for differences between blocks. If there is a significant difference between block means, then the reduction of SSW_1 to $(SSW_2 + SSB_{row})$ is valid. Therefore proceed to test for difference between the columns means.

If the difference between block means is not significant, then the reduction of SSW_1 to SSW_2 is not valid. Further independent samples should be selected as only a one-way ANOVA is valid.

Test statistics and degrees of freedom for two-way ANOVA

Test statistic for difference between blocks $F_{v_3, v_2} = \dfrac{s_{row}^2}{s_{w_2}^2}$: $v_3 = (n-1)$; $v_2 = (n-1)(c-1)$

Test statistic for difference between columns $F_{v_1, v_2} = \dfrac{s_B^2}{s_{w_2}^2}$: $v_1 = (c-1)$; $v_2 = (n-1)(c-1)$

The two-way ANOVA table is given in Table 10.11.

Table 10.11 Two-way ANOVA table

SS	df	$MSS = \dfrac{SS}{df}$	$F = \dfrac{MSSB}{MSSW_2}$	Degrees of freedom for sample-F
SSB_{col}	$c-1$	$MSSB_{col} = \dfrac{SSB_{col}}{c-1}$	$\dfrac{MSSB_{col}}{MSSW_2}$	$v_1 = (c-1)$, $v_2 = (c-1)(n-1)$
SSB_{row}	$n-1$	$MSSB_{row} = \dfrac{SSB_{row}}{n-1}$	$\dfrac{MSSB_{row}}{MSSW_2}$	$v_3 = (n-1)$, $v_2 = (c-1)(n-1)$
SSW_2	$(c-1)(n-1)$	$MSSW_2 = \dfrac{SSW_2}{(c-1)(n-1)}$		
TSS	$nc-1$			

Worked Example 10.7: Two–way ANOVA (for Overtime 3)

(a) (i) Calculate the two-way ANOVA table for the data in Table 10.9.
    (ii) Test the claim that there is no difference in the overtime hours worked by those in (i) different age groups; (ii) different departments.
(b) This data was originally presented in Table 10.1 when the HR assistant was unaware that age was an implicit blocking factor. In Worked Example 10.1, the null hypothesis that the mean overtime hours was the same in all departments was accepted.
    (i) Calculate a one-way ANOVA table with departments as the column factor.
    (ii) Confirm the results obtained in WE 10.1: $s_W^2 = MSSW_1$: $s_B^2 = MSSB_{col}$.
    (iii) Confirm the identity: $SSW_1 = SSB_{bl} + SSW_2$.

Solution

10. (a) (i) Method

Step 1. State c, the number of samples and n, the number of blocks. Calculate the block means, the column means and overall mean.
Step 2. Calculate TSS, SSB_{col}, SSB_{row}. Calculate SSW_2 from (10.18).

Step 3. Calculate the degrees of freedom.
Step 4. Calculate the ANOVA table.
Step 5. Test the equality of block means, using the step-by-step method for testing hypothesis.
Step 6. Test the equality of column means, using the step-by-step method for testing hypothesis.

Step 1. Set out the data as in Table 10.9(b). Three samples ($c = 3$) and five blocks ($n = 5$). Calculate the block means, column means and overall mean.

Table 10.12 Block and column means for hours overtime   ⟨File: Overtime 3⟩

Age	Marketing	Accounting	Personnel	Block means
20 < 25	14	15	13	14
25 < 30	12	13	10	11.67
30 < 40	11	13	9	11
40 < 50	9	10	9	9.33
50 < 60	7	8	6	7
Column means	10.6	11.8	9.4	10.6

Step 2. Calculate the Sums of Squares
Use the calculator to calculate the Sums of Squares, as described in section 10.2

$$\textbf{TSS} = \sigma_{All}^2 \times n \times c = 6.64 \times 5 \times 3 = 99.6$$
$$\textbf{SSB}_{col} = \sigma_{col}^2 \times n \times c = 0.96 \times 5 \times 3 = 14.4$$
$$\textbf{SSB}_{row} = \sigma_{row}^2 \times n \times c = 5.48 \times 5 \times 3 = 82.27$$
$$\textbf{SSW}_2 = \textbf{TSS} - \textbf{SSB}_{col} - \textbf{SSB}_{row} = 99.6 - 14.4 - 82.27 = 2.93$$

Step 3. Calculate the degrees of freedom ($n = 5$, $c = 3$)
The degrees of freedom associated with the $TSS = nc - 1 = 14$ and $SSB_{col} = c - 1 = 2$ (v_1)
The degrees of freedom associated with the SSB_{row} $v_3 = (n - 1) = 4$.
The degrees of freedom associated with the SSW_2 are $v_2 = (n - 1)(c - 1) = (4)(2) = 8$
The two-way ANOVA table is completed in Table 10.9(c)

Table 10.13 Two-way ANOVA table for Overtime 3

	SS	df	MSS	F-ratio	Degrees of freedom for F-ratio
$SSB_{col} =$	14.4	2	$MSSB_{col} = 7.2$	19.64	$v_1 = 2, v_2 = 8$
$SSB_{row} =$	82.27	4	$MSSB_{row} = 20.57$	56.09	$V_3 = 4, v_2 = 8$
$SSW_2 =$	2.93	8	$MSSW_2 = 0.37$		
$TSS =$	99.6	14			

Step 5. Test the equality of block means (see Table below)

Step 6. Test the equality of column means (see Table below)

(a) (ii)

i. Test for difference between blocks

1. H$_0$: $\mu_{1\bullet} = \mu_{2\bullet} = \mu_{3\bullet} = \mu_{4\bullet} = \mu_{5\bullet}$: Row (block) means are equal

2. H$_1$: At least one mean is different:

3. α: 5 %, 1 %

4. Test Statistic: F-ratio

$$F_{v_3, v_2} = \frac{\text{MSSB}_{\text{row}}}{\text{MSSW}_2} = \frac{s_{\text{row}}^2}{s_{w_2}^2}$$

5. Decision Rule:
Reject H$_0$ if the sample-F exceeds the critical-F

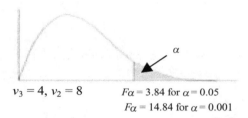

$v_3 = 4$, $v_2 = 8$ $F\alpha = 3.84$ for $\alpha = 0.05$
$F\alpha = 14.84$ for $\alpha = 0.001$

6. Calculations: From ANOVA table: Sample $F = 56.09$

7. Conclusion:

At $\alpha = 5\%$ and 0.1 % reject H$_0$ since the sample-F is greater at the than the critical-F. Reject H$_0$

Conclude there is evidence that the hours overtime is different for different ages

ii. Test for difference between columns

1. H$_0$: $\mu_{\bullet 1} = \mu_{\bullet 2} = \mu_{\bullet 3}$: Column means are equal

2. H$_1$: At least one mean is different:

3. α: 5 %, 1 %

4. Test Statistic: F-ratio where

$$F_{v_1, v_2} = \frac{\text{MSSB}_{\text{col}}}{\text{MSSW}_2} = \frac{s_B^2}{s_{w_2}^2}$$

5. Decision Rule:
Reject H$_0$ if the sample-F exceeds the critical-F

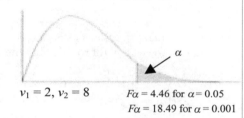

$v_1 = 2$, $v_2 = 8$ $F\alpha = 4.46$ for $\alpha = 0.05$
$F\alpha = 18.49$ for $\alpha = 0.001$

6. Calculations: From ANOVA table: Sample $F = 19.64$

7. Conclusion:

At $\alpha = 5\%$ and 0.1 % reject H$_0$ since the sample-F is greater than the critical-F.

Conclude there is evidence that the hours overtime is different for departments

(b)   (i) One-way ANOVA table
      Calculate the Sums of Squares
      Use the calculator to calculate the Sums of Squares, as described in section 10.2

$$\textbf{TSS} = \sigma_{All}^2 \times n \times c = 6.64 \times 5 \times 3 = 99.6$$
$$\textbf{SSB}_{\text{col}} = \sigma_{col}^2 \times n \times c = 0.96 \times 5 \times 3 = 14.4$$
$$\textbf{SSW}_1 = TSS - SSB_{\text{col}} = 99.6 - 14.4 = 85.2$$

The one-way ANOVA table is completed in Table 10.9(d).

Table 10.14   One-way ANOVA table for Overtime 3

	SS	df	MSS	F-ratio	Degrees of freedom for F-ratio
$SSB_{col} =$	14.4	2	$MSSB_{col} = 7.2$	1.014	$v_1 = 2, v_2 = 12$
$SSW_1 =$	85.20	12	$MSSW_1 = 7.1$		
$TSS =$	99.6	14			

(ii) In Worked Example 10.1, $s_W^2 = 7.1 = MSSW_1$: $s_B^2 = 7.2 = MSSB_{col}$
(iii) From Table 10.13 $SSB_{row} + SSW_2 = 82.97 + 2.93.$ ⎫ Hence
      From Table 10.14 $SSW_1 = 85.20$ ⎬ $SSB_{row} + SSW_2 = SSW_1$
                                                              ⎭

Skill Development Exercise SK10_4: Two-way ANOVA

Random samples of six households are selected from areas A, B and C. Their electricity expenditure is given in Table SK10_4.1.

(a) Carry out a one-way ANOVA to test whether expenditure is the same in all areas.
(b) It is thought that 'house type', e.g., apartment, detached, terraced, etc, effects expenditure on electricity, thus explaining some of the variation 'within' samples. Therefore in Table SK10_4.1 electricity expenditure for households of the same 'house type' is arranged in blocks. Carry out a two-way ANOVA to test whether expenditure on electricity is the same for all (i) house types (ii) areas

Table SK10_4.1   Expenditure on electricity classified by area and house type   File: Electricity 2

House type	Area A	Area B	Area C	Row (block) means
1	780	700	756	745.33
2	720	690	696	702.00
3	588	580	580	582.67
4	703	616	694	671.00
5	652	583	615	616.67
6	550	508	520	526.00
Column means	665.50	612.83	643.50	Overall mean 640.61

Answers

(a) One-way ANOVA table

Source	Sum of Squares	df	Mean Sum of Squares	Sample-F
SSB_{col}	8 396.44	2	4198.22	0.62
SSW_1	101 595.84	15	6773.06	
Total	109 992.28	17		

The degrees of freedom the sample-F for column means is $v_1 = 2$, $v_2 = 15$.

Difference between columns (Areas): the sample-F (0.62) is within the acceptance region and less than the critical $F = 6.36$ at $\alpha = 1\%$, so accept H_0. Conclude that the expenditure on energy is the same for all areas.

(b) Two-way ANOVA table

Source	Sum of Squares	df	Mean Sum of Squares	Sample-F
SSB_{col}	8 396.444	2	4 198.222	12.27709
SSB_{row}	98 176.28	5	19 635.26	57.42045
SSW_2	3 419.558	10	341.9558	
TSS	109 992.3	17		

The degrees of freedom the sample-F for row means is $v_1 = 5$, $v_2 = 10$
The degrees of freedom the sample-F for column means is $v_1 = 2$, $v_2 = 10$

Write out the hypothesis test, testing for difference between row means first; Difference between row factor (house types): the sample-F (57.42) exceeds the critical $F = 5.64$ at $\alpha = 1\%$, so reject H_0. Conclude that the expenditure on energy is different for at least one house type.

Difference between column factor (Areas): the sample-F (12.28) exceeds the critical $F = 7.56$ at $\alpha = 1\%$, so reject H_0. Conclude that the expenditure on energy is different for at least one area.

Comment: In the one-way ANOVA, no difference between areas was detected. But, in the two-way ANOVA house types explained a large part of the variation within samples. As a result the SSW was reduced and a difference between expenditures in areas was detected.

10.3.4 Simultaneous confidence intervals for two-way ANOVA

Simultaneous confidence intervals may be used to detect which means are different in a two-way ANOVA table.

The Bonferroni method, section 10.2.5 Recall the number of comparisons for c columns is given by
$$M = \frac{c(c-1)}{2}.$$
The simultaneous confidence interval for the difference between column means, such as ($\mu_{col\ A} - \mu_{col\ B}$) are given by the formula

$$\overline{x}_{col\ A} - \overline{x}_{col\ B} \pm t_{df(SSW_2),\alpha/(2M)} \sqrt{MSSW_2 \left(\frac{1}{n} + \frac{1}{n} \right)} \qquad (10.16)$$

...where $MSSW_2$ is the mean Sum of Squares within samples in the two-way table and $t_{df(SSW_2),\alpha/2M}$ is the t percentage point (critical value) for a tail area of $\dfrac{\alpha}{2M}$ and the degrees of freedom associated with $MSSW_2$: n is the column sample size (hence the number of rows in the table).

Similarly, the number of comparisons for n rows (or blocks) is $M = \dfrac{n(n-1)}{2}$ etc.

The simultaneous confidence intervals for the difference between population means ($\mu_{Row,A} - \mu_{Row\,B}$) are given by the formula

$$\overline{x}_{Row\,A} - \overline{x}_{Row\,B} \pm t_{df(SSW_2),\alpha/(2M)} \sqrt{MSSW_2 \left(\frac{1}{c} + \frac{1}{c}\right)} \qquad (10.17)$$

...where c is the number of elements in the row (the row 'sample size').

Scheffé's method

The simultaneous confidence intervals for differences between column means and row means in the two-way ANOVA is

$$\overline{x}_{col\,A} - \overline{x}_{col\,B} \pm \sqrt{MSSW_2(c-1)F_{\alpha,c-1,(n-1)(c-1)}\left(\frac{1}{n} + \frac{1}{n}\right)} \text{ and}$$

$$\overline{x}_{row\,A} - \overline{x}_{row\,B} \pm \sqrt{MSSW_2(n-1)F_{\alpha,c-1,(n-1)(c-1)}\left(\frac{1}{c} + \frac{1}{c}\right)} \text{ respectively} \qquad (10.18)$$

Individual confidence intervals: difference between means for paired samples

If individual confidence intervals are required for two-way ANOVA remember they are based on 'paired' samples for row or column means as follows:

$$\overline{d} \pm t_{n-1}\frac{S_d}{\sqrt{n}} \quad \text{or} \quad \overline{d} \pm t_{c-1}\frac{S_d}{\sqrt{c}} \qquad (10.19)$$

10.4 ANOVA and design of experiment

So far in this chapter, the researcher randomly selects samples from well defined populations and observes or records the response of intereset from each subject. For example in Table 10.1, subjects were randomly selected from different departments. The subject of the research is overtime; hence the recorded response is the number of overtime hours in the month. In section 10.2, the experiment is described as a completely randomised observational study since subjects are randomly selected. In section 10.3, an attempt was made to explain some of the variation in overtime hours within departments by 'pairing' or blocking subjects according to age. In practise, members of departments could be stratified by age, then a stratified random sample selected from each department. This type of study is described as an observational randomised block design. In both of these experiments the researcher observes and records the responses from the randomly selected subjects.

In a designed experiment the researcher has control over the variables of interest (also called treatments or factors) and the levels of the factors that affect a response in the subjects (people) or experimental units (things). For example, an experiment is carried out to investigate the affect of temperature on the drying time of paint. A selected brand of paint is applied to 20 one metre squares of plywood. The squares are randomly divided into four groups of five; each group is placed in a controlled temperature environment of 15°C; 20°C; 25°C and 30°C.

The drying time for the squares in each group is recorded in Table 10.15:

Table 10.15   Completely randomised design: drying time for paint at different temperatures | File: Paint 1 |

	Drying temperature			
	15°C	20°C	25°C	30°C
Drying time (hours)	6.5	5.3	5.2	4.2
	7.4	5.2	5.5	4.6
	4.6	4.2	3.8	3.5
	3.5	3.5	3.3	4.2
	5.4	4.1	3.4	2.2
Sample means	5.48	4.46	4.24	3.74

In this designed experiment the treatment (or factor) is temperature: the different temperatures are called 'levels' of the treatment. The response variable is drying time. The painted squares are the experimental units. A one-way ANOVA is used to test whether the mean drying times are different at different temperatures. The experiment is described as completely randomised since five squares were randomly assigned to each temperature.

A completely randomised block design is an experiment in which an attempt is made to explain some of the variation within the samples (treatments) by a second factor (blocking factor) as discussed in section 10.3: In Table 10.16 the blocking factor is the humidity **applied** at five different levels: To set up the experiment, four squares are randomly assigned to each of the five blocks (levels of humidity), then one unit from each block is randomly assigned to each treatment (drying temperature). The drying time (response) is recorded for each treatment–block combination in Table 10.16:

The results of this experiment can be analysed by a two-way ANOVA to test for a difference in drying times due to (a) different humilities and (b) different temperatures.

Table 10.16   Randomised block design: Drying time at all temperature-humidity combinations

File: Paint 2

| | | Drying temperature | | | | |
		15°C	20°C	25°C	30°C	Block means
Humidity	10%	6.5	5.3	5.2	4.2	5.3
	15%	7.4	5.2	5.5	4.6	5.675
	20%	4.6	4.2	3.8	3.5	4.025
	25%	3.5	3.5	3.3	4.2	3.625
	30%	5.4	4.1	3.4	2.2	3.775
	Column means	5.48	4.46	4.24	3.74	

Skill Development Exercise 10.5: Designed experiment

(a) Calculate a one-way ANOVA table for the data in Table 10.15. Test whether drying times are different at different temperatures
(b) Calculate a two-way ANOVA table for the data in Table 10.16.
   Test the claims that (i) humidity (ii) temperature has no effect on drying time.
   If the claims in (b) are not accepted, determine which means are different.
(c) **Comment** on the results in (a) and (b)

Answer

(a) One-way ANOVA table

Source of Variation	SS	df	MS	F	P-value	5% F crit
Temperatures	8.028	3	2.676	2.175168	0.130794	3.238867
Within Groups (Error)	19.684	16	1.23025			
Total	27.712	19				

(b) Two-way ANOVA table

Source of Variation	SS	df	MS	F	P-value	5% F crit
Humidities	14.142	4	3.5355	7.655359	0.002645	3.25916
Temperatures	8.028	3	2.676	5.794298	0.010962	3.4903
Error	5.542	12	0.461833			
Total	27.712	19				

(b) Both temperature and humidity affect drying time.
Drying time is longer for humidities of 10 % and 15 % compared to the others. Drying time is longer for a temperature of 15°C compared to 30°C.

Comment: the data in Table 10.15 and 10.16 is exactly the same! The one-way analysis does not detect a significant difference in drying times at different temperatures. In the two-way analysis, the blocking factor, humidity explains a significant difference in drying times, hence reduces the unexplained variation within the temperature groups: a difference in drying times at temperatures is detected and $a = 0.05$.

Progress Exercises 10.2: Two-way ANOVA

1. (a) Explain why the paired t-test is a more powerful test for the difference between two population means than the independent t-test.
   (b) See question 1, PE10.1. The bank's technical support team claim that the down-time has decreased over the years. As evidence, the team retrieved the records for six randomly selected branches, A, B, C, D, E, F, for the years 2005 and 2007. The down-times are given in Table PE10.2 Q1:

Table PE10.2 Q1 Down-time for computer systems | File: Downtime 2

Branch	Down-time (hours) 2005	2007
A	40	30
B	54	50
C	32	22
D	38	36
E	80	68
F	44	40

Test the claim made by the technical support team at the 5 % level of significance by (i) an independent t-test (ii) paired t-test. Is the independent t-test valid for this data?
Explain why these results in (i) and (ii) are different.
State and justify your conclusions
2. The two-way ANOVA for the data in question 1 is given below

Table PE10.2 Q2 2-way ANOVA table for data in question 1

Source of Variation	SS	df	MSS	F	F critical
Branches		5	554.2		5.0503
Years	147			17.0930	6.6079
Error	43		8.6		
Total		11			

(a) Complete the ANOVA table.

(b) Test the claim that there is a difference in down-time between (i) branches; (ii) years.

(c) Show that the F-ratio and the critical F-value in the ANOVA table are the same as the squared values of the test statistic and critical t-value in the paired t test in question 1.

3. Suppose a particular bank believes the downtime has not decreased progressively over several years. The records for six randomly selected branches are given in Table PE10.2 Q2.

Table PE10.3 Q3   Down-time in hours for six branches   File: Downtime 3

Branch	Down-time 2004	2006	2007	Block (row) means
A	56	40	30	42
B	61	54	50	55
C	54	32	22	36
D	40	38	36	38
E	71	80	68	73
F	48	44	40	44
Column means	55	48	41	48

(a) Calculate the two-way ANOVA table for the data.

(b) Test the claim that there is no difference in down-time between (i) branches; (ii) years.

4. (a) What is the purpose of 'blocking' in the randomised block design?

(b) In Quality control all measuring systems must be carefully monitored to ensure that the measured responses are not affected by either the measuring device or the operator using the device. To test the measurement system for long jumps, three operators measured the length of the jump (meters) for four athletes. Results are given in Table PE10.2 Q4:

Table PE10.2 Q4   Diameters of four tubes measured by each operator   File: Athletes

Athlete	Operator OP 1	OP 2	OP 3
A	5.5	5.5	7.9
B	4.7	5.9	7.4
C	6.6	5.8	8.3
D	4.5	5.7	7.5

Test whether there is a difference between (i) operators; (ii) athletes.

What conclusion do you reach about differences between operators? Is the measurement system satisfactory?

5. The average gross weekly earnings (not seasonally adjusted) for England, Wales, Scotland and Northern Ireland are given in Table PE10.2 Q5:

Table PE10.2 Q5   Gross weekly earnings (£) | File Earnings |

	England	Wales	Scotland	N. Ireland
Summer 2003	373	340	314	298
Spring 2004	375	327	346	328
Autumn 2004	387	339	345	344
Spring 2005	404	333	365	346

Source: Office for National Statistics. Crown copyright material is reproduced with the permission of the Controller of HMSO. Labour Force Survey

(a) Calculate a two-way ANOVA table for the data.
(b) Test the claim that there is no difference between the four areas.
(c) Calculate simultaneous confidence intervals to determine which areas are different.
(d) The data in Table PE10.2 Q5 was used in PE9.4 Q4 to test the differences between mean earnings for England and Scotland; England and Northern Ireland; Wales and Scotland.

Compare these results with those in (b) and (c) above. Which tests would you consider to be the most useful?

6. See question 5, PE9.4. The gross household disposable income (£000 m) for four regions in the UK is given in Table PE10.2 Q6 for years 1990, 1995 and 1999.

Table PE10.2 Q6   Gross disposable household income | File: Disposable income |

	1991	1995	1999
East	37.602	47.373	57.647
London	55.093	70.785	88.930
South East	55.978	72.840	89.299
South West	31.681	41.542	49.718

Source Office for National Statistics. Crown copyright material is reproduced with the permission of the Controller of HMSO.

(a) Calculate a two-way ANOVA table for the data.
(b) Test the claim that there is no difference between (i) regions; (ii) years.
(c) Calculate simultaneous confidence intervals to determine which regions are different. Compare this analysis with that in question 5, PE9.4.

7. See question 6, PE9.4.
(a) What are the conditions necessary for a two-way analysis of variance to be valid?
(b) The data in Table PE10.2 Q7 gives the daily revenue from parking fines (£,000) in three different areas collected by five traffic wardens:

Wardens	Table PE10.2 Q7 Revenue from parking fines (£, 000)			File: Parking fines
	Main street	Shop street	Old Market	
T	2.5	3.4	5.3	
R	1.2	1.0	3.4	
A	1.1	1.8	2.0	
F	1.5	1.8	2.9	
C	1.3	1.5	0.5	

(i) Test the claim that there is no difference between the values of the fines revenue collected by the five traffic wardens

(ii) Test the claim that there is no difference between the fines revenue collected from the different areas.

(iii) Construct a 90 % confidence interval for the difference between the income from fines collected in Main street and Old market.

Based on these results, write a brief report on your findings.

(c) Discuss the inference from the results above compared to the inference from Q6, PE9.4 which was based on the same data.

10.5 Excel and Minitab for ANOVA

Excel and Minitab software both produce ANOVA tables. The researcher must use the output from these software to formally set out the test of hypothesis for ANOVA, stating null and alternative hypotheses (H_0: all population means are equal vs. H_1: at least one mean is different), decision rule and conclusion.

In addition, the software, particularly Minitab provides other useful information that allows the researcher to check the model assumptions.

10.5.1 One-way ANOVA in Excel

The data are entered in an Excel worksheet in tabular form. For example, the data from Table PE10.1 (hours down-time for banks in 2003–5).

From the menu bar, select **Tools → Data Analysis** as illustrated in Figure 10.7.

From the drop down menu for 'Data Analysis' select **Anova: Single Factor**: see Figure 10.7.

The selection of Anova: Single Factor produces the dialogue box in Figure 10.8.

You are now required to enter the following:

1. The address for the data for the ANOVA, hence highlight the table (including labels).
2. Indicate whether the samples are presented in columns or rows.
3. Indicate whether the labels were included in 1.
4. State the level of significance of the ANOVA test.

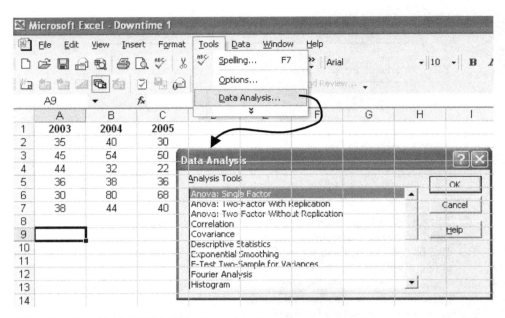

Figure 10.7   Data from Table PE10.1. (File: Downtime 1).

⑤. Indicate where you require the output. Usually it is convenient to have the output in the same worksheet as the data. In this case, click on the cell (or type the cell reference) of the top right-hand corner of the section of the worksheet where the results will be displayed. The other options for the location of output is a new worksheet or even a new workbook.

⑥. Click OK.

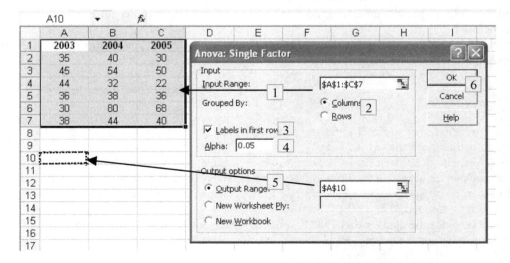

Figure 10.8   Give the location of data, level of significance and location of results of the analysis.

Finally, click OK. The output is given in Figure 10.9.

The output first gives the mean and sample standard deviation for each sample. In Excel the level of significance is $\alpha = 0.05$. In this example, the variance of the first sample is considerable smaller that the others. The user can test the equality of two variances by the F-test. In addition to the ANOVA table, the software gives p-values, the critical F-values.

f_x Anova: Single Factor

	A	B	C	D	E	F	G
1	**2003**	**2004**	**2005**				
2	35	40	30				
3	45	54	50				
4	44	32	22				
5	36	38	36				
6	30	80	68				
7	38	44	40				
8							
9							
10	Anova: Single Factor						
11							
12	SUMMARY						
13	*Groups*	*Count*	*Sum*	*Average*	*Variance*		
14	2003	6	228	38	32.4		
15	2004	6	288	48	299.2		
16	2005	6	246	41	263.6		
17							
18							
19	ANOVA						
20	ce of Varia	SS	df	MS	F	P-value	F crit
21	Between G	316	2	158	0.796371	0.469135	3.682317
22	Within Gro	2976	15	198.4			
23							
24	Total	3292	17				

Figure 10.9   Excel output: ANOVA table and other statistics for Downtime 1.

10.5.2   Two-way ANOVA in Excel

The steps for two-way ANOVA are similar to those described above and illustrated by entering the data in file: Overtime 3. Overtime hours for individuals classified by department and age (File: Overtime 3) are entered on a blank Excel worksheet in Figure 10.10.

Select ☐1 **Tools** → **Data Analysis** → ☐2 **Anova: Two-Factor Without Replication** ☐3 and ☐4 supply the input and output information. See Figure 10.10.

The results of the analysis are given in Figure 10.11.

It is vitally important that the researcher explains the results by setting out the hypotheses tests as given in Worked Example (a) (ii). How can a null hypothesis be accepted or rejected without defining the null hypothesis!

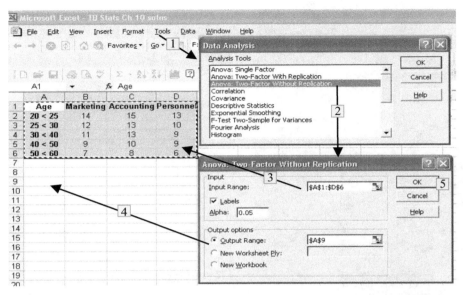

Figure 10.10   Excel: Two-way ANOVA in Excel for data in file: Overtime 3.

	A	B	C	D	E	F	G	H
1	Age	Marketing	Accounting	Personnel				
2	20 < 25	14	15	13				
3	25 < 30	12	13	10				
4	30 < 40	11	13	9				
5	40 < 50	9	10	9				
6	50 < 60	7	8	6				
7								
8								
9	Anova: Two-Factor Without Replication							
10								
11	SUMMARY	Count	Sum	Average	Variance			
12	20 < 25	3	42	14	1			
13	25 < 30	3	35	11.66667	2.333333			
14	30 < 40	3	33	11	4			
15	40 < 50	3	28	9.333333	0.333333			
16	50 < 60	3	21	7	1			
17								
18	Marketing	5	53	10.6	7.3			
19	Accounting	5	59	11.8	7.7			
20	Personnel	5	47	9.4	6.3			
21								
22								
23	ANOVA							
24	ce of Varia	SS	df	MS	F	P-value	F crit	
25	Rows	82.26667	4	20.56667	56.09091	6.83E-06	3.837854	
26	Columns	14.4	2	7.2	19.63636	0.00082	4.458968	
27	Error	2.933333	8	0.366667				
28								
29	Total	99.6	14					

Figure 10.11   Excel output for two-way for data in file: Overtime 3.

10.5.3   ANOVA in Minitab

Use Minitab to check the assumptions for ANOVA

Minitab not only produces the ANOVA tables, but also other statistics and charts that provides the user with an initial visual check on whether the ANOVA assumptions are satisfied.

Recall that the assumptions for the one-way ANOVA are

1. The samples are independent.
2. The samples are from Normal populations.
3. All populations have equal variances $\sigma_1^2 = \sigma_2^2 = \sigma_3^2 = \cdots = \sigma_c^2$ ($\sigma^2 =$ common variance).

Assumption 1: Residuals* vs. observation order should be random.

Assumption 2: the Normality of the data may be viewed by (i) histograms of the residuals, which should be reasonably bell-shaped (ii) Normal probability plots of residuals: the points should be close to the straight line in the Normal probability plot (iii) the box plots should be symmetrical with the mean close to the median and no outliers.

Assumption 3: a graphical confirmation of Normality and equal variance is given by box plots for each sample.

Minitab check for the assumptions for two-way ANOVA.

Recall, the assumptions for two-way ANOVA.

1. The probability distributions for all column-block combinations are Normal.
2. The variances of all distributions are equal.
3. There is no interaction between the row and column variables.

For two-way ANOVA, the residual* plots (Minitab) should be random, normally distributed, with no outliers. A curvilinear pattern of residuals against fitted values indicate that there may be interaction between block and treatment effects. Interaction may be flagged by plotting the line graphs for the values in each block against the treatments or vice-versa (use line graphs in Excel). If there is 'no interaction' the line graphs should be more or less parallel: remember, the graphs are based on sample data – so allow for sampling error.

*See Appendix L: a residual, $\varepsilon_i = x_i - \hat{x}_i$, is the difference between an observed value and the fitted value in the ANOVA table. If the model is appropriate, the residuals should be Normally distributed: $\varepsilon_i \sim N(0, \sigma^2)$.

One-way ANOVA tables in Minitab

Data may be entered column by column on a Minitab worksheet. See Figure 10.11 where the worksheet contains the data for the drying time for paint at different temperatures (File: Paint 1).

From the menu-bar, select **Stat → ANOVA → One-Way (Unstacked)...**

Graphical checks for the assumptions of normality and equal variance are produced by ticking the box for **Box plots**. Also tick the box **Three in One** under ***Residual Plots**: See Figure 10.13. The residual plots also provide visual inference about the model assumptions: residuals should be approximately normal, random, have no outliers, etc.

The output is given in Figure 10.14. The p-value, 0.131 or 13.1 % is sufficiently large to conclude that the null hypothesis (all population means are equal) should not be rejected. This conclusion is supported by the fact that all the individual 95 % confidence intervals overlap.

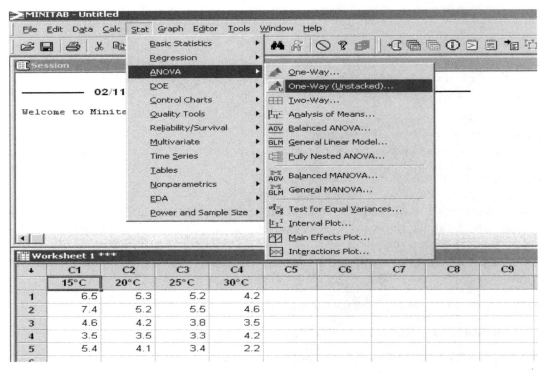

Figure 10.12   One-way ANOVA in Minitab: data in Table 10.12 (drying time for paint vs. temperatures, File: Paint 1).

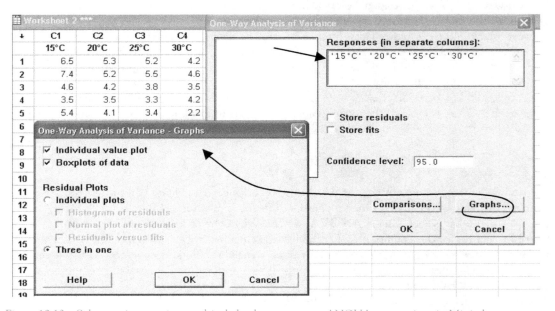

Figure 10.13   Select options to give graphical checks on one-way ANOVA assumptions in Minitab.

One-way ANOVA: 15°C, 20°C, 25°C, 30°C

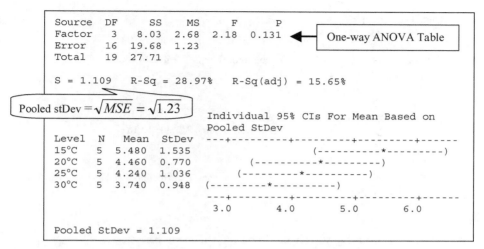

```
Source   DF     SS    MS     F      P
Factor    3   8.03  2.68  2.18  0.131  ◄──── One-way ANOVA Table
Error    16  19.68  1.23
Total    19  27.71

S = 1.109    R-Sq = 28.97%    R-Sq(adj) = 15.65%
```

Pooled stDev $= \sqrt{MSE} = \sqrt{1.23}$

```
                              Individual 95% CIs For Mean Based on
                              Pooled StDev
Level  N   Mean   StDev    ---+---------+---------+---------+------
15°C   5  5.480   1.535                        (----------*---------)
20°C   5  4.460   0.770               (----------*---------)
25°C   5  4.240   1.036            (---------*----------)
30°C   5  3.740   0.948     (---------*----------)
                         ---+---------+---------+---------+------
                          3.0       4.0       5.0       6.0

Pooled StDev = 1.109
```

Figure 10.14   Minitab output for data in Table 10.12. (File: Paint 1).

The box plots, Figure 10.15, show that the data for 15°C appears to be Normally distributed, but not the remaining samples, particularly for the sample at 30°C. In these three samples the mean is different from the median. The population variances are probably equal: remember that the F-ratio for two sample variances must be large (greater than 2.98 for $\alpha = 5\%$ and $v_1 = 10$, $v_2 = 10$ df) to reject Ho: $\sigma_a^2 = \sigma_b^2$.

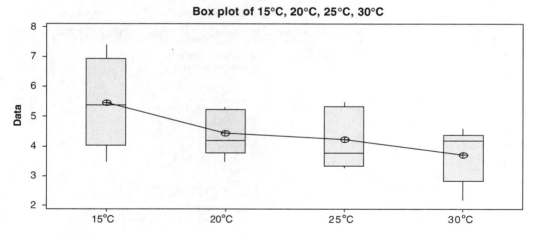

Box plot of 15°C, 20°C, 25°C, 30°C

Figure 10.15   Minitab output: box plots for data in Table 10.12. (File: Paint 1).

Two-way ANOVA in Minitab

For two-way ANOVA in Minitab, the data must be entered in three columns where the columns contain the row (block) references: column (treatment) references and the responses. For example the data for

Table 10.16 (File: Paint 2, drying time in terms of temperature and humidity) is entered as follows:

Humidity	Temp	Drying time	
10 %	*15*	*6.5*	
15 %	*15*	*7.4*	
20 %	*15*	*4.6*	Data from column 1, Table 10.16
25 %	*15*	*3.5*	
30 %	*15*	*5.4*	
10 %	20	5.3	
15 %	20	5.2	
20 %	20	4.2	Data from column 2, Table 10.16
25 %	20	3.5	
30 %	20	4.1	
10 %	*25*	*5.2*	Similarly for data from columns 3 and 4, Table 10.16
15 %	*25*	*5.5*	

Figure 10.16   Entering data (Paint 2) for two-way ANOVA in Minitab.

Having entered the data into the worksheet, then from the menu bar select

Stats → Anova → Two-way Anova

Into the 'Two-Way Analysis of Variance' dialogue box, enter the response, row factor, column factor as illustrated in Figure 10.17.

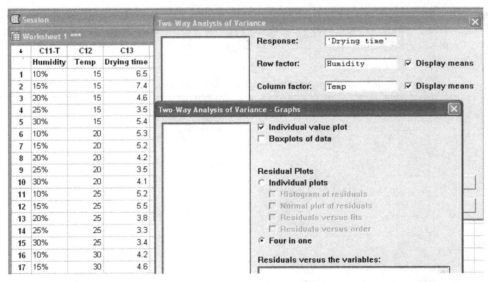

Figure 10.17   Minitab: Into the 'Two-Way Analysis of Variance' enter the response variable, row and column factors.

Also, click on the 'Graphs' button then select 'Four in one' for the plots to graphically check the ANOVA assumptions.

Recall: the assumptions are satisfied if the residuals are normally distributed, random and there are no outliers. Figure 10.18 shows that the assumptions are satisfactory in this case.

The Minitab output is given in Figure 10.19. The output gives the two-way ANOVA table and the individual confidence intervals for each row and column variable.

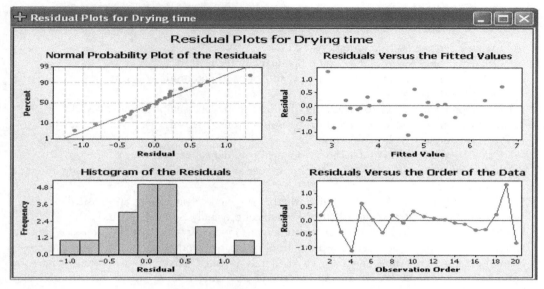

Figure 10.18   Minitab residual plots for two-way ANOVA, data in Table 10.15. (File: Paint 2).

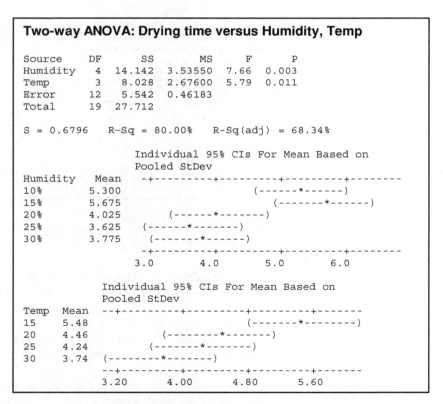

Figure 10.19   Minitab output for two-way ANOVA for data in Table 10.15. (File: Paint 2).

Progress Exercises 10.3: ANOVA with Excel, Minitab or otherwise

1. In Progress Exercises 10.1, for questions 2 to 7
   (a) Calculate the ANOVA tables and individual confidence interval.
   (b) If the means are different plot appropriate graphs to determine which means are different from the others.
   (c) Plot appropriate graphs to determine whether the model assumptions are satisfied.
   Write a brief report describing verbally the results of the analysis in (a), (b) and (c).
2. In Progress Exercises 10.2, for questions 1 to 7:
   (a) Calculate the ANOVA tables and individual confidence interval.
   (b) If the means are different plot appropriate graphs to determine which means are different from the others.
   (c) Plot appropriate graphs to determine whether the model assumptions are satisfied.
   Write a brief report describing verbally the results of the analysis in (a), (b) and (c).
3. Health and Safety officials set up an experiment to investigate whether the time drivers take to read road traffic signs is related to the level of lighting and/or the distances of the driver from the signs. The time taken by 48 randomly selected drivers to read traffic signs for various levels of light (A to F) and at distances ranging from 0.5 m to 4 m are given in Table PE10.2 Q3:

Table PE10.3 Q3   Time taken to read standard road traffic signs for light quality A to F at distances 0.4 to 4 m   File: Traffic signs

	Response time (sec)							
	0.5 m	1 m	1.5 m	2 m	2.5 m	3 m	3.5 m	4 m
A	29.8	29.1	30.7	30.8	35.3	36.4	37.1	42.1
B	29.6	31.7	28.4	30.6	34.1	37.2	36.5	38.5
C	30.2	31.1	30.9	30.4	35.6	39.2	36.7	35.8
D	36.1	38.7	37.4	38.9	39.1	38.6	36.2	38.6
E	28.4	28.7	30.8	31.3	32.9	37.9	38.2	39.2
F	25.2	32.7	28.7	36.3	35.7	31.2	33.7	38.1

   (a) In Excel calculate a two-way ANOVA table to test whether the time taken to read traffic signs is affected by the factors (i) level of light; (ii) distance from sign.
   (b) Carry out a two-way ANOVA in Minitab.
   (c) Does the Minitab printout confirm (i) the results in (a) (ii) the model assumptions.
   (d) Carry out any further analysis that could give useful information the factors that affect the time taken to read traffic signs.
   (e) Based on the results in (a) and (b) above write a brief report on the effect of lighting and distance from sign on the time taken to read notices. In the report indicate, with reasons, the factors and levels that have a significant effect on the response variable (time to read traffic signs).
4. In a double blind, randomised block experiment participants of similar typing skills were given asked to type a document after consuming a drink containing alcohol. Two document types were used in the experiment: technical (Doc A) and poetry (Doc B). The number of typing errors made by participants is given in Table PE10.16:

Table PE10.3 Q4   Typing errors   File: Type alcohol

Alcohol	Doc A	Doc B
0 units	2	4
1.0 units	6	7
1.5 units	3	4
2.0 units	7	12
2.5 units	5	16
3.0 units	9	12
3.5 units	12	14
4.0 units	7	14
4.5 units	9	17
5.0 units	13	5
5.4 units	18	3
5.5 units	16	2

(a) Why would the researcher use a 'double blind experiment?'
(b) Explain how participants are assigned alcohol-document combinations in this experiment. Why is it called 'completely randomised block' design?
(c) Carry out a two-way analysis of variance to determine whether there is a difference in the number of errors due to (i) level of alcohol (ii) document type.
(d) Are the model assumptions satisfied?
(e) In Excel plot line graphs for typing errors against units of alcohol. Comment on the trends. Do you think there is interaction between document type and level of alcohol?

5. Three groups of students are given the same assessment at the end of a 10 week course in typing. Group B attended the morning course; group B attended the 5–7 pm course while group C attended the 7–9 pm course (maximum score in the assessment is 200, File: Practical scores).

The Minitab printout for a one-way ANOVA is given in Table PE10.3Q5.

(a) Describe the Minitab printout and explain each of the calculations given.
(b) Write out the hypotheses test for the ANOVA. Write a report, comparing the performance of the three groups in the assessment.

Table PE10.3 Q5   Minitab printout for one-way ANOVA for practical scores for groups A, B and C

One-way ANOVA: C, B, A

Source	DF	SS	MS	F	P
Factor	2	7739	3870	3.64	0.031
Error	82	87144	1063		
Total	84	94884			

```
S = 32.60      R-Sq = 8.16%      R-Sq(adj) = 5.92%

                                Individual 95% CIs For Mean Based on
                                Pooled StDev
Level   N     Mean    StDev    --+---------+---------+---------+-------
C       23    107.48  22.78                       (--------*--------)
B       31    95.61   32.28              (-------*-------)
A       31    83.39   38.50    (-------*------)
                               --+---------+---------+---------+-------
                                75        90        105       120

Pooled stDev = 32.60
```

6. See question 5. The box plots for the data in question 5 is given in Figure PE10.3 Q6:

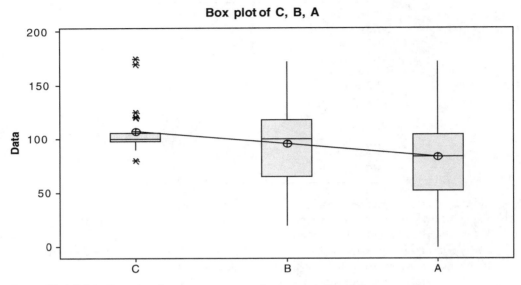

Figure PE10.3 Q6   Box plots for assessment scores for groups A, B and C.

Explain the box plots. Do you think the assumptions for one-way ANOVA are satisfied?

7. Compound X is a viscous liquid that is used extensively as a lubricant in the electronics industry. An experiment is carried out to determine the effect of band frequency (Freq) and film thickness (Th) on the optical density of compound X. The optical density is measured at twelve frequency-thickness combinations.

The Excel printout for a two-way ANOVA is given in Table PE10.3 Q7:

(a) State the row and column factors in the experiment. Give an outline of the contingency table from which the ANOVA table was calculated, filling in the row and column totals and averages.

(b) Show that the values given for the Sums of Squares for rows and columns in the ANOVA table are given by the formulae

$$SSB_{col} = \sigma^2_{col} \times n \times c \text{ and } SSB_{row} = \sigma^2_{row} \times n \times c$$

(c) Test, at the 5 % level of significance, whether (i) frequency (ii) film thickness have an effect on optical density.

8. Table PE10.3 Q8a is the Minitab printout for a two-way ANOVA for the time taken to commute to work by a number of motorists on different days.

(a) State the response variable, the number of motorists, the number of days and total number of observations in the source data.

(b) Explain the print out and the conclusions that may be made from this analysis.

(c) The 4 in 1 plot for this data is given in Table PE10.3 Q8c:

Do these plots indicate that the model assumptions are satisfied?

Table PE10.3 Q7   Excel printout for two-way ANOVA for optical density of Compound X

Anova: Two-Factor Without Replication

SUMMARY	Count	Sum	Average	Variance
Freq A	3	0.627	0.209	0.038437
Freq B	3	0.496	0.165333	0.026372
Freq C	3	1.162	0.387333	0.094721
Freq D	3	2.832	0.944	0.019812
Th (1.22)	4	2.433	0.60825	0.114419
Th (0.73)	4	1.745	0.43625	0.143384
Th (0.36)	4	0.939	0.23475	0.153418

ANOVA

Source of Variation	SS	df	MS	F	P-value	F crit
Rows	1.154564	3	0.384855	29.19226	0.000563	4.757055
Columns	0.279585	2	0.139792	10.60363	0.010725	5.143249
Error	0.079101	6	0.013183			
Total	1.513249	11				

Table PE10.3 Q8a   is the Minitab printout

Two-way ANOVA: Time versus Motorist, Day

Source	DF	SS	MS	F	P
Motorist	3	827.2	275.733	6.54	0.007
Day	4	1797.7	449.425	10.65	0.001
Error	12	506.3	42.192		
Total	19	3131.2			

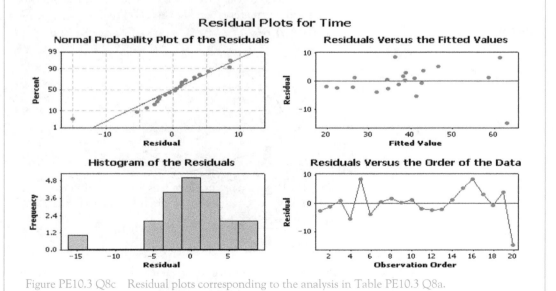

Figure PE10.3 Q8c   Residual plots corresponding to the analysis in Table PE10.3 Q8a.

Summary and overview analysis of variance

10.1 Analysis of variance

This is a technique for testing the equality of several means.

$H_0: \mu_1 = \mu_2 = \ldots\ldots = \mu_c$

H_1: at least one mean is different from the others

10.2 One-way analysis of variance

Summary of rational for one-way ANOVA

H_0 is True then $\mu_1 = \mu_2 = \mu_3$

The assumptions 1. Independent samples 2. Normal distributions 3. $\sigma_1^2 = \sigma_2^2 = \sigma_3^2$

Estimates of the common variance σ^2

Method A. Average sample variances $s_W^2 = \dfrac{s_1^2 + s_2^2 + s_3^2 + \ldots + s_c^2}{c}$

Method B. Use sample means to calculate $s_{\bar{x}}^2$. Then $s_B^2 = n \times s_{\bar{x}}^2$

If H_0 is true and the assumptions are true there is only one Normal population. Estimates of variance by different methods (A and B) will be approximately the same.	**If H_0 is NOT true**; at least one mean is different (but the assumptions are true) there is more than one Normal population. Estimates of variance by different methods will different The average of the sample variances (Method A) will be the same, whether Ho is true or not The estimate of variance based on sample means (Method B) will be greater when Ho is NOT True.

Ho true

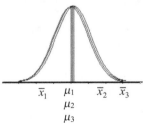

$$\bar{x}_1 \quad \mu_1 \quad \bar{x}_2 \quad \bar{x}_3$$
$$\mu_2$$
$$\mu_3$$

Ho NOT true

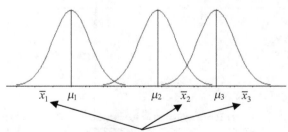

$$\bar{x}_1 \quad \mu_1 \qquad \mu_2 \quad \bar{x}_2 \quad \mu_3 \quad \bar{x}_3$$

B $\quad s_B^2 = n \times s_{\bar{x}}^2$

A $\quad s_W^2 = \dfrac{s_1^2 + s_2^2 + s_3^2}{3}$

B $\quad s_{\bar{x}}^2$ is larger since sample means are more dispersed

Hence $s_B^2 = n \times s_{\bar{x}}^2$ is larger when H_0 is NOT true

A $\quad s_W^2 = \dfrac{s_1^2 + s_2^2 + s_3^2}{3}$ is the same whether H_0 is true or not

So, if H_0 is NOT true then $s_B^2 > s_W^2$

Hence the test statistic, $F = \dfrac{s_B^2}{s_W^2}$ may be sufficiently large to reject H_0

Hence the test follows the familiar seven steps

1. H_0: $\mu_1 = \mu_2 = \ldots\ldots = \mu_c$ (hence $\sigma_B^2 = \sigma_W^2$)
2. H_1: at least one mean is different from the others (hence $\sigma_B^2 > \sigma_W^2$)
3. α: 5 %
4. **Test Statistic** F-ratio where $F_{v_1, v_2} = \dfrac{s_B^2}{s_W^2}$.
5. **Decision Rule**: If the sample-F exceeds the tables-F, (critical-F) reject H_0.
6. **Calculations**: Calculate the test statistic or sample $F = \dfrac{s_B^2}{s_W^2}$
7. **Compare the test statistic with the decision rule**. If the test statistic exceeds the critical F, then there is insufficient evidence to accept H_0.

The one-way ANOVA table

The ANOVA table is based on the separation of the estimates of variance in their simplest form into numerator and denominator:

$$s^2 = \frac{numerator}{denomintor} = \frac{sums\ of\ squares}{degrees\ of\ freedom}$$

... and the Sums of Squares identity, where the Sums of Squares and corresponding degrees of freedom total as in (10.6) and (10.7).

$$\text{Total Sum of Squares (TSS)} = \text{SS Between (SSB}_{col}) + \text{SS Within (SSW}_1)\ldots \qquad (10.6)$$

$$df(\text{Total}) = df(\text{Between column means}) + df(\text{Within columns})\ldots \qquad (10.7)$$

Table 10.3   One way ANOVA table

Source of Variation	Sum of Squares	Degrees of freedom	Mean Sum of Squares	Sample-F (Test Statistic)
Between Samples (or between Columns)	SSB$_{col}$	$c - 1$ (v_1)	MSSB$_{col} = \frac{SSB_{col}}{c-1}$ (s_B^2)	$\dfrac{MSSB_{column\ means}}{MSSW}$
Within Samples (or within Columns)	SSW$_1$	$c \times (n - 1)$ (v_2)	MSSW$_1 = \frac{SSW_1}{c(n-1)}$ (s_W^2)	$v_1 = c - 1\ v_2 = c(n - 1)$
Total	TSS	$nc - 1$		

Note: make appropriate adjustments for unequal sample sizes.

The ANOVA table simply sets out the calculations required to arrive at the test statistic $F_{v_1, v_2} = \dfrac{s_B^2}{s_W^2}$.

It is essential to set out the null hypothesis, alternative hypothesis, decision rule, etc.

Calculation of Sums of Squares on the calculator

$$\mathbf{TSS} = \sigma^2_{All} \times n \times c \quad \mathbf{SSB_{col}} = \sigma^2_{col} \times n \times c$$
Then, from (10.6) $\mathbf{SSW_1} = \mathbf{TSS} - \mathbf{SSB_{col}}$

10.3 One-way ANOVA with unequal sample sizes

Calculate TSS as $\mathbf{TSS} = \sigma^2_{All} \times$ number of data (the same as for equal sample sizes)

Calculate SSB_{col} directly from the formula $\mathbf{SSB_{col}} = \sum_{j=1}^{j=c} n_j \times (\overline{x}_j - \overline{\overline{x}})^2$.

10.4 Two-way ANOVA

A blocking factor is used to explain variation within treatments (columns/samples)

Sum of Squares for one-way ANOVA: $TSS = SSB_{col} + SSW_1$

Sum of Squares for two-way ANOVA: $TSS = SSB_{col} + \overbrace{SSB_{row} + SSW_2}$ (10.14)

where SSB_{row} is the Sum of Squares between row or 'block' means

Assumptions for two-way ANOVA

1. The probability distributions for all column-block combinations are Normal.
2. The variances of all distributions are equal.
3. There is no interaction between treatment and block variables.

Calculation of Sums of Squares

(Assume all tests are based on c treatments and n blocks: $n \times c$ table).

TSS and SSB_{col} is the same as that for one-way ANOVA. The SSB_{row} is calculated from the row (block) means:

$$SSB_{row} = c \times \sum_{i=1}^{n} (\overline{x}_{i \bullet} - \overline{\overline{x}})^2 \text{ On the calculator, } SSB_{row} = \sigma^2_{row} \times n \times c$$

$$\mathbf{SSW_2} = TSS - SSB_{col} - SSB_{row}$$

Table 10.11   Two-way ANOVA table

SS	df	$MSS = \dfrac{SS}{df}$	$F = \dfrac{MSSB}{MSSW_2}$	Degrees of freedom for sample-F
SSB_{col}	$c-1$	$MSSB_{col} = \dfrac{SSB_{col}}{c-1}$	$\dfrac{MSSB_{col}}{MSSW_2}$	$v_1 = (c-1),\ v_2 = (c-1)(n-1)$
SSB_{row}	$n-1$	$MSSB_{row} = \dfrac{SSB_{row}}{n-1}$	$\dfrac{MSSB_{row}}{MSSW_2}$	$v_3 = (n-1),\ v_2 = (c-1)(n-1)$
SSW_2	$(c-1)(n-1)$	$MSSW_2 = \dfrac{SSW_2}{(c-1)(n-1)}$		
TSS	$nc-1$			

Simultaneous confidence intervals: confidence intervals for multiple comparisons such that the overall probability of Type I error is at most α. There are several formulae for multiple comparisons: in this text formulae are given for Bonferroni (a set M comparisons) and Scheffé (experiment-wise comparisons).

10.5 Observational experiment and designed experiment

Further applications in which one-way and two-way ANOVA are used to test for significant differences between treatments.

10.6 Excel and Minitab

ANOVA calculations are produced by these software packages. The user must interpret the output. The model assumptions may also be checked.

ANOVA tables Excel and Minitab both calculate ANOVA tables: the researcher must then write out the test of hypothesis and use the test statistic (F-ratio) given in the table.

Checking model assumptions

Excel gives all sample mean and variances. The sample variances may be used to test the assumption of equal variances by the F-test. Line graphs for block or treatment variables may be used to flag violations of the 'no interaction' assumption in the two-way additive model.

   Minitab gives:

   (i) Individual confidence intervals for each variable (using the 'pooled' variance or $MSSW$).
  (ii) Box-plots: these provide useful visual indicators for the assumptions of normality and equal variances.
 (iii) Residual plots: residual plots should be random and normally distributed if the model assumptions are satisfied.

CHI-SQUARED TESTS

11

Chapter Objectives

Having carefully studied this chapter and completed the exercises you should be able to do the following

- Define and give examples of contingency tables
- State and give examples of contingency tables to test the following
  1. independence or 'no association' between two categorical variables
  2. that two or more populations are homogeneous
  3. the equality of several proportions
  4. the goodness of fit of a set of data to a given probability distribution.
- State the null hypothesis to test each of the tests in 1, 2, 3 and 4
- Carry out χ^2 tests for each of the tests in 1, 2, 3 and 4 and interpret the results
- Identify sources of association in a contingency table
- Carry out a χ^2 tests in Excel and Minitab and interpret the results

11.1 Introduction

The equality of two proportions was tested in Chapter 7 using the Z-test where the underlying assumption was that the proportions were Normally distributed. All the tests covered in this text so far are called **parametric tests**: the test statistics was calculated (i) on the basis that certain assumptions were true, for example that samples were selected from Normal populations (ii) that measurements were interval or ratio (for example, weights, heights, time etc.). The Chi-squared tests require no assumptions about the underlying probability distribution and hence are called **non-parametric tests**.

The Chi-squared distribution is symbolised χ^2 (where χ, pronounced 'kye', is the Uppercase Greek letter) and will be used for the following tests:

1. Independence (or no association) between two variables
2. The equality of several proportions
3. The homogeneity of two or more populations
4. The 'goodness of fit' of sample data to a given probability distribution

11.2 The χ^2 probability distribution

The critical values for the tests in this chapter will be given by the upper tail of χ^2 probability distribution.

The χ^2 **probability distribution** is defined in terms of one parameter: v, its degrees of freedom. As with all probability distributions, the total area under the probability curve is unity: the values of the χ^2 random variable range from 0 to infinity. The shape of the probability distribution changes for each value of v. As v increases the shape of the χ^2 distribution approaches that of a Normal distribution, with mean $= v$ and variance $= 2v$. See Figure 11.1.

The χ^2 percentage point tables give the critical values of χ^2 for selected upper tail areas (α) and degrees of freedom v. The layout of the χ^2 tables is similar to that of the t-tables: degrees of freedom are given in the left-hand margin (column): the levels of significance (upper tail areas) are given in the top

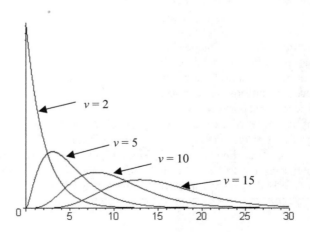

Figure 11.1 χ^2 distribution for 2, 5, 11 and 15 degrees of freedom.

Table 11.1   Extract of χ^2 tables for tail areas α and df 1 to 11

| Areas in the upper tail of the χ^2 distribution | | | | | | | |

df	$\alpha = 0.3$	0.25	0.2	0.15	0.1	0.05	0.025
1	1.0742	1.3233	1.6424	2.0722	2.7055	3.8415	5.0239
2	2.4079	2.7726	3.2189	3.7942	4.6052	5.9915	7.3778
3	3.6649	4.1083	4.6416	5.3170	6.2514	7.8147	9.3484
4	4.8784	5.3853	5.9886	6.7449	7.7794	9.4877	11.1433
5	6.0644	6.6257	7.2893	8.1152	9.2363	11.0705	12.8325
6	7.2311	7.8408	8.5581	9.4461	10.6446	12.5916	14.4494
7	8.3834	9.0371	9.8032	10.7479	12.0170	14.0671	16.0128
8	9.5245	10.2189	11.0301	12.0271	13.3616	15.5073	17.5345
9	10.6564	11.3887	12.2421	13.2880	14.6837	16.9190	19.0228
10	11.7807	12.5489	13.4420	14.5339	15.9872	18.3070	20.4832
11	12.8987	13.7007	14.6314	15.7671	17.2750	19.6752	21.9200

row of the table. Each column gives the critical value of χ^2 at various degrees of freedom. An extract of these tables is given in Table 11.1. For example, the critical value of χ^2 is 12.5916 (highlighted) for 5 % level of significance (a tail area, $\alpha = 0.05$) and $v = 6$ df. The calculation of the degrees of freedom is given in the Worked Examples that follow.

11.3   Contingency tables

The calculations for the χ^2 tests are based on contingency tables: a contingency table is a two-way cross-tabulation for every possible combination of levels for two variables: the sample data for a contingency tables is always *count data* (contingency tables have already been encountered in Chapter 4, section 4.3, Joint, Conditional and Marginal probability distributions). Table 11.2 is an example of a contingency: The two variables are the theory and practical results for a random sample of 600 candidates who

Table 11.2a   Performance of 600 individuals in theory and practise elements of driving test

		Practical driving test		
		A	Pass	Fail
Theory test	100 %	56	49	65
	80 % (pass)	63	120	127
	<80 % (fail)	21	41	58

underwent the driving test. Since each test is graded in three levels the contingency table must have three rows (for theory grades) and three columns (for the practical grades).

In a contingency table the sum of the row totals and the sum of the column totals must each give the overall total, as shown in Table 11.2b:

Table 11.2b   The sum of the Row Totals (and column totals) is the overall total

	Practical driving test				
	A	Pass	Fail	Row totals	
100 %	56	49	65	170	A row total (r)
80 % (pass)	63	120	127	310	
<80 % (fail)	21	41	58	120	Overall total (*Total*)
Column totals	140	210	250	600	

A column total (c)

The numbers in each cell (from sample data) are called the 'observed frequencies (f_o)'. For example, the observed frequency for the first cell is 56; 56 candidates were awarded 100 % in the theory test **and** a grade A in the practical test.

In the following Worked Examples, the term expected frequencies will feature the 'expected frequencies (f_e)' are the numbers expected in each cell if a given null hypothesis were true. If the observed frequencies (f_o) are close to the expected frequencies then the sample data may provide evidence to support the null hypothesis. On the other hand, when observed frequencies are considerably different from the expected frequencies then the stated null hypothesis may not be true. The calculation of the χ^2 test statistic is based on the difference between the observed and expected frequencies. If the χ^2 test statistic is so large that it exceeds the critical value of χ^2 then the null hypothesis is rejected (there is insufficient evidence to support the null hypothesis). See Figure 11.2.

11.4 χ^2 tests for independence (no association)

The first χ^2 test is a test for the independence of two variables which is also described as a test for no association between two variables. For example, in Table 11.2 independence would mean that a candidate's performance in the theory element of the driving test would give no indication (so is

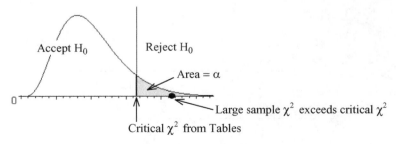

Figure 11.2   All χ^2 in this chapter for significant levels α are right-tailed tests.

independent) of their performance in the practical test. Stated another way, 'their performance in one test has no association with their performance in the other'. The sample data in Table 11.2 will be used as an example to discuss tests for independence.

The null hypothesis. The null hypothesis in tests for independence is **always** 'variables (*name the variables*) are independent' or there is 'no association between the variables' while the alternative hypothesis states that the variables are dependent or there is association.

H_0: The performance in the theory test is independent of performance in the practical test of the driving test (or there is no association between performance in the theory and practical tests for the driving test).
H_1: The performance in the theory test and practical test are dependent (performance in the theory test is related to (associated with) performance in the practical test).

To calculate the test statistic it is necessary to begin by setting up the contingency table of observed frequencies.

The contingency table of observed frequencies

The sample data given in Table 11.2b is a contingency table as described above. The 'expected frequencies' are then calculated from the contingency table of observed frequencies assuming the null hypothesis is true.

Calculation of expected frequencies

The expected frequencies are calculated on the assumption that the null hypothesis is true, hence performance in theory is independent of performance in practical. Independence may be expressed in terms of probability – recall equation (4.5): $P(AB) = P(A)P(B)$ if events A and B are independent.

In Table 11.2b, let $A =$ performance in theory: $B =$ performance in practical.

Then independence for the first cell means (1,1)
$P(100\% \text{ in theory } \textbf{and } A \text{ in practical}) = P(100\% \text{ in theory}) P(A \text{ in practical})$

$$\downarrow \qquad\qquad\qquad \downarrow \qquad \downarrow$$

$$\frac{56}{600} \qquad\qquad \frac{170}{600} \times \frac{140}{600}$$

...where probabilities are estimated by relative frequencies

Obviously $\frac{56}{600} \neq \frac{170}{600} \times \frac{140}{600}$ i.e. $0.093 \neq 0.066$. This difference may be due to sampling error or it may be so large that it indicates that the variables are not independent.

Therefore we will calculate the expected frequency (f_e) for each cell so that independence is satisfied and test for a significant difference between the observed and expected frequencies. The expected frequencies are calculated as follows:

	Practical driving test			
A		Pass	Fail	Row totals
100 %	f_e	49	65	170 = row 1 total
80 % (pass)	63	120	127	310
<80 % (fail)	21	41	58	120
Column totals	140 = column i total	210	250	Total = 600

Let (f_e) = the expected frequency for the first cell.

$P(AB) = P(A)P(B)$ ... for independence

$$\frac{f_e}{600} = \frac{170}{600} \times \frac{140}{600}$$ ... for f_e to satisfy independence

$$\frac{f_e}{Total} = \frac{row\ total}{Total} \times \frac{column\ total}{Total}$$ ... in general terms, f_e for any cell ....

$$f_e = \frac{row\ total \times column\ total}{Total}$$ ... solving for f_e

The expected frequency for each cell, assuming independence, is "(row total) multiplied by (column total) divided by the (overall total)."

$$f_e = \frac{(row\ total) \times (column\ total)}{Total} = \frac{r \times c}{Total} \tag{11.1}$$

The expected frequencies for Table 11.2b are given in Table 11.2c:

Table 11.2c   Calculation of the expected frequencies for the sample in Table 11.2a

Theory test	Practical driving test			
	A	Pass	Fail	Row totals
100 %	$f_o = 56$ $f_e = (170 \times 140)/600$ $= 39.67$	$f_o = 49$ $f_e = (170 \times 210)/600$ $= 59.50$	$f_o = 65$ $f_e = (170 \times 250)/600$ $= 70.83$	170
Pass	$f_o = 63$ $f_e = (310 \times 140)/600$ $= 72.33$	$f_o = 120$ $f_e = (310 \times 210)/600$ $= 108.50$	$f_o = 127$ $f_e = (310 \times 250)/600$ $= 129.17$	310
Fail	$f_o = 21$ $f_e = (120 \times 140)/600$ $= 28.00$	$f_o = 41$ $f_e = (120 \times 210)/600$ $= 42.00$	$f_o = 58$ $f_e = (120 \times 250)/600$ $= 50.00$	120
Column totals	140	210	250	600

The χ^2 test statistic

If the null hypothesis is true then the difference between observed and expected frequencies: $(f_o - f_e)$, should be small (allow small differences due to sampling error).

Therefore, *for each cell* calculate $(f_o - f_e)^2$ and divide the result by f_e

$$\dfrac{(f_o - f_e)^2}{f_e} \longleftarrow \left\{ \text{Should be small if } H_0 \text{ is true} \right.$$

$$\left\{ \begin{array}{l} \text{Division by } f_e \text{ gives a measure of the relative} \\ \text{importance of the difference } (f_o - f_e)^2 \end{array} \right.$$

The χ^2 test statistic is the sum of the $\dfrac{(f_o - f_e)^2}{f_e}$ for each cell.

The χ^2 test statistic, $T = \displaystyle\sum_{\text{all cells}} \left(\dfrac{(f_o - f_e)^2}{f_e} \right)$ (11.2)

The calculation of the test statistic is given in Table 11.2d.

Table 11.2d Calculation of the χ^2 test statistic

f_o	f_e	$(f_o - f_e)$	$\dfrac{(f_o - f_e)^2}{f_e}$
56	39.67	16.33	6.73
49	59.50	−10.50	1.85
65	70.83	−5.83	0.48
63	72.33	−9.33	1.20
120	108.50	11.50	1.22
127	129.17	−2.17	0.04
21	28.00	−7.00	1.75
41	42.00	−1.00	0.02
58	50.00	8.00	1.28
	Totals	0.00	14.57

$$T = \sum_{\text{all cells}} \frac{(f_o - f_e)^2}{f_e}$$

The individual χ^2's for each cell are summed (column 4) to give the test statistic (T).

$$\text{T(sample } \chi^2) = \sum_{\text{all cells}} \frac{(f_o - f_e)^2}{f_e} = \frac{(56 - 39.67)^2}{39.67} + \frac{(49 - 59.50)^2}{59.50} + \cdots + \frac{(58 - 50.00)^2}{50.00}$$

$$= 14.572$$

To determine the point at which the test statistic becomes too large to attribute the differences to chance, the test statistic is compared to the critical values for the upper tail in the χ^2 distribution that has $(r - 1)(c - 1)$ degrees of freedom, where r = number of rows and c = number of columns in the Table.

If the test statistic (sample χ^2) falls in the critical region for a given level of significance, there is insufficient evidence to support H_0, hence the null hypothesis is rejected.

The test of hypothesis is set out in the usual step by step format with the calculations outlined in Chart 11.1.

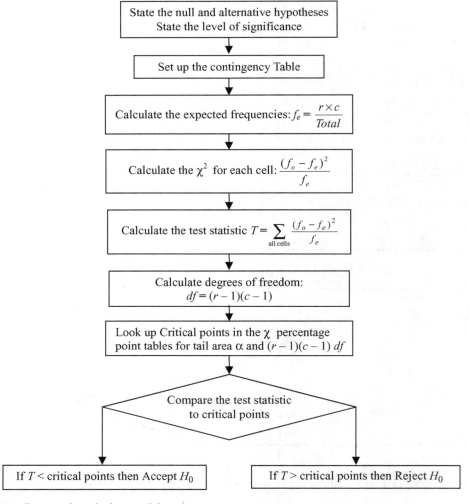

Chart 11.1   Steps in the calculation of the χ^2 test statistic.

Note

1. The sum: $\sum (f_o - f_e)$ is always zero. This is illustrated in column 3, Table 11.2d.
2. The minimum value of any χ^2 test statistic is zero. This occurs when the observed frequencies are identical to the expected frequencies.
3. The greater the difference between f_o and f_e, the greater the value of the χ^2 test statistic.

4. If the expected frequency is small, particularly if $f_e < 1$, the value $\dfrac{(f_o - f_e)^2}{f_e}$ may be unduly inflated. A general rule of thumb is that f_e should be 5 or more. However many statisticians will work with $f_e \geq 1$. If $f_e < 1$ then it is necessary to merge rows (or columns).

The test is demonstrated through Worked Example 11.1.

Worked Example 11.1: χ^2 Test for no association (independence)

(a) Test the claim that there is no association between a candidate's performance in the theory and practical elements of the driving test using the random sample of 600 given in Table 11.2.
(b) If association is found identify the possible source of the association in the contingency table.
(c) Give an example of a contingency table where there is no association between theory and practical results.

Solution

(a) Set out the test and the calculations as summarised in Chart 11.1.

H_0: there is no association between the performance of individuals in the practical and theory elements of the driving test.
H_1: there is an association
$\alpha = 5\,\%$ and $0.5\,\%$

Test Statistic: $\chi^2 = \displaystyle\sum_{\text{all cells}} \dfrac{(f_o - f_e)^2}{f_e}$

Decision Rule:
Reject H_0 if the sample χ^2 exceeds the critical χ^2 value from the χ^2 percentage point tables
The degrees of freedom, $v = (r - 1)(c - 1) = (3 - 1)(3 - 1) = 4$

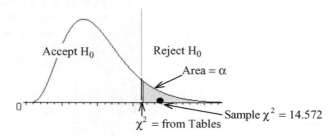

From tables: for $\alpha = 0.05$ the critical $\chi^2 = 9.4877$: for $\alpha = 0.005$ the critical $\chi^2 = 14.8602$

Calculations: calculate the test statistic $T = \displaystyle\sum_{\text{all cells}} \dfrac{(f_o - f_e)^2}{f_e} = 14.572$ as outlined in Tables 11.2c and 11.2d.

Conclusion: The results are not conclusive: when $\alpha = 0.05$, the sample χ^2 (14.572) falls in the reject region, so the null hypothesis is rejected and we conclude that there is association between practical and theory results.
But when $\alpha = 0.005$ the sample χ^2 (14.572) falls in the acceptance region, hence conclude that there is no association at this level of significance.
The p-value is between 0.05 and 0.005.

(b) To identify the likely association in the table (at $\alpha = 0.05$), look at the χ^2 values calculated for each cell in Table 11.2d. The largest value of χ^2 is the first cell where $\dfrac{(f_o - f_e)^2}{f_e} = 6.73$. In this cell the observed frequency, 58, is much larger than the expected frequency, $f_e = 30.67$. Hence there is an association here – those who do well in the theory test tend to do well in the practical, or vice-versa.

(c) There are numerous examples of tables showing no association: two examples are given in Tables 11.3a and 11.3b.

Table 11.3a   This table shows no association between theory and practical tests

		Practical driving test		
		A	Pass	Fail
	100 %	80	80	0
Theory test	Pass	80	80	0
	Fail	80	80	0

In Table 11.3a, 160 candidates were awarded 100 % in the theory test, 160 candidates were awarded a Pass and 160 failed. But irrespective of how they performed in the theory, 80 were awarded an A, 80 were awarded a Pass and none failed the practical. Hence there is no association between theory grade and practical grade. Alternatively, for **each practical grade** the numbers who are awarded 100 % pass and fail in the theory test is the same.

Similarly, in Table 11.3b the pass rate in the practical is the same, 50 % grade A, 25 % Pass and 25 % fail – **irrespective of theory grade**. Alternatively, for **each practical grade**, 2/3 of candidates are awarded 100 % in the theory test, 1/6 is awarded a pass and 1/6 of the candidates failed.

Table 11.3b   This table shows no association between theory and practical tests

		Practical driving test		
		A	Pass	Fail
	100 %	200	100	100
Theory test	Pass	50	25	25
	Fail	50	25	25

Row and column profiles

When numbers in groups are not as convenient as those in Table 11.3a and 11.4b, calculate the row profiles, that is, percentage of each row total that falls into the various column categories (similarly for column) totals. For example, Table 11.3c:

Table 11.3c   In this table it is easier to see no association by calculating the row profiles

| | | Practical driving test | | | |
		A	Pass	Fail	Row total
	100 %	90	45	30	165
Theory test	Pass	60	30	20	110
	Fail	150	75	50	275

| | | Practical driving test | | | |
		A	Pass	Fail	Row %
	100 %	55 %	27 %	18 %	100 %
Theory test	Pass	55 %	27 %	18 %	100 %
	Fail	55 %	27 %	18 %	100 %

Skill Development Exercise SK11_1: χ^2 Test for independence (no association)

(a) Test the claim that there is no association between a candidate's performance in Accounting and Mathematics given in Table 4.1, Chapter 4.
(b) If association is found, identify where the association exists in the Table.
(c) Give an example of a Table where there is no association between Mathematics and Accounting results.

Table 4.1   Examination results for students taking Mathematics and Accounting

| | | Mathematics | | | |
		Merit	Pass	Fail	Row totals
	Merit	120	10	0	130
Accounting	Pass	100	190	30	320
	Fail	20	10	20	50
	Column totals	240	210	50	500

Answers

(a) In Table SK11_1.1 the first and second entry in each cell is the observed and expected frequency.
The cell $\chi^2 = \dfrac{(f_o - f_e)^2}{f_e}$ is the third cell entry.

Table SK11_1.1   Calculation of expected frequencies and χ^2 test statistic

		Mathematics			Row totals
		Merit	Pass	Fail	
	Merit	$f_o = 120$ $f_e = 62.20$ $\chi^2 = 53.169$	$f_o = 10$ $f_e = 54.60$ $\chi^2 = 36.432$	$f_o = 0$ $f_e = 13.00$ $\chi^2 = 131.000$	130
Accounting	Pass	$f_o = 100$ $f_e = 153.60$ $\chi^2 = 18.704$	$f_o = 190$ $f_e = 134.40$ $\chi^2 = 23.001$	$f_o = 30$ $f_e = 32.00$ $\chi^2 = 0.125$	320
	Fail	$f_o = 20$ $f_e = 24.00$ $\chi^2 = 0.667$	$f_o = 10$ $f_e = 21.00$ $\chi^2 = 5.762$	$f_o = 20$ $f_e = 5.00$ $\chi^2 = 45.000$	50
Column totals		240	210	50	500

The test statistic, $\chi^2 = \sum\limits_{\text{all cells}} \dfrac{(f_o - f_e)^2}{f_e} = 195.86$. The critical χ^2 (5 % and 4 df) is 9.4877
The null hypothesis is rejected.

(b) There are associations in the Table, except for two groups 'a merit in maths and a fail in accounting' and 'a fail maths and a pass accounting.

(c) Give examples similar to those in Worked Example 11.1.

11.5 χ^2 test for homogeneous populations

Populations are described as homogeneous if they consist of the same subgroups (or strata) with the same percentage of the population in each stratum. For example, in Table 11.4, the numbers of male

Table 11.4   Homogeneous populations: percentage distribution of male applicants per faculty is the same as the percentage distribution of female applicants per faculty

	Faculty		
	Engineering	Science	Art
Male	200	100	100
Female	50	25	25

and female applicants for courses in three faculties are given. In this Table, 50 % of male applicants apply to the faculty of Engineering, 25 % to Science and 25 % to Arts. This is exactly the same as the percentage distribution for female applicants to each faculty. Therefore the distributions of male and female applicants in the three faculties are described as 'homogeneous'.

Table 11.4 can also be described as a Table in which there is no association between choice of faculty and gender of applicant. It follows that the test for homogeneity of populations may be described as a test for no association (or independence).

Worked Example 11.2: Test that two populations are homogeneous

Three hundred prospective students are asked about their intended field of study. The results are given in Table 11.5:

Table 11.5 Applicants for courses in three faculties classified by gender

	Faculty		
	Engineering	Science	Art
Male	37	41	44
Female	35	72	71

(a) Test the hypothesis that the populations are homogeneous (there is no association between gender and faculty) at the 10 % level of significance.
(b) Would your conclusion be different at the 5 % level of significance?

Solution

(a) Set out the test of hypothesis:

H_0: Both populations follow the same distribution. no association between faculty and gender.
H_1: The population distributions are different – there is association between gender and faculty
   $\alpha = 10\%$

Test Statistic: $\chi^2 = \sum_{\text{all cells}} \dfrac{(f_o - f_e)^2}{f_e}$

Decision Rule:
Reject H_0 if the sample χ^2 exceeds the critical χ^2 from the Tables for $v = (r-1)(c-1)df$
The degrees of freedom, $v = (r-1)(c-1) = (2-1)(3-1) = 2$
Hence, reject H_0 if sample χ^2 exceeds 4.605

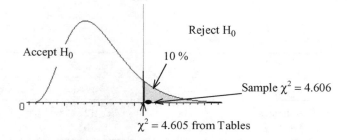

Calculations: Calculate the expected frequencies by $f_e = \dfrac{r \times c}{Total}$ (11.1).

	Faculty			Row totals
	Engineering	Science	Art	
Male	$f_o = 37$ $f_e = 29.28$	$f_o = 41$: $f_e = 45.95$	$f_o = 44$ $f_e = 46.77$	122
Female	$f_o = 35$ $f_e = 42.72$	$f_o = 72$ $f_e = 67.05$	$f_o = 71$ $f_e = 68.23$	178
Column totals	72	113	115	300

$$\chi^2 = \sum_{\text{all cells}} \frac{(f_o - f_e)^2}{f_e} = 2.035 + 0.54 + 0.164 + 1.395 + 0.366 + 0.112 = 4.606$$

Conclusion: The sample $\chi^2 = 4.606$, falls in the rejection region (only just!). Therefore reject the null hypothesis at the 10 % level of significance. Conclude that the populations are not homogeneous. The percentage of males who intend to apply to each faculty is different from the percentage of females who intend to apply to each faculty.

For example, for Engineering, the actual (or observed) number of males that intend to apply is 37, greater than the numbers expected (29), but the observed numbers of females is 35, less then the 43 expected, assuming H_0 is true.

(b) At the 5 % level of significance, the critical χ^2 value $= 5.991$. Hence, according to this decision rule, the test statistic (4.606) falls in the acceptance region. The conclusion is that the percentage of students that intend to apply to each faculty is the same for both genders.

11.6 χ^2 tests for the equality of several proportions

In Chapter 7 the Z-test was used to test the equality of two proportions. In this chapter, the χ^2 test is used to test the equality of two or more proportions. A plausible explanation is given for the χ^2 test statistic and its calculation.

Table 11.6a gives the test results for three random samples of candidates who took their test at Centres A, B and C. In this case we are interested in testing whether the proportion passing is the same at each Centre.

Table 11.6a Numbers of candidates who pass the driving test at various attempts

	Test Centre		
	Centre A	Centre B	Centre C
Number of candidates	200	126	74
Number of passes	130	88	62

Table 11.6b   Contingency table for the test of equality of proportions

| | Test Centre | | | |
	Centre A	Centre B	Centre C	Row totals
Numbers who passed	130	88	62	280
Numbers who failed	70	38	12	120
Number of candidates (column totals)	200	126	74	400

The null hypothesis

For this test the null hypothesis is *always* that all proportions are equal.

The contingency table of observed frequencies

Table 11.5b is not a contingency table: it does not contain every possible outcome (number of 'fails' is not included), hence the rows and columns do not add up to the overall total. The completed contingency table, with row and column totals, is given in Table 11.6b. The 'expected frequencies' are then calculated from the contingency table by $f_e = \dfrac{r \times c}{Total} \ldots (11.1)$.

Explanation for the formula for expected frequencies

The expected frequencies are calculated on the assumption that the null hypothesis is true (that all proportions are equal). Hence the expected proportions that pass (and fail) should be the same for all Centres.

The common proportion that pass is estimated from the data; $\dfrac{\text{Total passes}}{\text{Total}} = \dfrac{280}{400} = 0.7.$

The proportion that fail is estimated from the data; $\dfrac{\text{Total fails}}{\text{Total}} = \dfrac{120}{400} = 0.3.$

(Or, quoted as a percentage, 70 % pass, 30 % fail.)
Hence the expected numbers (frequencies) that pass and fail from each Centre are obtained by calculating 70 % and 30 % of the number in that Centre who took the test, respectively. See Table 11.6c.

Table 11.6c   Calculate expected frequencies

| | Test Centre | | | |
	Centre A	Centre B	Centre C	Row totals
Numbers passed	$f_o = 130$ $f_e = 200 \times 0.7 = 140$	$f_o = 88$ $f_e = 126 \times 0.7 = 88.2$	$f_o = 62$ $f_e = 74 \times 0.7 = 51.8$	280
Numbers failed	$f_o = 70$ $f_e = 200 \times 0.3 = 60$	$f_o = 38$ $f_e = 126 \times 0.3 = 37.8$	$f_o = 12$ $f_e = 74 \times 0.3 = 22.2$	120
Column totals	200	126	74	400

Careful examination of the calculations in Table 11.6c reveals the expected frequency for each cell is '(row total) multiplied by (column total) divided by the (overall total)' This is the formula for expected frequency given in $f_e = \dfrac{r \times c}{Total}$ ... (11.1).

Hence, applying formula (11.1) to the data in Table 11.6b gives

Table 11.6d   Use formula (11.1) to calculate expected frequencies

| | Test Centre | | | |
	Centre A	Centre B	Centre C	Row totals
Expected numbers passed	$f_e = \dfrac{280 \times 200}{400} = 140$	$f_e = \dfrac{280 \times 126}{400} = 88.2$	$f_e = \dfrac{280 \times 74}{400} = 51.8$	280
Expected numbers failed	$f_e = \dfrac{120 \times 200}{400} = 60$	$f_e = \dfrac{120 \times 126}{400} = 37.8$	$f_e = \dfrac{120 \times 74}{400} = 22.2$	120
Column totals	200	126	74	400

(The same results as in Table 11.6c above).

The calculation of the χ^2 test statistic for the data in Table 11.6a is given in Table 11.6e:

Table 11.6e   Calculations for the χ^2 test statistic

f_o	f_e	$(f_o - f_e)$	$(f_o - f_e)^2$	$\dfrac{(f_o - f_e)^2}{f_e}$
130	140.0	−10.0	100	0.71
88	88.2	−0.2	0.04	0.00
62	51.8	10.2	104.04	2.01
70	60.0	10.0	100	1.67
38	37.8	0.2	0.04	0.00
12	22.2	−10.2	104.04	4.69
Totals		0.00		9.08

The χ^2 test for the equality of several proportions is set out, step-by-step in Worked Example 11.3.

$$T = \sum_{all\,cells} \frac{(f_o - f_e)^2}{f_e}$$

Worked Example 11.3: χ^2 test for the equality of several proportions

Table 11.6a gives the numbers of candidates who took the driving test and the numbers who succeeded at Test Centres A, B and C.

(a) Calculate the proportions that pass at each Test Centre.

(b) Test the claim that the proportion of candidates that pass is the same at each Centre.

(c) If the proportions are different, determine which proportion(s) is different.

(d) Construct a 95 % confidence interval for the difference between the proportion that pass the test in Centre C and B. Interpret the interval. Does the interval support the sources of association identified in (c)?

Solution

(a) The proportions that pass are as follows:

	Test Centre			
	Centre A	Centre B	Centre C	Totals
Numbers of passes	130	88	62	280
Number of candidates	200	126	74	400
Proportion of passes (sample proportions)	0.65	0.70	0.84	0.70

(b) H_0: The proportions that pass are the same for each Centre.
H_1: At least one proportion is different
$\alpha = 5\%$

Test Statistic (sample χ^2): $T = \sum\limits_{\text{all cells}} \dfrac{(f_o - f_e)^2}{f_e}$

Decision Rule:
Reject H_0 if the sample χ^2 exceeds the critical χ^2 from the Tables.
The degrees of freedom, $v = (r - 1)(c - 1) = (2 - 1)(3 - 1) = 2$.
Reject H_0 if sample χ^2 exceeds 5.9915.

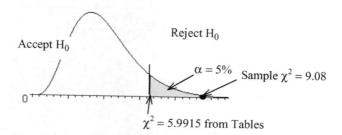

Calculations: From Tables 11.6d and 11.6e, the sample $\chi^2 = \sum\limits_{\text{all cells}} \dfrac{(f_o - f_e)^2}{f_e} = 9.08$

Conclusion: The sample $\chi^2 = 9.08$ falls in the reject region for H_0. Therefore reject the null hypothesis at the 5 % level of significance. Conclude that in at least one Test Centre the proportion of passes is different from the others.

(c) **Which proportion(s) is different?**
One way to identify the proportions that are different is to examine the χ^2's for individual cell. If a cell χ^2 is large, then there is a possible relationship between the variables. In Table 11.6e,

the individual cell χ^2's are given in the last column. The two largest individual cell χ^2's are 4.69 and 2.01: both from Centre C, where the observed number of passes (62) is higher than the expected number of passes (51.8), and consequently the observed number of fails (12) is lower than expected (22.2).

	f_o	f_e	$(f_o - f_e)$	$(f_o - f_e)^2$	$\dfrac{(f_o - f_e)^2}{f_e}$
A	130	140.0	−10.0	100	0.71
B	88	88.2	−0.2	0.04	0.00
C	62	51.8	10.2	104.04	2.01
A	70	60.0	10.0	100	1.67
B	38	37.8	0.2	0.04	0.00
C	12	22.2	−10.2	104.04	4.69
Totals	400	400	0.00		9.08

$$\sum (f_o - f_e) = 0$$

In this case the 95 % confidence interval for the difference in the proportions passing in Centres B and C is

$$p_C - p_B \pm z_{\alpha/2} \sqrt{\frac{p_C(1 - p_C)}{n_C} + \frac{p_B(1 - p_B)}{n_B}}$$

$$(0.85 - 0.70) \pm 1.96 \sqrt{\frac{(0.84)(0.16)}{74} + \frac{(0.7)(0.3)}{126}}$$

$$0.15 \pm 0.114$$

$$0.035 < \pi_C - \pi_B < 0.264$$

So, we are 95 % confident that the proportion that passes the test in Centre C is greater than the proportion that passes in Centre B by between 0.035 and 0.264 (between 3.5 % and 26.4 %).

The confidence interval does not span zero, hence there is a difference between the proportions passing in Centres B and C.

This confidence interval confirms the answer in (c) where Centre C appeared to have a larger proportion of passes than the others. The difference between the proportions A and C is even greater than between B and C, hence C is also different from A. If in doubt, construct confidence intervals for the difference between proportions for C and A.

Similarly it can be shown that there is no significant difference between A and B at the 5 % level of significance.

Skill Development Exercise SKD11_2: χ^2 tests for differences between proportions

A survey is carried out to determine whether the proportion of employees in favour of a new flexible working day is the same for all age groups. The results of the survey are as follows:

Table SKD11_2.1 Results of a survey on views of employees on flexible working day by age group

Age group	18–24	25–35	35–50	Over 50
No. in favour	20	62	40	40
No. surveyed	52	112	86	120

(a) Test the claim that there is no difference between the proportions in each age group that are in favour of a flexible working say at the 5 % level of significance.
(b) If the null hypothesis in (a) is rejected, identify proportions that are likely to be different.
(c) Construct a 95 % confidence interval for the difference in proportions between 18–14 and 25–35 year olds that are 'in favour' of a new flexible working day.

Outline Solution

(a) First, set up the contingency table to include the numbers 'not in favour' so that all row and column total. See Table SKD11_2.2.

Table SKD11_2.2 Contingency table for views of employees on flexible working day by age group

Age group	18–24	25–35	35–50	Over 50	Column totals
No. 'in favour'	20	62	40	40	162
No. 'not in favour'	32	50	46	80	208
No. surveyed	52	112	86	120	370

Hence the calculation of expected frequencies $f_e = \dfrac{(\text{row total}) \times (\text{column total})}{T}$

Test statistic $\chi^2 = \displaystyle\sum_{\text{all cells}} \dfrac{(f_0 - f_e)^2}{f_e}$.

$$= 0.366 + 3.426 + 0.146 + 2.993 + 0.262 + 2.669 + 0.114 + 2.331$$

$$= 12.278$$

The null hypothesis is rejected at the 5 % level of significance.

(b) From the individual χ^2's and the respective sample proportions 'in favour': 0.38; 0.55; 0.47; 0.33. The 25–35 and over 50 age groups are the most likely to be significantly different from each other. Row profiles are also useful to identify sources of association. To confirm that differences are significant, it is necessary to test appropriate hypotheses and/or construct confidence intervals for differences.

(c) 95 % confidence interval for the difference between the proportions of 18–24 and 25–35 year olds is 0.0091 to 0.3309.

Progress Exercises 11.1: χ^2 tests

1. (a) Explain and give an example of the following terms:
   (i) a contingency table
   (ii) observed frequency
   (iii) expected frequency
   (iv) degrees of freedom for a χ^2 test.

   (b) In the pharmaceutical industry, process control and quality assurance are of paramount importance. The successful implementation of a company's quality assurance depends on all staff in the organisation being conversant with quality assurance practices. At the end of their first six months in the company a group of 400 new staff are required to attend training courses on statistical process control (SPC) theory and methods. Attendees were graded on their performance at work and on the training course – see Table PE11.1 Q1.

   File: Quality work

   Table PE11.1 Q1   Grades on performance at work and on Statistical Process Control theory and methods

Performance at work	Performance on course		
	Above average	Average	Below average
Above average	24	32	38
Average	78	76	46
Below average	46	40	20

   (i) Test the hypothesis that there is no association between performance rates at work and on the course.
   (ii) If you find that there is a significant association, explain the likely sources of the associations.
   (iii) Suggest further test(s) that could be used to confirm the source of association in a table.

2. (a) Explain why the following Table shows 'no association' or independence between rows and columns:

   Table PE11.1 Q2a   No association or independence

	A	C	D	E
Group 1	40	80	120	160
Group 2	10	20	30	40

(b) Evidence from Labour Force Surveys indicates that the numbers employed in various sectors has changed in recent years. The county manager in a rural county council is particularly concerned about a shift in employment in between sectors. The results of a preliminary search on records from randomly selected areas in the county on numbers employed in agriculture, fisheries and forestry (AFF) and construction are given in Table PE11.1 Q2b:

Table PE11.1 Q2b   Survey of numbers employed in AFF and construction   File: AFF

	2000	2001	2002	2003	2004
AFF	22	19	18	17	76
Construction	26	32	36	44	138
Totals	48	51	54	61	214

(i) State the null and alternative hypothesis for a χ^2 test to determine whether there is a difference in the numbers employed in AFF and in construction from 2000 to 2004 at the 5 % level of significance.

(ii) The test statistic for this data was calculated as 3.968. Use this test statistic to test the hypothesis stated in (i).

(iii) Calculate the confidence interval for the difference between the proportions employed in construction in 2000 and 2004.

Based on the results above can you conclude that the numbers employed in these sectors have changed?

3. (a) A newsagent surveys 500 customers on their choice of daily paper. The survey results are summarised in Table PE11.1 Q3a.

Table PE11.1 Q3a   Choice of daily paper for 500 customers   File: Daily choice

	Male		Female		
	20–40	Over 40	20–40	Over 40	Totals
Business News	12	48	26	24	110
Squared Times	32	77	43	78	230
Sun & Stars	46	65	29	20	160
Totals	90	190	98	122	500

A χ^2 test is carried out to test whether there is any association between the age and gender of customers and choice of daily paper.

In Table PE11.1 Q3b, the expected frequencies are printed underneath the observed frequencies and the χ^2 contribution is printed underneath the expected frequencies.

Table PE11.1 Q3b   Expected frequencies and the χ^2 contribution

	Male		Female		
	20–40	Over 40	20–40	Over 40	Totals
Business News	12	48	26	24	110
(expected frequencies)	??	41.8	21.56	26.84	??
(χ^2 contribution)	3.076	??	0.914	0.301	
Squared Times	32	77	43	78	230
(expected frequencies)	41.4	87.4	45.08	??	
(χ^2 contribution)	2.134	1.238	0.096	??	
Sun & Stars	46	65	29	20	160
(expected frequencies)	28.8	60.8	31.36	39.04	
(χ^2 contribution)	10.272	0.29	0.178	9.286	
Totals	??	??	??	??	

(i) Fill in the missing cells indicated by (??).
(ii) State the null and alternative hypothesis to test the claim that males and females in different age brackets have different preferences for daily papers than males.
(iii) From the information given in Table 11.3b, confirm that the test statistic is 37.321. Use this test statistic to carry out the test at the 5 % level of significance.
(iv) State your conclusions carefully. If the null hypothesis is rejected, indicate which variables are likely to be associated.
(b) Carry out a χ^2 test for the equality of the proportions that read the Business News. Briefly describe the results as you would to a non-statistician.
4. (a) State the null and alternative hypotheses for equality of proportions. Explain the term 'expected frequency' for proportions.
(b) A pilot survey of 400 workers in the hospitality industry revealed that 130 were part-time and temporary. The breakdown by gender is given in Table PE11.1 Q4:

Table PE11.1 Q4   Survey of 400 workers in the hospitality industry

	Male	Female
Total surveyed	220	180
Part-time temporary	63	69

(i) It is claimed that there is a difference between the proportions of male and female part-time temporary workers. State the null and alternative hypotheses to test this claim.
(ii) Calculate the expected frequency for the test in (i).
(iii) Carry out a Chi-squared test ($\alpha = 5\%$) to ascertain whether there is a difference between the proportion of males and females who work in part-time temporary employment.

5. **See question 4.**
   (a) Calculate a 95 % confidence interval for the difference in proportions of male and female part-time temporary workers.
   (b) Test the hypothesis that the proportion of females exceeds the proportion of males who work part-time temporary.
   Explain the results in (a) and (b) with those in question 4.
6. (a) Give an example of a contingency table that shows (i) no relationship (ii) a relationship between two variables.
   (b) Two materials are tested to determine whether they can withstand extremely high temperatures. Specimens of each material are tested at a temperature of 500°C. The number of specimens that disintegrated were recorded in Table PE11.1 Q6.

Table PE11.1 Q6   Specimens that disintegrate at 500°C

	Material A	Material B
Disintegrated	26	41
Total	40	50

   (i) Use the χ^2 test to test the claim that the proportion disintegrated is the same for both materials at the 5 % level of significance. State your conclusions carefully.
   (ii) Test the claim that the proportion disintegrated is greater for material B at the 5 % level of significance.
   Compare the conclusions from the tests in (i) and (ii).
7. The number of public housing units and total housing units completed in each quarter of 2003 is given in Table PE11.1 Q7.

Table PE11.1 Q7   Total and public completed housing units in 2003   File: Public private housing

2003	Quarter 1	Quarter 2	Quarter 3	Quarter 4
Public housing units	1155	1470	1764	1744
Total housing units	14864	17379	19904	22805

Source: CSO.

Write down the proportion of public housing units completed each quarter.
A politician claims that the differences in proportions are simply due to chance and that the proportion of completed public housing units is the same for each quarter.
(a) State the null and alternative hypotheses to test the politician's claim.
(b) Set up a contingency for the test.
(c) Calculate the expected frequencies.
(d) Test the claim at the 5 % level of significance.
Give a verbal, not statistical, report on the results of the test.

8. A Chi-squared test is carried out to ascertain whether there is an association between hours sleep (0–2, 3–5, 6–8 and over 8 hours) and the number of mistakes made in a series of simple mathematical calculations (1–3, 4–6, 7–9 and 10 or more mistakes). The sample Chi-square was calculated as 15.36. Is this result significant?

9. Table PE11.1 Q9 summarises the examination results and number of hours per week that students spend in part-time employment:

Table PE11.1 Q9   Examination results and part-time hours worked   File: Exams Part time

| | Part-time hours work per week | | | | |
Exam result	0	1–5	6–10	>10	Totals
Distinction	32	14	11	5	62
Credit	18	20	12	8	58
Pass	31	28	21	34	114
Fail	10	8	12	21	51
Column totals	91	70	56	68	285

(a) Test the claim that there is no association between exam results and the number of part-time hours worked per week.

(b) Calculate a 90 % confidence interval between the percentage of distinctions for those who work between 1 and 10 hours and those who work more than 10 hours weekly. Interpret this interval. How does this result relate to that in (i)?

10. A company compares the quality of raw material from two suppliers. The materials are classified as 'good', 'acceptable' or 'reject'.

Explain the statement 'the quality of raw material from the two suppliers is homogeneous'. The results for random samples from each supplier are summarised in Table PE11.1 Q10:

Table PE11.1 Q10   Quality of raw material from two suppliers

	Good	Acceptable	Reject	Totals
Supplier A	390	34	76	500
Supplier B	238	20	42	300

Test the claim that there is no association between the quality of raw materials and the supplier. State your conclusions carefully.

11. The number of barrels of draught G sold in three bars on Monday, Tuesday and Wednesday are given in Table PE11.1 Q11:

Table PE11.1 Q11   Number of barrels of draught G sold

	Monday	Tuesday	Wednesday
Bar AA	10	13	10
Bar Ba	10	12	34
Bar X	14	19	44

(a) Does this data provide evidence to support the claim that there is no association between bars and days? Explain you conclusion verbally.

(b) Would the result of your analysis be different if the volume of ale sold was recorded in litres instead of barrels (assume 40 litres per barrel)?

12. The importation of raw cows' milk has been recorded since 1998. Table PE11.1 Q12 summarises the volume of milk produced domestically and the volume imported (the numbers have been rounded to the nearest million litres).

Table PE11.1 Q11 — Volume of cows, milk (m litres) produced domestically and imported | File: Milk |

Year	2002	2003	2004	2005
Domestic	5033	5157	5116	4913
Imported	279	349	377	550

Source: CSO.

(a) Set up a contingency table for the data.

(b) Carry out a test to determine whether the proportion of imported milk is the same for the years 2002 to 2005. Give a verbal description of your findings.

Would your results be affected if volumes were given in litres instead of thousands of litres?

13. The Office for National Statistics reported that the Labour Force Survey (Winter 2005) reported '3.1 % of women employees interviewed and 2.4 % of the men had been absent from work due to illness for at least one day in the previous week'. A large banking organisation carried out their own follow-up survey by selecting a random sample of weekly staff records during the period October to December. See Table PE11.1 Q12.

Table PE11.1 Q13 — Numbers absent for at least one day in a selected week | File: Absent age |

Age	Total Men	Men Absent	Total Women	Women Absent
16–24	45	3	68	7
25–34	56	3	65	8
35–49	45	4	74	8
50–59	42	5	34	5
60–65	4	0	5	1

(a) Calculate the percentage absent for each age group, men and women. Comment on these figures. Do you think the differences could be due to chance?

(b) Set up a contingency table for the data in Table PE11.1 Q13. Calculate the expected numbers of absentees. Why is it necessary to combine the two older age groups?

(c) Test whether there is a significant difference between the proportions absent for all age brackets and gender.

Give a verbal description of the results.

(d) Calculate the 95 % confidence interval for the difference between the proportion of male and female absentees. Interpret this result and compare it with the result in (c).

14. The distances travelled to work by the residents of neighbourhoods from Brentwood (B 005A) and London (Lon 001A) was given in Table PE1.4Q5 (File: Distance to work). Test the claim

that there is no association between neighbourhood and the distance travelled to work for those who travel from less than 2 km to 40 km or more. Would you describe the distribution of distance travelled to work for neighbourhoods Brentwood (B 005A) and London (Lon 001A) as homogeneous?

15. Table PE7.2 Q1 (File: Bus) gave the survey results on the views of commuters on a certain bus route. Carry out a χ^2 test to determine whether some of the groups of commuters feel differently about the bus services. Describe your results verbally, indicating the source of any association.

11.7 Goodness of fit tests

Recall that in Chapters 9 and 10 hypotheses testing and calculating confidence intervals were valid provided that the data is 'Normally distributed'. The χ^2 goodness of fit tests are used to test whether a set of data (a sample) fits a stated probability distribution. In this section we will test whether sample data fits a uniform distribution, an empirical probability distribution or a Poisson probability distribution. The assumption of Normality is so important that numerous tests for Normality have been devised, many more suitable than the χ^2 tests.

The χ^2 goodness of fit tests differs from the previous tests in the following ways:

1. The expected frequencies are calculated assuming the probability distribution stated in the null hypothesis is true.
2. The degrees of freedom are given by formula (11.3).

Degrees of freedom for goodness of fit

$$v = c - 1 - k \qquad (11.3)$$

where c is the number of classifications; k is the number of population parameters that must be estimated from the data in order to calculate the expected frequencies

---

11.7.1 Goodness of fit for empirical probability distributions

The L‿earner driving school claims that 80 % of their students pass the test on the first attempt, 15 % pass on the second attempt and the remaining 5 % on the third attempt. This claim defines the probability distribution given in Table 11.8a:

Table 11.7a Attempts to pass driving test as claimed by L‿earner school

	Pass driving test on		
	First attempt	Second attempt	Third attempt
Percentage passing	80	15	5
Probability of passing	0.8	0.15	0.05

To test this claim a random sample of 160 test candidates who attended the L_earner school is selected. The attempt on which they passed is given in Table 11.8b:

Table 11.7b   Numbers of L_earners who passed the driving test on attempt 1, 2 and 3

| | Pass driving test on | | | |
	First attempt	Second attempt	Third attempt	Total
Numbers	88	34	38	160

The χ^2 test will be used to test whether the observed data in Table 11.8b fits the empirical distribution in Table 11.8a in Worked Example 11.4.

Worked Example 11.4: χ^2 goodness of fit for empirical probability distributions

In Table 11.8a, the driving school claimed that 80 % of its students passed the test on the first attempt, 15 % passed on the second attempt and 5 % passed on the third attempt.
The number of attempts taken to pass the test is given in Table 11.8b for a random sample of 160 L_earners who attended the driving school.

(a) Test the claim made by the driving school.
(b) If there is association found, identify its source.
(c) Plot a bar chart to show the differences between the observed and expected frequencies.

Solution

(a) The null hypothesis states that the percentage of L_earners who pass the driving test on the first, second and third attempt are 80 %, 15 % and 5 %. Assuming the null hypothesis is true then 80 % of the 160 L_earners should pass on the first attempt, 15 % of the 160 should pass on the second attempt and 5 % of the 160 should pass on the third or subsequent attempt. These are the expected frequencies. With the values of f_o and f_e known the test statistic is calculated as in previous examples, formula 11.2. See Table 11.6c.

1. **H_0:** 80 % of L_earners pass the driving test on the first attempt, 15 % pass on the second attempt and 5 % on the third or subsequent attempt. See Table 11.6.
2. **H_1:** The distribution is not the same as that given in Table 11.6.
3. $\alpha = 5\,\%$
4. **Test Statistic:** $\chi^2 = \sum_{\text{all cells}} \dfrac{(f_0 - f_e)^2}{f_e}$

5. Decision Rule:
Reject H_0 if the sample χ^2 exceeds the critical χ^2 from the Tables.
The degrees of freedom, $\nu = c - 1 - k \ldots (11.3)$.
Hence $\nu = 2 - 1 - 0 = 2$. Reject H_0 if sample χ^2 exceeds 5.9915.

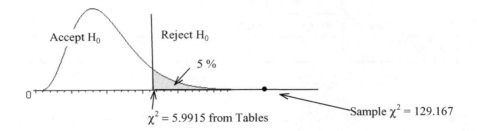

6. Calculations: calculate $\chi^2 = \sum\limits_{\text{all cells}} \dfrac{(f_0 - f_e)^2}{f_e}$. See Table 11.8c.

Table 11.7c   Calculation of test statistic for goodness of fit (attempts to pass driving test)

	Pass driving test on			
	First attempt	Second attempt	Third attempt	Totals
Numbers f_o	88	34	38	160
Expected frequency f_e	$160 \times \dfrac{80}{100} = 128$	$160 \times \dfrac{15}{100} = 24$	$160 \times \dfrac{5}{100} = 8$	160
$(f_o - fe)$	$(88 - 128) = -40$	$(34 - 24) = 10$	$(38 - 8) = 30$	0
$\dfrac{(f_o - f_e)^2}{f_e}$	12.5	4.167	112.5	129.167

Note: in Table 11.8c $\sum f_o = \sum f_e$. This should always be the case.

7. Conclusion: The sample $\chi^2 = 129.167$ is highly significant. Reject the null hypothesis and conclude that the distribution is not the same as that claimed in H_0.

(b) To find the source of the association check the individual sample χ^2's in Table 11.8c. The largest is 112.5. The reason for this large value is that $f_o = 38$ but the expected frequency (f_e) is only 8. For those who pass on the first attempt, $f_o = 88$ and $f_e = 128$ giving a sample $\chi^2 = 12.5$.
Row profiles are also useful to identify where categories do not fit the hypothesised distribution.

(c) The bar chart for the differences between the observed and expected frequencies is given in Figure 11.3. Relative to the expected frequency, the largest squared difference is for the third attempt.

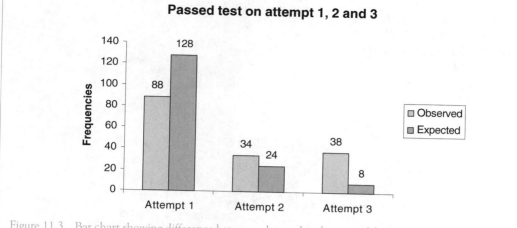

Figure 11.3 Bar chart showing difference between observed and expected frequencies.

Skill Development SKD11_3: χ^2 goodness of fit test

A company packs and distributes food products: The company states that a box of mixed nuts consists of 30 % peanuts; 20 % hazelnuts; 10 % pine nuts; 40 % cashew nuts.

A box of nuts contains the following numbers of each type:

Table SK11_3.1 Numbers of each type of nut in a randomly selected box

Type of nut	Peanut	Hazelnut	Pine	Cashew	Total
Number of nuts	210	100	80	330	720

(a) Is the company's claim justified at the 5 % level of significance?
(b) If the company's claim is rejected explain why.

Answer

If the packer's claim is true, then the number of each type of nut expected should be as follows:

Table SK11_3.2 Expected numbers of each type of nut in a randomly selected box

Type of nut	Peanut	Hazelnut	Pine	Cashew	Total
Expected number	$\frac{30}{100} \times 720 = 216$	$\frac{20}{100} \times 720 = 144$	$\frac{10}{100} \times 720 = 72$	$\frac{40}{100} \times 720 = 288$	720

Set out the test of hypothesis:
H_0: The distribution is 30:20:10:40
H_1: The distribution is not as stated in H_0
The degrees of freedom, $\nu = c - 1 - k$
Hence $\nu = 4 - 1 - 0 = 3$.

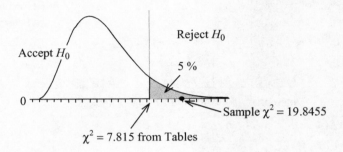

Reject H_0

Accept H_0

5 %

0

Sample $\chi^2 = 19.8455$

$\chi^2 = 7.815$ from Tables

Conclusion: the sample $\chi^2 = 19.8455$ falls in the rejection region for H_0. Therefore reject the null hypothesis: the contents are not the same as that claimed by the packer.

Skill Development Exercise SK11_4: Goodness of fit for a uniform distribution

In a discrete uniform probability distribution each outcome occurs with the same probability. An experiment to test whether a die is fair involved throwing a die 180 times. The number of times each face showed was recorded in Table SKD11_4.1:

Table SDK11_4.1   Frequency of faces on 180 throws of a die

Face of die	1	2	3	4	5	6
Number of throws	28	36	36	30	27	23

Does the data provide evidence that the die is fair? Plot a histogram to compare the observed and expected frequencies.

Solution

If the die is fair, the distribution will be uniform; that is the number of times each face shows (the expected frequency, fe) will be the same. Therefore six equally likely outcomes divided into 180 throws will result in each face showing $180/6 = 30$. So, 30 is the expected frequency in each case, as shown below.

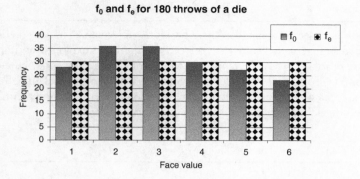

f$_0$ and f$_e$ for 180 throws of a die

Reject H$_0$ if sample χ^2 exceeds 11.07 (5 % 5 *df*)

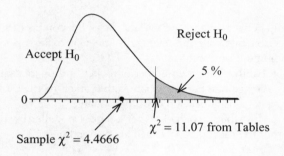

Conclusion: The sample $\chi^2 = 4.4666$ falls within the acceptance region for H$_0$. Therefore accept the null hypothesis that the die is fair.

11.7.2 Goodness of fit for a Poisson probability distribution

The Poisson probability distribution introduced in Chapter 5 section 5.3 modelled situations where random events occurred at a uniform rate over given interval of time or space (area of opportunity). To test whether sample data is modelled by a Poisson distribution proceed as follows:

1. If the average rate, λ, is not given then estimate the average from the sample data.

2. Use the average, λ, to calculate the probabilities, $P(X = x) = \dfrac{\lambda^x e^{-\lambda}}{x!}$
   For $x = 0, 1, 2, 3, \ldots$

3. Calculate the expected frequencies.
   Since the probability $P(X = x) = \dfrac{f_x}{n}$ then $f_x = n \times P(X = x)$
   Hence f_x is the expected frequency.

4. Carry out a χ^2 test for the difference between the observed and expected frequencies. The degrees of freedom are $v = c - 1 - k \ldots (10.3)$ where c is the number of categories, $k = 1$ if the average, λ is estimated from the sample, otherwise $k = 0$.

Worked Example 11.5: χ^2 goodness of fit for a Poisson distribution

In 1944, London was subjected to bombardment by pilotless German VI rockets (called doodlebugs). These rockets had sufficient fuel to reach London, then fell and exploded on impact. It was not clear whether these rockets were developed to the point where they could be aimed at specific targets or at London in general. In order to determine whether the rockets fell randomly R.D. Clarke divided an area of 144 square kilometres into 576 equal squares and recorded the number of bombs that fell in each square. The results are given in Table 11.8a:

Table 11.8a   Observed numbers of rockets falling per square

Rockets per square	0	1	2	3	4	5	Total (n)
No. of squares (frequency)	229	211	93	35	7	1	576

(a) State the relationship between mean and variance for a Poisson probability distribution. From the data in Table 11.8a calculate the average number of rockets per square. Calculate the sample variance. Comment on these values.
(b) (i) Calculate the probability that the number of rockets falling in a square is 0, 1, 2, 3, 4 and 5, assuming the average number per square is that given by the sample average in (a).
     (ii) Calculate the expected number of rockets per square.
(c) Test the claim that the distribution of rockets per square is modelled by the Poisson distribution.
(d) Can you conclude that the rockets were aimed at specific targets?

Solution

(a) The mean and variance are equal for a Poisson probability distribution.
    Calculate the mean for grouped data in Table 11.8b.

Table 11.8b   Calculation of mean and variance for number of rockets per square

Rockets per square	No. of squares (frequency)	fx	$f(x - \bar{x})^2$
0	229	0	197.55
1	211	211	1.07
2	93	186	106.71
3	35	105	150.15
4	7	28	66.03
5	1	5	16.57
Total	576	535	538.08

Average, $\lambda = \dfrac{\sum fx}{\sum f} = \dfrac{535}{576} = 0.9288.$

Sample variance, $s^2 = \dfrac{\sum f(x - \bar{x})}{n - 1} = \dfrac{538.08}{576} = 0.9333.$

Comment: The values of the mean and variance are very close. This indicates a possible Poisson distribution.

(b) (i) The calculation of the Poisson probabilities is given in Table 11.8c. Alternatively, you could use the Poisson probability tables or Excel.

    (ii) The expected frequencies are calculated by multiplying the probability by the total number of observations – remember the different ways to state probability: proportion or percentage – so we could say 'no rocket fell in 39.46 % of squares' etc.

Table 11.8c   Probabilities and expected frequencies for the numbers of rockets

Rockets per square	0	1	2	3	4	5	Total (n)
No. of squares (observed frequency)	229	211	93	35	7	1	576
(i) Poisson probability	$\dfrac{\lambda^0 e^{-\lambda}}{0!} =$	$\dfrac{\lambda^1 e^{-\lambda}}{1!} =$	$\dfrac{\lambda^2 e^{-\lambda}}{2!} =$	$\dfrac{\lambda^3 e^{-\lambda}}{3!} =$	$\dfrac{\lambda^4 e^{-\lambda}}{4!} =$	$\dfrac{\lambda^5 e^{-\lambda}}{5!} =$	0.9996
Mean, $\lambda = 0.93$	0.3946	0.3669	0.1706	0.0529	0.0123	0.0023	
(ii) Expected frequency $= n*$Poisson probability	228	211	98	30	7	1	576

(c) Set out the χ^2 test as follows:

    1. H_0: The sample data is modelled by a Poisson distribution.

    2. H_1: The sample data is NOT modelled by a Poisson distribution.

    3. $\alpha = 5\%$

    4. Test Statistic: $\chi^2 = \displaystyle\sum_{\text{all cells}} \dfrac{(f_0 - f_e)^2}{f_e}$

    5. Decision Rule:

    Reject H_0 if the sample χ^2 exceeds the critical χ^2 from the Tables.

    The degrees of freedom, $v = c - 1 - k \dots (11.3)$

    Hence $v = 6 - 1 - 6 = 24$. Reject H_0 if sample χ^2 exceeds 9.488.

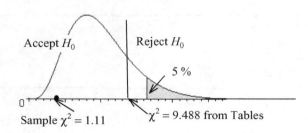

6. Calculations: calculate $\chi^2 = \displaystyle\sum_{\text{all cells}} \dfrac{(f_0 - f_e)^2}{f_e} = 1.11.$ See Table 11.8d.

Table 11.8d   Calculation of the Chi-squared test statistic

Rockets per square	0	1	2	3	4	5	Total
f_o: Observed frequency	229	211	93	37	7	1	576
f_e: Expected frequency	227	211	98	30	7	1	576
$(f_o - fe)$	$(229 - 227)$	$(211 - 211)$	$(93 - 98)$	$(37 - 30)$	$(7 - 7)$	$(1 - 1)$	0
	$= 2$	$= 0$	$= -5$	$= 7$	$= 0$	$= 0$	
$\dfrac{(f_o - f_e)^2}{f_e}$	0.02	0.00	0.26	0.83	0.00	0.00	1.11

7. The test statistic (sample χ^2) is in the acceptance region. Therefore accept the null hypothesis that the data is modelled by a Poisson distribution.

(d) Since the data is modelled by a Poisson distribution, the rockets fall randomly at a uniform rate. Hence the rockets are not aimed at specific targets.

Skill Development Exercise SKD11_5: χ^2 goodness of fit for a Poisson

The number of accidents per hour during peak periods on a stretch of motorway is given in Table SKD11_5.1.

(a) State the mean and variance for a Poisson probability distribution. Calculate the average number of accidents per hour. Calculate the sample variance. Comment on these values.
(b) (i) Calculate the probability that the number of accidents per hour is 0, 1, 2, 3, 4 and 5 assuming the average number per hour is given by the sample average in (a);
    (ii) Calculate the expected number of accidents per hour.
(c) Test the claim that the number of accidents per hour is modelled by the Poisson distribution.

Table SKD11_5.1   Number of accidents per hour during peak time

Accident per hour	0	1	2	3	4	5	6
Frequency	56	76	53	32	19	3	1

Answers

(a) Mean $= 1.5625$. Variance $= 1.6878$

(b)

(b)(i) Poisson prob.	0.21	0.33	0.26	0.13	0.05	0.02	0.01
(b)(ii) fe = N*prob	50.31	78.60	61.41	31.98	12.49	3.90	1.30
$\dfrac{(f_o - f_e)^2}{f_e}$	0.64	0.09	1.15	0.00	3.39	0.21	0.07

(c) Test statistic or Sample $\chi^2 = 5.55$. Degrees of freedom $= 5$. Critical $\chi^2 = 11.07$ at $\alpha = 5\%$ The null hypothesis is accepted.

Progress Exercises 11.2: Chi-squared goodness of fit

1. (a) Why is the Chi-squared test referred to as a non-parametric test?
   (b) In a small Lotto (digits 0 to 9 only), the number of times digits 0 to 9 turned up in a run of 300 selections are given in Table PE11.2 Q1.

Table PE11.2 Q1   Frequency of digits 0 to 9 in 300 selections

Digit	0	1	2	3	4	5	6	7	8	9
Frequency	24	22	25	32	32	35	37	26	32	35

Test the claim that the Lotto is fair, i.e., that the distribution is uniform.

2. In a discussion on pay and conditions, the managers of a taxi service claim that number of calls for the service is constant from Monday to Saturday.
   A survey by drivers recorded the number of calls received daily from Monday to Saturday in Table PE11.2 Q3.

Table PE11.2 Q3   Number of calls received per day

Day	Mon.	Tue.	Wed.	Thu.	Fri.	Sat.
Number of calls	72	63	71	110	126	128

(a) Calculate the expected number of calls per day from Monday to Saturday assuming the number of calls per day is constant.
(b) Carry out a χ^2 test to ascertain whether the company is justified in claiming that it receives the same number of service calls each day. If the claim is rejected, which days are likely to be different from the others?

3. The distributor of ethnic foods claims that the content per bag of Mixed Nut Special is 12 % brazil nuts; 25 % peanuts; 20 % cashew nuts; 28 % hazelnuts; 15 % walnuts.
   Several bags were selected; the total numbers of each type of nut were counted, Table PE11.2 Q3:

Table PE11.2 Q3   Numbers of each type of nut

	Type of Nut					
	Brazil	Peanut	Cashew	Hazelnut	Walnut	Total
Observed number of each type of nut	146	315	??	287	170	1130
Expected number of each type of nut	??	??	??	??	??	??

(a) Fill in the blank cells (??) in Table PE11.2 Q3.

(b) Calculate the χ^2 statistic for the data in Table PE11.2 Q3. Hence test the claim about the content of the mixture at the 5 % level of significance.

(c) Calculate the observed percentage and expected percentage of each nut (called *a row profiles*). By comparing the observed and expected percentages describe verbally how the contents of Mixed Nut Special differs from that given by the distributors.

4. Traditionally, 40 % of the demand for milk in a large food store was for the full cream milk, and 15 % for each of low fat milk, fortified milk, skim milk and soya milk. To assess whether the demand for each type is changing, the store manager surveys 200 customers on the type of milk purchased. The results are given in Table PE11.2 Q4.

Table PE11.2 Q4   Type of milk purchased by 200 customers

Type of milk	Full cream	Low fat	Fortified	Skim	Soya
Demand	65	40	15	20	60

Use the χ^2 test to determine whether the demand for each type of milk has changed.

5. Penalty points for driving offences were introduced in October 2002 in a major initiative to reduce the number of accidents and deaths on Irish roads. Between (i) November 2002 and December 2003 a total of 93,364 drivers (ii) November 2002 and March 2004 a total of 128, 966 drivers had incurred penalty points as follows:

Table PE11.2 Q5a   Drivers who received between 2 and 10 penalty points

Penalty points	2 points	4 points	6 points	8 points	10 points	12 points	Total drivers
(i) Percentage of drivers	94.55 %	4.98 %	0.40 %	0.06 %	0.01 %	0.00 %	100.00 %
(ii) Percentages of drivers	92.54 %	6.70 %	0.65 %	0.09 %	0.02 %	0.00 %	100.00 %

Source: Garda Siochana.

Suppose a survey of drivers who incurred penalty points in the province of Munster is given in Table PE11.16b:

Table PE11.2 Q5b   Survey from Munster: drivers who received between 2 and 10 penalty points

Penalty points	2 points	4 points	6 points	8 points	10 points	12 points	Total drivers
Numbers of drivers	2435	170	16	6	3	0	2630

Carry out a χ^2 test to determine whether the distribution of penalty points in Munster differs from the national distribution for (i) November 2002 to December 2003 (ii) November 2002 to March 2004. Write a brief report, explaining your results.

6. 'Neighbour Statistics' include statistics on the economic activities of 16–74 year olds (http:// neighbourhood.statistics.gov.uk).

Suppose follow-up surveys on economic activity are conducted in year in 2006. See Table PE11.2 Q6:

Table PE11.2 Q6   Economic activities of 16–74 year olds

Numbers	Census 2001	Sample 2004
Economically active	682	292
Economically active: Employed	551	206
Economically active: Self-employed	89	59
Economically active: Unemployed	22	15
Economically active: Full-time students	20	12
Economically inactive	386	145

(a) Calculate the percentage (according to Census 2001) that should be in each category: economically active (employee, self-employed, unemployed, full-time students) and economically inactive.
(b) Hence calculate the expected frequencies (according to Census 2001) for the 2004 data.
(c) Carry out a χ^2 test to determine whether the distribution of economic activity/inactivity has changed between 2001 and 2004.
(d) If the χ^2 test gives a significant result determine the categories that have changed since 2001.

7. See question 6.

Combine the category 'economically active: full-time students' in Table PE11.2 Q6 with (i) Economically active: Employee; (ii) Economically active: Self-employed.

Repeat question 6 for each case, (i) and (ii).

Explain why these results are different from those in question 6.

8. See question 8, PE5.3.

The number of car rentals per hour is recorded for 125 hourly intervals. The number of times (frequency) that 0, 1, 2....11, 12 cars were rented per hour is given in Table PE5.3 Q8:

File: Rented cars

Table PE5.3 Q8   Number of times that 0, 1, 2...,11 ,12 cars were rented per hour for 125 hours

Cars rented per hour	0	1	2	3	4	5	6	7	8	9	10	11	12
Frequency		1	3	4	10	15	20	24	19	13	10	4	1	1

(a) Calculate the mean number of rentals per hour.
(b) Assuming a Poisson process, use the mean number of rentals per hour in (a) to calculate
  (i) the probability that 0, 1, 2, 3..., 11, 12 cars are rented per hour;
  (ii) the expected frequencies f_e.
(c) Test the claim that the sample data is modelled by a Poisson probability distribution.

9. See question 10 Table PE11.2 Q9.

A company records the number of absentees per day for a period of 80 days in Table PE11.2 Q9.

Table PE11.2 Q9   Absentees per day for a period of 80 days File: Absentees per day

Absentees per day	0	1	2	3	4	5	6	7	8	9	10
Days		6	9	12	18	18	9	4	0	3	0	1

(a) Calculate the mean number of absentees per day.
(b) Use the average in (a) calculate the Poisson probabilities for 0 to 10 absentees per day. Hence calculate the expected numbers of absentees per day for 80 days.
(c) Test the claim that the sample data is modelled by a Poisson probability distribution.

11.8 χ^2 tests in Minitab and Excel

The calculations for the χ^2 test statistic involve a large amount of tedious arithmetic. The formula function in Excel may be used to deal with the arithmetic. A single command in Minitab will produce all the calculations – expected frequencies, individual cell χ^2's and the χ^2 test statistic. But remember that the packages do not declare the null and alternative hypotheses nor do they arrive at a conclusion or suggest follow up tests to determine why hypotheses are rejected. It is essential that the reader state the null and alternative hypotheses, sets out the decision rule and explains the conclusion in the context of the question.

11.8.1   Use Excel to do the calculations

The formula facility in Excel (described in Chapter 2, section 2.7) is very useful for doing the arithmetic required to calculate the expected frequencies and the χ^2 test statistic. To illustrate, consider the data in Table 11.2b entered in an Excel worksheet as shown in Figure 11.4. The expected frequency for the first cell is calculated by placing the cursor in the cell where the result is required (C9 in Figure 11.4) and entering the symbol '=' to begin a formula. The expected frequency is then calculated by clicking on the cells containing the required row total, column total and overall total as demonstrated below. In a similar manner, calculate the remaining expected frequencies or autofill the remaining eight cells from the formula in cell C9.

With the table of expected frequencies calculated, the formula for each individual χ^2, $\dfrac{(f_0 - f_e)^2}{f_e}$ is entered as shown in cell c14. In fact, if the worksheet is set out neatly with the tables of observed frequencies, expected frequencies and individual χ^2 in column order (or row order) the calculations of the table of individual χ^2 involves entering the formula in the first cell (cell C14 in Figure 11.5), then 'auto fill' the remainder of the table. The test statistic is the sum of the χ^2's for each cell.

Alternatively, the observed and expected frequencies may be arranged in columns as in Figure 11.6 (a similar layout to Table 11.2d in Worked Example 11.1). The formulae for the differences between observed and expected frequencies are entered into the first cell in column 3 and the formula for

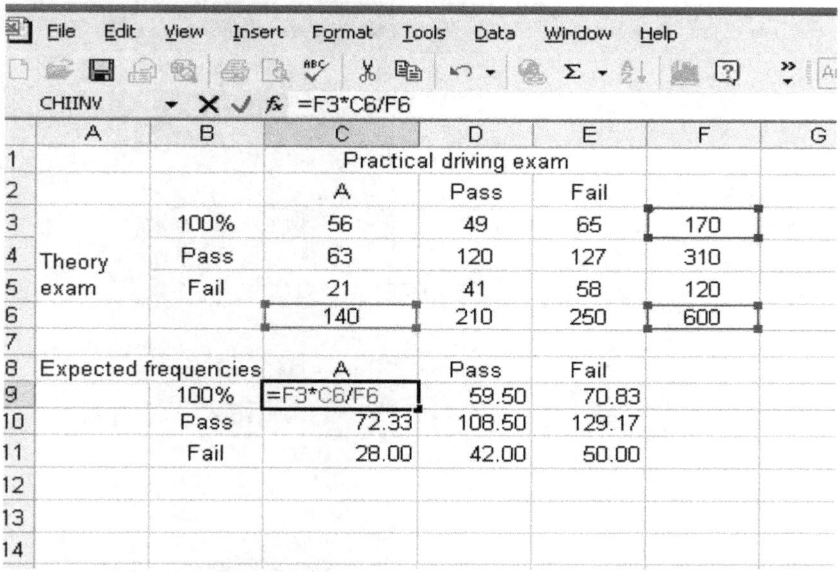

CHIINV =F3*C6/F6

	A	B	C	D	E	F	G
1			Practical driving exam				
2			A	Pass	Fail		
3		100%	56	49	65	170	
4	Theory	Pass	63	120	127	310	
5	exam	Fail	21	41	58	120	
6			140	210	250	600	
7							
8	Expected frequencies		A	Pass	Fail		
9		100%	=F3*C6/F6	59.50	70.83		
10		Pass	72.33	108.50	129.17		
11		Fail	28.00	42.00	50.00		
12							
13							
14							

Figure 11.4 Use formula in Excel to calculation of expected frequencies.

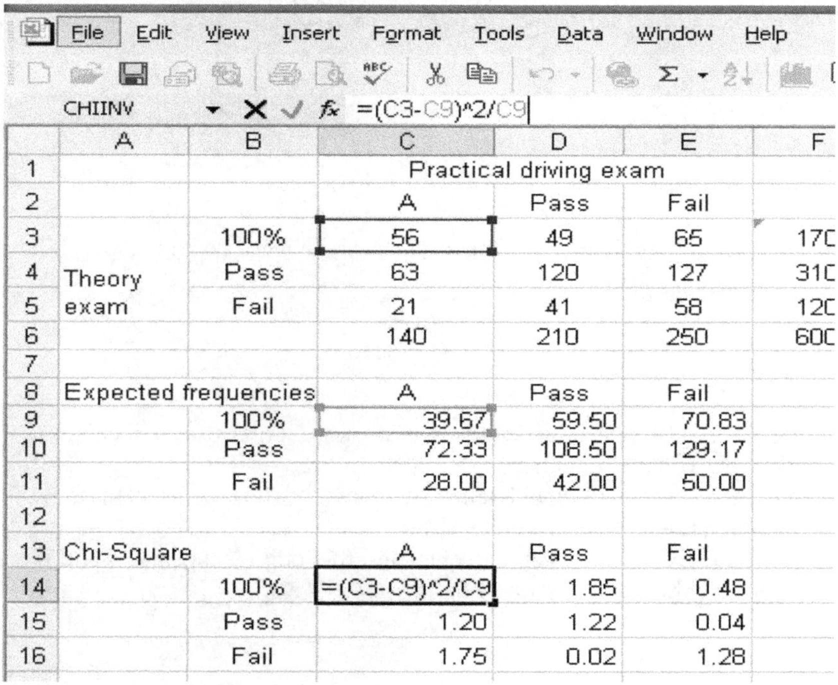

CHIINV =(C3-C9)^2/C9

	A	B	C	D	E	F
1			Practical driving exam			
2			A	Pass	Fail	
3		100%	56	49	65	170
4	Theory	Pass	63	120	127	310
5	exam	Fail	21	41	58	120
6			140	210	250	600
7						
8	Expected frequencies		A	Pass	Fail	
9		100%	39.67	59.50	70.83	
10		Pass	72.33	108.50	129.17	
11		Fail	28.00	42.00	50.00	
12						
13	Chi-Square		A	Pass	Fail	
14		100%	=(C3-C9)^2/C9	1.85	0.48	
15		Pass	1.20	1.22	0.04	
16		Fail	1.75	0.02	1.28	

Figure 11.5 Calculation of χ^2 for cell (1,1) by formula in Excel.

	A	B	C	D	E	F
12						
13	Chi-Square		A	Pass	Fail	
14		100%	6.73	1.85	0.48	
15		Pass	1.20	1.22	0.04	
16		Fail	1.75	0.02	1.28	
17						
18		f_o	f_e	$(f_o - f_e)$	$\frac{(f_o - f_e)^2}{f_e}$	
19		56	39.67		=(B19-C19)^2/C19	
20		49	59.50	-10.50	1.85	
21		65	70.83	-5.83	0.48	
22		63	72.33	-9.33	1.20	
23		120	108.50	11.50	1.22	
24		127	129.17	-2.17	0.04	
25		21	28.00	-7.00	1.75	
26		41	42.00	-1.00	0.02	
27		58	50.00	8.00	1.28	
28			Column totals	0.00	14.57	
29						

Figure 11.6 Alternative arrangements for the calculation of sample χ^2 in Excel.

$\frac{(f_o - f_e)^2}{f_e}$ in the first cell in column 4. These formulae are dragged down to 'auto fill' the table with the remainder of the calculations.

11.8.2 Minitab for χ^2

Using Minitab for χ^2 tests of independence is straightforward. Enter the data, column by column, in the Minitab worksheet. To carry out the χ^2 test click on **Stat** → **Tables** → χ^2 **tests** as directed in Figure 11.7. From the pop-up menu double-click the names of the column containing the data into the 'columns containing the table' box, Figure 11.8.

The Minitab printout gives the expected frequencies, the individual cell χ^2's and the test statistic as well as the p-value for the test in Figure 11.9.

Minitab printout

These same techniques may be applied to test the equality of several proportions *provided the full contingency table* is entered in the Minitab work sheet.

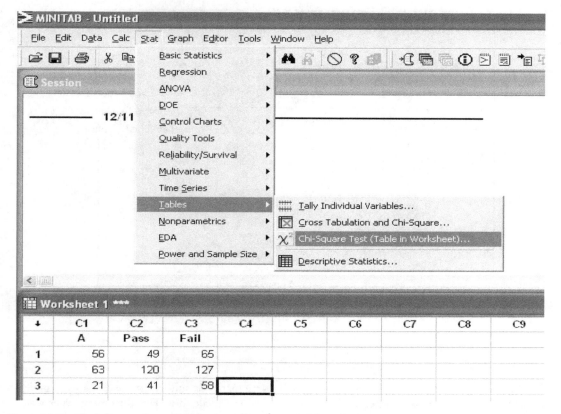

Figure 11.7   Worksheet and steps in carrying out a χ^2 test in Minitab.

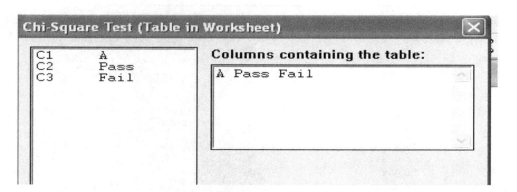

Figure 11.8   Double click on the columns in the left-hand box required for the χ^2 analysis.

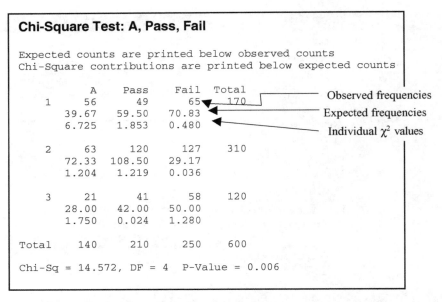

Chi-Square Test: A, Pass, Fail

```
Expected counts are printed below observed counts
Chi-Square contributions are printed below expected counts

           A     Pass    Fail  Total
    1     56       49      65    170
        39.67    59.50   70.83
        6.725    1.853   0.480

    2     63      120     127    310
        72.33   108.50   29.17
        1.204    1.219   0.036

    3     21       41      58    120
        28.00    42.00   50.00
        1.750    0.024   1.280

Total   140      210     250    600

Chi-Sq = 14.572, DF = 4   P-Value = 0.006
```

Observed frequencies
Expected frequencies
Individual χ^2 values

Figure 11.9   Minitab printouts for χ^2 test for independence.

Skill Development Exercise11_4: χ^2 tests with Minitab

In a double blind, randomised block experiment participants of similar typing skills were given asked to type a document after consuming a drink containing alcohol. Two document types were used in the experiment: technical (Doc A) and poetry (Doc B). The number of typing errors made by participants are given in Table PE10.3 Q4:

Table PE10.3 Q.4   Typing errors (File: alcohol)   File: alcohol

Alcohol	Doc A	Doc B
0 units	2	4
1.0 units	6	7
1.5 units	3	4
2.0 units	7	12
2.5 units	5	16
3.0 units	9	12
3.5 units	12	14
4.0 units	7	14
4.5 units	9	17
5.0 units	13	5
5.4 units	18	3
5.5 units	16	2

(a) Carry out a Chi-squared test to determine whether there is any relationship between document type and the units of alcohol consumed.

(b) If a significant relationship between document and alcohol consumption exists, describe this relationship verbally. Plot a suitable chart to illustrate the relationship.

(c) Delete the last two rows of Table PE10.3 Q4. Carry out the Chi-squared test on this Table. What do you conclude?

Outline solution

(a) In Minitab, the sample $\chi^2 = 38.466$, $df = 11$, p-value $= 0.000$.
There is an association between levels of alcohol and number of errors.

(b) The strongest association is that higher levels of alcohol are associated with lower numbers of errors for Doc B (poetry) and high number of errors for Doc A (technical).

(c) When the two highest levels of alcohol are deleted, then there is no significant association between the type of document and levels of alcohol.

The sample $\chi^2 = 11.516$, $df = 9$, p-value $= 0.242$.

Progress Exercises 11.3: χ^2 tests (general) with Excel, Minitab or otherwise

1. The number of typing errors made in typing documents A (technical) and B (poetry) by typists who had consumed between 0 and 3 units of alcohol are given in Table PE11.3 Q1:

Table PE11.3 Q1   Typing errors made under the influence of various units of alcohol

Units of alcohol	Doc A (Technical)	Doc B (Poetry)
0	11	15
1	21	40
2	28	45
3	47	10

Carry out a χ^2 test to determine whether there is any association between the number of errors in Documents A and B and the units of alcohol consumed by the typist.

2. An automobile emergency service is concerned about overstaffing on certain shifts. The number of emergency calls over a period of time is given in Table PE11.3 Q2:

Table PE11.3 Q2   Emergency calls per shift, Monday to Friday

Break-down calls	Shifts			Row totals
	8–4 pm	4–12 am	12–8 am	
Monday	16	32	12	60
Tueday	21	41	20	82
Wednesday	18	46	21	85
Thursday	23	45	38	106
Friday	42	44	46	132
Column totals	120	208	137	465

Test whether there is an association between shifts and days. If an association is found, describe this verbally. If the same number of staff had been scheduled for each shift, Monday to Friday what adjustments would you recommend?
3. Explain the limitations of the χ^2 tests.

Five brands of car tyre (A, B, C, D and E) are tested on 935 new cars. If any one of the set of four tyres is replaced within the guaranteed 5000 miles the life of the set of tyres is classified as 'short'. If any tyre is replaced within 5000 to 10 000 miles the life is classified as 'standard' and sets of tyres lasting longer than 10 000 miles as 'long'. See Table PE11.3 Q3.

Table PE11.3 Q3   Life length for tyres brand A, B, C, D, E   File: Tyres

	Short	Standard	Long
A	15	123	18
B	68	432	54
C	23	98	22
D	10	54	12
E	1	5	0

(a) Test the claim that there is no association between the brand of tyre and length of life.
(b) Calculate the row profiles for each brand of tyre.
Explain the difference between the results in (a) and (b).
4. See questions 6 and 7 PE11.2. A detailed breakdown of economic activity of the 16–74 year olds is given in Table PE11.3 Q4:

File: Economically active

Table PE11.3 Q4   Economic activity of 16–74 year olds according to census 2001 and subsequent surveys

	Census 2001	Sample 2002	Sample 2004	Sample 2006
All persons (16–74 years)	1068			
Economically active	682	286	292	284
Economically active: Employee	551	220	206	166
Economically active: Employee, part-time	118	50	56	85
Economically active: Employee, full-time	433	170	150	81
Economically active: Self-employed with employees	29	16	17	15
Economically active: Self-employed with employees, part-time	4	6	5	10
Economically active: Self-employed with employees, full-time	25	10	12	5
Economically active: Self-employed without employees	60	30	42	38
Economically active: Self-employed without employees, part-time	15	8	12	28
Economically active: Self-employed without employees, full-time	45	22	30	10
Economically active: Unemployed	22	10	15	10
Economically active: Full-time students	20	10	12	55
Economically inactive	386	142	145	66

Carry out appropriate χ^2 goodness of fit tests to determine whether the numbers in each category of activity or inactivity has changed with reference to distribution in Census 2001.

Summary and overview of χ^2 tests

In this text, the (non-parametric) χ^2 tests are applied to the following

1. H_0: There is no association between 2 categorical variables.
2. H_0: All populations are homogeneous.
3. H_0: The equality of several proportions.
4. H_0: The goodness of fit of a set of data to a given probability distribution.

The steps in the tests are as follows
Set up the contingency tables r rows $\times c$ columns
Calculate

Step 1: Expected frequencies: $f_e = \dfrac{r \times c}{T}$ and degrees of freedom.

Step 2: The χ^2 for each cell is $\dfrac{(f_o - f_e)^2}{f_e}$

Step 3: The χ^2 test statistic is $T = \sum\limits_{\text{all cells}} \dfrac{(f_o - f_e)^2}{f_e}$

Step 4: Degrees of freedom
$df = (r - 1)(c - 1)$ for tests 1, 2 and 3

For Goodness of Fit tests step 2 and step 4 are as follows

Step 2: $f_e = $ *Total number of outcomes $\times$ Probability of each outcome*
Step 4: $df = c - k - 1$ for test 4, where $c = $ number of classifications; $k = $ number of population parameters estimated from sample.

If the test statistic exceeds the critical point from the χ^2 percentage point tables for a given level of significance, α and df reject the null hypothesis.

To find the source of the association

Look for all individual χ^2 values that are close to or exceed the critical χ^2 percentage point.
Calculate the row and/or column profiles.

Limitations of χ^2 tests

Tests must be based on count data, set out in a contingency table.
Expected frequencies must exceed 1 but preferably exceed 5.

REGRESSION ANALYSIS

12

Chapter Objectives

Having carefully studied this chapter and completed the exercises you should be able to do the following

- Describe the linear regression model and state its assumptions
- Calculate standard error for residuals s_e
- State the sampling distribution for slope and calculate its standard error, s_b
- Calculate confidence intervals and test hypothesis for slope
- Calculate the standard errors for $\overline{y}_{x=x_0}$ and $y_{x=x_0}$
- Calculate confidence intervals for $\overline{y}$ when $x = x_0$
- Calculate prediction intervals for y when $x = x_0$
- Check the assumptions of the model graphically by residual plots in Excel
- Check the assumptions of the model graphically by residual plots in Minitab
- Describe the multiple linear regression model and state its assumptions
- Describe and analyse a Minitab printout for multiple linear regression

12.1   The simple linear regression model

The reader should find it helpful to review Chapter 3. The simple linear regression model describes a linear trend (fitted by the method of least squares) to a population of paired data $\{(x_1, y_2) \ldots (x_N, y_N)\}$.

As an example of such a population consider the pay per hour (£Y) and the corresponding hours worked per day (X hours) for an entire population. The scatter plot and the population least-squares line are given in Figure 12.1:

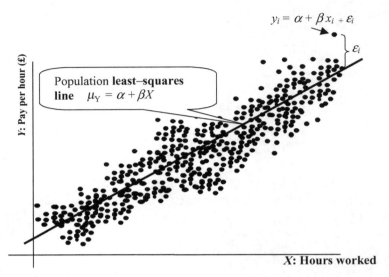

$$y_i = \alpha + \beta x_i + \varepsilon_i$$

ε_i

Y: Pay per hour (£)

Population **least–squares line** $\mu_Y = \alpha + \beta X$

X: Hours worked

Figure 12.1   Scatter diagram for rates of pay per hour vs. hours worked per day.

The scatter plot displays a general linear trend for which the pay per hour increases as the number of hours worked per day increases. The fact that the data points are scattered about the population least-squares line is not surprising – it is common knowledge that the hourly rate of pay varies, even for those who work the same number of hours per day. The intercept and slope of the population least-squares line are denoted by the Greek letters, α and β. However, **the population least-squares line is written as** $\mu_Y = \alpha + \beta X$ **and not '$Y = \alpha + \beta X$'.** The reason for using μ_Y and not Y is that the line represents the mean value for the points that are scattered above and below it. **An individual observation in the population is expressed as** (x_i, y_i) **where** $y_i = \alpha + \beta x_i + \varepsilon_i$ **:** ε_i is the random error in Y from the population line for observation i. See Figure 12.1.

The sample least-squares line $y = a + bx$ is calculated from a random sample, selected from a population (see Chapter 3). A sample least-squares line is an estimate of the population least-squares line $\mu_Y = \alpha + \beta X$, where

- the constant, a, in the sample least-squares line is an estimate of α the constant term for the population least-squares line;
- the slope, b, in the sample least-squares line is an estimate of β the slope for the population least-squares line.
- y is an estimate of μ_Y. y is sometimes referred to as $\hat{y}$.

For example, Figure 12.2 illustrates the sample least squares line $y = -15 + 5x$ that was calculated from the sample data given in Table 3.4.

REGRESSION ANALYSIS

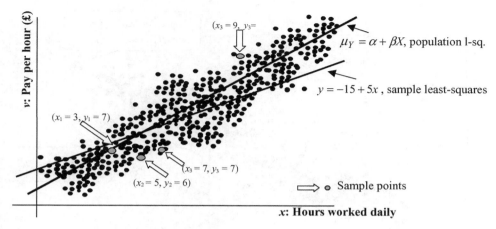

Figure 12.2   A sample least squares line (*LSL*), $y = a + bx$.

Since we rarely have data for an entire population, the objective in regression analysis is to make inference about the population least-squares line from a sample. The Simple Linear Regression model sets out the conditions under which such inferences are valid.

The **simple linear regression model assumes**

(a) the observations, Y_i are independent
(b) the probability distribution of Y for each $X = X_i$ is **Normal,** with its **mean value** given by the population least-squares line: $\mu_{Y_i} = \alpha + \beta X_i$ and **variance,** σ_ε^2.
(c) the variance, σ_ε^2 is the same for each value of X.

The property of 'equal variance' is referred to as homogeneous variance or **homoscedasticity**.

Independent observations (Y_i) means that the observations are not related in any way. For example, the hourly rate of pay (Y) for one person is not related in any way to the hourly rate for another person. Dependent observations frequently arise in time-related events. For example, the number of apartments constructed in any quarter will be related to the numbers in previous quarters since the construction projects are continuous over time and related to demand and other factors.

Normality: the assumption that the observations, Y_i are Normally distributed with mean $\alpha + \beta X_i$ and variance σ_ε^2, is illustrated graphically in Figure 12.3, where the probability distribution of $Y \sim N(\mu = \alpha + \beta X_1, \sigma_\varepsilon^2)$ for $X = X_1, Y \sim N(\mu = \alpha + \beta X_2, \sigma_\varepsilon^2)$ for $X = X_2$, etc.

Equal variance. For each X, the variance about the least-squares line is the same. For example, the variance for the hourly rates of pay is the same whether the hours worked per day are $X = 2$ hours or $X = 10$ hours in Figure 12.3. Visually, equal variance could be thought of as viewing a curved tunnel that has the same curvature, width and height, extending the length of the least-squares line. Cross-sections of such a tunnel curve are given by the Normal curves in Figure 12.3. The variance

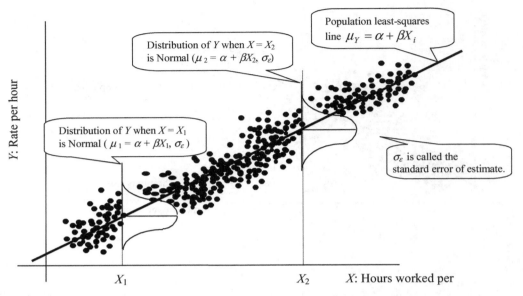

Figure 12.3 Simple Linear Regression model.

about the least-squares line, σ_ε^2 is called the variance of estimate: the standard deviation σ_ε is called the **standard error of estimate**:

The model assumptions may also be stated in terms of ε_i, the random error terms about the population line as follows:

1. ε_i are independent
2. $\varepsilon_i \sim N(\text{mean} = 0, \text{variance} = \sigma_\varepsilon^2)$
3. The variance σ_ε^2 is the same for each value of X.

Terminology reminder:
The variable, X, may be referred to as **independent, explanatory, predictor, exogenous (in economics)**. The random variable, Y, may be referred to as the **dependent, response, regressor endogenous (in economics)**.

12.1.1 The estimate of σ_ε, the standard error of estimate

The variance of the observations, Y_i about the population least-squares will feature in all of the regression analysis that follow, but σ_ε^2 must be estimated from the sample data (x_i, y_i).

The unbiased estimate of σ_ε^2 given in equation (12.1) is the sum of the squared deviations of the sample values, y_i from the corresponding mean value $\hat{y}_i$ divided by the sample size less two. *Recall,* the least-squares line $\hat{y}_i = a + bx_i$ is an estimate of μ_{Y_i}, the mean value of Y_i according to the simple linear regression model.

$$s_e^2 = \frac{\sum (y_i - \hat{y}_i)^2}{n - 2} = \frac{\sum \{y_i - (a + bx_i)\}^2}{n - 2}, \text{ where } n \text{ is the sample size.} \tag{12.1}$$

Note:

1. The divisor in the estimate of variance is $(n - 2)$: because 2 parameters (α and β) are estimated from the sample.
2. $e_i = y_i - \hat{y}_i$ is an estimate of the error term, ε_i and is called a RESIDUAL (see Chapter 3).

The residual $e_i = (y_i - \hat{y}_i)$ is an estimate of the error term, ε_i
s_e^2 is an estimate of the variance of estimate σ_ε^2

Computationally more convenient formulae (particularly if you are using a calculator) for s_e^2 are given in (12.2a) and (12.3).

$$s_e^2 = \frac{n}{n - 2} \left(\sum (y_i - \bar{y})^2 - b^2 \sum (y_i - \bar{y})^2 \right) \tag{12.2a}$$

$$s_e^2 = \frac{\sum y^2 - a \sum y - b \sum xy}{n - 2} \tag{12.3}$$

Note: (12.2a) may be further simplified for calculator computations by using the identity given in Chapter 10

$$\sum (x_i - \bar{x})^2 = n\sigma_x^2 \text{ and } \sum (y_i - \bar{y})^2 = n\sigma_y^2 \tag{10.8}$$

Hence

$$s_e^2 = \frac{n}{n - 2} \left(\sigma_y^2 - b^2 \sigma_x^2 \right) \tag{12.2b}$$

where σ_x and σ_y may be read from the calculator
OR

using $\sum (x_i - \bar{x})^2 = (n - 1)s_x^2, \sum (y_i - \bar{y})^2 = (n - 1)s_y^2$

$$s_e^2 = \frac{n - 1}{n - 2} \left(s_y^2 - b^2 s_x^2 \right) \tag{12.2c}$$

See section 10.2.3 for calculating Sums of Squares on the calculator.

12.2 Inference about the population slope (rate of change)

Inference about population slope β is made from the sample slope, b. It can be proved (mathematically) that the sampling distribution of slopes is Normal:

$$b \sim N\left(\mu_b = \beta, \sigma_b = \frac{\sigma_\varepsilon}{\sqrt{\sum(x_i - \bar{x})^2}}\right).$$

Since the sampling distribution is Normal, the methods described in Chapters 8, 9, and 10 for the calculation of confidence intervals and testing hypothesis may also be applied to calculate confidence intervals and test hypothesis for slope.

12.2.1 Sampling distribution for slope

To illustrate the concept of the sampling distribution for slope consider a random sample selected from the population in Figure 12.2, where the arrows point to the sample data. For example, in Figure 12.2 the sample least-squares line, $y = -15 + 5x$ was calculated from the sample data is given in Table 3.4.

Table 3.4   Rate of pay per hour for four randomly selected employees

X: Hours worked per day	3	5	7	9
Y: Rate of pay per hour (£)	7	6	7	40

Further random samples will yield different sample least-squares lines as illustrated in Figure 12.4: $y = a_1 + b_1 x : y = a_2 + b_2 x : y = a_3 + b_3 x : y = a_4 + b_4 x$, etc.

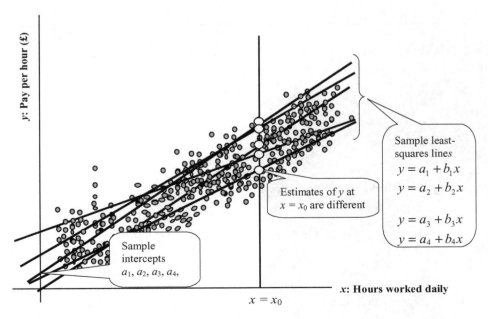

Figure 12.4   Some sample least-squares lines.

The coefficients a, b and the estimates of y at $x = x_0$ vary from sample to sample, hence a, b and the estimates of y at $x = x_0$ are sample characteristics and as such, have sampling distributions. The sampling distribution for slope is Normal, the mean $= \beta$ (the population slope) and variance $= \sigma_b^2$. See Figure 12.5:

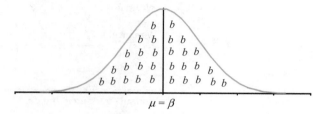

$\mu = \beta$

Figure 12.5 The sampling distribution for sample slopes, b is $N(\beta, \sigma_b)$.

The standard error for slope σ_b must be estimated from sample data. The sample standard error for slope is calculated by formula (12.5):

The sample standard error is for slope is given by the formula

$$s_b = \frac{s_e}{\sqrt{\sum (x_i - \overline{x})^2}} \tag{12.4}$$

Since σ_b is unknown the sampling distribution for slope is a t-distribution with $n - 2$ degrees of freedom. The confidence interval and test statistic are calculated by the formulae in (12.7) and (12.8):

Confidence interval for slope:

Point estimate $\pm$ (confidence coefficient) $\times$ (standard error)

$$b \pm (t_{n-2, \alpha/2}) \times (s_b) \tag{12.5}$$

Test Statistic for slope: $T = \dfrac{\text{estimate} - \text{claim made in } H_0}{\text{standard error}} = \dfrac{b - \beta_{H_0}}{s_b}$ $\tag{12.6}$

The comparison between the methods used to calculate confidence intervals and test hypotheses for population means and population slopes is given in Table 12.1.

Recall

For large samples ($n > 30$), the Z percentage (critical) points may be used to approximate t percentage points.

Table 12.1   Inference about μ from $\bar{x}$: inference about β from b

Review of inference about μ from $\bar{x}$	Inference about β from b
Sampling distribution for $\bar{x}$ Normal Variance $= \sigma_{\bar{x}}^2$ $\mu_{\bar{x}} = \mu$	Sampling distribution for b: Normal Variance $= \sigma_b^2$ $\sigma_b = \beta$
Formula for sample standard error for means: $s_{\bar{x}} = \dfrac{s}{\sqrt{n}}$	Formula for sample standard error for slopes $s_b = \dfrac{s_e}{\sqrt{\sum(x-\bar{x})^2}}$ (12.4) where $s_e = \sqrt{\dfrac{n\left(\sigma_y^2 - b^2\sigma_x^2\right)}{n-2}}$ (12.2b)
Confidence interval for μ $\bar{x} \pm Z_{\alpha/2} s_{\bar{x}}$ for $n > 30$ $\bar{x} \pm t_{(n-1),\alpha/2} s_{\bar{x}}$ for $n < 30$, (σ unknown)	Confidence interval for β $b \pm Z_{\alpha/2} s_b$ for $n > 30$ $b \pm t_{(n-2),\alpha/2} s_b$ for $n < 30$, (σ_b unknown) (12.5)
Test Statistic: $T = \dfrac{\text{estimate} - (H_0 \text{ claim})}{SE} = \dfrac{\bar{x} - (\mu_{H_0})}{s_{\bar{x}}}$	Test Statistic: $T = \dfrac{\text{estimate} - (H_0 \text{ claim})}{SE} = \dfrac{b - (\mu_{H_0})}{s_b}$ (12.6)

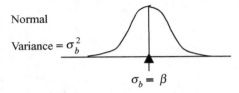

$(n-2)$ degrees of freedom because the 2 population parameters, α and β are estimated

---

Worked Example 12.1: Inference about slope

Consider the data in Table 3.4, Chapter 3 (reproduced below).

Table 3.4   Rate of pay per hour for four randomly selected employees

x: Hours worked per day	3	5	7	9
y: Rate of pay per hour (£)	7	6	7	40

(a) Calculate a 95 % confidence interval for slope.
(b) Test the hypothesis that the slope of the least-squares line is zero.
(c) Test the hypothesis that the rate of pay per hour increases by at least £ 5 when the hours worked per day increases by one.

When answering parts (a), (b) and (c), give a verbal explanation of your results and state your conclusions carefully.

It is recommended that you use your calculator to read off the inbuilt calculations for

$$\bar{x}, \bar{y}, \sum x, \sum y, \sum xy, \sum x^2, \sum y^2, \sigma_x, \sigma_y, s_x, s_y, n, a, b \text{ and } r$$

Solution

In order to illustrate the methods for confidence intervals and tests of hypothesis clearly certain **essential calculations** such as standard errors will be carried out in advance. In this example calculate (i) the equation of the least-squares line and (ii) the standard errors s_e and s_b.
Hence,

(i) The equation of the least-squares line $y = -15 + 5x$ was calculated in W.E 3.1, Chapter 3.
(ii) To calculate the standard errors s_e (12.3) and s_b (12.6), use your calculator to read off population standard deviation, σ. Then square σ to get variance. Hence $\sigma_x^2 = 5, \sigma_y^2 = 208.5$.
Substitute $\sigma_x^2, \sigma_y^2, n = 4$ and b into formulae (12.3b) and (12.6).

Variance of estimate: $s_e^2 = \dfrac{n \left(\sigma_y^2 - b^2\sigma_x^2\right)}{n-2} = \dfrac{4 \left(208.5 - (5)^2(5)\right)}{4-2} = 167$, by (12.3b).

Standard error of estimate, $s_e = \sqrt{167}$.

Standard error for slope s_b: substitute $s_e = \sqrt{167}, n = 4$ and $\sum (x_i - \bar{x})^2 = 20^*$ into (12.6).

Hence $s_b = \dfrac{s_e}{\sqrt{\sum (x_i - \bar{x})^2}} = \dfrac{\sqrt{167}}{\sqrt{20}} = \sqrt{\dfrac{167}{20}} = 2.889637$, by (12.6).

*Recall, $\sum (x_i - \bar{x})^2 = n\sigma_x^2 = (4)(5) = 20$ by (10.8).

Solution

(a) Confidence for slope, β
Point estimate $\pm$ (confidence coefficient) $\times$ (standard error)
$b \pm (t_{n-2,\alpha/2}) \times (s_b)$ From Tables: $t_{df=n-2,\alpha/2=2.5\%} = 4.303$
$5 \pm (4.303)(2.889637) \rightarrow 5 \pm 12.43411$
We are 95 % confident that the slope is between $- 7.43411$ to 17.43411
Since 0 is within the interval, then the claim $\beta = 0$ is plausible on evidence from this sample

(b) **Test the hypothesis that the slope $\beta = 0$**
Set out the hypothesis test as follows:
1. H_0: $\beta = 0$
2. H_a: $\beta \neq 0$ (recall, this is a 2-sided test)
3. $\alpha = 0.05$ (5 %)
4. **Test statistic:** $T = \dfrac{\text{estimate} - \text{claim made in } H_0}{\text{standard error}} = \dfrac{b - \beta_{H_0}}{s_b}$

> In fact, this sample provides evidence to support any claim for $\beta = -7.43$ to $\beta = 17.43$ Such imprecise intervals are of no practical value

5. **Decision Rule:** Write down the critical t values for $(n-2)$ degrees of freedom. Hence for $df = (4-2) = 2$, critical $t_{0.025,2} = 4.303$.
H_0 is rejected if the sample T is outside the critical points.

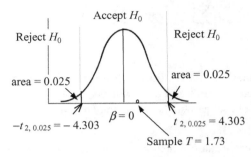

Reject H_0
area = 0.025
$-t_{2,\,0.025} = -4.303$

Accept H_0

Reject H_0
area = 0.025
$t_{2,\,0.025} = 4.303$

$\beta = 0$
Sample $T = 1.73$

Figure 12.6   Decision rule for a two-sided test for slope.

6. **Calculations:** $T = \dfrac{b - \beta_{H_0}}{s_b} = \dfrac{5 - 0}{2.889637} = 1.73$

7. **Conclusion:** since T is in the acceptance region, accept H_0: accept that slope is zero; hence there is no relationship between pay per hour and hours worked per day.
Note: the confidence interval supported the claim that $\beta = 0$.

8. **p-value:** 0.226

(c) **Test the hypothesis that the slope $\beta \geq 5$.** This statement is H_0 since the equality, '=' is in the statement 'β is greater than **or equal to 5**'

1. **H_0:** $\beta \geq 5$ —— $\boxed{\beta \text{ is at least 5}}$
2. **H_a:** $\beta < 5$ (one sided, left tail test)
3. $\alpha = 0.05$ (5%)
4. **Test statistic:** $T = \dfrac{\text{estimate} - \text{claim made in } H_0}{\text{standard error}} = \dfrac{b - \beta_{H_0}}{s_b}$

5. **Decision Rule:** Write down the critical t values for t percentage points, $(n-2)$ degrees of freedom. Hence for $df = (4-2)$, left tail critical: $-t_{2,.05} = -2.920$.
H_0 is rejected if the sample T is less than the critical point.
H_0 is accepted if the sample T is equal to or greater than the critical point.

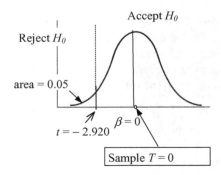

Reject H_0

area = 0.05

Accept H_0

$t = -2.920$

$\beta = 0$

$\boxed{\text{Sample } T = 0}$

6. **Calculations:** $T = \dfrac{b - \beta_{H_0}}{s_b} = \dfrac{5 - 5}{2.889637} = 0$

7. Conclusion: since T is in the acceptance region, accept H_0: so slope is at least 5, hence hourly rate increases by at least £5 per hour when the total hours per day increases by one.
   Note: the confidence interval also supported the claim that $\beta = 5$.
8. **p-value:** 0.5

The calculations for confidence intervals and test statistic for slope are summarised in Chart 12.1.

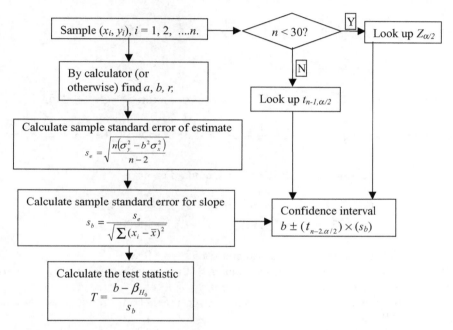

Chart 12.1   Calculation of confidence interval and test static for slope.

Skill Development Exercise 12_1: Inference about slope

The Engineers Profit when the number of PCs sold were as follows

Table SK_12.1   Engineers profit vs. PCs sold

PCs sold	1	3	5	9	12
Profit (£ 00)	−2	8	4	38	18

(a) Carry out the background calculations: $y = a + bx$, σ_x^2, σ_y^2, s_e, $\sum (x_i - \bar{x})^2 = n\sigma_x^2$, s_b
(b) Construct a 95 % confidence interval for slope.
(c) Test the hypothesis that the profit increases by at least £200 for each additional PC sold.

Describe the results of (b) and (c) verbally. Is there any evidence to support the claim that there is a relationship between PCs sold and profit?

Answer

Background calculations

The least-squares line is $\hat{y} = -2.1 + 2.55x$: $\sigma_y^2 = 196.16$: $\sigma_x^2 = 16$

The standard error of estimate

The variance of estimate, $s_e^2 = \dfrac{n}{n-2}\left(\sigma_y^2 - b^2\,\sigma_x^2\right) = 153.53$

N.B Don't forget to take the square root to get the standard error: $s_e = \sqrt{s_e^2} = \sqrt{153.53} = 12.3909$

The standard error for slope, $s_b = \dfrac{s_e}{\sqrt{\sum (x_i - \bar{x})^2}} = 1.3853$ where $\sum (x - \bar{x})^2 = n\sigma_x^2 = 80$

(b) Point estimate $\pm$ (confidence coefficient) $\times$ (standard error)

$$b \pm (t_{n-2,\alpha/2}) \times (s_b) \quad \text{From Tables,} \quad t_{df=3,\alpha/2=2.5\%} = 3.182$$
$$2.55 \pm (3.182)(1.3853) \quad \text{or} \quad 2.55 \pm 4.4081$$

We are 95 % confident that the slope is between -1.8581 to 6.9581.

Therefore, since the interval contains the value $\beta = 0$, the confidence interval above supports the claim that slope is zero or $\beta = 0$. Therefore at the 5 % level of significance, the rate at which profit increases for each additional PC sold is zero is acceptable hence profit is not related to the number of PCs sold!

(c) Test the hypothesis that the slope $\beta \geq 2$

1. H_0: $\beta \geq 2.00$
2. H_1: $\beta < 2.00$
3. $\alpha = 0.05$ (5 %)
4. Use the test statistic, T(since $n \leq 30$) to test the claim
5. H_0 is rejected if the T is less than $t_{3,0.05} = 2.353$. Note $d.f. = (5-2)$
6. Calculate the test statistic: $T = \dfrac{b - \beta_{H_0}}{s_b} = \dfrac{2.55 - 2.00}{(1.3853)} = -0.3970$

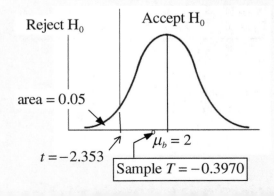

7. Conclusion: since the test statistic, T, is in the acceptance, accept H_0: the increase in profit per additional PC is £200 or more (profit is given in units of £00).

Note: the confidence interval supported the claim that β is within the range -1.8581 to 6.9581: this range includes $\beta = 0$ and also $\beta = 2$. The one-sided test of hypothesis supports the claim that $\beta \geq 2$.

Progress Exercises 12.1: The linear regression model and inference about slope

---

It is recommended that you use your calculator to read off the inbuilt calculations for

$$\bar{x}, \bar{y}, \sum x, \sum y, \sum xy, \sum x^2, \sum y^2, \sigma_x, \sigma_y, s_x, s_y, n, a, b \text{ and } r$$

1. See question 1, PE3.3 (File: Jerseys).
  (a) Plot the scatter diagram and calculate the equation of the least-squares line for the number of jerseys demanded (y) against price per jersey (x).
  (b) Write down the values for $\bar{x}, \bar{y}, \sum y, \sum xy, \sum y^2, \sigma_y^2, \sigma_x^2$ and sample size, n.
    Hence
    (i) Calculate the standard error of estimate, s_e by formulae (12.3) and (12.4).
    (ii) Calculate $\sum (x_i - \bar{x})^2$ by (12.5a) and (iii) the standard error for slope, s_b by formula (12.5).
  (c) Calculate the 95 % confidence interval for slope. Describe this interval verbally.
  (d) Test the hypothesis that the slope is zero at the 5 % level of significance.
  Can you conclude that the demand for jerseys is related to price? Use the results in (c) and (d) to justify your answer.
2. See question 5, PE3.3 (File: Weather).
  (a) Plot the scatter diagram and calculate the equation of the least-squares line for temperature (y) against hours of sunshine (x).
  (b) Calculate (i) the standard error of estimate (ii) the standard error for slope.
  (c) Calculate the 95 % confidence interval for slope. Describe this interval verbally.
  (d) Test the hypothesis that the slope is zero at the 5 % level of significance.
    Can you conclude that 'hours of sunshine' is an accurate predictor of temperature? Refer to the results in (c) and (d) to justify your answer.
3. See question 6, PE3.3 (File: Qualif_Gender).
  (a) Plot the scatter diagram and calculate the equation of the least-squares line for percentage of males (y) against percentage of females (x) that achieve five or more grades A to C in GCSE or SCE.
  (b) Calculate (i) the standard error of estimate (ii) the standard error for slope.
  (c) Calculate the 95 % confidence interval for slope. Describe this interval verbally.
  (d) Test the hypothesis that the slope is zero at the 5 % level of significance.
  What do you conclude about the relationship between the performance of males and females in GCSE/SCE?
    Use the results in (c) and (d) to justify your conclusions.
4. See question 8, PE3.3 (File: Lungs and sulphur 1).
  (a) Plot the scatter diagram and calculate the equation of the least-squares line for lung diseases (y) against SO_2 emissions (x).
  (b) Calculate (i) the standard error of estimate (ii) the standard error for slope.

(c) Calculate the 99 % confidence interval for the increase in the number of people suffering from lung disease (bronchitis/emphysema/asthma, all ages per 100 000) for each additional kg in SO_2 emissions per capita per year. Describe this interval verbally.

(d) Test the hypothesis that the slope is greater than one at the 1 % level of significance. Based on the results in (c) and (d), explain your conclusions about the relationship between lung disease and SO_2 emissions.

5. Selected vital statistics and GDP per capita for selected countries are given in Table PE12.1 Q5.

> File: UNESCAP popl

Table PE12.1 Q5   Selected vital statistics and GDP per capita for selected countries

Country	Crude Birth rate per 1000	Crude Death rate per 1000	Total fertility per woman	Age spec. rate 15–19 per 1000	Life expec. at birth male	Life expec. at birth Female	Mortality under 5 per 1000	% aged 0–14	% aged 60+	GD per capita
China	13.4	7	1.7	5	70	74	37	21	11	5003
Hong Kong China	7.2	5.4	0.9	3	79	85	5	14	16	21179
Indonesia	17.7	7.2	2.2	53	66	70	45	28	8	3361
Singapore	10.1	4.3	1.2	5	77	81	4	19	13	24481
Vietnam	18	5.9	1.9	25	69	73	24	29	7	2490
Russian Federation	11	15.8	1.4	29	59	72	21	15	17	9230
French Polynesia	17.7	5	2.3	37	71	76	11	27	8	
New Zealand	14.1	6.8	2	27	76	81	7	21	17	22582

Source: The United Nations Economic and Social Commission for Asia and the Pacific

If possible use Excel to answer the following:
(a) Plot a scatter diagram and give the equation of the least-squares line and correlation coefficient for the following sets of data:
   (i) Life expectancy at birth for females (y) against males (x).
   (ii) Percentage of the populations ages 0–14 (y) against the percentage aged 60+ (x).
   (iii) Percentage of the populations ages 0–14 (y) against crude birth rate.
   (iv) Crude birth rate against crude death rate.
   (v) GDP per capita against life expectancy for females.
   (vi) GDP per capita against life expectancy for males.
   In each case describe the relationship and comment on the strength of the relationship.
(b) For each set of data (i) to (vi) in part (a) above, use either a confidence interval or a test of hypothesis to determine whether the given variables are related at the 5% level of significance.
(c) In part (a) (i) remove the point (59, 72): life expectancies for males and females in the Russian Federation. Recalculate the correlation coefficient.
   Would you consider the point (59, 72) an outlier?

12.3 Confidence intervals and prediction intervals at $x = x_0$

In section 12.2, it was explained that each random sample (sample size n) is an estimate of the population least-squares line. Figure 12.4 illustrated that estimates of y at $x = x_0$ vary from sample to sample and constitute sampling distributions. Two sampling distributions arise:

(i) the sampling distribution for $\overline{Y}$, the mean value of Y at $X = x_0$ is given as $\overline{y}_{x_0} \sim N(\alpha + \beta x_0, \sigma_{\overline{y}/x_0})$. The estimates of the mean and variance of the sampling distribution are given in (12.7).

The mean value of Y at $X = x_0$ is estimated by the sample least-squares line $\overline{y}_0 = a + bx_0$

The sample variance is estimated by the formula $s^2_{\overline{y}/x_0} = s^2_e \left(\dfrac{1}{n} + \dfrac{(x_0 - \overline{x})^2}{\sum (x - \overline{x})^2} \right)$ (12.7)

(ii) the sampling distribution for the value of Y at $X = x_0$ is given as $\hat{y}_{x_0} \sim N(\alpha + \beta x_0, \sigma_{\hat{y}/x_0})$. The estimates of the mean and variance of the sampling distribution are given in (12.8):

The mean value for Y at $X = x_0$ is given by the sample least-squares line $\hat{y}_0 = a + bx_0$

The sample variance for the predicted y at $x = x_0$ is $s^2_{\hat{y}/x_0} = s^2_e \left(1 + \dfrac{1}{n} + \dfrac{(x_0 - \overline{x})^2}{\sum (x - \overline{x})^2} \right)$ (12.8)

Since the sampling distributions are Normal, the format for confidence intervals and test statistics are the same as those for means and proportions in Chapters 7, 8 and 9.

12.3.1 Confidence interval for the mean value of y, $(\overline{y})$ when $x = x_0$

Since $\sigma^2_{\overline{y}/x_0}$ is unknown and sample size, $n < 30$, then sampling distribution of $\overline{y}$ is a t-distribution with $(n - 2)$ degrees of freedom.

Hence, the confidence interval for the mean value of y, $(\overline{y})$ when $x = x_0$ is given in (12.9).

Point estimate $\pm$ (percentage point) $\times$ (standard error)

Confidence interval for the mean value of y, $(\overline{y})$ when $x = x_0$ $\longleftarrow$ This is referred to as the 'prediction interval' at $x = x_0$

$$(a + bx_0) \pm (t_{n-2,\alpha/2}) \times s_e \sqrt{\dfrac{1}{n} + \dfrac{(x_0 - \overline{x})^2}{\sum (x - \overline{x})^2}}$$ (12.9)

Where the standard error, $s_{\overline{y}/x_0} = s_e \sqrt{\dfrac{1}{n} + \dfrac{(x_0 - \overline{x})^2}{\sum (x - \overline{x})^2}}$

12.3.2 Confidence interval for y when $x = x_0$ (a prediction interval)

Since $\sigma^2_{\hat{y}/x_0}$ is unknown and $n < 30$, the sampling distribution of y_{x_0} is a t-distribution with $(n-2)$ degrees of freedom. Hence, the confidence interval is calculated as follows:

$$\left\{ \begin{array}{l} \text{Point estimate} \pm \text{(confidence coefficient)} \times \text{(standard error)} \\ (a + bx_0) \pm t_{(n-2),\alpha/2} \, s_{\hat{y}/x_0} \end{array} \right.$$

Prediction interval for y when $x = x_0$ ⟵ This is referred to as the 'prediction interval' at $x = x_0$

$$(a + bx_0) \pm (t_{n-2,\alpha/2}) \times s_e \sqrt{1 + \frac{1}{n} + \frac{(x_0 - \bar{x})^2}{\sum (x - \bar{x})^2}} \qquad (12.10)$$

The confidence interval for individual values of y when $x = x_0$ is called the **prediction interval**: The relationship between the confidence and the prediction interval is illustrated in Figure 12.7:

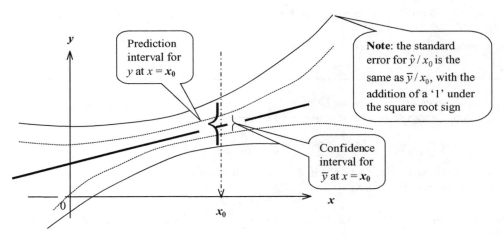

Figure 12.7   Confidence limits (inner curves) and prediction limits (outer curves).

Note: The difference between the confidence and prediction intervals is due to the standard errors. The relationship between the standard errors may be explained by the rule stated in Chapter 5: 'the sum of Normal independent random variables is also Normally distributed with the variance equal to the sum of the separate variances'. Hence, for linear regression, the variance for individual values of Y is the sum of the variance for the mean value of Y (given by the least-squares line) and the variance about the least-squares line:

$$s^2_{\hat{y}/x=x_0} = s^2_e + s^2_{\bar{y}/x=x_0} = s^2_e + s^2_e \left(\frac{1}{n} + \frac{(x_0 - \bar{x})^2}{\sum (x - \bar{x})^2} \right) = s^2_e \left(1 + \frac{1}{n} + \frac{(x_0 - \bar{x})^2}{\sum (x - \bar{x})^2} \right) \qquad (12.11)$$

Worked Example 12.2: Confidence intervals for y and $\bar{y}$ at a specified value of $x = x_0$

From the data in Table 3.4, calculate and describe verbally

(a) a 99% confidence interval for the average hourly rate for a person who works 6 hours daily.
(b) a prediction interval for hourly rate for a person who works 6 hours daily.

Solution

Background calculations: calculate the equation of the least-squares line and the standard errors:
From W.E 3.1, the equation of the least-squares line is $y = -15 + 5x$ and $\bar{x} = 6$
From W.E 12.1 $s_e = 12.92285$

The standard **error for $\bar{y}$ when $x = 6$** is calculated by (12.9):

> **Note:** when $x_0 = \bar{x}$, the standard error for $\bar{y}$ is $\dfrac{s_e}{\sqrt{n}}$

$$s_{\bar{y}/x_0} = s_e \sqrt{\frac{1}{n} + \frac{(x_0 - \bar{x})^2}{\sum (x - \bar{x})^2}} = 12.92285 \sqrt{\frac{1}{4} + \frac{(6-6)^2}{20}} = 12.92285\sqrt{0.25} = 6.46143$$

The standard **error for $\hat{y}$ when $x = 6$** is $s_e \sqrt{1 + \dfrac{1}{n} + \dfrac{(x_0 - \bar{x})^2}{\sum (x - \bar{x})^2}} \ldots (12.10)$:

$$s_{\hat{y}/x_0} = 12.92285 \sqrt{1 + \frac{1}{4} + \frac{(6-6)^2}{80}} = 12.92285\sqrt{1 + 0.25} = 14.44819$$

The estimate of $\hat{y}$ (and $\bar{y}$) when $x = 6$ is calculated from the equation of the least-squares line:
$y = -15 + 5(6) = 15$

Hence the required confidence intervals are

(a) Confidence for the mean value of y, ($\bar{y}$) when $x = 6$
Point estimate $\pm$ (Z or t percentage point) $\times$ (standard error)

$$15 \pm (4.303) \times 6.46143$$
$$15 \pm 27.8035$$
$$-12.8035 < \bar{y} < 42.8035$$

We are 95% confident that the average rate per hour for those who work 6 hours per day is somewhere between £0 and £42.8 (negative rates of pay are not meaningful!).

(b) Confidence for $\hat{y}$ when $x = 6$ (prediction interval)
Point estimate $\pm$ (Z or t percentage point) $\times$ (standard error)

$$15 \pm (4.303) \times 14.44819$$
$$15 \pm 62.17056$$
$$-47.17056 < y < 77.17506, \text{ that is } 0 < y < 77.17506$$

We are 95% confident that the rate per hour for those who work 6 hours per day is between £0 and £77.18 (negative rates of pay are not meaningful!).

Skill Development Exercise SK12_2: Confidence and prediction intervals

For the Engineers profit for PCs sold in Skill Development Exercise SK12_1, Table SK_12.1:

(a) Construct a 95% confidence interval for the average profit 10 PCs are sold.
(b) Construct a 95% confidence interval for the profit when 10 PCs are sold.

Answer

Background calculations given in SK12_1.

The least-squares line is $\hat{y} = -2.1 + 2.55x$ and $\bar{x} = 6$

The standard error for $\bar{y}$ when $x = 10$ is calculated as $s_e \sqrt{\dfrac{1}{n} + \dfrac{(x_0 - \bar{x})^2}{\sum (x - \bar{x})^2}}$ (12.9)

$$s_{\bar{y}/x_0} = 12.3909 \sqrt{\frac{1}{5} + \frac{(10-6)^2}{80}} = 12.3909 \sqrt{0.2 + 0.2} = 7.8367$$

The standard error for y when $x = 10$ is calculated as: $s_e \sqrt{1 + \dfrac{1}{n} + \dfrac{(x_0 - \bar{x})^2}{\sum (x - \bar{x})^2}}$ (12.10)

$$s_{y/x_0} = 12.3909 \sqrt{1 + \frac{1}{5} + \frac{(10-6)^2}{80}} = 12.3909 \sqrt{1 + 0.2 + 0.2} = 12.3909 \sqrt{1.4} = 14.6611$$

The estimate of $\hat{y}$ (and $\bar{y}$) when $x = 10$ is calculated from the equation of the least-squares line: $y = -2.1 + 2.55(10) = 23.4$

(a) **Confidence for the mean value of y, ($\bar{y}$) when $x = 10$**
    Point estimate $\pm$ (confidence coefficient) $\times$ (standard error)
    $23.4 \pm (3.182) \times (7.8367)$
    23.4 ± 24.9364
(b) Confidence for $\hat{y}$ when $x = x_0$ (the prediction interval)
    $23.4 \pm (3.182) \times 14.6611$
    23.4 ± 46.6517

Progress Exercises 12.2: Regression analysis

1. **See question 1, PE12.1 (File: Jerseys)**
   (a) Given that the price per jersey is € 15, calculate the standard error for (i) the mean value of y (the average number of jerseys demanded); (ii) the value of y (the number of jerseys demanded).

(b) Calculate the 95% confidence interval for the mean number of jerseys demanded when the price per jersey is € 15.

(c) Calculate the 95% prediction interval for the number of jerseys demanded when the price per jersey is € 15.

Explain the intervals in (b) and (c) verbally as you would to a non-statistician.

2. See question 2, PE12.1 (File: Weather).

(a) Calculate the standard errors for (i) the confidence interval (ii) prediction interval when there is 5 hours of sunshine per day.

(b) Calculate the 90% confidence interval for the mean temperature when there is 5 hours of sunshine per day.

(c) Calculate the 90% prediction interval for the temperature when there is 5 hours of sunshine per day.

Would you recommend the use of the regression line to predict temperature from hours of sunshine? Use the results of (b) and (c) to justify your answer.

3. See question 3, PE12.1 (File: Qualif_Gender).

(a) Calculate the 95% confidence interval for the mean percentage of males that achieve five or more grades A to C in GCSE or SCE when 40% of females achieve five or more grades A to C in GCSE or SCE.

(b) Calculate the 95% prediction interval for the mean percentage of males that achieve five or more grades A to C in GCSE or SCE when 40% of females achieve five or more grades A to C in GCSE or SCE.

Would you recommend the use of the regression line to predict performance of male students in GCSE or SCE ? Use the results of (a) and (b) to justify your answer.

4. See question 4, PE12.1 (File: Lungs and sulphur 1).

(a) Plot the scatter diagram and calculate the equation of the least-squares line the number of people suffering from lung disease (bronchitis/emphysema/asthma, all ages per 100 000) (y) against SO_2 emissions, kg per capita is emitted per year (x).

(b) Calculate the standard errors for (i) confidence interval (ii) the prediction interval when 50 kg of SO_2 per capita is emitted per year.

(c) Calculate (i) the 99% confidence interval; (ii) the 99% prediction interval when 50 kg of SO_2 per capita is emitted per year.

Describe each interval verbally.

5. In the context of linear regression, explain the following:
(i) Residual variance; (ii) Standard error of the residuals

(a) The data in Table PE12.2 Q5 was provided by a haulage company for the cost of (in € 000) delivering loads of vegetables along with the weight of the load and the distance transported.

File: Vegs

Table PE12.2 Q5   Cost of delivering vegetables of various weights and distance transported

Weight (1000 kg)	25	54	47	23	29	32	83	62	78	38
Distance (km)	52	65	58	26	150	120	98	180	155	240
Cost (€ 000)	2.6	2.8	3.6	2.5	5.8	7.5	5.5	15.3	16.4	21.5

(i) Plot the scatter diagram with distance as the explanatory variable and cost as the response.

(ii) The haulage company claims that the cost of delivery is at least € 1000 per km. Test this claim at the 5% level of significance.

(iii) Calculate a 95% confidence interval for the average cost of delivery when the distance is 50 km.

(b) Repeat part (a) above for weight as the explanatory variable and cost as the response. Explain the results of (a) and (b). What do you conclude about the cost of delivering vegetables?

6. See question 2 PE3.3. The number of apartments given as a percentage of all domestic dwelling sold since 2000 is given in Table PE3.8: (File: % Apartments)

(a) Plot the scatter diagram and number of apartments given as a percentage of all domestic dwelling sold (y) against years since 2000 (x).

(b) Calculate the standard errors for (i) confidence interval (ii) the prediction interval $x = 4$ years since 2000.

(c) Calculate (i) the 95% confidence interval; (ii) the 95% prediction interval when $x = 4$. Describe each interval verbally.

Do you think the assumptions of the linear regression model are satisfied in this example?

12.4 Checks on the model assumptions based on residuals plots

In section 12.1, **the assumptions of the Simple Linear Regression model** were stated in terms of the random errors ε_i where $\varepsilon_i \sim N(\mu_\varepsilon = 0, \text{variance} = \sigma_\varepsilon^2)$.

Recall, that the residuals, $e_i = y_i - \hat{y}_i$ are estimates of the random errors ε_i. Hence, residual plots will be used as graphical checks that the model assumptions are satisfied, namely:

(a) Linearity
(b) Independence
(c) Equal variance (homoscedasticity)
(d) Normality

Calculation of residuals

For example, the residuals for the data in Table 3.1, Skill Development SK3_1 (File: Savings vs. income 1) are calculated in Table 12.2.

The estimated value of y is calculated from the equation of the least-squares line in column 3. The residual is the difference between the observed and estimated value of y as given in column 4.

File: Saving vs. income 1

Table 12.2   Calculation of residuals for data savings vs. income in Table 3.1

Col. 1	Col. 2	Col. 3	Col. 4
(x) Income $\times$ €10 000	Observed y. Saving $\times$ €10	Predicted € $\hat{y} = 83.873 + 52.327x$	Residual $e_i = y - \hat{y}$
2	240	188.527	51.473
3	250	240.854	9.146
4	364	293.181	70.819
5	350	345.508	4.492
6	272	397.835	−125.835
7	422	450.162	−28.162
8	410	502.489	−92.489
9	495	554.816	−59.816
10	687	607.143	79.857
11	750	659.47	90.53

The XY (scatter) plot for the least-squares line is given in Figure 12.8, and the plot of residuals vs. x is given Figure 12.9. Notice: the residuals are simply the data points with the least-squares line removed.

If the assumptions of the linear regression model are satisfied then the residuals should be randomly scattered with mean value of zero and the same variance throughout the range for x. Hence residuals plots may be used as graphical checks for the assumptions as follows:

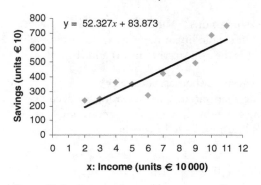

Figure 12.8   Scatter plot and least-squares line.   Figure 12.9   Plot of residuals vs. x.

(a) Graphical check on the Linearity Assumption

Plot the residuals vs. x on a scatter plot as in Figure 12.9. If a linear relationship is appropriate, the residuals will form a random pattern with no obvious trend.

In Figure 12.9, there is no obvious trend, so the linear model is the most appropriate.

What does a residual plot look like when the linear model is not appropriate? The residuals form a pattern. For example, the residual plot in Figure 12.10 is a U-shaped curve an: this suggests a quadratic model for the data.

Note: Figure 12.10 is the residual plot in for the data in Table SK3_4.3, Figure 3.1(c), (File: Savings vs. income) depicts a curvilinear trend.

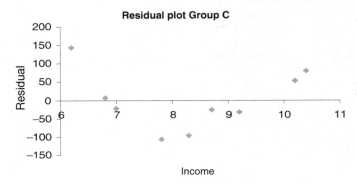

Figure 12.10 Linear model is not appropriate: A quadratic would be better. See Skill Development Exercise SK3_4 Chapter 3.

(b) Graphical check for the independence assumption

This test is particularly important if observed data are collected over consecutive periods of time where there is a risk that the response in any time period is related to that in previous period(s). Examples include monthly sales data; unemployment data; passenger numbers etc. See also question 6, PE12.2; the number of apartments given as a percentage of all domestic dwelling in any quarter must be related to the numbers in previous quarters. A residual plot for the order in which the observed data were obtained provides a graphical check for independence. If there is a pattern in the residuals, then consecutive residuals are related to the independent variable.

Another test (given in Minitab) to graphically check the assumptions (a) linearity and (b) homoscedasticity is the plot of residuals vs. fitted values ($\hat{y}$). Since, in simple linear regression, the fitted values ($\hat{y}$) are directly related to x the interpretation of this plot is the same as that in (a) and (b).

(c) Graphical check for the assumption of equality of variance (homoscedasticity)

Plot the residuals vs. x. If there is a 'fanning in or out' effect, then variance in increasing (or decreasing) as x increases. Then we conclude that variance is not the same for all x, hence not constant.

In Figure 12.9, constant variance about the least-squares line appears to be satisfied.

(d) Graphical check on the assumption of Normality

The assumption of normality may be checked graphically as follows:

 (i) A histogram of the residuals (if there are sufficient data points!). A reasonable bell-shaped histogram indicates normality assumption is satisfied.
 (ii) A box plot: the box should be symmetrical, with mean close to the median and contain no outliers.
(iii) A plot on normal probability graph paper should form a reasonable straight line. Such a plot is given in Figure 12.14.
(iv) Standardised residuals. A residual is standardised when it is divided by its standard error: $\dfrac{e_i}{s_e}$ where

$$s_e = \sqrt{\frac{\sum (y_i - \hat{y}_i)^2}{n - k - 1}}, k = \text{number of independent variables in the model, hence } k = 1 \text{ for simple}$$

regression. If the assumption of Normality is satisfied, the standardised residuals are distributed as $N(0, 1)$ a standard Normal distribution, mean equal to zero and variance equal to one. Hence, since the standard error is one, then 68% should fall within ± 1 and 94.5% within ± 2, etc.

For small sample sizes the histogram in (i) give little information. In test (ii) a single extreme outlier can affect the value of the mean: *outliers should always be investigated*. Test (iii) is the most reliable for small data sets.

Tests (i); (ii) and (iii) are readily executed with a click on the mouse in Minitab. Results for this data set are given in Figure 12.15. The value of the standardised residual for any unusual observation is given in Minitab printouts. However, even a simple calculation of the residual mean, median, with the coefficient of skewness can give an initial indication of Normality. In Table 12.3, the residuals have a mean $= 0.0015$ and a median, $Q_2 = 6.819$. However, the difference between $\overline{x}$ and Q_2 can be misleading if the variance is large (in this case it is 75!) Pearson's coefficient of skewness is calculated as $\dfrac{3(\overline{x} - Q_2)}{\sigma} = -0.2861$. The small coefficient of skewness suggests that the assumption of Normality may be satisfied.

In the next section, you will be given instructions on how to use Minitab to produce residual plots as well as Normal probability plots, histograms of the residuals and box plots. The 'Help' feature in Minitab gives detailed explanations about how to detect departures from the model assumptions

12.5   Regression analysis in Minitab and Excel

To illustrate regression analysis in Minitab the data for savings vs. income, (File: savings vs. income 1), is used.

Open Minitab. Enter data in columns (or cut and paste from an Excel worksheet).

Then click the **Stat** button the menu-bar as shown in Figure 12.11.

Select '**Regression**' from the first and second drop down menus.

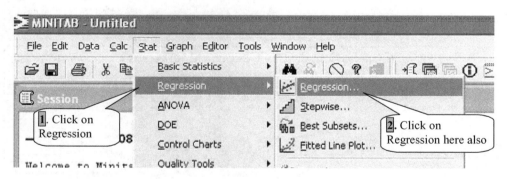

Figure 12.11   Minitab for regression.

In the regression analysis dialogue box, Figure 12.12

1 select the response variable (y) and 2 the predictor variable (x). Double click to enter these in the response and predictor boxes. This is sufficient to produce the 'regression analysis' that will be displayed in the Minitab 'session' window.

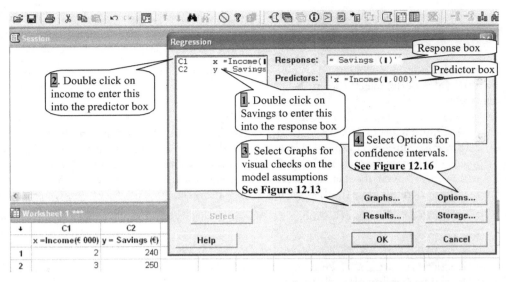

Figure 12.12   Minitab; selecting the response variable (y) and the predictor variable(s), (x).

The regression analysis is displayed in the 'Session' window

1. Click on 'Window' in the menu bar and select 'session' to display the regression analysis in Figure 12.13. The output gives
2. The equation of the least-squares line
3. The coefficients, a and b of the least squares line, with
4. the standard errors (called SE Coef) for a and b.
5. The standard error of estimate, s_e.
6. **The test statistic for slope** $T = \dfrac{b - \beta_{H_0}}{s_b}$

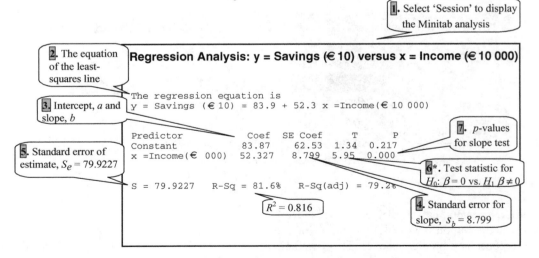

Figure 12.13   Minitab 'Session' window, displaying the regression analysis.

The Minitab output in regression analysis (see Figure 12.13) gives the statistics: b, s_e and s_b and T, test statistic for H_o: $\beta = 0$. Hence the test statistic is $T = \dfrac{b - \beta_{H_0}}{s_b} = \dfrac{52.325 - 0}{8.799} = 5.95$.

*However, the values given in the printout for b and s_b may be used to calculate the test statistic (T) for any other value of β, such as H_o: $\beta = 40$.

$$T = \frac{b - \beta_{H_0}}{s_b} = \frac{52.325 - 40}{8.799} = 1.4007$$

Confidence intervals for slope are readily calculated by using the statistics: b and s_b given in the Minitab printout:

$$b \pm (t_{n-2, \alpha/2}) \times (s_b) \quad \text{where } t_{df=8, \alpha/2=2.5\%} = 2.7515$$
$$52.3 \pm (2.7515)(8.799) \rightarrow 52.3 \pm 24.2104$$

A visual check the model assumptions from the 'Four in one' plots.

Click on the 'Graphs' icon in Figure 12.12. The 'Graphs' dialogue box is given in Figure 12.14.

1. Select 'Four in one' to graphically test the model assumptions.
2. Click OK to finish.

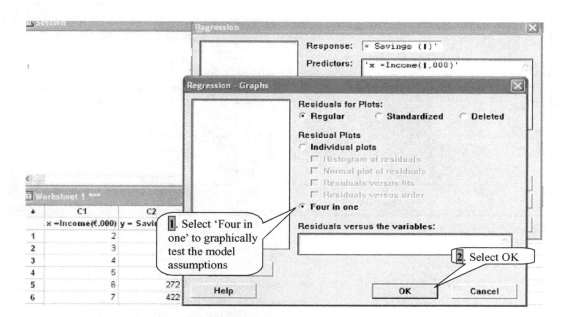

Figure 12.14   Select 'Graphs' to test the model assumptions.

The Minitab output for the 'Four in one' plots is given in Figure 12.15.

Refer back to section 12.4 for more detail and the interpretation of residual plots. The 'Help' on Minitab also gives very useful information and guidelines. Comments on the model assumptions are briefly summarised below.

(a) **Linearity Assumption:** The residuals vs. fitted values are random; hence a linear model is appropriate.
(b) **Constant variance Assumption:** Residuals vs. fitted values are random with no outliers; hence residual variance is likely to be constant.
(c) **Independence Assumption:** Residuals vs. order (observation order) are random, so there is no apparent relationship between successive residuals suggesting that the assumption of independence is satisfactory.
(d) **Normality Assumption:** the Normal probability plot is fairly linear, indicating that the assumption of normality is plausible. However, there are too few point to read anything from the histogram.

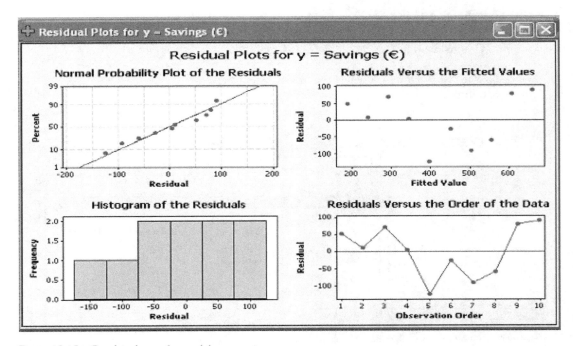

Figure 12.15   Graphical tests for model assumptions.

Confidence and prediction intervals at a given value of x:

The confidence and prediction intervals are obtained following the selection of 'Options' in Figure 12.12. The 'Options' dialogue box id given in Figure 12.16:

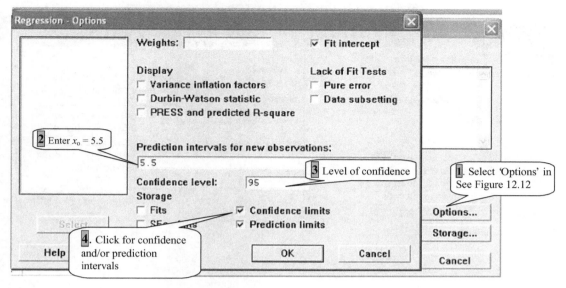

Figure 12.16   Confidence and prediction intervals at $x = 5.5$.

In the 'Options' dialogue box, Figure 12.16, enter the value of x_0 at which the confidence and prediction intervals are required (2). Then select the level of confidence (3) and indicate whether you require either or both intervals (4).

The Minitab output is displayed in the session window with the regression analysis, see Figure 12.17

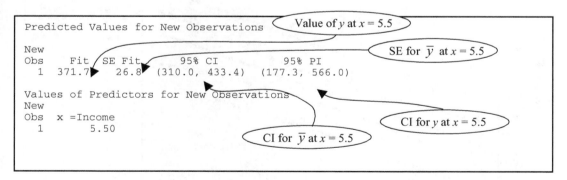

Figure 12.17   Confidence and prediction intervals at $x = 5.5$ (extract of Minitab regression analysis).

Note: the standard error $s_{\hat{y}/x=x_0}$ for prediction interval is not given. It may be may be calculated from standard errors, s_e and $s_{\bar{y}/x_0}$ (S and SE Fit, see Figures 12.13 and 12.17) that are given using the equation $s_{\hat{y}/x=x_0}^2 = s_e^2 + s_{\bar{y}/x=x_0}^2$

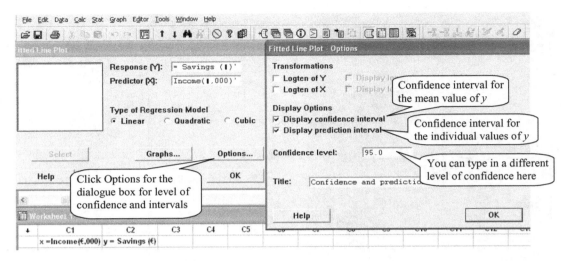

Figure 12.18 Setting up the data for confidence intervals in regression.

Plot the confidence and prediction intervals for the entire range of *x* given in the data: Return to the menu bar and Select: Stat > Regression > Fitted line plot. Then select the response and predictor variables as shown in Figure 12.18. Next, click on 'Options'. Then select confidence and prediction intervals and level of confidence as required. The graph is given in Figure 12.19.

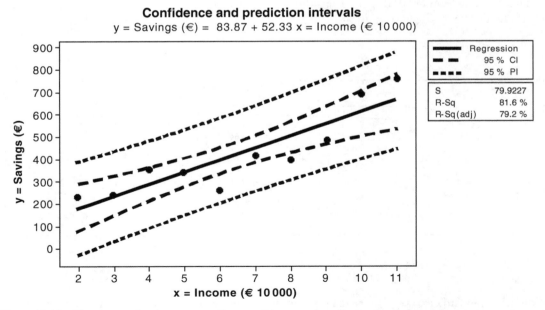

Figure 12.19 The scatter plot, least-squares line, confidence and prediction intervals.

Why are the confidence intervals above at their narrowest when $x = \bar{x}$?

Regression analysis in Excel is similar to Minitab in many of its features; hence it is unnecessary to cover it in detail here. The 'built in' regression analysis in Excel may be accessed by the usual steps Tools → Data analysis → Regression analysis.

Review graphs, XYscatter plots and regression in Chapter 3, section 3.3.

Worked Example 12.3: Regression analysis Minitab and calculator

The discussion above gives the Minitab analysis for the data in Table 3.1 (File: Savings vs. income 1).

x = Income (€ 10 000)	2	3	4	5	6	7	8	9	10	11
y = Savings €	240	250	364	350	272	422	410	495	687	750

(a) Use the calculator (or manually, if you prefer) to calculate the least-squares line, correlation coefficient, standard errors (i) for slope (ii) for the confidence and the prediction intervals at $x = 5.5$.

　　Compare the results with the Minitab printout given above.
(b) **Test the hypothesis that the slope $\beta \geq 40$** (since at least € 40 means € 40 or more) and give the 95 % confidence interval for slope. What do you conclude?
(c) Calculate a 95% confidence interval for average savings when income is 5.5 (i.e. € 55 000) Calculate a 95% confidence interval for savings when income is € 55 000 (called a prediction interval).

Solution

(a) By direct calculations with calculator

$$y = 83.873 + 52.327x : \bar{x} = 6.5\sigma_x = 2.8732; \sigma_y = 166.4326; n = 10;$$

Substitute these into formulae (12.3b), (12.6), (12.9) and (12.10)

Same as
Minitab. See
Figures 12.15
and 12.17

$$s_e = \sqrt{\frac{n}{n-2}\left(\sigma_y^2 - b^2\sigma_x^2\right)} = \sqrt{\frac{10}{8}(27\,699.8 - (238.1435)(8.25))} = 79.9227$$

(12.2b)

Recall $\sum (x - \bar{x})^2 = n\sigma_x^2 = 10(8.25) = 82.5$

$$s_b = \frac{s_e}{\sqrt{\sum (x_i - \bar{x})^2}} = \frac{79.9227}{\sqrt{10(8.25)}} = 8.799$$

(12.6)

$$s_{\bar{y}/x_0} = s_e\sqrt{\frac{1}{n} + \frac{(x_0 - \bar{x})^2}{\sum (x - \bar{x})^2}} = 79.9227\sqrt{\frac{1}{10} + \frac{(5.5 - 6.5)^2}{10(8.25)}} = 26.7617$$

(12.7)

Note: $s_{\bar{y}/x_0}$ is called the standard error of fit (SE Fit) in Minitab.

The standard error for the prediction interval $s_{\hat{y}/x_0}$ is not given in the Minitab print-out but is easily calculated by substituting $x = 5.5$, s_e and $\sum (x - \bar{x})^2 = n\sigma_x^2$ into equation (12.8):

$$s_{\hat{y}/x_0} = s_e\sqrt{1 + \frac{1}{n} + \frac{(x_0 - \bar{x})^2}{\sum (x - \bar{x})^2}} = 79.9227\sqrt{1 + \frac{1}{10} + \frac{(5.5 - 6.5)^2}{10(8.25)}} = 84.2842 \qquad (12.8)$$

OR use (121.13), the sums of variances: $s_{\hat{y}/x=x_0}^2 = s_e^2 + s_{\bar{y}/x=x_0}^2 = (79.9227)^2 + (26.7617)^2 = 7103.8266$. Taking the square root of 7103.8266 gives $s_{\hat{y}/x_0} = 84.2842$
When $x = 5.5$, the fitted value, $\hat{y} = 83.873 + 52.327(5.5) = 371.7$,
These results agree with those given in Figures 13.13 and 12.17.

(b) **Test the hypothesis that the slope $\beta \geq 40$**
   1. H_0: $\beta \geq 40$
   2. H_a: $\beta < 0$ (one-sided left tail test)
   3. $\alpha = 0.05(5\%)$
   4. Test statistic: $T = \dfrac{\text{estimate} - \text{claim made in } H_0}{\text{standard error}} = \dfrac{b - \beta_{H_0}}{s_b}$
   5. The sampling distribution is t with $(n - 2) = (10 - 2) = 8$ degrees of freedom.
      H_0 is rejected if the T less than the left tail 5% point: from Tables $t_{8,0.05} = -1.860$.

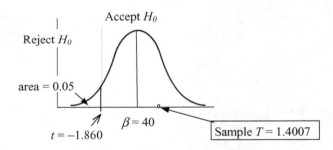

   6. Calculate $T = \dfrac{b - \beta_{H_0}}{s_b} = \dfrac{52.325 - 40}{8.799} = 1.4007$.
   7. Conclusion: since T is in the 'Accept' region, accept H_0: the slope is at least 40; savings increase by at least € 40 × 10 when income increases by € 10 000.

The 95% confidence interval for slope is
$b \pm (t_{n-2,\alpha/2}) \times (s_b)$, where $t_{df=8,\alpha/2=2.5\%} = 2.306$
$52.3 \pm (2.306)(8.799) \rightarrow 52.3 \pm 20.291$
This interval includes $\beta = 40$, hence supports the hypothesis $\beta \geq 40$

(c) (i) $a + bx_0 \pm (t_{n-2,\alpha/2}) \times (s_{\bar{y}/x_0}) \rightarrow 371.7 \pm 2.307(26.76) \rightarrow (310.0, 433.4)$
   (ii) $a + bx_0 \pm (t_{n-2,\alpha/2}) \times (s_{\hat{y}/x_0}) \rightarrow 371.7 \pm 2.307(84.28) \rightarrow (177.3, 566.0)$

Progress Exercises 12.3: Regression analysis with Minitab

1. See question 1 (File: Jerseys), 2 (File: Weather), 3 (File: Qualif_Gender), 3 and 4 (File: Lungs and Sulphur 1), 2PE12.1 and PE12.2.

   Carry out a regression in (a) Minitab and (b) Excel, for the data in the files listed above. Compare the results with those in PE12.1 and PE12.2.

2. See question 5, PE12.2 (File: Vegs).

   The Minitab output for the regression analysis for cost as the response variable and distance as the predictor variable is given below. The Confidence intervals and prediction intervals are for $x = 50$ km.

```
Regression Analysis: Cost (€ 000)_1 versus Distance (km)

The regression equation is
Cost(€ 000) = - 2.36 + 0.0936 Distance (km)

Predictor        Coef   SE Coef      T      P
Constant        -2.361    1.865   -1.27  0.241
Distance (km)   0.09363  0.01424   6.58  0.000
S = 2.87010     R-Sq = 84.4%    R-Sq(adj) = 82.4%

Unusual Observations
      Distance    Cost
Obs     (km)    (€ 000)  Fit   SE Fit  Residual  St Resid
  5      150     5.800  11.683  1.040   -5.883    -2.20R

R denotes an observation with a large standardized residual.

Predicted Values for New Observations
New
Obs   Fit    SE Fit       95% CI              95% PI
  1   2.320  1.290   (-0.655, 5.296)    (-4.936, 9.577)


Values of Predictors for New Observations
New   Distance
Obs   50(km)
  1      50.0
```

Figure PE12.3 Q2a   Minitab output: Regression analysis for File: Vegs.

   (a) Explain each line and explain the calculation in the Minitab output in Figure PE12.3 Q2a.
   (b) Can you conclude that there is a relationship between the cost and distance transported?
3. The residual plots for the data in question 2 (File: Vegs) are given in Table PE12.3 Q3.
   (a) Explain the residual plots. Do the plots indicate that all the model assumptions are satisfied?
   (b) The haulage company checked out the 'unusual' observation (150, 5.8) and found that cost of 5.8 (€ 000) had been incorrectly recorded. Therefore remove this point and repeat the Minitab analysis. Has the removal of this point altered the conclusions in questions 2 and 3?

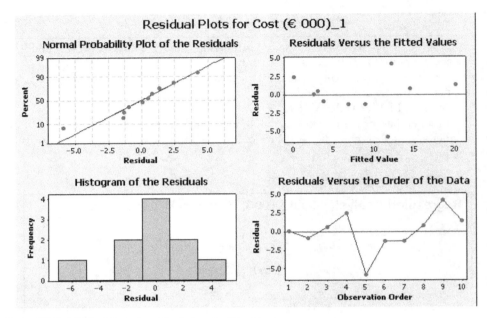

Figure PE12.3 Q3   Minitab residual plots for File: Vegs.

4. Table PE12.3 Q4 summaries the unemployment rates against highest qualification:

Table PE12.3 Q4   Percentage unemployed by highest qualification Spring 2003 | File: Source ONS |

| | Percentage unemployed by highest qualification Spring 2003 | | | | | |
	No qualif.	Degree or equivalent	GCE A Level or equivalent	GCSE grades A* – C or equivalent	Other qualif.	Total (thousands)
United Kingdom	8.9	2.7	3.8	5.5	6.9	1409
North East	11.9		4.1	9.3	8.7	74
North West	8.7		3	5.5	8.3	153
Yorkshire and the Humber	9	2.9	4.2	5.5	8.3	128
East Midlands	8		3.1	4.4	4.4	83
West Midlands	10.2	2.5	3.8	5.5	7.4	141
East	5.5	2	3	4.4	6.2	107
London	12	3.9	6.1	8.1	8.9	248
South East	7.1	2.7	2	3.8	4.8	153
South West	7.2	2.5	3.1	4.3	4.1	87

(a) Plot a scatter diagram with the percentages with A level or equivalent as the response variable and the percentages with no qualification as the predictor variable. Give the regression equation and correlation coefficient.

(b) Carry out a Minitab regression analysis for the data in (a) and include confidence and prediction intervals when the unemployment rate for those with no qualification is 10%.

(c) Is there any evidence that there is a relationship between the percentages unemployed with no qualification and those with A levels or equivalent? Use the results in (a) and (b) to justify your answers.

(d) Test the model assumptions.

5. **See question 5, PE12.2 (File: % Apartments).** The Minitab output for the regression analysis is given:

```
Regression Analysis: % Apartments (y) versus Years since 2000 (x)

The regression equation is
% Apartments (y) = 20.4 + 1.65 Years since 2000 (x)

Predictor                Coef   SE Coef     T      P
Constant               20.400    2.573   7.93  0.004
Years since 2000 (x)   1.6500    0.5252  3.14  0.052

S = 3.32165   R-Sq = 76.7%   R-Sq(adj) = 68.9%

Predicted Values for New Observations
New
Obs    Fit   SE Fit      95% CI           95% PI
  1  61.65   11.13   (26.23, 97.07)   (24.69, 98.61)XX

XX denotes a point that is an extreme outlier in the predictors.
```

Figure PE12.3 Q5   Minitab output for regression analysis for File: % Apartments.

Explain this output.

Is there any evidence that the percentage of apartments increased between 2000 and 2004?

6. See question 5. The residual plots for this data are given in Table PE12.4 Q6. Explain the residual plots below. Do these plots indicate that the model assumptions are satisfied?

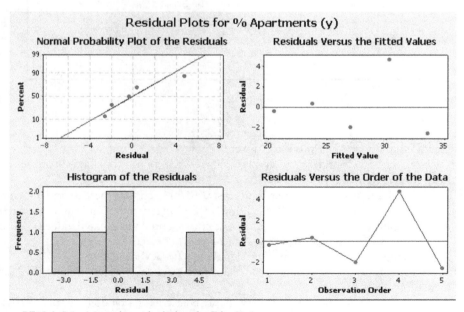

Figure PE12.3 Q6   Minitab residual plots for File: % Apartments.

7. The levels of temperature and gas flow rate (l/h) are thought to be related to yield in the manufacture of paper pulp. A sample of 26 yields was recorded at various temperatures and gas flow rates, see Table PE12.3 Q7:

Table PE12.3 Q7   Yield of paper pulp vs. gas flow rate and temperature | File: Paper pulp |

Gas flow rate (i/h)	1	1	1.1	1.1	1.2	1.2	1.2	1.2	1.3	1.2	1.3	1.3	1.4
Temp	35	35	40	40	45	45	45	45	50	50	50	50	60
Yeild	31.1	32.3	29.8	28.2	27.9	26.3	25.9	28.9	28.6	29.8	27.2	25.2	26.2
Gas flow rate (i/h)	1.2	1.4	1.4	1.3	1.5	1.2	1.6	1.6	1.6	1.7	1.7	1.8	1.8
Temp	60	60	60	65	65	65	70	70	70	80	90	90	90
Yield	28.4	24.5	26.4	28.5	26.2	28.4	24.1	23.7	23.6	22.8	24.3	23.5	19.5

An extract from the Minitab output for yield vs. temperature is given in Figure PE12.3 Q7

```
Regression Analysis: Yield versus Temp
The regression equation is
Yield = 34.8 - 0.141 Temp

Predictor       Coef      SE Coef         T        P
Constant      34.835        1.297     26.85    0.000
Temp         -0.14059      0.02132     -6.59    0.000

S = 1.75544    R-Sq = 64.4%    R-Sq(adj) = 62.9%

Fit      SE Fit        95% CI              95% PI
27.805    0.391    (26.999, 28.611)   (24.093, 31.517)
```

Figure PE12.3 Q7   Minitab output for regression analysis: File: Paper pulp.

Use the information in the printout to answer the following:
(a) Calculate the 95% confidence interval for slope. What do you conclude about the relationship between yield and gas flow rate?
(b) Test the claim that yield decreases by at least 0.10 for each unit increase in temperature.
(c) Explain the terms 'Fit', 'SE Fit', '95% CI' and '95% PI' given in the printout for a temperature of 50°C. Confirm by manual calculation, the corresponding calculated values: '27.805', '0.391', '(26.999, 28.611)' and '(24.093, 31.517)'.
(d) Test the model assumptions by using residual plots from Minitab.
Do these results in (a) and (b) support the claim that yield decreases by at least 0.10 for each unit increase in gas flow rate?

8. The energy conservation team in electricity supply board takes a random sample of 20 domestic dwellings. The monthly electricity consumption (kWh) is recorded along with the energy efficient rating and the number of hours that the dwelling was occupied in the last month in Table PE12.3 Q8:

Table PE12.3 Q8   Electricity consumption, EER and hours
occupied for 20 domestic dwellings

File: EER house

x_1	x_2	y
Hours	EER	Electricity kWh
75	85	525
95	80	678
130	74	950
65	65	778
154	63	1125
82	64	925
175	52	1508
210	43	1755
125	45	1275
175	60	1250
163	35	1875
92	48	1125
225	38	1580
240	50	1550
293	45	1540
165	68	980
96	65	890
120	54	1084
122	46	1235
164	42	1560

(a) Plot a scatter diagram for electricity consumption against EER. Write down the equation of the regression line and coefficient of determination. Give a verbal description of the slope and intercept.

(b) Carry out a regression analysis to determine whether EER is significant predictor of electricity consumption.

(c) Carry out a residual analysis to determine whether the model assumptions are satisfied.

Repeat this question for electricity consumption against hours occupied

9. The number of marriages and divorces per 1000 persons in 30 European countries in 2003 is given in (File: Marriage Divorce 2003), and an extract is given in Table PE12.3 Q9.

(a) On the same diagram, plot line graphs comparing the marriage and divorce rates for the 30 countries. Would you expect a strong correlation between marriages and divorces?

(b) Plot the data for marriage and divorce rates on a scatter diagram. Explain the inference displayed in the diagram.

(c) Carry out a regression analysis with divorce rate as the response variable. What can you conclude about the relationship between the marriage and divorce rates?

Table PE12.3 Q9   Marriage and divorce rates per 1000 persons in 30 countries (extract only given here)

| File: Marriage Divorce |

Country	Marraiges/1000 persons	Divorces/1000 persons
Belgium	4	3
Czech Republic	4.8	3.2
Denmark	6.5	2.9
Germany	4.6	2.6
Estonia	4.2	2.9
Greece	5.5	1.1
Spain	5	1.1
Ireland	5.1	0.7
Italy	4.5	0.8
Norway	6.3	3.1
Switzerland	4.9	2.4

Source: Eurostat.

10. Following a brainstorming session it was decided to investigate the relationship between the Wear of tyres (y mm), vehicle Speed (x_1 km per hour) and the Load (x_2 units of 100kg). The following results were recorded for a fleet of 8 identical pick-up trucks tested under identical conditions. See Table PE12.3 Q9.

Table PE12.3 Q9:   Wear of tyres (y), vehicle Speed (x_1) and the Load (x_2). File: Wear Tyres

| File: Wear Tyres |

Wear (y)	30	56	73	92	56	97	82	95
Speed (x_1)	70	80	95	70	95	50	44	94
Load (x_2)	150	160	200	240	140	290	210	256

(a) Plot a scatter diagram with Wear as the response variable and Load as the explanatory variable.
   Write down the equation of the least-squares line and the correlation coefficient.
(b) Carry out a regression analysis in Minitab. Use the residual plots to determine whether the model assumptions are satisfied.

Repeat the above analysis with Wear as the response variable and Speed as the explanatory variable.

Based on the results above write a brief report explaining whether the Speed and/or Load are related to the Wear of tyres.

12.6 Multiple regression

In the previous sections, the simple linear regression model, gave the value of Y in terms of one independent or explanatory variable X. However, in most real-world situations the value of Y is dependant on not just one explanatory variable but several. To return to the examples on savings (Y) vs. income (X), it is common knowledge that income is only one factor that explains the amount of savings. Some other explanatory variable could include loan repayments (X_2), family size (X_3), cost of education (X_4), holiday expenses (X_5), etc.

The multiple regression model for savings in terms of the explanatory (or independent) variables is expressed as $Y = \beta_0 + \beta_1 X_1 + \beta_2 X_2 + \beta_3 X_3 + \beta_4 X_4 + \dots \varepsilon$ where the constant term is β_0 and the regression (slope) coefficients for income, family size, education, holidays are $\beta_1, \beta_2, \beta_3. \beta_4$, etc.

The method of least square is used to derive formulae for the estimators of regression coefficients; $b_0, b_1, b_2, b_3 \dots \dots$ Hence the sample least-squares estimator of the multiple regression model is written

$$y = b_0 + b_1 x_1 + b_2 x_2 + b_3 x_3 + b_4 x_4 + \dots$$

Note: the constant term is called b_0 in multiple regression instead of 'a' in simple regression.

Interpreting the coefficients in multiple regression

In the savings vs. income example above, the explanatory variables and regression coefficients are interpreted as follows:

If X_1 = income, then β_1 is the change in savings per unit increase in income, assuming all other variable are constant (do not change).

If X_2 = loan repayments then β_2 is the change in savings per unit increase in loan repayments, assuming all other variable are constant (do not change).

If X_3 = family size then β_3 is the change in savings per unit increase in family size, assuming all other variable are constant (do not change).

In this text all the multiple regression calculations will be carried out in Minitab and/or Excel. The test of hypothesis for each slope coefficient $\beta_0, \beta_1, \beta_2, \beta_3 \dots$ is $H_0: \beta_i = 0$, as in simple regression.

Worked Example 12.4: Multiple Regression

The data in Table 12.3 gives savings in terms income and personal loan repayments for 10 randomly selected individuals.

File: Saving vs. income 2

Table 12.3   Savings, income and car loan repayments. File: Savings vs. income 2

($\times$ € 10 000) x_1 (Income)	$\times$ € 10 y (Savings)	$\times$ € 10 x_2 (Car repayments)
2	240	124
3	250	118.5
4	364	110.2
5	350	215.7
6	272	322.6
7	422	245
8	410	336.2
9	495	452.2
10	687	417.6
11	750	350

The Minitab printout (partial) of the multiple linear regression is given below:

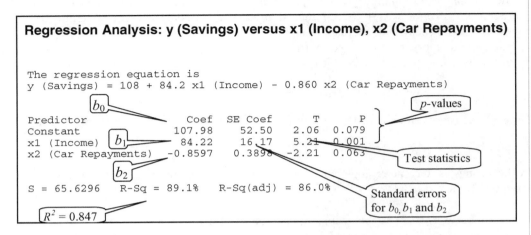

Figure 12.20   Minitab output for multiple regression. File: Savings vs. income 2.

The equation for savings in terms of income and car repayments is

$$y = b_0 + b_1 x_1 + b_2 x_2$$
$$y = 108 + 82.2 x_1 - 0.860 x_2$$

This equation is stated, with variable names, in the Minitab printout in Figure 12.20.

$$y(\text{Savings}) = 108 + 82.2 x_1 (\text{Income}) - 0.860 x_2 (\text{Car repayments})$$

The regression coefficients are described verbally as follows

- $b_0 = 108$:
  This means that average savings $= 108$ when income, $x_1 = 0$ and repayments, $x_2 = 0$.
- $b_1 = 82.2$:
  This means that average savings increase by € 82.2 × 10 = € 822 when income (x_1) increases by 1 unit and car repayments, x_2 do not change (1 unit of income = € 10 000).
- $b_2 = -0.860$:
  This means that average savings **decrease** by € 0.860 × 10 = € 8.6 when car repayments, x_2 increases by 1 unit and income, x_1 does not change (1 unit of car repayments = € 10).

Testing the significance of the slope coefficients

The test for slope follows the same steps as that for simple regression. The standard errors given in the Minitab printout in Figure 12.20 above.

Income

(i) Test the hypothesis that the slope $\beta_1 = 0$
1. $H_0: \beta_1 = 0$
2. $H_1: \beta_1 \neq 0$　　(2-sided test)
3. $\alpha = 0.05$ (5%)

4. Test statistic: $T = \dfrac{b_1 - \beta_{H_0}}{s_{b1}}$

5. Decision Rule:
The sampling distribution is t with $(n - k - 1)$ df. Since $k = 2$, $df = (n - 3) = (10 - 3) = 7$ H_0 is rejected if the T is less than the percentage point in the Tables $t_{0.025,7} = 2.365$

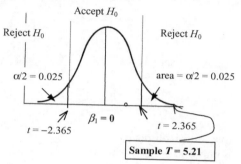

Sample $T = 5.21$

6. Calculations
$$T = \frac{b_1 - \beta_{H_0}}{s_{b_1}} = \frac{84.22}{16.17} = 5.21$$
7. Conclusion: since T is in the reject region, hence reject H_0. Conclude that there is a relationship between savings and income at the 5% level of significance

Car repayments

(ii) Test the hypothesis that the slope $\beta_2 = 0$
1. $H_0: \beta_2 = 0$
2. $H_0: \beta_2 \neq 0$　　(2-sided test)
3. Same

4. Same

5. Same

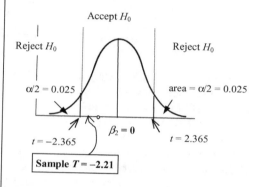

Sample $T = -2.21$

6. Calculations
$$T = \frac{b_2 - \beta_{H_0}}{s_{b_2}} = \frac{-0.8597}{0.3898} = -2.21$$
7. Conclusion: since T is in the accept region, hence accept H_0. Conclude that there is NO relationship between savings and car repayments at the 5% level of significance

Note: the p-value for each of the above coefficients: for savings against income the p-value is 0.001, hence H_0 is rejected while the p-value $= 0.063$ for savings against car repayments is greater than 0.05 (the level of significance of the test) and Ho is accepted – but only just!

Notes on multiple regression

Before making inference in multiple regression it is necessary to state the assumptions of the model:

Assumptions

1. The explanatory variables, X may be fixed or random variables
2. For each explanatory variable, X, the observations Y are independent, Normally distributed, mean value: $Y = \beta_0 + \beta_1 X_1 + \beta_2 X_2$, and variance σ^2
   OR, expressed in terms of the residuals (e), for each X, $\varepsilon \sim N(0, \sigma^2)$
3. For each explanatory variable, X, the variance of the residuals, σ^2 is constant (homoscedasticity).

4. The explanatory variables are not linearly related.

The first 3 assumptions are much the same as those for the simple linear regression model. Assumption 4 applies for multiple regression only. To get some idea of its importance, it is necessary to examine the formulae for the multiple regression coefficients for the model above with two independent variables (explanatory or predictor variables):

$$y = b_0 + b_1 x_1 + b_2 x_2$$

$$b_1 = \frac{s_y(r_{x_1 y} - r_{x_1 x_2} r_{x_2 y})}{s_{x_1}(1 - r_{x_1 x_2}^2)}$$

$$b_2 = \frac{s_y(r_{x_2 y} - r_{x_1 x_2} r_{x_2 y})}{s_{x_2}(1 - r_{x_1 x_2}^2)}$$

$$b_0 = \bar{y} - b_1 \bar{x}_1 - b_2 \bar{x}_2$$

where

$s_y : s_{x_1} : s_{x_2}$ are the sample standard deviations for y, x_1 and x_2 respectively.
$r_{x_1 y}$: correlation between x_1 and y.
$r_{x_2 y}$: correlation between x_2 and y.
$r_{x_1 x_2}$: correlation between x_1 and x_2.

Correlation between independent variables is referred to as **multicollinearity**.
   If the independent variables are highly correlated it is impossible to say which variable predicts changes in the dependent variable, y. Consider the two extreme cases for the independent variables, x_1, x_2.

If $r_{x_1 x_2} = 0$, the formulae for b_1 and b_2 are the same as those in simple regression
For example, $b_1 = \dfrac{s_y(r_{x_1 y} - 0 \times r_{x_2 y})}{s_{x_1}(1 - 0)} = \dfrac{s_y(r_{x_1 y})}{s_{x_1}}$

If $r_{x_1 x_2} = 1$, the coefficients above are not defined because of division by zero
For example, $b_1 = \dfrac{s_y(r_{x_1 y} - 1 \times r_{x_2 y})}{s_{x_1}(1 - 1)} = \dfrac{s_y(r_{x_1 y} - r_{x_2 y})}{s_{x_1}(0)}$, not defined

This means that if x_1 and x_2 are perfectly correlated it will be impossible to say which variable predicts the change in y.
   The following 'rule of thumb' is useful:

Rule of thumb : if $|r_{x_1 x_2}| > 0.7$, then the model suffers from multicollinearity; it is unclear whether either explanatory variable is a significant predictor of y, even if the t-test concludes that the regression coefficient is significant.

The Adjusted R^2.
In simple regression the coefficient of determination was defined in equation (2.3) as

$$R^2 = \frac{\text{Explained variation}}{\text{Total variation}} = \frac{\sum(\hat{y}_i - \bar{y})^2}{\sum(y_i - \bar{y}_i)^2}$$

In multiple regression the value R^2 may be inflated artificially by the number of explanatory variables, particularly if these variables are not significant. The following formula makes adjustments for the number of independent explanatory variables in the model

$$R_{Adj}^2 = 1 - (1 - R^2) \times \left[\frac{n - 1}{n - k - 1} \right]$$
(12.12)

R^2 is the coefficient of determination calculated according to equation (12.12)
n is the number of data points
k is the number of independent explanatory variables

It is particularly important to use the Adjusted R^2 when comparing two or more models that predict the same dependent variable but have different numbers of explanatory variables.
   In the Minitab printout Figure 12.20, the adjusted R^2 is calculated as follows:

$$R_{Adj}^2 = 1 - (1 - 0.847) \times \left[\frac{22 - 1}{22 - 2 - 1} \right] = 1 - (0.169) = 0.831$$

Note: if $R^2 = 1$, then the adjusted R^2 is also 1.

Progress Exercises 12.4: Multiple Regression

1. **See question 5, PE12.2, File: Vegs.**
   This file gives the cost of delivery for loads of vegetables in terms of distance and weight. The Minitab multiple regression printout is given below:

Regression Analysis: Cost (€ 000) versus Distance (km), Weight (1000 kg)

```
The regression equation is
Cost (€ 000) = - 3.75 + 0.0910 Distance (km) + 0.0359 Weight (1000 kg)

Predictor          Coef      SE Coef        T        P
Constant         -3.753       2.631     -1.43    0.197
Distance (km)    0.09103      0.01500     6.07    0.001
Weight (1000 kg) 0.03587      0.04653     0.77    0.466

S = 2.94576    R-Sq = 85.6%    R-Sq(adj) = 81.5%
```

   (a) Write down the multiple regression equation for cost in terms of the predictor variables, distance and weight. Explain each coefficient in terms of cost, distance and weight.
   (b) Explain each part of the Minitab printout.
   (c) Is there a significant relationship between the cost of delivery in terms of distance and weight at the 1% level of significance?
   (d) Calculate the correlation coefficient between distance and weight. Does the result have an effect on the assumptions for the multiple regression model in (c)?

2. **See question 5, PE12.1, File: UNESCAP popl**
   This file gives selected vital statistics for nine countries.
   (a) Carry out a multiple regression analysis for the crude birth rate as the response variable with life expectancy for females and % aged 0–14 as predictor variables.
   (b) Explain the Minitab printout. Justify your conclusion using the analysis form he printout.

3. **See question 2, PE12.1, File: UNESCAP popl**
   Carry out a multiple regression analysis for the life expectancy of males as the response variable with crude birth rate and crude death rate as predictor variables.

(a) Explain the Minitab printout.

(b) Carry out a simple regression analysis for crude birth rate as the response variable and crude death rate as the predictor variable.

Do you think the assumption that the independent variables are uncorrected in part (a) is satisfied?

4. See question 7, PE12.3, File: EER house.

This file gives electricity consumption, EER and hours occupied for private dwellings.

The Minitab multiple regression analysis is given below in Figure PE12.4 Q4:

```
Regression Analysis: Electricity kWh versus Hours occupied, EER

The regression equation is
Electricity kWh = 1890 + 2.28 Hours occupied - 18.2 EER

Predictor          Coef   SE Coef     T      P
Constant         1889.7     191.6   9.86  0.000
Hours occupied   2.2787    0.5484   4.16  0.001
EER             -18.151     2.348  -7.73  0.000

S = 119.169   R-Sq = 90.7%    R-Sq(adj) = 89.6%
```

Figure PE12.4 Q4   Minitab multiple regression analysis for electricity consumption. File: EER house.

(a) Write down the equation relating electricity consumption to EER and hours occupied. Calculate the electricity consumption for a dwelling that is occupied for 10 hours and has an EER of 50.

(b) Explain the adjusted R^2. Does the multiple regression model suffer from multicollinearity?

(c) Test the claim that electricity consumption increases by at least 3 kWh or each hour that the dwelling is occupied. Assume the EER is constant.

(d) Test the claim that electricity consumption decreases by 20 kWh for each unit increase in EER. Assume the hours occupied is constant.

5. See question 6, PE12.3, File: Paper pulp.

(a) In Minitab, produce a multiple regression for the yield of paper pulp in terms of the predictor variables, temperature and gas flow rate.

(b) From the printout, test the claim that (i) temperature; (ii) gas flow rate are related to yield.

(c) Give simple regression analysis for yield in terms of gas flow rate. Calculate the 90% confidence interval for the average yield when the gas flow rate is 1.5.

6. The penalty points system is designed to reduce fatalities and injuries caused by road traffic accidents. Table PE12.4 Q6:

Table PE12.4 Q6   Statistics on road traffic 2003 Source: Irish road safety   File: Road deaths

Year	Deaths	Monthly average	No. Vehicles	No. Licence holders
1998	458	38	1510853	1943184
1999	413	34	1608156	2039509
2000	415	34	1682221	2014296
2001	411	34	1769684	2036624
2002	376	31	1850046	2110666
2003	335	28	1937429	2217076
2004	374	31	2036307	2285323
2005	399	33	2138680	2352540

(a) Give the equation of multiple regression line. Test whether there is a significant relationship between the numbers of deaths and (i) the number of vehicles; (ii) the number of licence holders.

(b) Give a simple regression analysis for the predictor variables: numbers of vehicles vs. licence holders. Is there a significant between them? Test the model assumptions.

(c) Carry out a simple regression analysis for the numbers of deaths as the response variable and the number of licence holders as the explanatory variable. State your conclusions clearly.

(d) Fit (i) a linear trend (a least-squares line); (ii) a suitable curvilinear trend for the numbers of deaths vs. the number of licence holders (use Excel or Minitab). In each case, give the equations and the coefficient of determination. Which model best fits the data?

7. Total energy consumption (KJ), passenger transport (Km driven per person) and non-dairy cattle (millions) are just three of climate change driving force indicators according to scientists and environmentalists. The projected values of these indicators for 2010 is given in Table PE12.4 Q7:

Table PE12.4 Q7   Projected data for selected climate change indicators 2010   File: Energy indicators

Country	Euro/capita GDP	Total KJ Total Energy	Km/capita Transport	Millions Non-dairy cattle
Austria	30 700	1 325	13 339	1.53
Belgium	29 513	2 556	13 979	2.34
Denmark	39 228	817	18 800	1.30
Finland	31 873	1 559	16 216	0.76
France	28 419	12 291	15 835	15.73
Germany	29 984	14 555	13 218	10.13
Greece	16 272	1 510	14 667	0.40
Ireland	38 846	748	14 654	5.87
Italy	25 395	7 683	17 421	5.10
Luxembourg	65 087	198	16 126	0.16
Netherlands	29 913	3 351	13 837	2.53
Portugal	15 026	1 227	14 074	1.07
Spain	19 867	6 146	14 558	4.87
Sweden	34 605	2 285	15 460	1.32
United Kingdom	32 628	9 734	14 999	8.97
Cyprus	17 339	120	12 441	0.03
Czech Republic	8 659	1 712	10 454	0.87
Estonia	7 234	202	9 124	0.14
Hungary	8 513	1 123	9 828	0.46
Latvia	5 780	162	9 170	0.19
Lithuania	5 861	281	6 352	0.46
Malta	13 821	48	9 967	0.01
Poland	7 381	4 039	7 181	2.58
Slovakia	5 998	736	8 029	0.39
Slovenia	14 651	304	15 480	0.26

(a) Do a multiple regression analysis for GDP as the response variable and total energy, transport and non-dairy cattle as predictor variables.

(b) Calculate the correlation coefficients between every possible pair of independent variables. Do these results give any indication that any of the model assumptions are violated? State your conclusions verbally.

(c) Repeat (a) for transport and non-dairy cattle as the two predictor variables. Have your conclusions changed?

Summary and overview of regression analysis

Regression analysis is concerned with making inference about a population linear trend from a sample least-squares line calculated from a random sample selected from that population.

The population least-squares line is written as $\mu_Y = \alpha + \beta X$.

An individual observation in the population is expressed as (x_i, y_i) **where** $y_i = \alpha + \beta x_i + \varepsilon_i$ where ε_i is the random error in Y from the population line for observation i.

The **simple linear regression model assumes**

(a) the observations, Y_i are independent
(b) the probability distribution of Y for each X is **Normal,** with its **mean value** given by the population least-squares line: $\mu_i = \alpha + \beta X_i$ and **variance,** σ_ε^2.
(c) the variance, σ_ε^2 is the same for each value of X.

Note: the model assumptions may also be stated in terms of ε_I, the random error terms about the population line as follows:

1. ε_i are independent
2. $\varepsilon_i \sim N(\text{mean} = 0, \text{variance} = \sigma_\varepsilon^2)$
3. The variance σ_ε^2 is the same for each value of X.

The residual $e_i = (y_i - \hat{y}_i)$ is an estimate of the error term, ε_i.
The variance about the population least-squares line is estimated from sample data by formula (12.1).

$$s_e^2 = \frac{\sum (y_i - \hat{y}_i)^2}{n - 2} = \frac{\sum \{y_i - (a + bx_i)\}^2}{n - 2}, \text{ where } n \text{ is the sample size.} \quad (12.1)$$

Alternative expressions for s_e^2 **are given by formulae (12.2), (12.3) and (12.2b)**

$$s_e^2 = \frac{n}{n - 2} \left(\sum (y_i - \overline{y})^2 - b^2 \sum (y_i - \overline{y})^2 \right) \quad (12.2a)$$

$$s_e^2 = \frac{\sum y^2 - a \sum y - b \sum xy}{n - 2} \quad (12.3)$$

$$s_e^2 = \frac{n}{n - 2} \left(\sigma_y^2 - b^2 \sigma_x^2 \right). \quad (12.2b)$$

Inference about population slope is based on the point estimate, b, the confidence interval and test statistic.

Confidence interval for slope:

Point estimate $\pm$ (confidence coefficient) $\times$ (standard error)

$$b \pm (t_{n-2,\alpha/2}) \times (s_b) \tag{12.4}$$

Test Statistic for slope:

$$T = \frac{\text{estimate} - \text{claim made in } H_0}{\text{standard error}} = \frac{b - \beta_{H_0}}{s_b} \tag{12.5}$$

where

$$s_b = \frac{s_e}{\sqrt{\sum (x_i - \overline{x})^2}} \tag{12.6}$$

Confidence interval for the mean value of y, $(\overline{y})$ **when** $x = x_0$

$$(a + bx_0) \pm (t_{n-2,\alpha/2}) \times s_e \sqrt{\frac{1}{n} + \frac{(x_0 - \overline{x})^2}{\sum (x - \overline{x})^2}} \tag{12.7}$$

Prediction interval for y **when** $x = x_0$

$$(a + bx_0) \pm (t_{n-2,\alpha/2}) \times s_e \sqrt{1 + \frac{1}{n} + \frac{(x_0 - \overline{x})^2}{\sum (x - \overline{x})^2}} \tag{12.8}$$

Regression analysis in Minitab gives the full analysis for slope and intercept with confidence and prediction intervals. In addition, Minitab also produces residual plots that may be used as visual checks that the model assumptions are satisfied.

Multiple regression. The multiple regression model for a dependent variable in terms of several the explanatory (or independent) variable is $Y = \beta_0 + \beta_1 X_1 + \beta_2 X_2 + \beta_3 X_3 + \beta_4 X_4 + \ldots \varepsilon$.

Multiple regression analysis is carried out in Minitab.

APPENDIX A

TECHNICALITIES AND CONVENTIONS FOR DEFINING CLASS INTERVALS, MID-INTERVAL, WIDTHS, LIMITS, BOUNDARIES

Definition of intervals must be clear and unambiguous. However, there are different conventions for defining intervals, two of which are illustrated in (a) and (b) below.

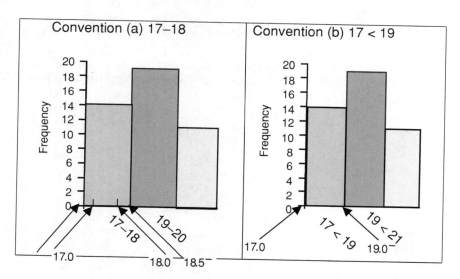

Detail on Class and Interval Widths, Boundaries, Limits and Mid-Interval Value (Skill Development Exercise SKD_1.2)

(a) **17–18:** according to this convention, all numbers that fall within the range 17.0 to 18.0 when rounded correct to one decimal place are included this interval.

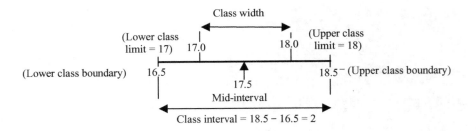

(b) **17 < 19:** according to this convention all numbers that fall within the range 17.0 and up to but not including 19.0 (written as 19.0^-) are included this interval.

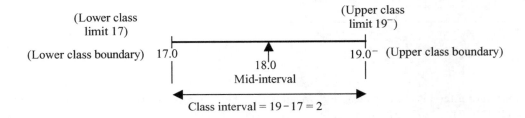

The convention 17 < 19 was used in Skill Development Exercise SK1_2 where the data was initially sorted into a stem-and-leaf plot.

Strictly speaking, for count data the class boundaries and class limits are the same since data can only assume discrete values: but since histograms cannot have 'gaps' the class boundaries must extend by 0.5 on either side of the stated limits.

APPENDIX B

FORMULAE FOR CALCULATING THE QUARTILES FOR GROUPED DATA

The formulae for calculating Q_1, Q_2 (median) and Q_3 for grouped data are given in Table 2.5

Table Appendix 2.1 Formulae for the calculation of Quartiles Q_1, Q_2, Q_3 from the frequency distribution table

Q_1 (Lower quartile)	Q_2 (Median or second quartile)	Q_3 (Upper quartile)
$$Q_1 = L_{Q1} + \left(\frac{\frac{N}{4} - cf}{f_{Q1}} \right)(w)$$	$$Q_2 = L_{Q2} + \left(\frac{\frac{N}{2} - cf}{f_{Q2}} \right)(w)$$	$$Q_3 = L_{Q3} + \left(\frac{\frac{3N}{4} - cf_3}{f_{Q3}} \right)(w)$$
(2.6)	(2.7)	(2.8)

Where

L_{Q1} = Lower limit of interval (or class) containing Q_i.

f_{Q1} = Frequency of the interval containing Q_i.

w = width of the interval containing Q_i.

cf = Cumulative frequency up to, but not including the Q_i interval

Example: Determine the Median and Quartiles from the Frequency Distribution Table

For the data given in Table 2.5 (hours worked by the 49 staff) calculate the values of the quartiles using the formulae in Table Appendix 2.1

Solution

The first step in calculating more accurate estimates for Q_1, Q_2 (median) and Q_3 for grouped data is to identify the intervals that contain Q_1, Q_2 and Q_3. See Table 2.4, Worked Example 2.3.

Now apply the formulae in Table Appendix 2.1.

The results are given in Table Appendix 2.2, below.

Table Appendix 2.2   Calculation of quartiles from grouped data

Q_1 (Lower quartile)	Q_2 (Median or second quartile)	Q_3 (Upper quartile)
$Q_1 = L_{Q1} + \left(\dfrac{\frac{N}{4} - cf}{f_{Q1}} \right)(w)$	$Q_2 = L_{Q2} + \left(\dfrac{\frac{N}{2} - cf}{f_{Q2}} \right)(w)$	$Q_3 = L_{Q3} + \left(\dfrac{\frac{3N}{4} - cf_3}{f_{Q3}} \right)(w)$
(2.6)	(2.7)	(2.8)
$Q_1 = 20 + \left(\dfrac{\frac{49}{4} - 3.5}{12} \right)5 = 22.58$	$Q_2 = 30 + \left(\dfrac{\frac{49}{2} - 24}{1} \right)5 = 30.5$	$Q_3 = 40 + \left(\dfrac{\frac{3(49)}{4} - 3.5}{10} \right)5 = 43.5$

Compare the results for the raw data calculated in Worked Example 2.1 where $Q_1 = 21.45$; $Q_2 = 30.2$, $Q_3 = 41.5$.

APPENDIX C

OUTLINE OF DERIVATION OF FORMULAE (3.2) AND (3.3) FOR THE SLOPE AND INTERCEPT OF THE LEAST-SQUARES LINE, $Y = a + bX$

The criterion for the best fit line is that the sum of the squared residuals should be a minimum. That is, $S = \sum (d_i)^2$ should be a minimum.

The residual for any point (x_i, y_i) is $d_i = y_i - \hat{y}_i$ where $\hat{y}_i = a + bx_i$

Hence $SS = \sum (d_i)^2 = \sum (y_i - \hat{y})^2 = \sum (y_i - a - bx_i)^2$

SS will be a minimum when $\dfrac{\partial SS}{\partial a} = 0$ and $\dfrac{\partial SS}{\partial b} = 0$

$$SS = \sum (y_i - a - bx_i)^2$$

$$\frac{\partial SS}{\partial a} = \sum -2(y_i - a - bx_i) = -2\left[\sum y_i - \sum a - b \sum x_i\right]$$

$$\frac{\partial SS}{\partial b} = \sum -2x_i(y_i - a - bx_i) = -2\left[\sum x_i y_i - \sum a x_i - b \sum x_i^2\right]$$

Equate the first derivatives to zero

$$\frac{\partial SS}{\partial a} = 0$$

$$\sum y_i - \sum a - b \sum x_i = 0$$

$$\text{Hence } \sum y_i = an + b \sum x_i = 0 \tag{C.1}$$

$$\frac{\partial SS}{\partial b} = 0$$

$$\sum x_i y_i - \sum a x_i - b \sum x_i^2 = 0$$

$$\text{Hence } \sum x_i y_i = a \sum x_i + b \sum x_i^2 = 0 \tag{C.2}$$

Solve equations (C1) and (C2) to find the values of a and b at whish SS is a minimum

$$\sum x_i \sum y_i = an \sum x_i + b \sum x_i \sum x_i = 0 \qquad \text{(C1)} \times \sum x_i$$

$$n \sum x_i y_i = an \sum x_i + bn \sum x_i^2 = 0 \qquad \text{(C2)} \times n$$

$$\sum x_i \sum y_i - n \sum x_i y_i = bn \sum x_i^2 - b \left[\sum x_i\right]^2 = b \left(n \sum x_i^2 - \left[\sum x_i\right]^2\right)$$

Hence $b = \dfrac{n \sum x_i y_i - \sum x_i \sum y_i}{n \sum x_i^2 - \left[\sum x_i\right]^2}$

Or, dropping the subscripts, $b = \dfrac{n \sum xy - \sum x \sum y}{n \sum x^2 - \left[\sum x\right]^2}$ \qquad (3.2)

Solve equation (1) for a

$$a = \frac{\sum y - b \sum x}{n} = \frac{\sum y}{n} - b \frac{\sum x}{n} = \bar{y} - b\bar{x} \qquad (3.3)$$

APPENDIX D

BRIEF REVIEW OF THE MATHEMATICS FOR THE BINOMIAL

In maths, a 'Binomial' is the sum or difference of 2 terms raised to a power:

$$(q + p)^0 = 1..\text{since (any real number)}^0 = 1$$
$$(q + p)^1 = q + p \quad ....$$
$$(q + p)^2 = q^2 + 2qp + p^2 ...$$
$$(q + p)^3 = q^3 + 3q^2 p + 3qp^3 + p^3$$
$$(q + p)^4 = q^4 + 4q^3 p + 6q^2 p^2 + 4qp^3 + p^4 \text{ etc.}$$

There is a very definite pattern developing in each of the expansions, $(q + p)^n$ above, for example, consider $(q + p)^4$: Start by looking at the powers on q in each term of the series

> Power on q **starts at $n = 4$**, then decreases by 1 in each term of the series from to $n = 0$

$$(q + p)^4 = q^{④} + 4 q^{③}p + 6q^{②}p^2 + 4q^{①}p^3 + q^{⓪}p4 \text{ ...writing out } q = q^1 \text{ and } 1 = q^0$$

The powers on p are going in the opposite way

$$(q + p)^4 = q^4 p^{⓪} + 4q^3 p^{①} + 6 q^2 p^{②} + 4q^1 p^{③} + q^0 p^{④}$$

> Power on p increases by 1 in each term of the series from $n = 0$ to $n = 4$

Now, the numbers that multiply each term of the series (called coefficients)

$$(q + p)^4 = \text{①}q^4 p^0 + \text{④}q^3 p^1 + \text{⑥}q^2 p^2 + \text{④}q^1 p^3 + \text{①}q^0 p^4$$

> Numbers multiplying each term (coefficients) are **1 4 6 4 1**
> from Pascal's triangle or ^nC_r formulae

The coefficients may be written down from Pascal's triangle:

Pascal's Triangle: coefficients for $(q + p)^n$

$(q + p)^0$								1								
$(q + p)^1$							1		1							
$(q + p)^2$						1		2		1						
$(q + p)^3$					1		3		3		1					
$(q + p)^4$				1		4		6		4		1				
$(q + p)^5$			1		5		10		10		5		1			
$(q + p)^6$		1		6		15		20		15		6		1		
$(q + p)^7$	1		7		21		35		35		21		7		1	
$(q + p)^8$	1	8		28		56		70		56		28		8		1

Any row of Pascal's triangle may be calculated by the formula

$$^nC_x = \frac{n!}{(n-x)!x!} \text{ where } n \text{ is the power on the binomial: } x = 0, 1, 2, 3, \ldots n$$

n! is called 'n factorial' and is evaluated as $n! = (n)(n-1)(n-2)\ldots.3.2.1,$
For example, $4! = 4 \times 3 \times 2 \times 1 = 24$.
   The formula ^nC_x (and $n!$) may be evaluated on most scientific calculators for positive integer values
of both n and x. They may also be calculated manually as follows

$$^4C_{0:} = \frac{n!}{(n-r)!r!} = \frac{4!}{(4-0)!0!} = \frac{4 \times 3 \times 2 \times 1}{(4 \times 3 \times 2 \times 1)(1)} = 1 \text{...where } 0! = 1$$

$$^4C_{1:} = \frac{n!}{(n-r)!r!} = \frac{4!}{(4-1)!1!} = \frac{4!}{(3)!1!} = \frac{4 \times 3 \times 2 \times 1}{(3 \times 2 \times 1)(1)} = 4$$

$$^4C_{2:} = \frac{n!}{(n-r)!r!} = \frac{4!}{(4-2)!2!} = \frac{4!}{(2)!2!} = \frac{4 \times 3 \times 2 \times 1}{(2 \times 1)(2 \times 1)} = 6$$

$$^4C_{3:} = \frac{n!}{(n-r)!r!} = \frac{4!}{(4-3)!3!} = \frac{4!}{(1)!3!} = \frac{4 \times 3 \times 2 \times 1}{(1)(3 \times 2 \times 1)} = 4$$

$$^4C_{4:} = \frac{n!}{(n-r)!r!} = \frac{4!}{(4-4)!4!} = \frac{4 \times 3 \times 2 \times 1}{(0)(4 \times 3 \times 2 \times 1)} = 1$$

Therefore, to write out the expansion of $(q + p)^n$ apply the above pattern to the powers on q and p and use the ^nC_x formula to calculate the

$$\text{coefficients} (q + p)^n = {}^nC_0 q^n p^0 + {}^nC_1 q^{n-1} p^1 + {}^nC_2 q^{n-2} p^2 + \ldots {}^nC_x q^{n-x} p^x + \ldots + {}^nC_n q^0 p^n$$

This is the probability that there will be x successes in a sample size n

Recall, in chapter 4, section 4.2, the number of different arrangements of n objects in n_1 were identical: n_2 were identical. . . . . n_k were identical is $\dfrac{n!}{n_1! n_2! \ldots n_k!}$

Hence the number of different arrangements of n objects in x were one type (the x successes) and the remaining $(n - x)$ were of another type (non- successes) is $\dfrac{n!}{x!(n - x)!}$

But this is identical to the formula: $^nC_x = \dfrac{n!}{(n - x)! x!}$

(**n!** is called 'n factorial' and is evaluated as $n! = (n)(n - 1)(n - 2) \ldots 3.2.1$).

APPENDIX E

THE NUMBER *e*

The Number *e*

The letter e represents a number which has an unending decimal part, just like the number π.

The number $\pi = \dfrac{22}{7} = 3.1415927\ldots\ldots$ arose naturally in circular measurements, the length of the circumference of a circle was shown to be 2π multiplied by the radius. Since the number π is an unending decimal it is represented by the letter π.

The number e arises when growth and decay in all types of systems are described mathematically, for example, the size of a population at time t is calculated from the formula $P = P_0 e^{rt}$ where P_0 is the population at $t = 0$ and r is the annual rate of growth.

The value of investments is given by the formula: $I = I_0 e^{rt}$: the amount of radioactive material remaining at time, t is calculated from the equation: $N = N_0 e^{-\lambda t}$, etc.

But, like π, the number e is represented by a letter, so that the user can evaluate it to any required number of decimal places.

For example, to evaluate e, evaluate e^1 on the calculator. You should get $e = 2.7182818$.

APPENDIX F

CALCULATION OF MEAN AND VARIANCE OF PROPORTIONS BY EXPECTED VALUES

The **expected value** of $g(x)$, a function of a random variable is $E(g(x)) = \sum g(x_i) \times P(X = x_i)$ ...(5.7).

Show that $E(aX) = aE(X)$

$E(X) = \sum xP(x)$

hence

$$E(aX) = \sum (ax) \times P(x)$$
$$= \sum ax P(x)$$
$$= a \sum x P(x)$$
$$= a E(x)$$

Show $V(aX) = a^2 V(X)$

$V(X) = E(X^2) - [E(X)]^2$

hence

$$V(aX) = E\{(aX)^2\} - [E(aX)]^2$$
$$= E(a^2X^2) - [aE(X)]^2$$
$$= a^2 E(X^2) - a^2 [E(X)]^2$$
$$= a^2 \left\{ E(X^2) - [E(X)]^2 \right\}$$
$$= a^2 V(X)$$

Sample proportion $p = \dfrac{X}{n}$, where X is $B(n, p)$. Hence $E(X) = n\pi; V(X) = n\pi(1 - \pi)$.

$$E(p) = E\left(\frac{X}{n}\right) = \sum \frac{x}{n} \times P(x)$$
$$= \frac{1}{n} \sum x P(x)$$
$$= \frac{1}{n} E(X)$$
$$= \frac{1}{n} \times n\pi$$
$$= \pi$$

$$V(p) = V\left(\frac{X}{n}\right) = E\left\{\left(\frac{X}{n}\right)^2\right\} - \left[E\left(\frac{X}{n}\right)\right]^2$$
$$= E\left(\frac{X^2}{n^2}\right) - \left[\frac{1}{n}E(X)\right]^2$$
$$= \frac{1}{n^2} E(X^2) - \frac{1}{n^2} [E(X)]^2$$
$$= \frac{1}{n^2} \left\{ E(X^2) - [E(X)]^2 \right\}$$
$$= \frac{1}{n^2} V(X) = \frac{1}{n^2} \times n\pi(1 - \pi)$$
$$= \frac{\pi(1 - \pi)}{n}$$

APPENDIX G

CONFIDENCE INTERVALS FOR MEANS AND PROPORTIONS

$$\varepsilon = Z_{\alpha/2}\sqrt{\frac{\overline{p}(1 - \overline{p})}{n}}$$

Hence $n = \left(\dfrac{Z_{\alpha/2}}{\varepsilon}\right)^2 \overline{p}(1 - \overline{p})$

$$n = \left(\frac{Z_{\alpha/2}}{\varepsilon}\right)^2 (\overline{p} - (\overline{p})^2)$$

n is a maximum when $\dfrac{dn}{d\overline{p}} = 0$ and $\dfrac{d^2n}{d\overline{p}^2}$ is negative

Hence find the first and second derivatives of n with respect to $\overline{p}$.

$$\frac{dn}{d\overline{p}} = \left(\frac{Z_{\alpha/2}}{\varepsilon}\right)^2 (1 - 2(\overline{p})) : \frac{d^2n}{d\overline{p}^2} = -2\left(\frac{Z_{\alpha/2}}{\varepsilon}\right)^2$$

To find the value of $\overline{p}$ for which n is a maximum, solve $\dfrac{dn}{d\overline{p}} = 0$

$$\frac{dn}{d\overline{p}} = \left(\frac{Z_{\alpha/2}}{\varepsilon}\right)^2 (1 - 2(\overline{p})) = 0$$

$$(1 - 2(\overline{p})) = 0 \rightarrow \overline{p} = 0.5$$

The second derivative is negative, hence n is a maximum when $\overline{p} = 0.5$.

APPENDIX H

DEGREES OF FREEDOM

The degrees of freedom are the number of independent pieces of information used to estimate variance (and hence standard error). To estimate of σ by the sample standard deviation, $s = \sqrt{\dfrac{\sum (x_i - \bar{x})^2}{n-1}}$ one must first calculate $\bar{x}$. $\bar{x}$ has $(n-1)$ independent pieces of information. For example, suppose $n = 4$ and $\bar{x} = 10$: then 3 of the sample data may assume any values but the fourth data must be such that the value of $\bar{x} = 10$ (if $x_1 = 2$; $x_2 = 24$; $x_3 = 5$ then x_3 must be 9 to give $\bar{x} = 10$). Hence there are 3 degrees of freedom.

APPENDIX I

NOTES ON SUMMATIONS AND DOUBLE SUMMATIONS

(a) Summing a constant, 'a' n times

$$\sum_{i=1}^{i=n} a = a + a + a + \dots + a = an$$

For Example, $\displaystyle\sum_{i=1}^{4} a = a + a + a + a = 4a$

(b) Double summing a constant.

For Example $\displaystyle\sum_{j=1}^{5} \left[\sum_{i=1}^{2} a \right] = \sum_{j=1}^{5} [a + a]$

> Write out the inner sum first, for $i = 1$ and $i = 2$

$$= \sum_{j=1}^{5} 2a$$

> Write out the outer sum next for $j = 1, 2, 3, 4$ and 5.

$$= 2a + 2a + 2a + 2a + 2a = 5(2a)$$

3. Summing Variables 3 from one dimensional lists, such as $x_1, x_2, x_3 \dots$ and $y_1, y_2, y_3 \dots$

For Example $\displaystyle\sum_{j=1}^{3} \left[\sum_{i=1}^{2} x_i y_j \right]$

$$\sum_{j=1}^{3} \left[\sum_{i=1}^{2} x_i y_j \right] = \sum_{j=1}^{3} [x_1 y_j + x_2 y_j]$$

> Write out the inner sum first, for $i = 1$ and $i = 2$

$$= \sum_{j=1}^{3} y_j [x_1 + x_2]$$

> Write out the outer sum next for $i = 1, 2, 3$.

$$= y_1[x_1 + x_2] + y_2[x_1 + x_2] + y_3[x_1 + x_2]$$
$$= (y_1 + y_2 + y_3)[x_1 + x_2]$$

Alternatively: this summation could also have been written as follows:

$$\sum_{j=1}^{3} y_j \sum_{i=1}^{2} x_i$$

> Since x is depends on i only and y depends on j only

Then summing over each index separately

$$\sum_{j=1}^{3} y_j \sum_{i=1}^{2} x_i = (y_1 + y_2 + y_3)[x_1 + x_2]$$

4. Summation of variables in a two dimensional table: n rows and c columns. For Example, the Data in Table 10.1 may be represented in five rows and three columns

Marketing	Accounts	Personnel
$14 = x_{1,1}$	$8 = x_{1,2}$	$13 = x_{1,3}$
$9 = x_{2,1}$	$13 = x_{2,2}$	$9 = x_{2,3}$
$11 = x_{3,1}$	$15 = x_{3,2}$	$6 = x_{3,3}$
$12 = x_{4,1}$	$10 = x_{4,2}$	$10 = x_{4,3}$
$7 = x_{5,1}$	$13 = x_{5,2}$	$9 = x_{5,3}$

Each datum is referred to as $x_{i,j}$ or $x_{row,\ column}$ where i and j give the row and column that contains $x_{i,j}$ respectively OR i and j give the position in the sample and the sample number for the value $x_{i,j}$:

$$\sum_{j=1}^{3} \left[\sum_{i=1}^{5} x_{i,j} \right] = \sum_{j=1}^{3} [x_{1,j} + x_{2,j} + x_{3,j} + x_{4,j} + x_{5,j}] \ \ldots i = 1 \text{ to } 5 \text{ for the inner sum}$$

Next write out the outer summation:

$$= [x_{1,1} + x_{2,1} + x_{3,1} + x_{4,1} + x_{5,1}] \quad \ldots \ldots J = 1, \text{ lists the five data in sample 1}$$
$$+ [x_{1,2} + x_{2,2} + x_{3,2} + x_{4,2} + x_{5,2}] \quad \ldots \ldots J = 2, \text{ lists the five data in sample 2}$$
$$+ [x_{1,3} + x_{2,3} + x_{3,3} + x_{4,3} + x_{5,3}] \quad \ldots \ldots J = 3, \text{ lists the five data in sample 3}$$

Mean Value Estimated from All the Data

The overall mean for all the data in the table is expressed as

$$\bar{\bar{x}} = \frac{\sum\limits_{j=1}^{3}\sum\limits_{i=1}^{5} x_{i,j}}{(3)(5)}$$

Variance Estimated From All the Data

The estimate of variance from all the data in the table may be expressed as

$$s^2 = \frac{\sum\limits_{j=1}^{3}\sum\limits_{i=1}^{5}(x_{i,j} - \bar{\bar{x}})^2}{(3)(5) - 1}$$

$$= \frac{\sum\limits_{j=1}^{3}\left[(x_{1,j} - \bar{\bar{x}})^2 + (x_{2,j} - \bar{\bar{x}})^2 + (x_{3,j} - \bar{\bar{x}})^2 + (x_{4,j} - \bar{\bar{x}})^2 + (x_{5,j} - \bar{\bar{x}})^2\right]}{(3)(5) - 1}$$

$$\ldots i = 1 \text{ to } 5\text{: each of sample is size } 5$$

$$= \frac{\left[(x_{1,1} - \bar{\bar{x}})^2 + (x_{2,1} - \bar{\bar{x}})^2 + (x_{3,1} - \bar{\bar{x}})^2 + (x_{4,1} - \bar{\bar{x}})^2 + (x_{5,1} - \bar{\bar{x}})^2\right]}{(3)(5) - 1}$$

$$\ldots \text{ squared deviations of each value in sample 1 from } \bar{\bar{x}}$$

$$+ \frac{\left[(x_{1,2} - \bar{\bar{x}})^2 + (x_{2,2} - \bar{\bar{x}})^2 + (x_{3,2} - \bar{\bar{x}})^2 + (x_{4,2} - \bar{\bar{x}})^2 + (x_{5,2} - \bar{\bar{x}})^2\right]}{(3)(5) - 1}$$

$$\ldots \text{ squared deviations of each value in sample 2 from } \bar{\bar{x}}$$

$$+ \frac{\left[(x_{1,3} - \bar{\bar{x}})^2 + (x_{2,3} - \bar{\bar{x}})^2 + (x_{3,3} - \bar{\bar{x}})^2 + (x_{4,3} - \bar{\bar{x}})^2 + (x_{5,3} - \bar{\bar{x}})^2\right]}{(3)(5) - 1}$$

$$\ldots \text{ squared deviations of each value in sample 3 from } \bar{\bar{x}}$$

APPENDIX J

EXPRESSING THE ESTIMATES OF VARIANCE AS SUMS OF SQUARES DIVIDED BY DEGREES OF FREEDOM IN THEIR SIMPLEST FORM

In order to generalise the estimates of variance s_T^2, s_B^2, s_W^2 arrange the c samples (each sample is size n) in columns as shown in Table 10.3. For each datum, $x_{i,j}$, the values of the index i range from 1 to n and give the position of the datum within a sample. The index j ranges from 1 to c and refers to the sample number.

Table 10.3 c samples, each of size n arranged in columns

	Column 1 (Sample 1)	Column 2 (Sample 2)	Column 3 (Sample 3)	—	Column c (Sample c)
Row 1	$x_{1,1}$	$x_{1,2}$	$x_{1,3}$	—	$x_{1,c}$
Row 2	$x_{2,1}$	$x_{2,2}$	$x_{2,3}$	—	$x_{2,c}$
Row 3	$x_{3,1}$	$x_{3,2}$	$x_{3,3}$	—	$x_{3,c}$
—	—	—	—	—	—
Row n	$x_{n,1}$	$x_{n,2}$	$x_{n,3}$	—	$x_{n,c}$
Sample mean	$\overline{x}_{\bullet,1}$	$\overline{x}_{\bullet,2}$	$\overline{x}_{\bullet,3}$	—	$\overline{\overline{x}}$ (overall mean)

..where the '$\bullet$' in place of subscript i means addition has taken place over all data, i in the column $j = 1$

Note, the mean of sample 1 is written $\overline{x}_{\bullet,1}$

hence $\overline{x}_{\bullet,1} = \dfrac{\sum\limits_{i=1}^{i=n} x_{i,1}}{n}$. Similarly, the mean of sample 2 is $\overline{x}_{\bullet,2} = \dfrac{\sum\limits_{i=1}^{i=n} x_{i,2}}{n}$, etc.

Hence

1. **Variance estimated form all the data (Total)**
   This estimate of variance is based on all the data from the c samples, each sample of size n

$$s_T^2 = \frac{(x_{1,1} - \bar{\bar{x}})^2 + (x_{1,2} - \bar{\bar{x}})^2 + (x_{1,3} - \bar{\bar{x}})^2 \Lambda\Lambda + (x_{n,c} - \bar{\bar{x}})^2}{nc - 1} = \frac{TSS}{df(Total)}$$

$$= \frac{\sum_{j=1}^{c} \sum_{i=1}^{n} (x_{i,j} - \bar{\bar{x}})^2}{nc - 1} = \frac{TSS}{df(Total)}$$

TSS is Total Sum of Squares

2. **Variance estimated from the difference between sample means** (Method B)

$$\frac{s_B^2}{n} = \frac{(\bar{x}_{\bullet,1} - \bar{\bar{x}})^2 + (\bar{x}_{\bullet,2} - \bar{\bar{x}})^2 + \Lambda\Lambda + (\bar{x}_{\bullet,c} - \bar{\bar{x}})^2}{c - 1} \ldots \text{estimate variance of means}$$

$$s_B^2 = \frac{n \times \lfloor (\bar{x}_{\bullet,1} - \bar{\bar{x}})^2 + (\bar{x}_{\bullet,2} - \bar{\bar{x}})^2 + \Lambda\Lambda + (\bar{x}_{\bullet,c} - \bar{\bar{x}})^2 \rfloor}{c - 1}$$

$$s_B^2 = \frac{n \times \sum_{j=1}^{j=c} (\bar{x}_{\bullet,j} - \bar{\bar{x}})^2}{c - 1} = \frac{SSB_{column\ means}}{df(between)}$$

SSB_{column means} Sum of Squares Between column or sample means

For different sample sizes $s_B^2 = \dfrac{\sum\limits_{j=1}^{j=c} n_j \times (\bar{x}_{\bullet,j} - \bar{\bar{x}})^2}{c - 1}$

… where n_j is the number of data in sample j.

3. **Variance estimated from within samples** (Method A)
   Recall s_w^2 is the average of the c independent sample variances

$$s_W^2 = \frac{s_1^2 + s_2^2 + s_3^2 + K + s_c^2}{c}$$

So, write out the calculation of each sample variance, assuming equal samples, size n

$$s_w^2 = \frac{1}{c} \left(\frac{\sum_{i=1}^{n} (x_{i,1} - \bar{x}_{\bullet,1})^2}{n - 1} + \frac{\sum_{i=1}^{n} (x_{i,2} - \bar{x}_{\bullet,2})^2}{n - 1} + \frac{\sum_{i=1}^{n} (x_{i,3} - \bar{x}_{\bullet,3})^2}{n - 1} - K + \frac{\sum_{i=1}^{n} (x_{i,c} - \bar{x}_{\bullet,c})^2}{n - 1} \right)$$

$$s_w^2 = \frac{\sum\limits_{i=1}^{n}(x_{i,1} - \overline{x}_{\bullet,1})^2 + \sum\limits_{i=1}^{n}(x_{i,2} - \overline{x}_{\bullet,2})^2 + \sum\limits_{i=1}^{n}(x_{i,3} - \overline{x}_{\bullet,3})^2 K + \sum\limits_{i=1}^{n}(x_{i,c} - \overline{x}_{\bullet,c})^2}{c(n-1)}$$

$$s_w^2 = \frac{\sum\limits_{j=1}^{c}\sum\limits_{i=1}^{n}(x_{i,j} - \overline{x}_{\bullet,j})^2}{n(c-1)} = \frac{SSW}{df\,(within)}$$

SSW is Sum of Squares Within samples

APPENDIX K

SUMS OF SQUARES FOR THE ONE-WAY ANOVA

The total deviation of any datum from the overall mean may be expressed as the sum of the deviation within the sample from the sample mean plus the sum of the deviation of the sample mean from the overall mean

$$x_{i,j} - \overline{\overline{x}} = (x_{i,j} - \overline{x}_{\bullet,j}) + (\overline{x}_{\bullet,j} - \overline{\overline{x}}) \qquad \text{(K10.3)}$$

Total deviation	Within column deviation	Deviation between column mean and overall mean

For example, the first entry in row 1 in Table 10.4 is

$$14 - 12.5 = (14 - 10.6) + (10.6 - 12.5)$$
$$1.5 = 3.4 + (-1.9)$$

Square each side of equation (K.10.3). Then sum over all rows and columns

$$\sum_{j=1}^{c}\sum_{i=1}^{n}(x_{i,j} - \overline{\overline{x}})^2 = \sum_{j=1}^{c}\sum_{i=1}^{n}\left[(x_{i,j} - \overline{x}_{\bullet,j}) + (\overline{x}_{\bullet,j} - \overline{\overline{x}})\right]^2$$

$$\text{TSS} = \sum_{j=1}^{c}\sum_{i=1}^{n}\left[(x_{i,j} - \overline{x}_{\bullet,j})^2 + 2(x_{i,j} - \overline{x}_{\bullet,j})(\overline{x}_{\bullet,j} - \overline{\overline{x}}) + (\overline{x}_{\bullet,j} - \overline{\overline{x}})^2\right]$$

> ...independent of i,
> there add n times

$$= \sum_{j=1}^{c}\sum_{i=1}^{n}(x_{i,j} - \overline{x}_{\bullet,j})^2 + \sum_{j=1}^{c}\sum_{i=1}^{n}2(x_{i,j} - \overline{x}_{\bullet,j})(\overline{x}_{\bullet,j} - \overline{\overline{x}}) + \sum_{j=1}^{c}\sum_{i=1}^{n}(\overline{x}_{\bullet,j} - \overline{\overline{x}})^2$$

$$= \sum_{j=1}^{c}\sum_{i=1}^{n}(x_{i,j} - \overline{x}_{\bullet,j})^2 + 2\sum_{j=1}^{c}(\overline{x}_{\bullet,j} - \overline{\overline{x}})\sum_{i=1}^{n}(x_{i,j} - \overline{x}_{\bullet,j}) + n\sum_{j=1}^{c}(\overline{x}_{\bullet,j} - \overline{\overline{x}})^2$$

> ...sum of deviation
> from the mean is zero

$$\text{TSS} = n\sum_{j=1}^{c}(\overline{x}_{\bullet,j} - \overline{\overline{x}})^2 + \sum_{j=1}^{c}\sum_{i=1}^{n}(x_{i,j} - \overline{x}_{\bullet,j})^2$$

$$\text{TSS} = \text{SSB}_{\text{column means}} + \text{SSW}_{\text{within columns (samples)}}$$

APPENDIX L

FITTED VALUES AND RESIDUALS

In two-way ANOVA
The fitted (predicted) value for any observations in two-way ANOVA is
Fitted value = overall mean + column effect + row effect
Fitted value = overall mean + column effect + row effect

$$\hat{x} = \overline{\overline{x}} + (\overline{x}_{col} - \overline{\overline{x}}) + (\overline{x}_{row} - \overline{\overline{x}}) = \overline{x}_{col} + \overline{x}_{row} - \overline{\overline{x}}$$

More formally,

$$\hat{x}_{i,j} = \overline{\overline{x}} + (\overline{x}_{\bullet,j} - \overline{\overline{x}}) + (\overline{x}_{i,\bullet} - \overline{\overline{x}}) = \overline{x}_{\bullet,j} + \overline{x}_{i,\bullet} - \overline{\overline{x}}$$

The residual is the (observed value – fitted value): unexplained difference between observed and the value predicted by ANOVA

In one-way ANOVA
Fitted value = overall mean + column effect = $\overline{\overline{x}} + (\overline{x}_{col} - \overline{\overline{x}}) = \overline{x}_{col}$

$$\hat{x}_{i,j} = \overline{\overline{x}} + (\overline{x}_j - \overline{\overline{x}}) = \overline{x}_j$$

The residual = Observed value – Fitted value = $x_{i,j} - \hat{x}_{i,j} = x_{i,j} - \overline{x}_j$

Note: the 'within column' deviation in equation (A 10.3) in Appendix 10.3 is actually the residual for one-way ANOVA

In two-way ANOVA, the
Fitted value = overall mean + column effect + row (block) effect

$$\hat{x}_{i,j} = \overline{\overline{x}} + (\overline{x}_{col} - \overline{\overline{x}}) + (\overline{x}_{row} - \overline{\overline{x}}) = \overline{x}_{col} + \overline{x}_{row} - \overline{\overline{x}}$$

The residual = Observed value − Fitted value = $x_{i,j} - \hat{x}_{i,j} = x_{i,j} - \overline{x}_{col} - \overline{x}_{row} + \overline{\overline{x}}$

APPENDIX M

SUMS OF SQUARES IDENTITY FOR TWO-WAY ANOVA

The total deviation of any datum from the overoll mean be expressed as the sum of column effect, block effect and residual

$$x_{i,j} - \bar{\bar{x}} = \alpha_j + \beta_I + \varepsilon_{i,j} \tag{M10.4a}$$

$$(x_{i,j} - \bar{\bar{x}}) = (\bar{x}_{\bullet,j} - \bar{\bar{x}}) + (\bar{x}_{i,\bullet} - \bar{\bar{x}}) + \left[x_{i,j} - (\bar{x}_{\bullet,j} - \bar{\bar{x}}) - (\bar{x}_{i,\bullet} - \bar{\bar{x}}) - \bar{\bar{x}}\right] \tag{M10.4b}$$

Total deviation	Deviation between column mean and overall mean	Deviation between row mean and overall mean	Residual for 2-way

Squaring and summing each side of (J 10.4) – use the abbreviated version in equation (J 10.4a) first to illustrate the method:

$$\sum_{j=1}^{c}\sum_{i=1}^{n}(x_{i,j} - \bar{\bar{x}})^2 = \sum_{j=1}^{c}\sum_{i=1}^{n}(\alpha_j + \beta_i + \varepsilon_{i,j})^2$$

$$= \sum_{j=1}^{c}\sum_{i=1}^{n}\left[(\alpha_j)^2 + (\beta_i)^2 + (\varepsilon_{i,j})^2 + 2\alpha_j\beta_i + 2\alpha_j\varepsilon_{i,j} + 2\beta_i\varepsilon_{i,j}\right]$$

On summing, all the cross terms will be zero: each will involve one summation of the form $\sum_{all\,i}(x_i - \bar{x}) = 0$

$$\sum_{j=1}^{c}\sum_{i=1}^{n}\alpha_j\beta_i = 0, \quad \sum_{j=1}^{c}\sum_{i=1}^{n}\alpha_j\varepsilon_{i,j} = 0, \quad \sum_{j=1}^{c}\sum_{i=1}^{n}\beta_i\varepsilon_{i,j} = 0. \quad \sum_{j=1}^{c}\sum_{i=1}^{n}\beta_i\varepsilon_{i,j}$$

$$\sum_{j=1}^{c}\sum_{i=1}^{n}(x_{i,j}-\overline{\overline{x}})^2 = \sum_{j=1}^{c}\sum_{i=1}^{n}\left[(\alpha_j)^2 + (\beta_i)^2 + (\varepsilon_{i,j})^2 + 2\alpha_j\beta_i + 2\alpha_j\varepsilon_{i,j} + 2\beta_i\varepsilon_{i,j}\right]$$

$$\sum_{j=1}^{c}\sum_{i=1}^{n}(x_{i,j}-\overline{\overline{x}})^2 = \sum_{j=1}^{c}\sum_{i=1}^{n}(\overline{x}_{\bullet,j}-\overline{\overline{x}})^2 + \sum_{j=1}^{c}\sum_{i=1}^{n}(\overline{x}_{i,\bullet}-\overline{\overline{x}})^2 + \sum_{j=1}^{c}\sum_{i=1}^{n}(x_{i,j}-\overline{x}_{\bullet,j}-\overline{x}_{i,\bullet}+\overline{\overline{x}})^2$$

$$= n\sum_{j=1}^{c}(\overline{x}_{\bullet,j}-\overline{\overline{x}})^2 + c\sum_{i=1}^{n}(\overline{x}_{i,\bullet}-\overline{\overline{x}})^2 + \sum_{j=1}^{c}\sum_{i=1}^{n}(x_{i,j}-\overline{x}_{\bullet,j}-\overline{x}_{i,\bullet}+\overline{\overline{x}})^2$$

for example

$$\sum_{j=1}^{c}\sum_{i=1}^{n}(\overline{x}_{i,\bullet}-\overline{\overline{x}})(x_{i,j}-\overline{x}_{\bullet,j}-\overline{x}_{i,\bullet}+\overline{\overline{x}}) = \sum_{i=1}^{n}\sum_{i=1}^{c}(\overline{x}_{i,\bullet}-\overline{\overline{x}})(x_{i,j}-\overline{x}_{\bullet,j}-\overline{x}_{i,\bullet}+\overline{\overline{x}})$$

$$= \sum_{i=1}^{n}(\overline{x}_{i,\bullet}-\overline{\overline{x}})\sum_{j=1}^{c}(x_{i,j}-\overline{x}_{\bullet,j}-\overline{x}_{i,\bullet}+\overline{\overline{x}})$$

$$= \sum_{i=1}^{n}(\overline{x}_{i,\bullet}-\overline{\overline{x}})\left[(x_{i,1}-\overline{x}_{\bullet,1}-\overline{x}_{i,\bullet}+\overline{\overline{x}}) + (x_{i,2}-\overline{x}_{\bullet,2}-\overline{x}_{i,\bullet}+\overline{\overline{x}}) + (x_{i,3}-\overline{x}_{\bullet,3}-\overline{x}_{i,\bullet}+\overline{\overline{x}})\right.$$

$$\left. + \ldots (x_{i,n}-\overline{x}_{\bullet,n}-\overline{x}_{i,\bullet}+\overline{\overline{x}})\right]$$

$$\sum_{i=1}^{n}(\overline{x}_{i,\bullet}-\overline{\overline{x}})\left[\left(\sum_{i=1}^{n}x_{i,j}-n\overline{\overline{x}}-n\overline{x}_{i,\bullet}+n\overline{\overline{x}}\right)\right]$$

$$\sum_{i=1}^{n}(\overline{x}_{i,\bullet}-\overline{\overline{x}})\left[\left(\sum_{i=1}^{n}x_{i,j}-n\overline{x}_{i,\bullet}\right)\right]$$

$$\sum_{i=1}^{n}(\overline{x}_{i,\bullet}-\overline{\overline{x}})\left[\left(\sum_{i=1}^{n}x_{i,j}-\overline{x}_{i,\bullet}\right)\right] = (0)(0) = 0$$

CHAPTER 1 SOLUTIONS

PE 1.1

2. Bias (see section 1.1). The sampling frame is a list of members of the target population from which the sample is to be selected. Omission of any key groups from the list will result in a sample that is biased in favour of the groups that are included in the sampling frame.

3. Response bias (error) arises when questions or terminology are misunderstood or answers are inaccurate.

4. See text, section 1.2 for (a) a simple random sample (b) a stratified random sample (iii) a cluster sample (iv) a multistage sample.

(b) (i) For views on statistics held by the current first year students attending the statistics course, use a simple random sample. The population is finite, a population list is available
    (ii) For views on support for a football team by the members of the football supporters club use a quota sample since no population list is available, etc.
    (iii) For the views of primary school children on homework could use a cluster sample or stratified cluster sample.

6. Possible contact methods include the following Contact those selected for (i) by email, mobile/landline or personal contact. Follow up by same contact those in (ii) by interviewer at matches for a cross-section of all supporters. Contact signed-up members of the supporters club by email, mobile/landline or letter and follow up by same and offer an incentive, such as free tickets. In (iii), depending on the sampling method, direct contact through schools is a possibility or personal interview.

8. Reasons include any one or all of the following: population list not available; all members of the population not accessible; survey results required quickly; reduce the cost of the survey.

PE 1.2

1.

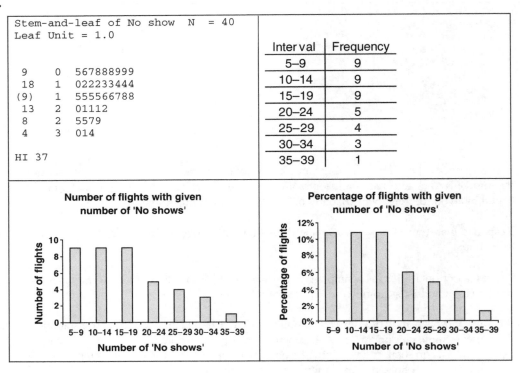

```
Stem-and-leaf of No show   N  = 40
Leaf Unit = 1.0


   9    0    567888999
  18    1    022233444
  (9)   1    555566788
  13    2    01112
   8    2    5579
   4    3    014


HI  37
```

Interval	Frequency
5–9	9
10–14	9
15–19	9
20–24	5
25–29	4
30–34	3
35–39	1

Number of flights with given number of 'No shows'

Percentage of flights with given number of 'No shows'

Distribution is skewed. The majority of flights (67.5%) have between 5 and 19 'no shows'. 20% have more than 25, with only two flights with 35 to 39 'no shows'. No flights have more than 39 'no shows'.

2.

Stem-and-Leaf Display: Exam marks	Interval (marks)	Frequency
Stem-and-leaf of Exam marks N = 60	6 < 10	1
Leaf Unit = 1.0 Stem unit = 10	10 < 20	2
	20 < 30	3
0 6	30 < 40	6
1 28	40 < 50	13
2 069	50 < 60	13
3 016689	60 < 70	14
4 0235578899999	70 < 80	4
5 0002234556788	80 < 90	3
6 00114555567899	90 < 100	1
7 1244		
8 012		
9 4		

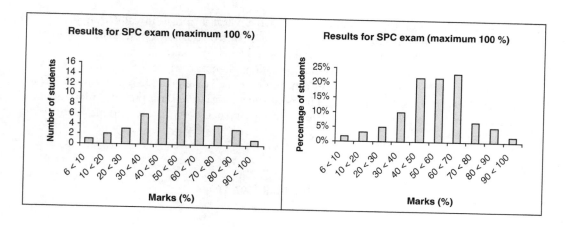

The distribution of marks is reasonably symmetrical about the three central groups. Two-thirds of students were awarded marks between 40 % and 70 %.

4.

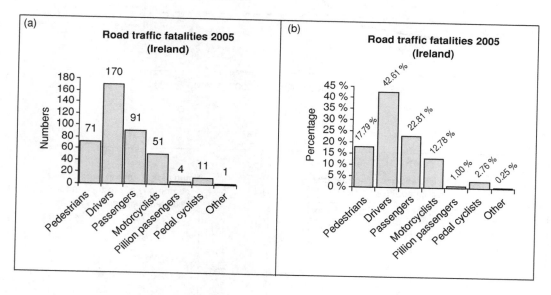

Almost half of all road fatalities were drivers (42.6 %) of vehicles. Passengers account for 22.8 % of fatalities. Seventy-one pedestrians (18 % of fatalities) were killed. Only four fatalities were pillion passengers, while 51 motorcyclists were killed.

6. (a) The only information from this chart is that 'whatever it is' ranges in value from 0.2 to 0.56. Every second bar relates to a country whose name is given on the horizontal axis. At a minimum, the chart requires a title, units on vertical axis and labels for each bar.

(b) The data in the file may be used to compare changes in cost over time and make comparisons of costs in the different countries over time. One such chart is given here for countries Spain, France, Greece, Ireland and Italy. The notes given with the data provide information on the type of call, duration range and time of day. Hence costs in the various countries are comparable.

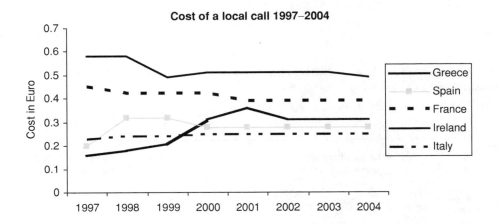

8.

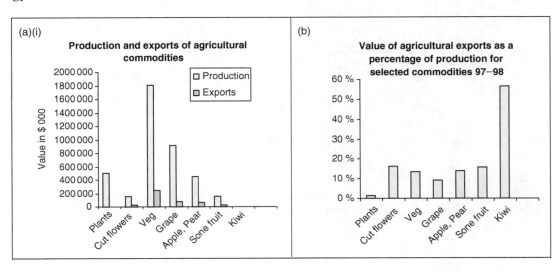

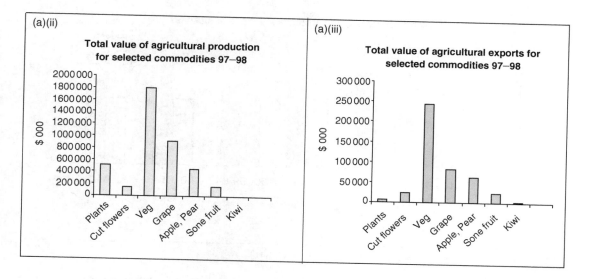

(a)(ii)

Total value of agricultural production for selected commodities 97–98

(a)(iii)

Total value of agricultural exports for selected commodities 97–98

Comparisons are difficult to 'see' in (a)(i) because some groups are extremely large and others extremely small. Plotting separate bar charts for production and exports is an improvement: it would appear that the greater the value of production the greater the value of export for all commodities except plants. In (b) the percentages of each product exported reveals that kiwis are produced mainly for export.

PE 1.3

Answers

1. Relative costs are similar over the years, with Ireland the most expensive, France next and Spain the least expensive.

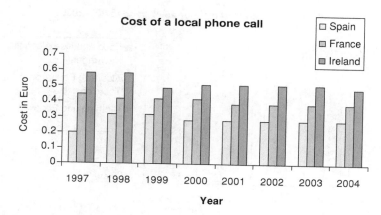

Cost of a local phone call

☐ Spain
▨ France
▨ Ireland

2. Comments to note (i) percentage of women is higher that men for all age groups except 60–65+ and it is higher for younger ages (ii) male absentee rate is lowest for the 25–34 group.

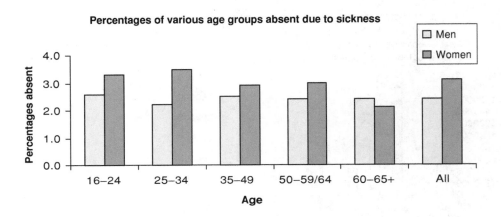

4.

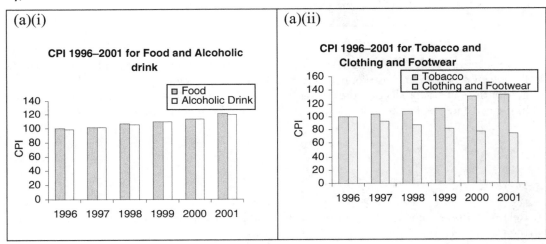

4. (b)

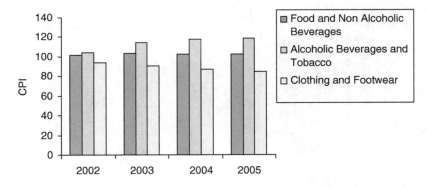

4. (c) No. Each series is reset to a base of 100 in mid-December 2001 and there was a change in the goods grouped together in the first two categories in 2002.

6. (a)

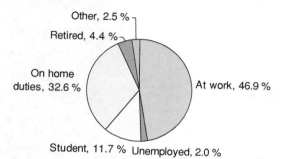

Percentage in each economic activity, Female

Other, 2.5 %

Retired, 4.4 %

On home duties, 32.6 %

At work, 46.9 %

Student, 11.7 %   Unemployed, 2.0 %

(b)

Economic activity by gender 2005

☐ Males
☐ Females

Numbers

1,200.00

1,000.00

800.00

600.00

400.00

200.00

0.00

At work   Unemployed   Student   On home duties   Retired   Other

Comments to include: there is a higher percentage of males at work, unemployed and retired. The numbers of males on home duties is almost insignificant compared to the percentage of females.

8. (a)

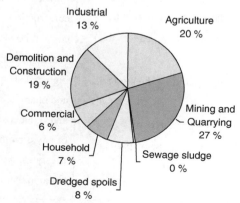

Total annual waste by source UK

Industrial
13 %

Agriculture
20 %

Demolition and
Construction
19 %

Commercial
6 %

Mining and
Quarrying
27 %

Household
7 %

Sewage sludge
0 %

Dredged spoils
8 %

Total annual waste by source UK

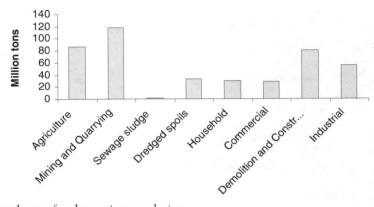

10. These are total costs for the entire population.

(a)

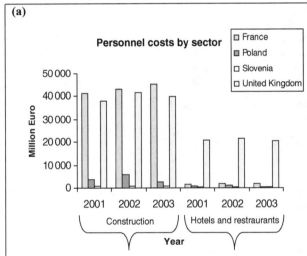

In this chart it is difficult to make comparisons between Poland and Slovenia whose total costs are much less than those of France and the UK.

Comparisons are made clearer by separating the high and lost cost countries.

It would be interesting to have costs per capita of economically active population or the cost per capita employed in the given industry.

(b)(i)

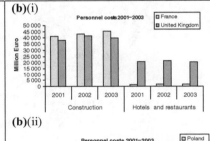

(b)(ii)

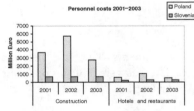

In **(b)(i)** Costs are similar for construction, but the UK has higher costs in hotels and restaurants than France. Possible explanations for higher costs may include the size and character of the industry, variable and fixed costs such as rents, rates of pay, raw materials, etc., as well as lifestyles, type of ownership, personnel employed, tax systems etc.

PE 1.4

Answers

1.

(a)

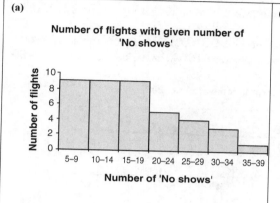

This data is not Normally distributed: it is slightly skewed to the right or positively skewed

(b)

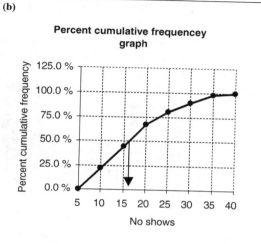

From the Ogive, 50 % of flights have 16 or fewer 'no shows'

2.

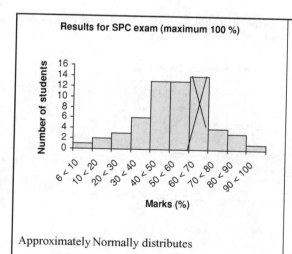

Approximately Normally distributes

From the Ogive, the lower 30 % of results received 45 marks or fewer: upper 10 % of results received 75 marks or more.

4. The bar charts in these questions were for categorical data.

6. (a) Total income before tax is for the lowest 1 %, 5 % 10 % ... 95 % and 99 % of taxpayers in the years 1990–1 and 2003–4. Similarly, total income after tax for the lowest percentages from 1 % to 99 % of taxpayers is also given.

(b) In the year 1990–1, the lower 10 % of incomes was £ 4650 or less before tax and £ 4300 or less after tax. In 2003–4, the lower 10 % of incomes was £ 7000 or less before tax and £ 6790 or less after tax.

In the year 1990–1, the lower 50 % of incomes was £ 10 600 or less before tax and £ 8980 or less after tax. In 2003–4, the lower 50 % of incomes was £ 16 000 or less before tax and £ 14 000 or less after tax.

(c)

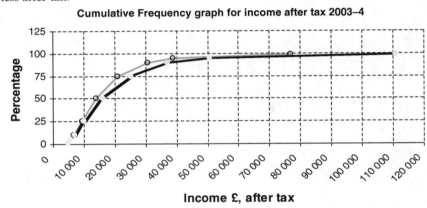

Cumulative Frequency graph for income after tax 2003–4

PE 1.5

1. Employment in Health is increasing steadily; employment in agriculture is declining while numbers employed in transport is increasing but levelling off.

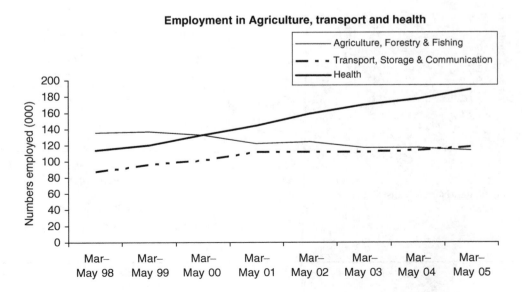

Employment in Agriculture, transport and health

2. (a)

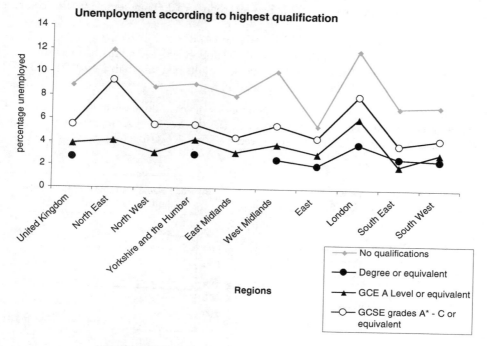

2. (b) In spite of regional differences in overall unemployment, in all regions the unemployment rate is higher for those with no qualification in all regions, while the rate is lowest for those with degrees (with the exception of those with A levels in the South East).

4.

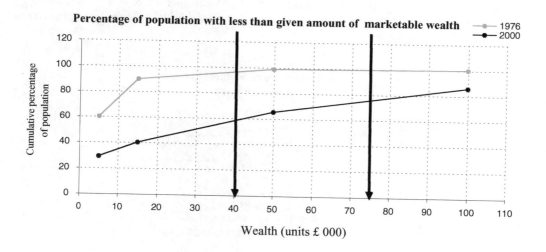

(b) From the graph, approximately 95 % had less than £ 40 000 marketable wealth in 1976, compared to 60 % in 2000. In 1976, 99 % had less than £ 75 000 marketable wealth compared to 75 % in 2000.

(c) The adult population was 40 486 000 in 1976 and 47 828 000 in 2000.
Hence 38 461 700 adults had less than £ 40 000 and 40 081 140 adults had less than £ 75 000 marketable wealth in 1976. In 2000, 28 696 800 adults had less than £ 40 000 and 35 871 000 adults had less than £ 75 000 marketable wealth.

6. (a) From the following graphs (i) the average hours for males in the basic metal industries is 45–50 per week compared to around 35 for women, (ii) the average earnings for males range from € 650 to € 700 from 2002 to 2005 compared € 350 to € 400 for women.

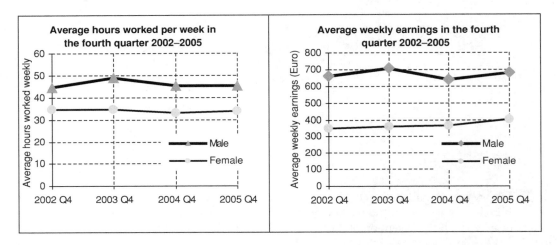

(b) The average hourly rate for males fell slightly between 2002 and 2004 before increasing to approximately € 15 in 2005, the average hourly rate for women increased steadily from € 10 in 2002 to € 12 in 2005 – hence the difference in rates for males and females decreased from € 4.8 to € 3 approximately.

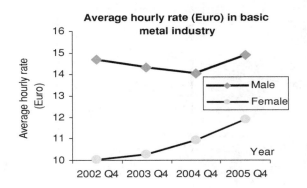

8.

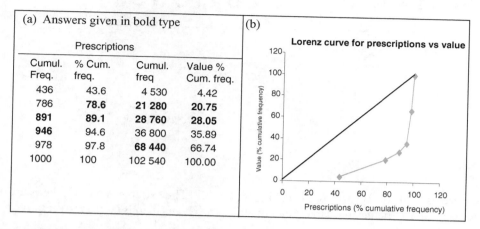

(a) Answers given in bold type

Prescriptions

Cumul. Freq.	% Cum. freq.	Cumul. freq	Value % Cum. freq.
436	43.6	4 530	4.42
786	**78.6**	**21 280**	**20.75**
891	**89.1**	**28 760**	**28.05**
946	94.6	36 800	35.89
978	97.8	**68 440**	66.74
1000	100	102 540	100.00

(b)

Lorenz curve for prescriptions vs value

Most prescriptions are relatively inexpensive. For example, from the graph in (b) approximately 80 % of all prescriptions account for only 20 % of the total value of all prescriptions.

CHAPTER 2   SOLUTIONS

PE 2.1

1. Arrayed data 14, 16, 16, 16, 16, 17, 17, 18, 18, 19, 108. The extreme value, 42, skews the data to the right; hence the mean is larger than the median. The median is more representative.

Mean	Mode	Q1	Q2	Q3
25	16	16	17	18

2.

Stem (tens)	Leaf (units)
0	0,0,0,2,4,5,5,9
1	0,0,0,0,3,5,5,5,5,5,5,5,5,5,8,8
2	0,0,1,2,2,3,3,4,5,5,5,5
3	0,0,2,5

Mean	Q1	Q2	Q3	Mode
16.275	10	15	23	15

4. (a)

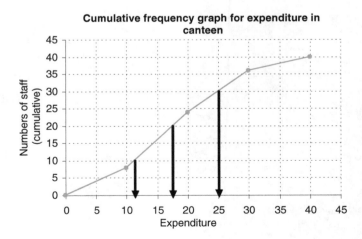

(b) Quartiles from the Ogive are Q1 = 11, Q2 = 15.5; Q3 = 25.

(c) From the Ogive, the expenditure for the lower 10 % point is 5, for the 90 % point is 30. Hence the middle 80 % spend between 5 and 30 per week.

6. (File: Exams) (a) From the raw data Q1 = 43.5, Q2 = 52.5, Q3 = 65, mode = 29, mean = 52.617
(b) From data grouped as in question 2, PE1.2, Q1 = 42, Q2 = 53, Q3 = 65, mode (from histogram, PE1.4, question 2) = 66, mean = 53.
The mean and median are almost identical; therefore the data should not be skewed.

8.

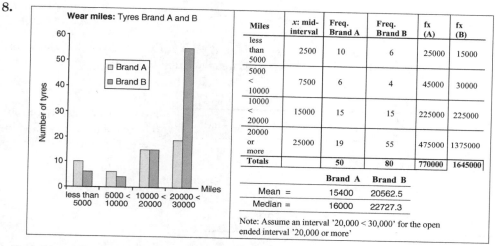

Miles	x: mid-interval	Freq. Brand A	Freq. Brand B	fx (A)	fx (B)
less than 5000	2500	10	6	25000	15000
5000 < 10000	7500	6	4	45000	30000
10000 < 20000	15000	15	15	225000	225000
20000 or more	25000	19	55	475000	1375000
Totals		50	80	770000	1645000

	Brand A	Brand B
Mean =	15400	20562.5
Median =	16000	22727.3

Note: Assume an interval '20,000 < 30,000' for the open ended interval '20,000 or more'

Both sets of data are skewed. The numbers of tyre that wear out are relatively small for low mileage, but increases rapidly for 20,000 miles or more, particularly brand B.
This is demonstrated by the bar chart and also by the large difference between the mean and median. For each brand use the median life as typical. Assume a lower and an upper limit of 0 and 30000 respectively for both sets of data.

10. Assume an upper limit 60km to less than 80km

Distance	x: Mid-interval	Frequency Brentwood	Frequency London	fx Brentwood	fx London
Less than 2km	1	94	539	94	539
2km to less than 5km	2.5	160	237	400	592.5
5km to less than 10km	7.5	57	80	427.5	600
10km to less than 20km	15	159	16	2 385	240
20km to less than 30km	25	85	9	2 125	225
30km to less than 40km	35	285	4	9 975	140
40km to less than 60km	50	22	13	1 100	650
60km and over	70	10	21	700	1470
Totals		872	919	17 206.5	4456.5

Brentwood	mean = 19.73	median = 20 approx	London	mean = 4.85	median = 1.7 approx

Assumption: assume an upper limit of 80 for the '60km and over' interval.
Comment. Distances traveled in London are shorter, but the relatively large value of the mean compared to the median implies that some individuals travel a longer distance than the majority. In a more rural community like Brentwood, people must travel greater distances to work; the average distance is 19.73 miles, which is close to the median. This implies that there are few, if any, who travel extremely long distances.

PE 2.2

1. $\sigma^2 = 690.55$ and $s^2 = 759.60$. $\sigma = 26.28$, $s = 27.56$. IQR $= 2$, QD $= 1$. Standard deviation is larger the QD because of the outlier, 108 – the squared deviation of this value from the mean is $(108 - 25)^2 = 6889!$

2.

Expenditure	f: frequency	x: mid interval	fx	f(x−mean)²
0 < 10	8	5	40	1352
10 < 20	16	15	240	144
20 < 30	12	25	300	588
30 < 40	4	35	140	1156
Totals	40	80	720	3240

From grouped data (see table) $\sigma^2 = 81$ and $s^2 = 83.08$. $\sigma = 9$, $s = 9.11$, IQR $= 13$, QD $= 6.5$

4. **Exam marks** (a) For raw data $\sigma^2 = 298.94$, $\sigma = 17.29$.
   IQR $= 65 - 44 = 21$, QD $= 10.5$
   (b) From grouped data (question 2 PE1.2) $\sigma^2 = 319.33$ and $\sigma = 17.87$, IQR $=$ Q3 $-$ Q1 $= 65 - 42 = 23$, QD $= 11.5$.
   The average and median marks are both close to 53, hence data is symmetrically distributed. The variation in marks is not great, $\sigma = 17.87$.

6. **Mobile calls** (a) $\sigma^2 = 113.77$ and $s^2 = 116.00$. $\sigma = 10.67$, $s = 10.77$. IQR $= 7.82$, QD $= 3.91$.
   (b) $\sigma^2 = 38.24$ and $s^2 = 38.99$. $\sigma = 6.18$, $s = 6.24$
   IQR $= 8.5$, QD $= 4.25$
   The variance for grouped data is smaller because an upper limit of 19.5 is assumed; hence the large values are not included in its calculation.
   Data is skewed to the right, mean $= 7.85$, median $= 3.19$. The presence of extreme values (large) in the raw data is reflected in the value of standard deviation, $s = 10.77$ which is greater than QD $= 3.91$. This is not the case for grouped data where the assumed upper limit hides the extremely high data.

8. (a) mean $= 19.73$, median $= 20$ for Brentwood; mean $= 4.85$, median $= 1.70$ (see question 10 PE2.1)
   (b) $\sigma^2 = 237.16$, $\sigma = 15.40$ for Brentwood, $\sigma^2 = 148.18$, $\sigma = 12.09$ for London
   There is a wider range of distances traveled by the residents of Brentwood than by the resident of the London district.

PE 2.3

1.

						Range	Variance (pop)	mean	median	CV	Sk
C	48	49	50	51	52	4	2	50	50	2.83	0
D	1	2	3	4	5	4	2	3	3	47.14	0
E	1001	1002	1003	1004	1005	4	2	1003	1003	0.14	0

(a) The variance is a measure of the dispersion of data from the mean value and is independent of location: the mean is a measure of location.

(b) The CV gives the magnitude of variance relative to the magnitude of the data values. Hence a variance of 2 is large relative to data values 1, 2, 3, 4 and 5 but small relative to data values 1001, 1002, 1003, 1004 and 1005.

(c) All the sets of data are perfectly symmetrical: mean value = median value, hence the coefficient of skewness is zero.

2. (a) CV = 55.57 (b) Sk = 0.42 (c)
See PE2.1 question 2

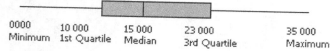

0000	10 000	15 000	23 000	35 000
Minimum	1st Quartile	Median	3rd Quartile	Maximum

Data is only slightly skewed to the right.

4. (a) CV = 33.1372 (b) Sk = 0.0201 (c) *
See PE2.1 question 6 and PE2.2 question 4.

6000		43 500	52 500	65 000	94 000
Minimum		1st Quartile	Median	3rd Quartile	Maximum

Data is almost symmetrical.

6. The larger negative coefficient of skewness for B is due to the small number of very low values relative to the remainder of the data. Hence the mean is less than the median and the skewness is negative.

	Brand A	Brand B
Mean	15400	20562.5
Median	16000	22727.27
s =	8797.031	7280.273
(a) CV =	0.571236	0.354056
(b) Sk =	−0.20461	−0.89204

Data skewed to right.

8. (a)(i)

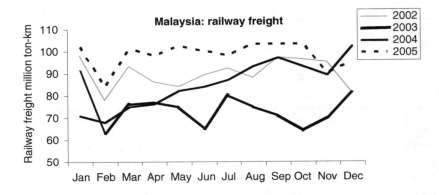

8. (a)(ii)

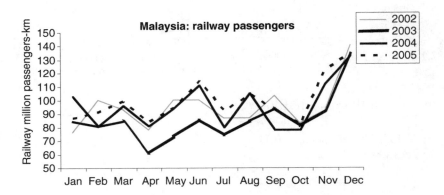

8. (b)

Freight million ton-km	Mean	Median	Stdev (s)
2002	89.75	90.50	6.52
2003	74.00	75.00	8.07
2004	84.75	85.50	10.69
2005	98.25	100.50	5.94
Passenger-km			
2002	94.92	92.50	16.78
2003	85.92	85.00	17.57
2004	96.00	95.00	17.42
2005	99.75	93.50	16.24

8. (c) Freight: The volume of freight was lowest in 2003, increasing somewhat in 2004, and highest in 2005. This is reflected in the values of the means and medians. Passenger volume is very seasonal, highest in December each year. The volume of passengers was lowest in 2003, but similar for the other years – the averages and medians are similar.

The standard deviations are only marginally different for both freight and passengers, with the exception of 10.69 and 5.94 for freight in 2004 and 2005.

PE 2.4

Answers to questions 1 to 5 given in previous exercises.

6. Descriptive Statistics: Beer, Cider & perry, Wine, Spirits

Variable	Count	Mean	StDev	Variance	Coef Var	Minimum	Q1	Median
Beer	14	60088	1947	3790633	3.24	56139	59220	59885
Cider & perry	14	5266	835	697495	15.86	3777	4485	5592
Wine	14	10276	2429	5897969	23.63	7342	8022	10045
Spirits	14	926.6	123.3	15198.9	13.31	810.0	835.8	887.5

Variable	Q3	Maximum	Skewness	Kurtosis
Beer	60511	64541	0.55	2.23
Cider & perry	5911	6291	-0.79	-0.85
Wine	12672	14795	0.40	-1.10
Spirits	980.3	1210.0	1.43	1.41

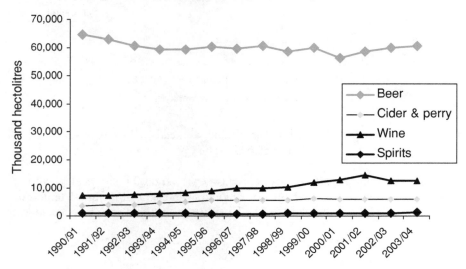

Beer has the highest consumption, smallest coefficient of variation with only a small positive skew since there has been relatively small changes over the years.

Cider & perry small average, high coefficient of variation and skewed to left.

Wine, second highest average, high coefficient of variation since volume of wine is increasing over the years.

Spirits have been relatively steady with the exception of increases in last two years, hence a positive skewness (1.43).

10. Cholesterol
Descriptive Statistics: LDL, HDL

Variable	Mean	StDev	Variance	Coef Var	Minimum	Q1	Median
LDL	5.0194	0.6475	0.4192	12.90	3.7200	4.5150	5.0350
HDL	0.9884	0.3033	0.0920	30.68	0.4500	0.7650	0.9650

Variable	Q3	Maximum	Skewness	Kurtosis
LDL	5.4625	6.6100	-0.11	-0.59
HDL	1.2200	1.6700	0.34	-0.65

Company B has a higher average sick days ranging from 1 to 63 compared to 6.25 for company A. Company B has some extreme high absentee rates ($Sk = 2.55$) and has a peaked distribution, kurtosis = 7.43.

12. (a) From the bar graph, the mode is within the range $40 < 50$ for both males and females.

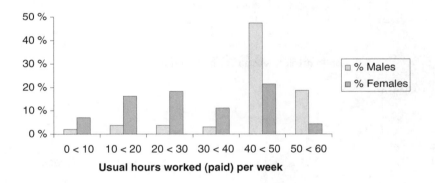

Distribution of usual weekly hours of paid work

(b) The mean and variance for the number of hours worked by (i) males is 40.73 and 102.04 and (ii) females is 30.06 and 163.46 respectively. The CV = 24.8 for males and 42.5 for females.

The majority of males work in paid employment between 40 and 60 hours weekly, with an average of 40.73 hours. Women work various hours in paid employment ranging almost uniformly from 10 to 50 hours weekly with an average of 30.6. The larger variation for women is reflected in the larger coefficient of variation.

CHAPTER 3   SOLUTIONS

PE 3.1

1. (a) intercept 2, slope 4 upward sloping (b) intercept −5 slope 8 upward sloping (c) intercept −2 slope 3 upward sloping (d) intercept 12 slope −1 downward sloping.

2.

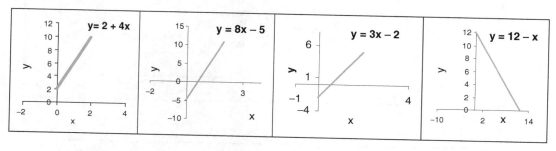

4.

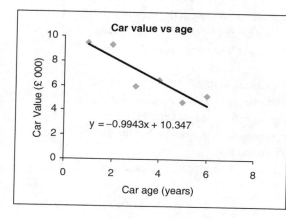

When a car is 4.5 years its value should be −0.9943(4.5) + 10.347 = 5.87265 i.e. £ 5872.65. In the scatter plot the points are scattered from the line, hence the answer should be accurate within a small margin of error.

6.

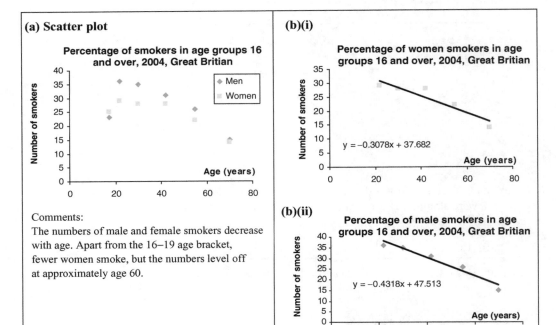

(a) Scatter plot

Comments:
The numbers of male and female smokers decrease with age. Apart from the 16–19 age bracket, fewer women smoke, but the numbers level off at approximately age 60.

(b)(i)

$y = -0.3078x + 37.682$

(b)(ii)

$y = -0.4318x + 47.513$

8.

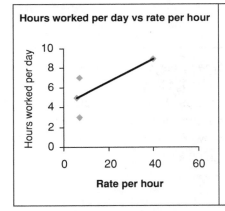

Hrs per day $= 0.1199 *$ rate per hour $+ 4.2014$
($x = 0.1199y + 4.2014$)
According to this equation, the hours worked per day increase by 0.1199 when the rate per hour increases by £ 1.

The regression in WE 3.1 $y = 5x - 15$, when rearranged is $x = 0.2y + 3$. Conclude that the regression for y vs x different to the regression line for x vs y (unless the line is a perfect fit).

10.

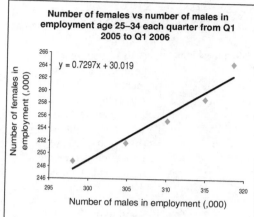

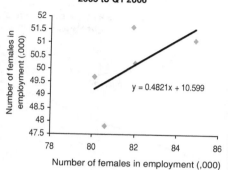

From the slope, for each additional male there is 0.73 females or when the numbers of males increase by 100, the number of females increases by 73 between Q1 2005 and Q1 2006 for the 25–34 age group. The trend is strong.

From the slope, for each additional male there is 0.48 females or when the numbers of males increase by 100, the number of females increases by 48 between Q1 2005 and Q1 2006 for the 55 to 59 age group. The trend is very weak.

PE 3.2

1. (i) (as shown on graphs) (a) $R^2 = 1, r = 1$ (b) $R^2 = 0.9932, r = -0.9661$ (c) $R^2 = r = 0$ (ii) Verbal explanations are given in the text following Figure 3.11

2. (a) $R^2 = 0.6824, r = 0.8261$ (b) $R^2 = 0.9992, r = 0.9996$

PE 3.3

1.

(a), (b) and (c)

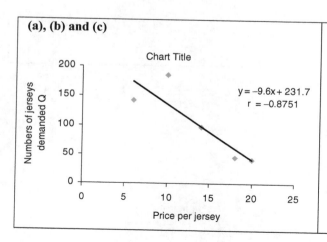

(d) when the price is 15, 237 jerseys will be demanded.
The points are scattered about the regression line, 76.58 % of the change is explained by the equation ($R^2 = 0.7658$), hence the estimated values will be accurate within a small margin of error.

2.

(b) (i)

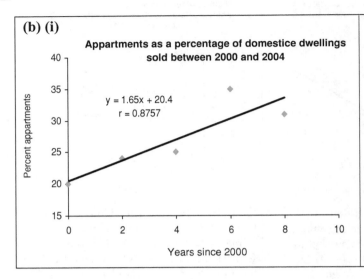

(b)(ii) Each year, the percentage of apartments increases by 1.65. The percentage of apartments sold is given by the regression equation.
(b)(iii) In 2003, $x = 3$, hence the percentage of apartments sold is $1.65(3) + 20.4 = 25.35$. 25.35 % of 280,000 is 70,980.

4.

(b) (i), (ii) and (iii)

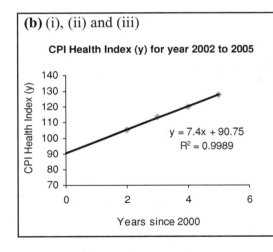

(b)(iv)

The CPI for health is rising very steadily ($r = 0.9995$) at the rate of 7.4 percentage points per year. This trend is unlikely to continue–otherwise health will become unaffordable.

6.

(a) $r = 0.9915$. The correlation coefficients is high and po a strong relationship between the percentage of males and females. The equation of the least squares line will make accurate estimates. **(b)**	**(b)** The percentage of females is given by the equation $y = 0.9925x + 11.005$, where x is the percentage of males. Hence the percentage of females is higher, but each 1 % increase for males is accompanied by a 0.9925 % increase in females. The regression equation may only be used to calculate the percentage of females when the percentage of males is between 40 and 51.5. **(c)** 50.5, 65.6, 85.4. No

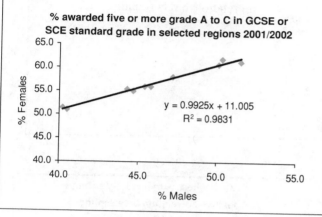

% awarded five or more grade A to C in GCSE or SCE standard grade in selected regions 2001/2002

$y = 0.9925x + 11.005$
$R^2 = 0.9831$

8.

(a), (b) 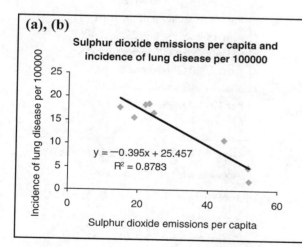 Sulphur dioxide emissions per capita and incidence of lung disease per 100000 $y = -0.395x + 25.457$ $R^2 = 0.8783$	$y = -0.395x + 25.457$ $R^2 = 0.8783$ There appears to be a reasonable strong negative linear, $r = -0.9372$: as the sulphur dioxide emissions increase by 1Kg per capita the incidence of lung disease decreases by 0.395 per 100 000.

10.

(a), (b) 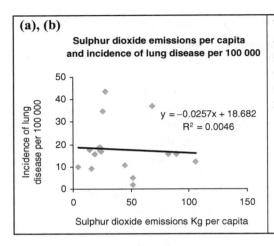	The relationship between SO_2 emissions and the incidence of lung disorder in question 8 is probably a coincident (spurious). When further data is included in questions 9 and 10, no relationship is found.

12.

(a) (i) 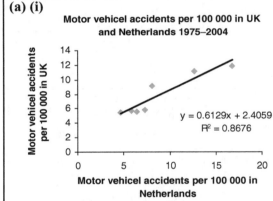	(a)(ii) There is moderately strong positive correlation between the accident rate in the UK and Netherlands $R = 0.9315$, $R^2 = 0.8676$ From the equation of the least-squares line, $y = 0.6129x + 2.4059$, the rate in the UK increases by 1 per 1000 000 the rate in the Netherlands increases by 0.6129 per 100 000 – See (c)
(a) (ii) 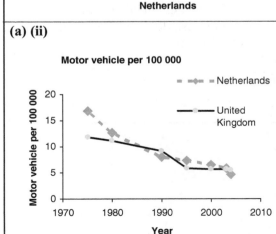	**Greece** $y = -0.0299x + 18.326$ $R^2 = 0.0186$ Accident rate is fairly constant **Iceland** $y = -0.2422x + 493$: $R_2 = 0.5433$ Accident rate is decreasing, but trend is weak **Netherlands** $y = -0.3701x + 746.31$ $R^2 = 0.9392$ Accident rate is decreasing, trend is strong **UK** $y = -0.2437x + 493.45$. $R^2 = 0.9407$ Accident rate is decreasing, trend is strong

14.

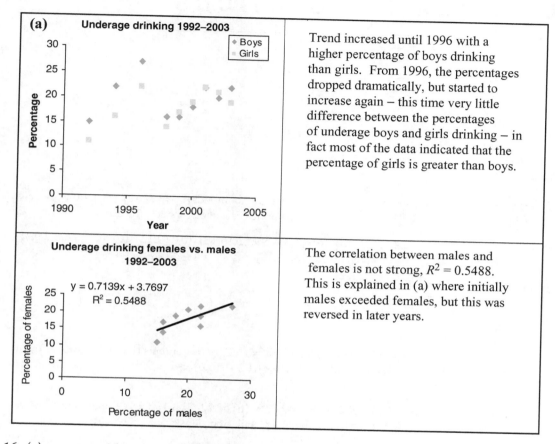

(a) Underage drinking 1992–2003

Trend increased until 1996 with a higher percentage of boys drinking than girls. From 1996, the percentages dropped dramatically, but started to increase again – this time very little difference between the percentages of underage boys and girls drinking – in fact most of the data indicated that the percentage of girls is greater than boys.

Underage drinking females vs. males 1992–2003

$y = 0.7139x + 3.7697$
$R^2 = 0.5488$

The correlation between males and females is not strong, $R^2 = 0.5488$. This is explained in (a) where initially males exceeded females, but this was reversed in later years.

16. (a) age group $25 - 34$, $r = 0.9724$, $R^2 = 0.9455$. There is a strong positive relationship – the number of males and females both increase over the years.
(b) age group $55 - 59$, $r = 0.6099$, $R^2 = 0.372$. There is a very weak positive relationship between the numbers of males and numbers of females in employment for this age group.

CHAPTER 4   SOLUTIONS

PE 4.1

2. (a) {list the 52 cards}, # = 52 (b) {list the hearts and diamonds}, # = 26 (c) {Ace of hearts, Ace of diamonds}, # = 2 (d) {list the hearts, the diamonds, the ace of spades, the ace of clubs}, # = 28

4. (a) {1,2,3,5,6,7} (b) {2,4,6,7,16,18} (c) {1,3,4,5,9,12,16,18} (d) {9,12} (e) {1,2,3,5,6,7,9,12} (f) {1,3,5} (g) {9,12} (h) {1,2,3,4,5,6,7,9,12,16,18}

6. (a) 4900 (b) 4450 (c) 400 (d) 5300

PE 4.2

2. (a) #S = 36, #E = 1, P(both six) = 1/36 (b) #E = 6, P(both numbers the same) = 6/36 = 1/6. (c) E = {(4,4), (4,5), (4,6), (5,4), (5,5), (5,6), (6,4), (6,5), (6,6)}, #E = 9, P(both greater than 3) = 9/36 = 1/4.

4. (a) (i) 0.4884 (ii) 0.4961 (iii) 0.9821 (b) (i) 49.77 % (ii) 50.22 % (iii) 98.5 %. Population density = 0.8063 (c) (i) 0.4923 (ii) 0.5057 (iii) 0.9791. Population density = 1.3996 (d) (i) 0.4289 (ii) 0.5711.

6. (a) {(H,1), (H,2), (H,3), (H,4),(H,5), (H,6), (T,1), (T,2), (T,3), (T,4),(T,5), (T,6)}
(b) (i) {(H,2),(H,2),(H,6)} (ii) {(H,1),(H,6)} (c) (i) 0.25 (ii) 2/12 = 0.0667.

8. (a) This is subjective probability and hence the answer will vary from 0 (pessimists) and 1 (supreme optimists).

10. Select the fist site. 80 × 0.95 = 0.76.76 % of the 80 % of sites that are commercially profitably will be workable.

PE 4.3

2. (a) independent (b) independent if the first card is replaced before selecting the second. If the first card is not replaced the events are dependent. (c) dependent-temperature is related to sunshine (d) independent – temperature is the same for all seasons (e) dependent.

4. (a) $\dfrac{^5C_2}{^{10}C_2} = 0.2222$ (b) $\dfrac{^5C_2 + {}^3C_2 + {}^3C_2}{^{10}C_2} = 0.3111$ (c) 1 – (b) = 0.689 (d) 0.9333

6. (a) $(0.15 \times 0.15) = 0.0225$ (b) $(0.15 \times 0.85) + (0.85 \times 0.15) = 0.2250$ (c) $(0.85 \times 0.85) = 0.7225$.

8. (a) True, Independent by Rule (4.5) (b) True, Independent by Rule (4.7) (c) False, they are mutually exclusive, Rule (4.6) (d) False, $P(AB) = 0$ means they are mutually exclusive, Rule (4.6).

10. (a) $9 \times 10 \times 10 \times 10 = 9000$ (b) 9000 (c) $9 \times 10 \times 10 \times 10 \times 10 \times 10 - 9000 = 891\,000$

PE 4.4

1. (a) Joint probability, $P($ fine issued in main St. AND by Tom$) = 120/3000$.
Marginal probability, $P($ issued given by Tom$) = 630/3000$.
Conditional probability, $P($ fine issued by Tom/Main St.$) = 120/840$

(b) The Joint probability distribution

Wardens	Main street	Shop street	Old Market
T	0.04	0.1	0.07
R	0.05	0.08	0.05
A	0.07	0.07	0.06
F	270	0.07	0.08
C	0.03	0.05	0.09

(c)

T	R	A	F	C
21 %	18 %	20 %	24 %	17 %
0.21	0.18	0.2	0.24	0.17

Row 2 gives the % of fines issued by each warden. Row 3 gives the probability that a fine was issued by each warden. These probabilities constitute the marginal distribution for wardens

(d) 0.08 (e) 90/840 (f) 210/720

2. (a) The marginal distribution for personal grade is given in the right margin while the marginal distribution for technical grades is given in the bottom row (margin).

	Technical Grades			
	A_T	B_T	F_T	Marginal for pers.
A_P	0.09	0.14	0.02	0.25
B_P	0.22	0.128	0.044	0.392
F_P	0.05	0.136	0.172	0.358
Marginal for tech	0.36	0.404	0.236	1

(b) The joint distribution if given in the body of the table (in bold).
(c) (i) 9 % grade A for both (ii) $110/196 = 56\%$ (iii) $64/202 = 32\%$
    (i) 0.09, joint (ii) 0.56, conditional (iii) 0.136, joint

4. (a) (i) If the events are independent then $P(A/B) = P(A)$ or $P(B/A) = P(B)$ or $P(AB) = P(A)P(B)$. Hence, in question 2, $P(A_P/A_T) = 45/180 = 0.25$, the same as $P(A_P) = 0.25$ etc., (ii) mutually exclusive events cannot occur together at the same time, such as a Head and a Tail on one toss of a coin (iii) events are dependent when the occurrence of one changes the probability of the occurrence of the other, such as sampling without replacement from a small population.

4. (b) (i) $\frac{3}{8} \times \frac{2}{7} \times \frac{1}{6}$ (ii) $\frac{3}{8} \times \frac{5}{7} \times \frac{4}{6}$ (iii) $3 \times \left(\frac{3}{8} \times \frac{2}{7} \times \frac{5}{6}\right)$ (iv) $\frac{5}{8} \times \frac{4}{7} \times \frac{3}{6}$

6.

Minutes late	5 < 15	15 < 30	30 < 45	45 or more	Total
(a) P(late/London)	0.3016	0.4127	0.1587	0.1270	1.0000
(b) P(late/Sydney)	0.1515	0.4343	0.3535	0.0606	1.0000

Given that the flight is from Sydney, the probability that it will be more than 30 minutes late is $0.3635 + 0.0606 = 0.4141$. Given the origin is London the probability that it will be at least 30 minutes late is 0.2875. This does not confirm the views of the travellers association.

8. (a)

	Vitamin C	Garlic	Placebo
Probability	0.333333	0.353333	0.313333

(b) (i) $36/600 = 0.06$ (ii) $448/600 = 0.7467$ (iii) $48/200 = 0.2400$ (iv) $120/448 = 0.2679$

PE 4.5

1. (b) (i)

	Supplier			
	A	B	C	Total
Probability from Supplier A_i	$P(A_1) = 0.60$	$P(A_2) = 0.30$	$P(A_3) = 0.10$	1
Probability defective, (given Supplier)	$P(D/A) = \frac{5}{100}$	$P(D/B) = \frac{8}{100}$	$P(F/C) = \frac{3}{100}$	
Probability 'from Supplier AND defective'	$0.2 \times \frac{5}{100} = 0.03$ $= P(A \text{ AND } D)$	$0.3 \times \frac{8}{100} = 0.024$ $= P(B \text{ AND } D)$	$0.1 \times \frac{3}{100} = 0.003$ $= P(C \text{ AND } D')$	$P(D)$ $= 0.057$
(ii) P(Source/D)	$\frac{P(AD)}{P(D)} = \frac{0.03}{0.057}$ $= 0.526$	$\frac{P(BD)}{P(D)} = \frac{0.024}{0.057}$ $= 0.420$	$\frac{P(CD)}{P(D)} = \frac{0.003}{0.057}$ $= 0.052$	

2. (b) (i) 0.082 (ii) 0.2866 (iii) 0.7134. The test is not satisfactory because in 71 % of cases the test is positive when the board is perfect.

4. (a) 0.040279 (b) 0.0072 (c) $0.999990622 = 1$. No, the test is not satisfactory because there is only a 0.72 % chance that the virus is present when the test is positive.

6. (a) (i) 0.914 (ii) 0.086 (b) (i) 0.5581 (ii) 0.4419 (c) (i) 0.9978 (ii) 0.0022
The outcome is satisfactory for those who are guilty; there is a high chance (0.9978) that they will be found guilty. For an innocent defendant the outcome is worrying – there is a 55.81 % chance that they will be found guilty.

CHAPTER 5   SOLUTIONS

PE 5.1

1. (a) X is the number that shows when the die is thrown: $x-$ 1,2,3,4,5,6. #S $= 6$

   (b) X is the number on the card that is selected: $x =$ 1,2,3,4,5,6,7,8,9,10. #S $= 10$

   (c) X is the number that shows on the first and second die when both are thrown: $x =$ (1,1), (1,2),(1,3),(1,4),(1,5),(1,6),(2,1),(2,2),(2,3)...(6,1),(6,2),(6,3),(6,4),(6,5),(6,6). #S$= 36$.

2. (a) X is the faces that show on each coin. $x =$ HH, HT, TH, TT

   (b) X is the number on the die and the face on the coin. $x =$ 1H,2H,3H,4H,5H,6H,1T,2T,3T, 4T,5T,6T.

   (c) X is the colour of the ball that is selected. $x =$ white, yellow

4. (b) Question 2 (a) outcomes are $x =$ HH, HT, TH, TT. The probability of each outcome is 0.25. Question 2 (b) outcomes are $x =$ 1H,2H,3H,4H,5H,6H,1T,2T,3T,4T,5T,6T. The probability of each outcome is 1/12 Question 2 (c) outcomes are $x =$ white, yellow. P(white) $= 0.4$, P(yellow) $= 0.6$

6. (a) Total number of sample $= 317$

Number of defectives	0	1	2	3	4	5	6	more than 6
Probability x defective parts	$\frac{80}{317}$	$\frac{125}{317}$	$\frac{64}{317}$	$\frac{21}{317}$	$\frac{15}{317}$	$\frac{3}{317}$	$\frac{5}{317}$	$\frac{4}{317}$
Number of samples	80	125	64	21	15	3	5	4

(b) Probabilities are given correct to four decimal places

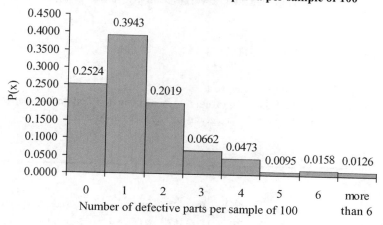

Probability histogram for defective part a per sample of 100

(c) Average defective parts per sample $= 1.416$ assuming 'more than 6' is '7'. (d) Since sample sizes are equal this is the proportion of defective parts for the entire data (proportion $= 449/31700$).

8. (a) $p = 0.11$ (b) 0.34 (c) 21 % (d) 2.24

PE 5.2 Binomial Probabilities

1. (a) 0.25, 0.50, 0.25

2. 1/8, 3/8, 3/8, 1/8

4. (a) 224 (b) 0.43 (c) (i) see Chapter 1 (ii) 0.1260

6. (a)(i) 0.4832 (ii) 0.6242 (b) (i) 0.2013 (ii) 0.3222

8. (a) 0.1298 (b) 0.8202 (c) 9.

10. (a) 0.8784 (b) 0.8670

12. (a) (i) 0.35, on-line (ii) 0.40, by phone (iii) 0.25, other. Not on-line 0.65
(b)(i) 0.1218 (ii) 0.000044 (iii) 0.9101 (iv) 0.0389

PE 5.3 Poisson Probabilities

1. (a) (i) 7.3891 (ii) 0.4060 (iii) 0.0410 (iv) 0.0161
(b) (i) $\dfrac{\lambda^4 e^{-\lambda}}{24}$ (ii) 0.1254

2. (a) 0.1465 (b) 0.2532 (c) 0.2424 (d) 0.2236

4. (a) mean $= 2$, variance, $\sigma^2 = 2$ (b) 0.8571 (c) mean and variance are the same – possible indication of a Poisson process.

6. (a) 0.3476 (b) 0.4966 (c) 0.0341 (d) 0.0009 (e) 0.0111

8. (a) the mean number of rentals is $735/125 = 5.88$
   (b) probabilities given in row four of the following table; number of times (expected frequencies) that there are 0, 1, 2, ...12 rentals per hour are given in the last row of the table.

Rentals per hour	0	1	2	3	4	5	6	7	8	9	10	11	12	Totals
Frequency (f_o)	1	3	4	10	15	20	24	19	13	10	4	1	1	125
fx	0	3	8	30	60	100	144	133	104	90	40	11	12	735
$P(x)$	0.00	0.02	0.05	0.09	0.14	0.16	0.16	0.13	0.10	0.06	0.04	0.02	0.01	0.992445
$fe = P(x)*125$	0	2	6	12	17	20	20	17	12	8	5	3	1	124.0557

Comparing the observed number of times and the expected number of times (assuming a Poisson with $\lambda = 2.88$) for 1, 2, 3 ... 12 rentals per hour – a Poisson process does appear appropriate.

10. (b) average $= 262/80 = 3.275$

Absentees per day	0	1	2	3	4	5	6	7	8	9	10	Totals
Frequency	6	9	12	18	18	9	4	0	3	0	1	80
fx	0	9	24	54	72	45	24	0	24	0	10	262
$P(x)$	0.0378	0.1239	0.2028	0.2214	0.1813	0.1187	0.0648	0.0303	0.0124	0.0045	0.0015	0.9994
$fe = P(x)*80$	3	10	16	18	15	9	5	2	1	0	0	80

Comparing the observed number of absentees and the expected number of absentees (assuming a Poisson with $\lambda = 3.275$) for 1, 2, 3 ... 12 – a Poisson process may be appropriate – there are differences, for example on six occasions there were no absentees but when modelled by a Poisson there should be only three days on which there are no absentees. (There are statistical tests for the goodness of fit for a Poisson process.)

PE 5.4 Normal probabilities

1. (a) 0.1587 (b) 0.1587 (c) 0.3414 (d) 0.6828 (e) 0.8664

2. (a) 36 $\pm$ 9.8694, i.e. between 26.1306 and 45.8694 minutes. (b) 36 $\pm$ 11.76 minutes
(c) 36 $-$ (0.5244)(6) i.e. 36 $-$ 3.1464 = 32.8536 minutes (d) 45.8694 minutes

4. (a) 550 $\pm$ 2.5758 mm (b) 0.6079 (c) 550.1776

6.

Grade	A	B	C	D	E
(a) Marks %	84 or more	72 < 84	48 < 72	36 < 48	36 or less
(b) Percentage awarded grade	2.275	13.595	68.26	13.595	2.275
(c) Number awarded	3	17	84	17	3

8. (a)(i) 0.0239 (ii) 0.2244 (iii) 0.2643 (b) (i) 293 (ii) 67 (iii) 221

10. (a) For 12 adults, weights are normally distributed, $\mu = 12 \times 69 = 828$, $\sigma = \sqrt{12 \times 8.5^2} =$ 29.44. Hence P (weight of 12 exceeds 900kg) = 0.00714.
(b) (i) 0.1423 (ii) 99 % of groups of 14 weigh 1040kg or less. 99 % of groups of 12 weigh 897kg or less (iii) a maximum of 12 persons will no breach health and safety limits. Also calculate probability of a group of 13 exceeding the limit.

12. 99.9 %

14. (a) From question 11, PE2.4, the mean and variance for the number of hours worked by (i) males is 40.73 and 102.04 and (ii) females is 30.06 and 163.46 respectively. (b) (i) 0.2148 (ii) 0.3785 (c) mean = 10.67, variance = 265.5 (i) 74.22 % (ii) 25.78 % (iii) 51.6 %.

PE 5.5

1. See section 5.5.1

2.

$x =$	0	1	2	3
$P(x) =$	0.064	0.288	0.432	0.216
$xP(x) =$	0	0.288	0.864	0.648
$E(x) =$	1.8			

4. $E(X) = \sum xP(x) = 1.4164$. $V(X) = E(X^2) - [E(X)]^2 = 3.9779 - (1.4164)^2 = 1.9717$. The same values for mean and variance are obtained by using the formulae for grouped data $\frac{\sum fx}{N} = \frac{449}{317} = 1.4164$ and $\frac{\sum fx^2}{N} - (\mu)^2 = \frac{1261}{317} - (1.4164)^2 = 1.9717$.

6. (b) (i) Binomial 0.2770, Poisson 0.2681 (ii) Binomial 0.2762, Poisson 0.2706 (iii) Binomial 0.1227, Poisson 0.1221 (iv) Binomial 0.9401, Poisson 0.9362.

8. (a) (i) $n = 20$, mean $= 1$, variance $= 0.95$ (ii) $n = 250$, mean $= 12.5$, variance $= 11.875$
   (b) (i) 0.0755 (ii) 0.2810 (using a Normal approximation to the Binomial, with continuity correction).

10. (a) mean $= 16.8$ lbs, variance $= 17.1727$lbs
   (b) (i) 8.85 % (ii) 72.89 %

12. (a) mean $= 82.4°$F, variance $= 10.4976°$F, standard deviation $= 3.24°$F
   (b) (i) 0.2877 (ii) 0.7704 (iii) 0.5788 (iv) 0.3567

CHAPTER 6 SOLUTIONS

PE 6.1

1. (a) $\mu = 9.25$, $\sigma = 1.9203$ (b) Samples of size two are $\{(10,8), (10,12), (10,7), (8,12), (8,7), (12,7)\}$. Sample means are $\{9,11,8.5,10,7.5,9.5\}$. The means and standard deviation of the sample means are 9.25 and 1.1087 respectively. (c) The mean of the sampling distribution of means is the same as the population mean, the standard deviation (standard error) is

$$\frac{\sigma}{\sqrt{n}}\sqrt{\frac{N-n}{N-1}} = \frac{1.9203}{\sqrt{2}}\sqrt{\frac{4-2}{4-1}} = 1.1087.$$

Comparing (a) and (b) the mean of all the sample means is the same as the population mean. The standard deviation of all the sample means = 1.1087 is the same as that calculated by (6.1).

2. (b) (i) $\mu = 11.5$, $\sigma = 1.7078$ (ii) $\bar{x} = 12$ and $s = 2$. Sampling errors for μ and σ are $|12 - 11.5| = |2|$ and $|2 - 1.7078| = |0.2922|$. $\mu_{\bar{x}}$ and $\sigma_{\bar{x}}$ are calculated from all the sample means – not a single mean.

4. (a) (i) $\sigma_{\bar{x}} = 42$ (ii) $\sigma_{\bar{x}} = 10.5$ (iii) $\sigma_{\bar{x}} = 8.4$. (c) (i) 320 ± 82.32 (ii) 320 ± 20.58 (iii) 320 ± 16.464 (c) (i) 36.88 % (ii) 94.26 % (iii) 98.268 %.

As sample size increases, sample means are more accurate – closer to population mean – hence the sampling error decreases. Hence as sample size increases, there is a larger percentage of sample means within €320 ± €20 (the population mean).

8. (a) Not necessarily – for example, stratified samples are more accurate when sampling from populations that consist of distinct strata (subgroups) of different sizes. Non-random samples such as quota samples may be representative but the standard error may only be estimated from a random sample. (b) False. Standard error is the standard deviation of all sampling errors- sampling error is the difference between a population parameter and an estimate of it. (c) True. A sampling distribution of means is a list of every possible sample mean for samples size n. Hence the probability of selecting a sample with a given mean value may be calculated. (d) False. The standard error of the mean is the standard deviation of all sample means.

10. (a) (i) 46.76 % (ii) 168 055 (b) (i) 0.0475 (ii) 0.9521 (iii) 0.0004

12. (a) True – but even when a population list is available a quota sample may be used, particularly if time and cost are important (b) False, the nature of randomness means that there is always a chance that a random sample may be biased. (c) The Central Limit Theorem applies for small samples only when sampling is from a Normal population and the population standard deviation, σ is known. (d) True, the standard error $\dfrac{\sigma}{\sqrt{n}}$ decreases as the sample size, n increases.

PE 6.2

1. (a) $\pi = 1/3$ (b) Samples $= \{(15, 21, 17, 24, 28), (15, 21, 17, 24, 33), (15, 17, 24, 28, 33), (15, 21, 17, 28, 33), (15, 21, 24, 28, 33), (21, 17, 24, 28, 33)\}$. Sample proportions of even numbers $= 0.4$, $0.2, 0.4, 0.2, 0.4, 0.4$. The mean and variance of all sample proportions $\mu_p = 1/3, \sigma_p = 0.0943$. (c) The formulae for the mean and variance of sample proportions are π and $\dfrac{\pi(1-\pi)}{n}\dfrac{N-n}{N-1}$. Hence mean $= 1/3$ and variance $= \dfrac{\frac{1}{3}\left(1-\frac{1}{3}\right)}{5}\dfrac{6-5}{6-1} = 0.00888888$ respectively. Hence $\sigma_p = 0.0943$, the same as that calculated directly from the list of all sample proportions.

4. (a) (i) 0.2611 (ii) 0.1587 (b) 0.3205 to 0.5795. (90 % of sample proportions will fall between 0.3205 and 0.5795).

6. (a) True. $0 \le \pi \le 1$, therefore $\pi(1 - \pi) < 1$ while the sample size, $n > 1$. Hence the standard error is $\sigma_p = \sqrt{\dfrac{\pi(1-\pi)}{n}} = \sqrt{\dfrac{\text{a number less than 1}}{\text{a number greater than 1}}} < 1$. The minimum value is zero, when $\pi = 0$ or $\pi = 1$. (b) False. Increasing the sample size decreases the value of $\sigma_p = \sqrt{\dfrac{\pi(1-\pi)}{n}}$, the standard error.
(c) Rarely true. Sample proportions are Normally distributed about the population proportion, therefore differ from the population proportion – there is a very, very small probability that a sample proportion is the same as the population proportion. (d) True. If a large number are polled, sampling error and hence standard error decreases – the sample proportion is closer to the population proportion.

8. (a) 0.6826 (b) (i) 384 (ii) 9587

10. (a) 0.0901 (b) 1.0000 (c) (i) 0.0257 (ii) 0.0000 (d) (i) 1937 (ii) 542

CHAPTER 7 SOLUTIONS

PE 7.1

2. (a) The margin of error for (i) a 95 % confidence interval is $\pm1.96\sigma_{\bar{x}}$ (ii) a 90 % confidence interval is $\pm1.6449\sigma_{\bar{x}}$. The formulae are derived on the assumption that $\bar{x} \sim N(\mu, \sigma_{\bar{x}})$ hence the margin of error is the intervals, symmetrical about the mean, which contains (i) 95 % of sample means (ii) 90 % of sample means.

(b) (i) 78 ± 11.76, i.e. 66.24kg to 89.76kg. We are 95 % confident that the population average weight is somewhere between 66.24kg and 89.76kg. The interval is too wide to be useful. (ii) when $n = 84$, the standard error is $\sigma_{\bar{x}} = \dfrac{\sigma}{\sqrt{n}} = \dfrac{36}{8} = 4.5$ compared to $\sigma_{\bar{x}} = 6$ when $n = 36$. Hence the interval will be more precise: $\pm1.96(4.5) = \pm8.82$.

4. (a) Target population depends on the intended market for the daffodils, for example a specific urban population or a national network of retail outlets etc. Quota controlled samples will be cheaper and faster than random samples with quotas for specific groups in the population. Cluster sampling could be used for retail outlets distributed nationally.

(b) The mean spending per household is between 8.55 ± 0.87 or £ $7.68 \le \mu \le £9.42$ with 95 % confidence. **(c)** £ 7219 to £ 8855

6. (a)

95 % confidence intervals	(i) Long haul	(ii) Package	(iii) Business
(a) Upper confidence limit	495.80	861.57	64.14
(a) Lower confidence limit	471.20	846.43	52.86
(b) Upper one-sided confidence limit	473.18	849.05	53.76

There is no overlap between the intervals; hence at 95 % confidence the mean amount spent by the three categories of customers is different. The CI for business customers may not be valid: $n < 20$, σ is unknown and no statement declaring that expenditure by these customers is Normally distributed.

(b) See above: we are 90 % confident that long haul customers spend up to € 473.18 etc.

8. From PE2.4, question 12, the mean and variance for the number of hours worked by (i) males is 40.73 and 102.04 and (ii) females is 30.06 and 163.46 respectively (a) 95 % confidence interval for males is 40.73 to 40.74, for females is 30.05 to 30.07. The precision is due to large sample sizes, 13432784 males and 11978863 females.

PE 7.2

1. (a) (i) 0.4 (ii) 0.6 (iii) 0.180 (iv) 0.250 (v) 0.220 (b) $0.4 \pm 0.0255 \rightarrow 0.3745$ to 0.4255 (c) 0.4341 to 0.4859. From (b) we are 90 % confident that between 37 % and 43 % of commuters are male, 50 % is not in the interval, hence does support the claim that $\pi = 50\%$. Similarly in (c) $\pi = 0.5$ or 50 % is the interval.

2. (a) 0.59 ± 0.0330 or 0.5570 to 0.6230. The interval contains 0.5, but only supports the claim that up to 62.3 % of female commuters are dissatisfied.

4. (a) False. The population proportion can be anywhere within the confidence interval (b) True. The standard error, $\sqrt{\dfrac{p(1-p)}{n}}$ decreases, hence the width of the interval, $Z_{\alpha/2}\sqrt{\dfrac{p(1-p)}{n}}$ also decreases as n increases (c) True, as $Z_{\alpha/2}$ increases, the width of the interval, $Z_{\alpha/2}\sqrt{\dfrac{p(1-p)}{n}}$ decreases. (d) True. The product $p(1-p)$ is largest for $p = 0.5$, hence the width of the interval, $Z_{\alpha/2}\sqrt{\dfrac{p(1-p)}{n}}$ is largest, assuming the sample size and level of confidence are fixed.

PE 7.3

2. (a) [27.31, 28.69] (b) 22.98, 24.5, 28.0. (c) 100, 36, 202.

4. (b) (i) [0.32, 0.48] (iii) 62, 144

6. 1692

PE 7.4

2. (a) No difference (b) There is a difference, $\mu_1 > \mu_2$ (c) There is a difference, $\mu_2 > \mu_1$

4. (a) 0.2088 (b) (i) 0.52 ± 0.4093 (ii) 0.52 ± 0.5378. Only the 99 % confidence interval contains zero. At the 99 % level of confidence, there is support for the claim that there is no difference between offices.

6. (a) 7 ± 1.0033 minutes for recovery times for existing – new medication. The interval indicates the mean recovery time is greater for the existing medication, therefore supports the claim that the new medication is faster.

8. (a) (i) 0.0464 to 0.1136 (ii) 0.0200 to 0.0810 (b) −0.0478 to 0.1068 (c) in (b) there is evidence that the difference between thermo A and thermo B is zero – don't change to the new thermostat.

10. (a) (i) 0.7326 (ii) 0.5224 (b) (0.1569 to 0.2635, hence a higher proportion moved to the competitor in the last month; the provider's alarm is justified.

12. (a) (i) 0.65 (ii) 0.72.95 % CI for difference $(-0.0373, 0.1717)$. (b) 95 % CI (women $-$ men) is 0.19 ± 0.0787. Hence there is a longer waiting for women at the 95 % confidence level (c) 4.84 ± 0.8375. Comments, women take longer to get to hospital after the initial attack. (d) At the 95 % confidence level, there is no difference in the proportions of men and women referred to hospital by their GPs. Women take longer to decide to go to hospital and wait longer to be treated when they get there.

14. From PE2.4, question 12, the mean and variance for the number of hours worked by (i) males is 40.73 and 102.04 and (ii) females is 30.06 and 163.46 respectively. There are 13 432 784 males and 11 978 863 females.

The 95 % confidence for the difference between the average hours worked by males and females is 10.67 ± 0.0090. There is evidence that the average weekly hours worked by males exceeds that of females by 10 hours per week.

CHAPTER 8 SOLUTIONS

PE 8.1

2. (a) $H_0 \mu \geq 16, H_1 \mu < 16$ (b) $H_0 \mu = 5, H_1 \mu \neq 6$ (c) $H_0 \mu \leq 45, H_1 \mu > 45$ (d) $H_0 \mu \leq 8, H_1 \mu > 8$

4. Test statistic (sample Z) $= -2.784$. Reject H_0 at the 5 % level of significance and conclude that the average amount of fill is not 16g.

6. $H_0 \mu \leq 25$. Reject H_0 if the test statistics (sample Z) exceeds 1.6449. Test statistic (sample Z) $= 3.56$. Conclusion, reject H_0 and conclude that the average is greater than 25.

8. (a) The test statistics, $T = -2.17$. $H_0: \mu = 20$, $H_1 \neq 20$. H_0 is rejected at the 5 % level of significance but accepted at 1 %. (b) The 95 % confidence interval does not include 20, hence does not support the claim that $\mu = 20$. The 99 % confidence interval is (18.5873, 20.1327) does contain 20, hence supports $H_0: \mu = 20$.

10. $H_0 \mu = 120$. Reject H_0 if the test statistics (sample Z) is less than -2.5758 or greater than 2.5758. Test statistic (sample Z) $= -2.65$ is less than -2.5758. Conclusion, reject H_0 and conclude that the average is not 120. The p-value $= 0.00403 \times 2 = 0.00804$ (Assume $n \geq 30$).

12. $H_0: \mu = 5$. $H_1: \mu \neq 5$. Given that $\sigma = 2$

Sample size, $n = 30$, $\bar{x} = 5.6$, $s = 6.26209$, SE mean $= \sigma_{\bar{x}} = \dfrac{\sigma}{\sqrt{n}} = \dfrac{5}{\sqrt{30}} = 0.91287$. The 95 %

CI is $\bar{x} \pm 1.96\sigma_{\bar{x}} = 5.6 \pm 1.96(0.91287)$. $Z = \dfrac{\bar{x} - \mu_{H_0}}{\sigma_{\bar{x}}} = \dfrac{5.6 - 5}{0.91287} = 0.6573$ or 0.66 correct to two decimal places. P-value $= 2 \times 0.2562 = 0.5125$. Slight differences between these answers and those given in the Minitab print out are due to rounding and use of more accurate Normal probability tables.

PE 8.2

2. (b) (i) $(0.3565, 0.4935)$ (ii) $H_0 \pi \leq 0.50$, $H_1 \pi > 0.50$. Test statistic (sample Z) $= -2.14$. Accept H_0. Conclude the support is at most 50 %. P-value $= 0.9838$.

4. $H_0 \pi \geq 0.20$, $H_1 \pi < 0.20$. Test statistic $= 2.08$. Accept H_0 at the 5 % and 0.5 % levels of significance and conclude that 20 % or more of accidents are due to untidy working conditions. This sample does not provide evidence to support the manager's claim. The p-value $= 0.9812$.

6. (a) $T = 2.53$, accept H_0, conclude that at least 50 % are satisfied/very satisfied. P-value $= 0.9943$
(b) H_0 $\pi = 0.54$, H_1 $\pi \neq 0.54$. $T = 3.36$, reject H_0. Hence the percentage of females aged 45 and younger that are not satisfied is not 54 %; p-value $= 0.0008$ (c) H_0 $\pi \leq 0.70$, H_1 $\pi > 0.70$. $T = -11.13$, accept H_0. P-value $= 1$. There is insufficient evidence to conclude that more than 70 % of males are satisfied.

PE 8.3

1. (a) $\overline{X} \sim N\left(\mu_{\bar{X}} = 9.10, \sigma_{\bar{X}} = \dfrac{1.87}{\sqrt{30}}\right)$, $\overline{Y} \sim N\left(\mu_{\bar{Y}} = 8.70, \sigma_{\bar{X}} = \dfrac{2.24}{\sqrt{45}}\right)$

(b) $(\overline{X} - \overline{Y}) \sim N\left(\mu_{\bar{X}-\bar{Y}} = 0.40, \sigma_{\bar{X}-\bar{Y}} = \sqrt{\dfrac{(1.87)^2}{30} + \dfrac{(2.24)^2}{45}}\right) = N(0.4, \ 0.4776)$

$P\,(\bar{x} - \bar{y}) > 0 = 0.7995$ (c) (i) H_0: $\mu_X \leq \mu_Y$ H_1: $\mu_X > \mu_Y$. Test statistics $= 1.88$. Reject H_0 at $\alpha = 0.05$. Conclude that the mean price for brand X is greater than brand Y. P-value $= 0.0301$. (ii) H_0: $\mu_X - \mu_Y = 1$. H_1: $\mu_X - \mu_Y \neq 1$. Test statistics $= -0.2094$. Accept H_0. The mean price for brand X is £ 1 more than the mean price for brand Y.

2. $\mu_1 = \mu_2$. Accept H_0 if the test statistic falls between -1.6449 and 1.6449 inclusive. Conclusion is to reject H_0. p-value $= 0.0644$. At $\alpha = 1\,\%$, H_0 would be accepted.

4. (a) SE $= 0.2088$ (b) Test statistic $= 2.49$. Reject H_0 at $\alpha = 5\,\%$ and conclude that the mean process times are different. Accept H_0 at $\alpha = 1\,\%$ and conclude that there is no difference between process times. (c) No, the test is for average process times – not individual times.

6. (a) Morning: $p \sim N(0.15, 0.03992)$ (ii) Evening: $p \sim N(0.25, 0.03953)$, P ('no shows' is at least 10 in mornings) $= 0.7357$, P ('no shows' is at least 10 in evenings) $= 1$.
(b) $p_1 - p_2 \sim N\left(\mu_{p_1-p_2} = \pi_1 - \pi_2, \ \sigma^2_{p_1-p_1} = \dfrac{\pi_1(1-\pi_1)}{n_1} + \dfrac{\pi_2(1-\pi_2)}{n_2}\right)$
$p_1 - p_2 \sim N(\mu_{p_1-p_2} = 0.10, \sigma^2_{p_1-p_1} = 0.0032)$. Hence $\sigma_{p_1-p_1} = 0.0562$. (i) 0.1867.
(c) $p_c = 0.12$, $s_{p_1-p_1} = 0.0469$ (i) Test statistic $= -4.26$. One cannot conclude that there at least 15 % more 'no shows' on evening flights. (ii) Test statistic $= -1.07$, so accept that the percentage of 'no shows' is the same.

8. (a) $T = 1.24$. Accept that there is no difference (b) do not replace existing model with the new design.

10. (a) SE $= 7.3701$, test statistic $= 2.78$. Accept H_0 and conclude that the mean expenditure by package holiday customers exceeds that by long-hauls by at least € 350. (b) Sample size is less than 30, the distribution of expenditure by business customers is not stated to be Normal and σ is not known.

12. (a) $p_c = 0.667$. Test statistic $= 1.31$. Conclude that the proportion of women referred is not greater than the proportion of men referred by GP. (b) SE $= 0.0402$. Test statistic $= 4.7264$. Conclude that the mean time before treatment is longer for women.

14. The smallest mean is 19.41, standard deviation $= 7.69$. The largest mean is 33.20, standard deviation $= 8.30$. Sample sizes are 30. The test statistic $= 6.68$. Conclude the mean amounts saved are different. No conclusion can be made about the other groups – each pair must be tested.

CHAPTER 9 SOLUTIONS

PE 9.1

1. (a) 98.268 % (b) 100 % (c) (i) 95 % CI (50.0879, 50.2721) (ii) $T = 3.83$, therefore reject H_0 and conclude that the average weight is not 50 mg. This agrees with the CI which does not contain the value, 50.

2. (a) $\bar{x} \pm 1.6449 \dfrac{0.15}{\sqrt{8}} = \bar{x} \pm 0.8723$ (b) (i) 90 % CI is (2.3665, 2.5410) (ii) $T = -0.87$, accept H_0 and conclude that the mean weight is 2.5kg. This agrees with the CI which contains the value, 2.5.

4. (a) 99 % CI is (9.8182, 12.2618) (b) $T = 2.19$, accept H_0 and conclude that the mean time is 10 min. This agrees with the CI which contains the value, 10.

6. (a) $s = 0.1647$ (b) $\bar{x} \pm 1.895 \dfrac{0.1647}{\sqrt{8}} = 2.4538 \pm 0.1104$ (c) $T = -0.79$, accept H_0 and conclude that the mean weight is 2.5kg. This agrees with the CI which contains the value, 2.5.

8. (a) $s = 3.3712$ (b) 95 % CI is (16.1141, 24.4859) (c) $T = 2.19$. The critical region is outside ± 2.776; therefore conclude that the true mean time is 17. In addition, 17 is within the 95 % confidence interval. The results in question 3 concluded that the mean was not 17. The assumed standard deviation in question 3 was 2.2 while in question 8 the estimated standard deviation was 3.3712. The larger value of standard deviation together with the use of $t_{0.025,4}$ instead of $z_{0.025}$ resulted gave a wider interval that included 17. Also the smaller value of T in question $8(T = 2.19)$ led to the acceptance of H_0: $\mu = 17$.

10. (a) 95 % CI for clinic 1 is $15.4 \pm 2.145 \dfrac{3.2}{\sqrt{15}} = 15.4 \pm 1.7723 = (13.6277, 17.1723)$

   95 % CI for clinic 2 is $14.6 \pm 2.069 \dfrac{2.5}{\sqrt{24}} = 14.6 \pm 1.0558 = (13.5442, 15.6558)$

   95 % CI for clinic 3 is $20.6 \pm 2.365 \dfrac{6.7}{\sqrt{8}} = 20.6 \pm 5.6022 = (14.9978, 26.2022)$.

   All three intervals overlap, suggesting that the true mean time could be the same for all clinics.
   (b) The test statistics for clinics 1, 2, and 3 are 0.4841, -0.7839 and 2.3641 respectively. In each case the claim that the mean time is 17 is accepted. Times for tests are assumed to be normally distributed.

PE 9.2

2. (a) (i) $s_p^2 = 3.11405$ (ii) $s_p^2 = 4.5276$ (iii) $\sigma_p^2 = 274.9167$ (b) (i) $s_{\bar{x}_1 - \bar{x}_2} = 0.7892$ (ii) $s_{\bar{x}_1 - \bar{x}_2} = 0.7930$ (iii) $\sigma_{\bar{x}_1 - \bar{x}_2} = 4.7864$ (c) 90 % confidence intervals for differences between the mean Hyper and Hipo prices are (i) $(-.1581, 3.0581)$ (ii) $(-3.8241, -0.5759)$ (iii) $(-9.3213, 9.4413)$. At the 5 % level of significance the mean price for Chablis is higher in the Hyper stores but there is no difference between the prices of strawberries or petrol.

3. (a) mean and sample standard deviation for Model 1.8 and Model 1.4 respectively are $\bar{x} = 7839, 6181$ and $s = 3038, 917$ (b) $s_p^2 = 5559115.1$ (c) 98 % CI for difference: $(-1230.88, 4547.83)$ (c) The test statistic $T = 1.48$ is less than the critical value, $t_{16, 0.01} = 2.583$. Hence we cannot conclude that the average miles for Model 1.8 is greater that for Model 1.4.

4. (a) 95 % CI for the difference in mean scores between the control group and the test group are: (i) $(-11.6998, 39.1998)$, before the course (ii) $(-32.761052, 31.761052)$ after the course.
(b) 95 % CI for the difference in the mean scores (after − before) the course (i) $(12.3894, 72.2773)$ for the control group and (ii) $(28.4439, 84.7228)$ for the test group There is no evidence that the test group performs better than the control group. In (a) there is no difference between the control and test groups, either before or after the course. In (b) the scores were higher after the course, but there is a large overlap between the confidence intervals, suggesting that the increase in scores is the same for both groups. The sample variances are all large (1052, 755, 1450, 1454) but of the same order of magnitude.

6. (a) 6 pairs ($^4C_2 = 6$) (b) 95 % CI for differences: (i) A and C $(0.60315, 3.39685)$ (ii) B and C $(3.41779, 6.74221)$ (iii) B and D $(-0.471774, 2.351774)$. There is evidence that the journey time is longer in the morning (both 6–8 am and 8–10 am) than during 'normal time', 10 am to 4 pm. There is no difference between the journey time during the morning or evening rush 'hour' at the 5 % level of significance.
(c) 95 % CI for difference between B and D when sample size for D is 9, $(-1.448284, 3.328284)$. The small sample size gives rise to a large standard error and a larger critical t-percentage point. The result is a wider or less precise confidence interval.

PE 9.3

1. (i) $F_{4,8,0.05} = 3.84$ (ii) $F_{2,10,0.05} = 4.10$ (iii) $F_{2,2,0.025} = 39.00$ (iv) $F_{5,3,0.025} = 14.88$

2. Strawberries. Test statistic $= 3.00$, $df = (9, 9)$. The critical (5 %) $F = 3.1789$. Chablis. Test statistic $= 1.79$, $df = (17, 11)$. The critical (5 %) $F = 2.6851$. No significant difference at 5 %.

3. Test statistic $= 10.97$, $df = (9, 7)$. The difference between variances is significant at 0.5 %. The critical $F = 3.6767$.

4. Before the course, Test statistic $= 1.39$. $df = (11, 11)$. The critical (5 %) $F = 2.8179$. No significant difference at 5 %.

6. (i) A and C. Test statistic $= 6.07$. $df = (23, 17)$. The critical (0.5%) $F = 3.53$. There is a significant difference at 0.5%. (ii) B and C. Test statistic $= 8.95$. $df = (24, 17)$. The critical (0.5%) $F = 3.51$. There is a significant difference at 0.5%. (iii) B and D. Test statistic $= 4.122$. $df = (24, 28)$. The critical (0.5%) $F = 2.79$. There is a significant difference at 0.5%. (iv) B and D. Test statistic $= 4.122$. $df = (24, 8)$. The critical (0.5%) $F = 6.50$. There is no significant difference at 0.5%.

8. (a) $n = 30$ for both samples. The sample means and standard deviations are 5.6, 12.4 and 6.25, 13.2 for companies A and B respectively. $SE = \sigma/\sqrt{n}$. The test is two-sided, H_0 $\mu_A - \mu_B = 0$ vs. H_0 $\mu_A - \mu_B \neq 0$. The test statistic, $T = -2.57$. The evidence from these samples (the test and the 95% CI) support the alternative hypothesis: $\mu_A - \mu_B \neq 0$.

8. (b) Both box plots reveal the presence of outliers, particularly B, hence Normality is not satisfied.

PE 9.4

1. Paired differences for (statistics – marketing) are 21, 8, −5, 23, 13, 8. $\bar{d} = 11.3333$, $s_d = 10.2111$, $s_{\bar{d}} = \dfrac{s_d}{\sqrt{n}} = 4.1687$. The 95% CI is $\bar{d} \pm t_{5,0.025}\dfrac{s_d}{\sqrt{n}} = 11.3333 \pm 2.571\dfrac{10.2111}{\sqrt{6}} = (0.6157$ to $22.051)$. There is a difference; average statistics mark is greater than average marketing mark at the 5% level of significance.

2. 95% CI for difference (statistics – marketing) is $(-13.7710, 36.4377)$. According to this CI, there is no difference between the marks.
No difference is detected in this test because the standard error is larger for independent samples (11.2670) than for paired samples (4.1687).

4. (a) (i) Test statistic for paired differences is $T = 6.38$. $df = 3$. Conclude that average earnings are higher in England than in Wales at the 5% and 1% levels of significance. (ii) $T = 7.79$. Conclude that average earnings are higher in England than in N Ireland at the 5% and 1% levels of significance.
(b) The 95% CI for mean differences in weekly earnings between Wales and Scotland is $(-47.33651, 31.83651)$.
Hence average earnings are higher in England than in N Ireland or Scotland but there is no difference between Wales and Scotland.

6. (a) The 90% CI for mean difference between wardens T and F is $(0.48256, 2.85077)$. Warden T collects more fines, on average, than warden F. (b) Use $H_0: \mu_{Old\ market} \leq \mu_{Shop\ street}$ $H_1: \mu_{Old\ market} > \mu_{Shop\ street}$. The test statistic, $T = 1.51$. Accept H_0, hence we cannot conclude that the revenue from Old market is greater than Main street or Shop street. (c) The Minitab printout tests the difference in average revenue from Main street and Shop street. The print gives sample size, sample mean, standard deviation and standard error for Main street, Shop street and the paired differences for (Main St. – Shop St.) The 95% confidence interval for the paired differences contains zero, hence supports the claim that revenue is the same for both streets. The two-sided test accepts H_0: mean revenue is the same from Main St and Shop St. The critical values for the test are ± 2.776. The test statistics, $T = -1.96$ is within the acceptance region. The p-value $= 0.121$ indicates a 12.1% chance of obtaining these samples when H_0 is true – hence supports H_0.

CHAPTER 10 SOLUTIONS

PE 10.1

1. (a) (i) 0.1271 (ii) 0.0212 (iii) 0.8517 (b) 99 % CI for the mean account is $(6239.15, 6530.85)$. The interval does not support the banks claim: 6554 is not in the interval.

2. (a) $(23.96, 58.04)$ (b) (ii) H_0: All population means are equal (hence $\sigma_B^2 \leq \sigma_W^2$) H_1: at least one mean is different from the others (hence $\sigma_B^2 > \sigma_W^2$) (iii) $s_W^2 = 198.4$, $s_B^2 = 158$ (iii) $F_{v_1, v_2} = \frac{s_B^2}{s_W^2} = 0.7963$, $df = (2, 15)$. There is no significant difference.

(c)

Source	DF	SS	MS	F	F-critical (5 %)
Between groups	2	316	158	0.80	3.682
Within groups	15	2976	198		
Total	17	3292			

The MS's (mean sum of squares), the SS/df (sums of squares divided be the corresponding degrees of freedom) are s_B^2 and s_W^2 respectively. F in the table is the test statistic or F-ratio, $F_{v_1, v_2} = \frac{s_B^2}{s_W^2} = \frac{MS(Factor)}{MS(Error)}$

4. (a) One-way ANOVA: Degree, Diploma, Trade, Professional

Source	DF	SS	MS	F	F-critical (5 %)
Qualif	3	2512.6	837.5	12.47	3.2387
Error	16	1074.4	67.2		
Total	19	3587.0			

Conclude that there is a difference between salaries.

(b) 95 % CI for Degree $(26.68, 41.72)$, Diploma $(24.81, 35.19)$, Trade $(28.00, 49.60)$, Professional $(44.58, 73.82)$.

(c) Using Scheeffe's method, the 95 % simultaneous CI are $\bar{x}_2 - \bar{x}_1 \pm 16.16$.
Using Bonferroni's method, the 95 % simultaneous CI are $\bar{x}_2 - \bar{x}_1 \pm 15.70$

(d) Hence there are significant difference between those professionally qualified and all the others.

6. (a) Accept the null hypothesis that population means are equal at $\alpha = 2.5\,\%$, but reject the null hypothesis at $\alpha = 5\,\%$.

ANOVA

Source of Variation	SS	df	MS	F	F critical
Between Groups	20766.28	4	5191.57	2.89	2.73 (5 %)
Within Groups	48457.94	27	1794.74		3.31 (2.5 %)
Total	69224.22	31			

(b) 95 % individual CI for the difference in means is between maths groups (i) OA – OD, (15.64, 131.70) (ii) OB – OD, (15.08, 88.75). These intervals indicate a difference between grades OD and OA/OB.

PE 10.2

1. (b) (i) When the data is treated as independent samples, the 95 % CI for difference for 2005– 2007 is $(-14.90908, 28.90908)$. The test statistic is $T = 0.72$. Conclude that there is no significant difference at the 5 % and 1 % levels.
(ii) For paired data, the 95 % CI for mean difference for 2005–2007 is $(2.6477, 11.3523)$. The test statistic is $T = 4.13$. Conclude that there is a difference at the 5 % and 1 % levels of significance.
Note. The independent t test is not valid since it is based on the assumptions that the samples are independent – in this example, the samples are related according to bank branch.

2. (a)

Source of Variation	SS	df	MSS	F	F critical (5 %)
Branches	2771	5	554.2	64.4419	5.0503
Years	147	1	147	17.0930	6.6079
Error	43	5	8.6		
Total	2961	11			

(b) The differences between (i) branches and (ii) years are significant at the 5 % levels. There is no significant difference between years at $\alpha = 0.1\,\%$.

(c) The square root of the critical F (5 %) $\sqrt{.6.6079} = 2.571$. The critical t (2.5 %) $= 2.571$
The square root of the test statistic in ANOVA (F-ratio) $= \sqrt{17.093} = 4.134$. This is the test statistic for the paired t-test.

4. (b) Two-way ANOVA

Source	DF	SS	MS	F
Athlete	3	1.8225	0.6075	2.43
Operator	2	13.8200	6.9100	27.64
Error	6	1.5000	0.2500	
Total	11	17.1425		

The difference between athletes is not significant. The difference between operators is significant. The critical $F = 17.00$ at $\alpha = 0.1\,\%$. Since the measured lengths vary, depending on the operator the measurement system is not satisfactory.

6. (a)

Source	DF	SS	MS	F	P
Region	3	2400.55	800.184	44.35	0.000
Year	2	1384.46	692.232	38.36	0.000
Error	6	108.27	18.044		
Total	11	3893.28			

(b) There difference between regions and between years are both significant.

(c) Simultaneous confidence intervals by (i) Bonferroni's method $\bar{x}_2 - \bar{x}_1 \pm 15.6915$ (ii) Scheeffe's method $\bar{x}_2 - \bar{x}_1 \pm 13.6239$.
The gross household disposable income is greater in London and the South East than in the other two regions.

PE 10.3

1. (a) and (b) will be left to the reader.
1. (c)
   PE10.1 Q2. The one-way ANOVA revealed no significant difference between mean downtimes at $\alpha = 5\,\%$. The box plots reveal an outlier in 2004 which violates the Normality assumption: this source of this outlier should be checked.
   PE10.1 Q3. A significant difference between the mean power-up times was found. The box-plots suggest that the variance of Brand 3 may be greater than the others and its Normality is in doubt: the mean and median are very different.
   PE10.1 Q4. A significant difference between salaries was found. The assumptions appear to satisfied but one could test the variance for the professional group (it appears large) against the others.
   PE10.1 Q5. No significant difference found between the strengths of the resins. The assumptions appear to satisfied.
   PE10.1 Q6. A significant difference between average points at $\alpha = 5\,\%$. The box plots suggest that the assumption of equal variance may not be satisfied and the Normality of the OD sample.
   PE10.2 Q7. A significant difference between average points at $\alpha = 5\,\%$. The equality of variances assumption is almost certainly violated and the Normality of the site 1 is in doubt.

2. Parts (a) and (b) will be left to the reader.
   PE10.2 Q3. (c) A significant difference between both branches and years is found. Assumptions appears to be satisfied.

PE10.2 Q4. (c) A significant difference between operators is found-operator 3 appears to produce higher readings that the others. The model assumptions appear satisfied.
PE10.2 Q5. (c) A significant difference between both areas and quarters is found. The model assumptions appear satisfied.

4. (c) From the ANOVA table, there is no evidence of differences between the average numbers of errors for (i) the type of document or (ii) the units of alcohol consumed by the typists

Source	DF	SS	MS	F	P
Alcohol Units	11	227.458	20.6780	0.60	0.797
Doc Type	1	0.375	0.3750	0.01	0.919
Error	11	381.125	34.6477		
Total	23	608.958			

(d) From Minitab, the box plots and 4 in 1 plots suggest that the residual variance is not constant: it increases as the levels of alcohol increase.
(e)

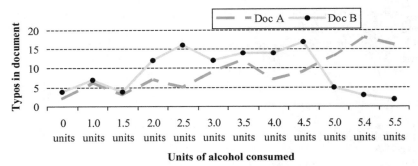

Number of Typos in Doc A and Doc B against units of alcohol consumed by typist

While the difference between the average number of typos in Doc A and B were not significant, the graph indicates a swing from high numbers of typos in Doc B to low numbers (and vice-versa for Doc A) at the three higher levels of alcohol, hence an interaction between document and units of alcohol.

6. The box plots for groups B and A suggest that the data are from Normal populations with equal variances. The box plot for group C, suggests that its variance is smaller than B or C and the data does not appear to be Normally distributed – contains extreme values.

8. (a) The response variable is time. The time was given for each of four motorists on five days. Hence there are 20 observations (b) The ANOVA indicates a significant difference between days and between motorists. (c) The data contains an outlier: the presence of the outlier is shown on all four graphs. From residual vs. order, it is the final observation (motorist 4, day 5) – Since the residual = observed – fitted value, the value of this observation is approximately 15 units less than the fitted value.

CHAPTER 11 SOLUTIONS

PE 11.1

1. (b)(i) H_0: there is no association between the performance at work and on the course H_1: there is an association. Use $\alpha = 5\%$

Test Statistic: $\chi^2 = \sum_{\text{all cells}} \frac{(f_0 - f_e)^2}{f_e}$

Decision Rule: Reject H_0 if the sample χ^2 exceeds the critical χ^2. The degrees of freedom, $\nu = (r-1)(c-1) = (3-1)(3-1) = 4$. From Tables the critical $\chi^2 = 9.4877$ and 14.8602 for $\alpha = 0.05$ and $\alpha = 0.005$ respectively.

Calculations: calculate the test statistic $T = \sum_{\text{all cells}} \frac{(f_0 - f_e)^2}{f_e} = 15.311$

```
Expected counts are printed below observed counts
Chi-Square contributions are printed below expected counts
```

		Course		
Work	Above		Below	
	average	Average	average	Total
Above	**24**	**32**	**38**	94
average	34.78	34.78	24.44	
	3.341	0.222	7.523	
	78	**76**	**46**	200
Average	74.00	74.00	52.00	
	0.216	0.054	0.692	
Below	**46**	**40**	**20**	106
average	39.22	39.22	27.56	
	1.172	0.016	2.074	
Total	148	148	104	400

```
Chi-Sq = 15.311, DF = 4
```

Conclusion. Reject H_0 and conclude that performance at work and on the course are related.
(b)(ii) The largest individual χ^2's are 7.523 and 3.341 for those who perform above average at work and below average on the course and those who perform above average on the course and below average at work.

(b)(iii) Calculate row profiles: the percentages in row 1 for above average and below average are dissimilar to those in rows 2 and 3.
Calculate the confidence interval for the difference between the proportions rated above average at work who performed above average and below average on the course. The 95 % CI for difference is (0.0933, 0.3132): the proportion that perform below average on the course is higher than those who perform above average.

2. (b)(ii) The degrees of freedom are $(5 - 1)(2 - 1) = 4$. Hence the critical $\chi^2 = 9.4877$ at the 5 % level of significance. Therefore H_0 is accepted.
(b)(iii) The 95 % CI for difference: $(-0.0516605, 0.258047)$.
Conclusion is that there is no difference in the numbers employed in these sectors over the years.

4. (b) (i) H_0: The proportions of part-time temporary male workers is the same as the proportions of part-time temporary female workers: $\pi_M = \pi_F$ H_1: the proportions are different: $\pi_M \neq \pi_F$.
(b) (ii) answers for fe are given in the Table below. (iii) the individual cell χ^2 and their sum (the test statistic) are given in the Table:

	Male	Female	Totals
Part-time temporary f_o	63	69	132
f_e	72.6	59.4	
Individual cell χ^2	1.269421	1.551515	
Not P/T temp f_o	157	111	268
f_e	147.4	120.6	
Individual cell χ^2	0.2304	0.2304	
Total surveyed	220	180	400
Test statistic $= 3.28174$		df $= 2$	Tables $= 5.9915$

The test statistic is less that the critical χ^2 at the 5 % level of significance – conclude that there is no difference between the proportion of part-time temporary female and male workers.

5. (a) The 95 % CI for difference in the proportion of female and male p/t temps is (0.00416217, 0.189777). This supports the claim that the proportion of females is greater than the proportion of males. (b) The test statistic is $T = 2.05$. Conclude that the proportion of females is greater. The parametric tests may be one-sided or two sided and are more powerful than the non-parametric χ^2 test.

6. (b)(i) Test statistics, $T = 3.376$ is less than the critical $\chi^2 = 5.991$ at the 5 % level of significance, $df = 1$. Hence conclude that there is no difference. (b)(ii) The 90 % CI for difference for the proportion of B – proportion of A disintegrating is (0.0171126, 0.322887).
    The test statistic for the one sided test, (H_1: $\pi_B > \pi_A$) is $T = 1.83$. Hence conclude that the proportion of B disintegrating is greater than the proportion of A disintegrating at the 5 % level of significance. The parametric tests may be one-sided or two sided and are more powerful than the non-parametric χ^2 test.

8. Degrees of freedom are $(4-1)(4-1) = 9$. The test statistic $T = 15.36$ exceeds the and critical χ^2 at the 10 % level of significance (where critical $\chi^2 = 14.6837$) or higher. The result is only marginally significant.

10. Chi-Square Test: Good, Acceptable, Reject
   Expected counts are printed below observed counts
   Chi-Square contributions are printed below expected counts

	Good	Acceptable	Reject	Total
A	390	34	76	500
	392.50	33.75	73.75	
	0.016	0.002	0.069	
B	238	20	42	300
	235.50	20.25	44.25	
	0.027	0.003	0.114	
Total	628	54	118	800

   Chi-Sq $T = 0.230$, $df = 2$, P-Value $= 0.891$
   Conclude that there is no difference in the quality of material from suppliers A and B

12. $T = 105.124$, $df = 3$, conclude that there is a difference. The volume of milk imported was relatively lower than expected in 2002 ($f_e = 379.36m$, $f_o = 279m$) and relatively greater than the expected in 2005 ($f_e = 390.14m$, $f_o = 550m$). The result is highly significant.

14. Chi-Sq $T = 782.517$, $df = 6$. The results are highly significant. The distances travelled to work by residents of the two neighbourhoods are not homogeneous.

PE 11.2

1. $T = 8.4$, $df = 9$. Critical $\chi^2 = 14.6837$ at $\alpha = 10$ %. Accept that the distribution is uniform, hence the lotto is fair.

2. $T = 46.3579$, $df = 5$. Critical $\chi^2 = 20.5147$ at $\alpha = 0.10$ %. Conclude that the distribution is not uniform. Thursday is the only day when the observed and expected frequencies are close $fo = 110$ and $fe = 95$.

4. $T = 46.9792$, $df = 4$. Critical $\chi^2 = 18.4662$ at $\alpha = 0.10$ %. Conclude that the distribution is not the same as that claimed. The observed demand for soya milk had double the expected demand.

6. $T = 22.4807$, $df = 5$. Critical $\chi^2 = 20.5147$ at $\alpha = 0.10$ %. Conclude that the distribution of economic activity is different than in 2001. The economically active self employed group has dropped from 59 to 37.

8. (a) mean = 5.88 (b) (i) and (ii)

Cars rented per hour	0 or 1	2	3	4	5	6	7	8	9	10	11	12 or more
f_o	4	4	10	15	20	24	19	13	10	4	1	1	125
Probability	0.02	0.05	0.09	0.14	0.16	0.16	0.13	0.10	0.06	0.04	0.02	0.02	1
f_e	2.4	6.04	11.84	17.40	20.46	20.05	16.84	12.38	8.09	4.76	2.54	2.19	125
Chisq =	1.07	0.69	0.29	0.33	0.01	0.78	0.28	0.03	0.45	0.12	0.94	0.65	5.62

(c) $T = 5.62$, $df = 10$. Critical $\chi^2 = 18.307$ at $\alpha = 5\%$. The result is not significant. Accept the claim that the data follows a Poisson distribution with a mean $= \lambda = 5.88$.

PE 11.3

1. Chi-Sq = 34.475, DF = 3, P-Value = 0.000.

2. Chi-Sq = 16.358, DF = 8, P-Value = 0.038.

4. There is no significant difference between Census 2001 and sample 2002 ($T = 9.206$). The difference between Census 2001 and sample 2004 is significant at $\alpha = 0.001$ ($T = 29.911$). The difference between Census 2001 and sample 2006 is highly significant. ($T = 241.899$).

CHAPTER 12 SOLUTIONS

PE 12.1

(a)

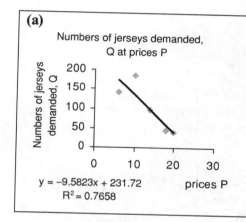

Numbers of jerseys demanded, Q at prices P

y = −9.5823x + 231.72
R² = 0.7658

prices P

(b)

$\bar{x} = 13.6, \bar{y} = 101.4, \sum y = 507, \sum xy, = 5638,$
$\sum y^2 = 67141, \sigma_y^2 = 3146.24, \sigma_x^2 = 26.24$
(i) $s_e = 35.04$ (ii) $\sum (x_i - \bar{x})^2 = 131.2$
(iii) $s_b = 3.059$
(c) The 95 % CI for slope is $-9.5823 \pm$
$(3.182)(3.059) = -9.5823 \pm 9.7337$
(d) Test statistic, $T = -3.13$.
Slope in not zero at $\alpha = 5$ %, hence demand is
related to price

2. (a) Mean Temperature $(y) = 16.9 - 1.54$ Hours of sunshine (x)
(b) $s_e = 0.5266$, $s_b = 0.5336$ (c) 95 % CI for slope is $-1.5443 \pm 2.306(0.5336)$ (d) Test statistic
$T = -2.89 < -2.305$ (critical t at $\alpha = 0.05$, $df = 8$). Conclude that there is a relationship between
hours of sunshine and temperature.

4. (a) The regression equation is Lung $= 25.5 - 0.395$ SO2
(b) $s_e = 2.43255$, $s_b = 0.06003$ (c) 95 % CI for slope is $-0.39496 \pm 2.447(0.06033)$ (d) Test
statistic $T = -6.58 \ll -1.943$ (critical t at $\alpha = 0.05$, $df = 6$).

6. The regression equation is % Apartments $(y) = 20.4 + 1.65$ Years since 2000 (x)
$s_e = 3.32165$, $s_b = 0.5252$ (b) H_0 $\beta = 2$. Test statistic $T = 0.67 < 3.353$ (critical t at $\alpha = 0.05$,
$df = 3$). Conclude that the percentage of apartment increases by 2 % annually.

PE 12.2

1. (a) (i) 16.2 (ii) 38.6273 (b) 95 % CI for the mean number of jerseys demanded when the price is
€ 15 is $88 \pm 3.182(16.2) = (36.3, 139.7)$ (c) The 95 % prediction interval for the number of jerseys
demanded when the price per jersey is € 15 is $88 \pm 3.182\ 36.3077 = (-34.9, 210.9)$

2. (a) (i) 0.452 (ii) 0.6940 (b) 90 % CI for the mean temperature when there is 5 hours of sunshine per day $9.215 \pm 1.860(0.452) = (36.3, 139.7)$ (c) The 90 % prediction interval for the temperature when there is 5 hours of sunshine per day $9.215 \pm 1.860(0.6940) = (7.924, 10.505)$.

4. (a) $y = 25.5 - 0.395x$ (b) (i) 1.409 (ii) 2.8112 (c) (i) The 99 % CI for the mean numbers (per 100 000) suffering from lung disease is $5.707 \pm 3.707(1.409) = (0.486, 10.931)$ (ii) The 99 % prediction interval for the numbers (per 100 000) suffering from lung disease is $5.707 \pm 3.707(2.8112) = (-4.713, 16.128)$.

6. (a) The regression equation is % Apartments $(y) = 20.4 + 1.65$ Years since 2000 (x) (b) (i) 1.49 (ii) 3.6405 (c) The 95 % confidence interval for the percentage apartments in 2004 is $27 \pm 3.182(1.49) = (22.27, 31.73)$ (ii) The 95 % prediction interval for the percentage apartments in 2004 is $27 \pm 3.182(3.6405) = (15.42, 38.58)$.

PE 12.3

2. **Line 1** is the equation of the least-squares line
   The regression equation is Cost($€$,000) $= -2.36 + 0.0936$ Distance (km)
   Lines 2, 3 and 4 give the coefficients, i.e. the constant and slope of the least-squares line, followed by the standard error, test statistic and p-value for test of hypothesis $H_0: \alpha = 0$ and $H_0: \beta = 0$.

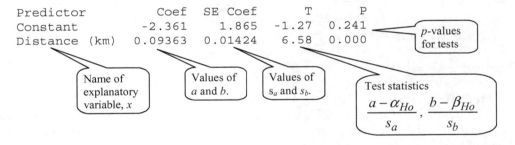

```
Predictor           Coef   SE Coef       T      P
Constant          -2.361     1.865   -1.27  0.241
Distance (km)    0.09363   0.01424    6.58  0.000
```

```
S = 2.87010 R-Sq = 84.4% R-Sq(adj)  = 82.4%
```

Line 5 gives the standard error of estimate, $S = S_e = 2.87010$; the coefficient of determination $R^2 = 0.844 \times 100$ and the adjusted $R^2 = 0.824 \times 100$.
Line 6–8 'Unusual Observations' gives statistics for an 'unusual' observation- the fifth observation, $x = 150$, $y = 5.800$. Fit, is its fitted value according to the least-squares line, $y = -2.36 + 0.0936 (150) = 11.68$.

Obs	Distance (km)	Cost ($€$,000)	Fit	SE Fit	Residual	St Resid
5	150	5.800	11.683	1.040	-5.883	-2.20R

SE Fit is the standard error for $\bar{y}$ when $x = 150$. Residual is the observed value of y minus its fitted value at $x = 150$. i.e. $5.8 - 11.683 = -5.883$.

Std Resid is the standardised residual

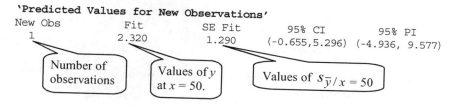

```
'Predicted Values for New Observations'
New Obs          Fit         SE Fit        95% CI            95% PI
   1            2.320        1.290      (-0.655,5.296)  (-4.936, 9.577)
```

Number of observations

Values of y at $x = 50$.

Values of $s_{\bar{y}/x = 50}$

95 % CI and 95 % PI are the 95 % confidence and prediction intervals for y at x = 50.

`Values of Predictors for New Observations` refers to the value of x (the predictor or explanatory variable) used to estimate the value of y (the response variable) for which the confidence and prediction intervals are calculated.

The equation of the least-squares line describes a positive relationship between distance and cost – for each additional km (unit of distance) cost increases by 0.0936 (€ 93.6). The test statistic for slope $T = 6.58$ with a p-value $p = 0.000$ means that the null hypothesis, H_0 $\beta = 0$ will be rejected. Hence there is a relationship between distance and cost since $\beta \neq 0$. However, the confidence interval for cost when distance, $x = 50$ ranges from −0.665 and 5.296 and contains zero, indicating a very weak or no relationship- the imprecision of this interval may be due to the fact that (i) $x = 50$ is close to one end of the range for x, see Figures 12.7 and 12.19 (ii) the data contains an 'unusual observation' as indicated on the Minitab printout (and also on the residual plots in question 3); such an observation will increase the values of standard errors and hence the widths of confidence/prediction intervals.

2. (b) There is a relationship between distance and cost: $T = 6.58 > t_{0.05\%, 8df}$ (5.041). This is confirmed by the Minitab printout which gives the p-value $= 0.000$.

4. (a)

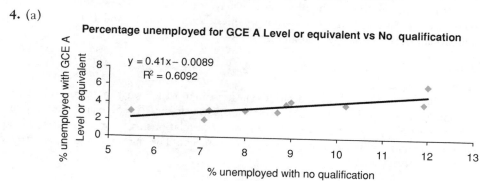

Percentage unemployed for GCE A Level or equivalent vs No qualification

$y = 0.41x - 0.0089$
$R^2 = 0.6092$

% unemployed with GCE A Level or equivalent

% unemployed with no qualification

(b) The regression equation is
GCE A Level or equivalent $= -0.01 + 0.410$ No qualifications
(c) To test whether there is a relationship, test H_0: $\beta = 0$ vs. H_1: $\beta \neq 0$. The test statistic is $T = 3.53$. The critical t-value $= 3.355$ at $\alpha = 0.01$ (1 %). Hence H_0: is rejected, conclude that slope is not zero and that these is a relationship between the % unemployed with A level and the % unemployed with no qualification. The relationship is more easily described through confidence intervals. When 10 % of those with no qualification are unemployed,

the prediction intervals for the percentages with A level that are unemployed are between 2.314 % and 5.865 %

Obs	Fit	SE Fit	95% CI	95% PI
1	4.092	0.265	(3.481, 4.702)	(2.315, 5.868)

(d) The model assumptions appear to be satisfied from the residual plots. The Minitab printout gives one unusual observation.

6. The sample size ($n = 5$) is small, hence information that may be deduced from the residual plots is limited. The fourth observation is highlighted as unusual and should be investigated. The Normal probability is almost linear. The histogram give no information (5 points are too few). However, the model assumption of 'independence' is not satisfied. The values of the response variable (% apartments) are time related; building projects are continuous over several quarters.

8. **Electricity consumption vs EER** (a) The regression equation is Electricity kWh = 2535 − 23.6 EER, a negative relationship in which electricity consumption decreases by 23.6 per unit increases in EER (b) Test statistic for slope is $T = -8.82$ and the p-value is 0. Hence reject H_0: $\beta = 0$ and conclude that there is a relationship between electricity consumption and EER (c) From the residual plots, the model assumptions of independence and homoscedasticity are satisfactory – as well as normality.

Electricity consumption vs hours occupied (a) The regression equation is Electricity kWh = 518 + 4.66 Hours, a positive relationship in which electricity consumption increases by 4.66 for each additional hour that the apartment is occupied (b) Test statistic for slope is $T = 4.98$ and the p-value is 0. Hence reject H_0: $\beta = 0$ and conclude that there is a relationship between electricity consumption and hours occupied (c) From the residual plots, the model assumptions of homoscedasticity appear satisfactory but the residual plot against order of observation is not random, hence the observations may not be independent. The normality of the residuals may not be satisfied – two extreme values (observation 11 and 15) are present in the normal probability plot and the histogram.

10. **Wear vs. load**
(a) The regression equation is Wear (y) = −9.3 + 0.398 Load (x2), a positive relationship in which wear increases by 0.398 for each additional unit of load. $R^2 = 0.822$ (b) Test statistic for slope is $T = 5.26$ and the p-value is 0.002. Hence reject H_0: $\beta = 0$ and conclude that there is a relationship between wear and load. The assumption of Normality does not appear to be satisfied. The data contains one extreme value (the first observation).

Wear vs. speed
(a) The regression equation is Wear (y) = 92.9 − 0.272 Speed (x1), a negative relationship in which wear decreases by 0.272 for each km/hour increase in speed. $R^2 = 0.053$ (b) Test statistic for slope is $T = -0.58$ and the p-value is 0.583. Hence accept H_0: $\beta = 0$ and conclude that there is no relationship between wear and speed. The assumptions of Normality and homoscedasticity appear satisfactory. The data contains one extreme value. Overall, there is a relationship between wear and load but no relationship between wear and speed. These conclusions are valid, provided the model assumptions are satisfied.

PE 12.4

1. (a) Cost (€,000) $= -3.75 + 0.0910$ Distance (km) $+ 0.0359$ Weight (1000 kg). Cost $= 03.75$ when distance and cost are both zero. Cost increases by 0.0910 (€ 91) when distance increases by one km, assuming weight is constant. Cost increases by 0.0359 (€ 35.9) when weight increases by one unit (1000 kg), assuming distance is constant. (b) The extract of the Minitab printout is similar to that for simple linear regression. Immediately beneath the regression equation, the printout gives a line for the value of each coefficient in the equation, with the corresponding standard error, test statistic and the p-value for the test that coefficient is zero. The final line of the printout gives the standard error of estimate (s_e), the coefficient of determination and the adjusted value of it. (c) From the test statistic and p-value for each coefficient, we see that cost is related to distance; $T = 6.07$, p-value $= 0.001$, but there is no relationship between cost and weight since $T = 0.77$, p-value $= 0.466$.
(d) $R^2 = 0.94$; $|r| = 0.9695 > 0.7$; the model suffers from multicollinearity.

2. The regression equation is
Crude Birth rate $= -0.0150$ (life expect at birth female) $+ 0.679$ (% aged 0 – 14)
From the printout, the percentage aged 0-14 is related to the crude birth rate ($T = 9.84$, p-value $= 0$). The life expectancy of females at birth is not a significant predictor of crude birth rate ($T = -0.74$, p value $= 0.485$).

3. (a) The regression equation is
Life expectancy at birth males $= 3.87$ Crude Birth rate $+ 1.91$ Crude Death rate
From the printout, male life expectancy at birth is related to the crude birth rate ($T = 3.44$, p-value $= 0.014$) but not to the crude death rate ($T = 0.95$, p-value $= 0.3754$).
(b) The regression equation is

Crude Birth rate per 1000 $= 14.9 - 0.169$ Crude Death rate per 1000

The relationship between crude birth rate and crude death rate is not significant since $T = -0.38$ and p-value $= 0.719$, $R^2 = 0.023$. The model does not suffer from autocorrelation. Observation 6 is unusual has a large influence.

4. (a) The regression equation is

$$\text{Electricity kWh} = 1890 + 2.28 \text{ Hours occupied} - 18.2 \text{ EER}$$
$$\text{Electricity kWh} = 1890 + 2.28 (10) - 18.2 (50) = 1002.8$$

(b) The adjusted R^2, $R^2_{Adj} = 1 - (1 - R^2) \times \left[\dfrac{n-1}{n-k-1}\right]$. No, $|r_{x_1,x_2}| = 0.5624 < 0.7$, hence the model does not suffers from multicollinearity.

(c) H_0 $\beta_1 \geq 3$ H_1 $\beta < 3$ Test statistic $T = \dfrac{b_1 - \beta_1}{s_{b_1}} = \dfrac{2.28 - 3}{0.5484} = -1.31$ The critical value is $t = -1.734$ ($\alpha = 5\%$). Hence accept the null hypothesis and conclude that the electricity consumption increases by at least 3 kWh for each additional hour that the dwelling is occupied.

(d) H_0 $\beta_2 = -20$ H_1 $\beta \neq -20$ Test statistic $T = \dfrac{b_2 - \beta_2}{s_{b_2}} = \dfrac{-18.151 - (-20)}{2.348} = 0.79$ The critical value is $t = 2.101$ ($\alpha = 5\%$). Hence accept the null hypothesis and conclude that the electricity consumption decreases by 20 kWh for unit increases in EER.

6. (a) The regression equation is
Deaths $= 587 - 0.000131$ No. Vehicles $+ 0.000023$ No. Licence holders
(i) The number of road deaths is not related to the number of vehicles at $\alpha = 0.015$ or lower. ($T = -0.58$, p-value $= 0.587$)
(ii) The number of road deaths is not related to the number of licence holders at $\alpha = 0.001$ or lower. ($T = 0.07$, p-value $= 0.949$)
(b) The regression equation is
No. Vehicles $= -1250363 + 1.44$ No. Licence holders. $R^2 = 0.94$.

Predictor	Coef	SE Coef	T	P
Constant	−1250 363	317 129	−3.94	0.008
No. Licence holders	1.4434	0.1489	9.69	0.000

Hence there is a relationship between the independent variables. This violates the model assumptions in (a): that the predictor variables are independent of each other.
(c) The regression equation is Deaths $= 750 - 0.000166$ No. Licence holders

Predictor	Coef	SE Coef	T	P
Constant	750.3	164.9	4.55	0.004
No. Licence holders	−0.00016596	0.00007744	−2.14	0.076

There is no significant relationship between the numbers of deaths and licence holders. The residual plot suggests a curvilinear relationship.
(d) The scatter plot with the equations for a linear trend and a curvilinear trend are given below. Use the curvilinear to model the data, $R^2 = 0.9033$ compared to $R^2 = 0.4336$ for the linear trend.

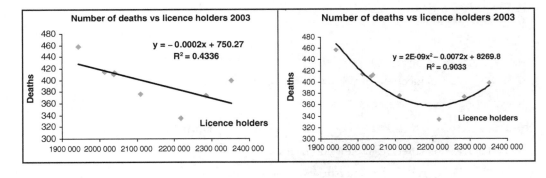

TABLES

Table 1. Cumulative Binomial Probability tables

These tables give the cumulative probabilities $P(x \le r)$

$= \sum_{x=0}^{x=r} P(x)$ and various values of p. For example, for a

sample of size $n = 5$, $p = 0.20$, the cumulative probability

for $r = 1$ is the probability one successes or fewer, $P(x \le$

$1) = \sum_{x=0}^{x=1} P(x) = 0.7373$

$P(x \le 1) = \sum_{x=0}^{x=1} P(x) = P(x = 0.3277 + 0.4096 = 0.7373$

Probability Histogram, n = 5, p = 0.2

Cumulative Binomial Probability tables $n = 5$

r	p = 0.01	p = 0.05	p = 0.10	p = 0.20	p = 0.30	p = 0.50	p = 0.60	p = 0.70	p = 0.80	p = 0.90	p = 095	p = 0.99
0	0.9510	0.7738	0.5905	0.3277	0.1681	0.0313	0.0102	0.0024	0.0003	0	0	0
1	0.9990	0.9774	0.9185	0.7373	0.5282	0.1875	0.0870	0.0308	0.0067	0.0005	0	0
2	1	0.9988	0.9914	0.9421	0.8369	0.5000	0.3174	0.1631	0.0579	0.0086	0.0012	0
3	1	1	0.9995	0.9933	0.9692	0.8125	0.6630	0.4718	0.2627	0.0815	0.0226	0.0010
4	1	1	1	0.9997	0.9976	0.9688	0.9222	0.8319	0.6723	0.4095	0.2262	0.0490

Cumulative Binomial Probability tables $n = 10$

r	p = 0.01	p = 0.05	p = 0.10	p = 0.20	p = 0.30	p = 0.50	p = 0.60	p = 0.70	p = 0.80	p = 0.85	p = 095	p = 0.99
0	0.9044	0.5987	0.3487	0.1074	0.0282	0.0010	0.0001	0	0	0	0	0
1	0.9957	0.9139	0.7361	0.3758	0.1493	0.0107	0.0017	0.0001	0	0	0	0
2	0.9999	0.9885	0.9298	0.6778	0.3828	0.0547	0.0123	0.0016	0.0001	0	0	0
3	1	0.9990	0.9872	0.8791	0.6496	0.1719	0.0548	0.0106	0.0009	0.0001	0	0
4	1	0.9999	0.9984	0.9672	0.8497	0.3770	0.1662	0.0473	0.0064	0.0014	0	0
5	1	1	0.9999	0.9936	0.9527	0.6230	0.3669	0.1503	0.0328	0.0099	0.0001	0
6	1	1	1	0.9991	0.9894	0.8281	0.6177	0.3504	0.1209	0.0500	0.0010	0
7	1	1	1	0.9999	0.9984	0.9453	0.8327	0.6172	0.3222	0.1798	0.0115	0.0001
8	1	1	1	1	0.9999	0.9893	0.9536	0.8507	0.6242	0.4557	0.0861	0.0043
9	1	1	1	1	1	0.9990	0.9940	0.9718	0.8926	0.8031	0.4013	0.0956

Table 1. Cumulative Binomial Probability tables (continued)

Cumulative Binomial Probability tables $n = 50$

	p = 0.01	p = 0.05	p = 0.10	p = 0.20	p = 0.30	p = 0.50	p = 0.60
0	0.6050	0.0769	0.0052	0	0	0	0
1	0.9106	0.2794	0.0338	0.0002	0	0	0
2	0.9862	0.5405	0.1117	0.0013	0	0	0
3	0.9984	0.7604	0.2503	0.0057	0	0	0
4	0.9999	0.8964	0.4312	0.0185	0.0002	0	0
5	1	0.9622	0.6161	0.0480	0.0007	0	0
6	1	0.9882	0.7702	0.1034	0.0025	0	0
7	1	0.9968	0.8779	0.1904	0.0073	0	0
8	1	0.9992	0.9421	0.3073	0.0183	0	0
9	1	0.9998	0.9755	0.4437	0.0402	0	0
10	1	1	0.9906	0.5836	0.0789	0	0
11	1	1	0.9968	0.7107	0.1390	0	0
12	1	1	0.9990	0.8139	0.2229	0.0002	0
13	1	1	0.9997	0.8894	0.3279	0.0005	0
14	1	1	0.9999	0.9393	0.4468	0.0013	0
15	1	1	1	0.9692	0.5692	0.0033	0
16	1	1	1	0.9856	0.6839	0.0077	0.0001
17	1	1	1	0.9937	0.7822	0.0164	0.0002
18	1	1	1	0.9975	0.8594	0.0325	0.0005
19	1	1	1	0.9991	0.9152	0.0595	0.0014
20	1	1	1	0.9997	0.9522	0.1013	0.0034
21	1	1	1	0.9999	0.9749	0.1611	0.0076
22	1	1	1	1	0.9877	0.2399	0.0160
23	1	1	1	1	0.9944	0.3359	0.0314
24	1	1	1	1	0.9976	0.4439	0.0573
25	1	1	1	1	0.9991	0.5561	0.0978
26	1	1	1	1	0.9997	0.6641	0.1562
27	1	1	1	1	0.9999	0.7601	0.2340
28	1	1	1	1	1	0.8389	0.3299

Cumulative Binomial Probability tables $n = 120$

r	p = 0.01	p = 0.05	p = 0.10	p = 0.20	p = 0.30
0	0.2735	0.0013	0	0	0
1	0.6299	0.0104	0	0	0
2	0.8602	0.0410	0.0001	0	0
3	0.9587	0.1092	0.0007	0	0
4	0.9901	0.2222	0.0028	0	0
5	0.9980	0.3709	0.0087	0	0
6	0.9997	0.5326	0.0221	0	0
7	0.9999	0.6822	0.0482	0	0
8	1	0.8022	0.0925	0	0
9	1	0.8872	0.1587	0	0
10	1	0.9408	0.2469	0.0001	0
11	1	0.9714	0.3530	0.0003	0
12	1	0.9872	0.4689	0.0008	0
13	1	0.9947	0.5847	0.0019	0
14	1	0.9979	0.6914	0.0042	0
15	1	0.9993	0.7823	0.0086	0
16	1	0.9997	0.8542	0.0164	0
17	1	0.9999	0.9074	0.0294	0
18	1	1	0.9441	0.0496	0
19	1	1	0.9680	0.0791	0
20	1	1	0.9825	0.1197	0.0001
21	1	1	0.9909	0.1724	0.0003
22	1	1	0.9955	0.2371	0.0006
23	1	1	0.9979	0.3123	0.0012
24	1	1	0.9990	0.3953	0.0023
25	1	1	0.9996	0.4825	0.0043
26	1	1	0.9998	0.5697	0.0079
27	1	1	0.9999	0.6528	0.0136
28	1	1	1	0.7285	0.0226

Table 2. Cumulative Poisson Probabilities $\sum_{x=0}^{x=r} P(x \le r) = \sum_{x=0}^{x=r} \dfrac{\lambda^x e^{-\lambda}}{x!}$

These tables give the cumulative probabilities $P(x \le r) = \sum_{x=0}^{x=r} P(x)$ and various values of p. For example, for $\lambda = 4$, sample size $n = 5$, $p = 0.20$, the cumulative probability for $r = 2$ is the probability two successes or fewer is

$$P(x \le 2) = \sum_{x=0}^{x=2} P(x) = 0.7373$$

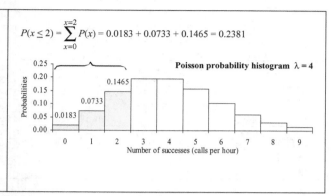

$$P(x \le 2) = \sum_{x=0}^{x=2} P(x) = 0.0183 + 0.0733 + 0.1465 = 0.2381$$

Cumulative Poisson Probability tables

r	λ = 0.0	λ = 0.1	λ = 0.2	λ = 0.3	λ = 0.4	λ = 0.5	λ = 0.6	λ = 0.7	λ = 0.8	λ = 0.9
0	1	0.9048	0.8187	0.7408	0.6703	0.6065	0.5488	0.4966	0.4493	0.4066
1	1	0.9953	0.9825	0.9631	0.9384	0.9098	0.8781	0.8442	0.8088	0.7725
2	1	0.9998	0.9989	0.9964	0.9921	0.9856	0.9769	0.9659	0.9526	0.9371
3	1	1	0.9999	0.9997	0.9992	0.9982	0.9966	0.9942	0.9909	0.9865
4	1	1	1	1	0.9999	0.9998	0.9996	0.9992	0.9986	0.9977
5	1	1	1	1	1	1	1	0.9999	0.9998	0.9997

r	1 λ = 1.0	1.1 λ = 1.1	1.2 λ = 1.2	1.3 λ = 1.3	1.4 λ = 1.4	1.5 λ = 1.5	1.6 λ = 1.6	1.7 λ = 1.7	1.8 λ = 1.8	1.9 λ = 1.9
0	0.3679	0.3329	0.3012	0.2725	0.2466	0.2231	0.2019	0.1827	0.1653	0.1496
1	0.7358	0.6990	0.6626	0.6268	0.5918	0.5578	0.5249	0.4932	0.4628	0.4337
2	0.9197	0.9004	0.8795	0.8571	0.8335	0.8088	0.7834	0.7572	0.7306	0.7037
3	0.9810	0.9743	0.9662	0.9569	0.9463	0.9344	0.9212	0.9068	0.8913	0.8747
4	0.9963	0.9946	0.9923	0.9893	0.9857	0.9814	0.9763	0.9704	0.9636	0.9559
5	0.9994	0.9990	0.9985	0.9978	0.9968	0.9955	0.9940	0.9920	0.9896	0.9868
6	0.9999	0.9999	0.9997	0.9996	0.9994	0.9991	0.9987	0.9981	0.9974	0.9966
7	1	1	1	0.9999	0.9999	0.9998	0.9997	0.9996	0.9994	0.9992
8	1	1	1	1	1	1	1	0.9999	0.9999	0.9998

r	λ = 2.0	λ = 2.1	λ = 2.2	λ = 2.3	λ = 2.4	λ = 2.5	λ = 2.6	λ = 2.7	λ = 2.8	λ = 2.9
0	0.1353	0.1225	0.1108	0.1003	0.0907	0.0821	0.0743	0.0672	0.0608	0.0550
1	0.4060	0.3796	0.3546	0.3309	0.3084	0.2873	0.2674	0.2487	0.2311	0.2146
2	0.6767	0.6496	0.6227	0.5960	0.5697	0.5438	0.5184	0.4936	0.4695	0.4460
3	0.8571	0.8386	0.8194	0.7993	0.7787	0.7576	0.7360	0.7141	0.6919	0.6696
4	0.9473	0.9379	0.9275	0.9162	0.9041	0.8912	0.8774	0.8629	0.8477	0.8318
5	0.9834	0.9796	0.9751	0.9700	0.9643	0.9580	0.9510	0.9433	0.9349	0.9258
6	0.9955	0.9941	0.9925	0.9906	0.9884	0.9858	0.9828	0.9794	0.9756	0.9713
7	0.9989	0.9985	0.9980	0.9974	0.9967	0.9958	0.9947	0.9934	0.9919	0.9901
8	0.9998	0.9997	0.9995	0.9994	0.9991	0.9989	0.9985	0.9981	0.9976	0.9969
9	1	0.9999	0.9999	0.9999	0.9998	0.9997	0.9996	0.9995	0.9993	0.9991
10	1	1	1	1	1	0.9999	0.9999	0.9999	0.9998	0.9998
11	1	1	1	1	1	1	1	1	1	0.9999

Table 2. Cumulative Poisson Probability tables (continued)

r	λ = 3.0	λ = 3.1	λ = 3.2	λ = 3.3	λ = 3.4	λ = 3.5	λ = 3.6	λ = 3.7	λ = 3.8	λ = 3.9
0	0.0498	0.0450	0.0408	0.0369	0.0334	0.0302	0.0273	0.0247	0.0224	0.0202
1	0.1991	0.1847	0.1712	0.1586	0.1468	0.1359	0.1257	0.1162	0.1074	0.0992
2	0.4232	0.4012	0.3799	0.3594	0.3397	0.3208	0.3027	0.2854	0.2689	0.2531
3	0.6472	0.6248	0.6025	0.5803	0.5584	0.5366	0.5152	0.4942	0.4735	0.4532
4	0.8153	0.7982	0.7806	0.7626	0.7442	0.7254	0.7064	0.6872	0.6678	0.6484
5	0.9161	0.9057	0.8946	0.8829	0.8705	0.8576	0.8441	0.8301	0.8156	0.8006
6	0.9665	0.9612	0.9554	0.9490	0.9421	0.9347	0.9267	0.9182	0.9091	0.8995
7	0.9881	0.9858	0.9832	0.9802	0.9769	0.9733	0.9692	0.9648	0.9599	0.9546
8	0.9962	0.9953	0.9943	0.9931	0.9917	0.9901	0.9883	0.9863	0.9840	0.9815
9	0.9989	0.9986	0.9982	0.9978	0.9973	0.9967	0.9960	0.9952	0.9942	0.9931
10	0.9997	0.9996	0.9995	0.9994	0.9992	0.9990	0.9987	0.9984	0.9981	0.9977
11	0.9999	0.9999	0.9999	0.9998	0.9998	0.9997	0.9996	0.9995	0.9994	0.9993
12	1	1	1	1	0.9999	0.9999	0.9999	0.9999	0.9998	0.9998
13	1	1	1	1	1	1	1	1	1	0.9999

r	λ = 4.0	λ = 4.1	λ = 4.2	λ = 4.3	λ = 4.4	λ = 4.5	λ = 4.6	λ = 4.7	λ = 4.8	λ = 4.9
0	0.0183	0.0166	0.0150	0.0136	0.0123	0.0111	0.0101	0.0091	0.0082	0.0074
1	0.0916	0.0845	0.0780	0.0719	0.0663	0.0611	0.0563	0.0518	0.0477	0.0439
2	0.2381	0.2238	0.2102	0.1974	0.1851	0.1736	0.1626	0.1523	0.1425	0.1333
3	0.4335	0.4142	0.3954	0.3772	0.3594	0.3423	0.3257	0.3097	0.2942	0.2793
4	0.6288	0.6093	0.5898	0.5704	0.5512	0.5321	0.5132	0.4946	0.4763	0.4582
5	0.7851	0.7693	0.7531	0.7367	0.7199	0.7029	0.6858	0.6684	0.6510	0.6335
6	0.8893	0.8786	0.8675	0.8558	0.8436	0.8311	0.8180	0.8046	0.7908	0.7767
7	0.9489	0.9427	0.9361	0.9290	0.9214	0.9134	0.9049	0.8960	0.8867	0.8769
8	0.9786	0.9755	0.9721	0.9683	0.9642	0.9597	0.9549	0.9497	0.9442	0.9382
9	0.9919	0.9905	0.9889	0.9871	0.9851	0.9829	0.9805	0.9778	0.9749	0.9717
10	0.9972	0.9966	0.9959	0.9952	0.9943	0.9933	0.9922	0.9910	0.9896	0.9880
11	0.9991	0.9989	0.9986	0.9983	0.9980	0.9976	0.9971	0.9966	0.9960	0.9953
12	0.9997	0.9997	0.9996	0.9995	0.9993	0.9992	0.9990	0.9988	0.9986	0.9983
13	0.9999	0.9999	0.9999	0.9998	0.9998	0.9997	0.9997	0.9996	0.9995	0.9994
14	1.0000	1.0000	1.0000	1.0000	0.9999	0.9999	0.9999	0.9999	0.9999	0.9998

r	λ = 5.0	λ = 5.1	λ = 5.2	λ = 5.3	λ = 5.4	λ = 5.5	λ = 5.6	λ = 5.7	λ = 5.8	λ = 5.9
0	0.0067	0.0061	0.0055	0.0050	0.0045	0.0041	0.0037	0.0033	0.0030	0.0027
1	0.0404	0.0372	0.0342	0.0314	0.0289	0.0266	0.0244	0.0224	0.0206	0.0189
2	0.1247	0.1165	0.1088	0.1016	0.0948	0.0884	0.0824	0.0768	0.0715	0.0666
3	0.2650	0.2513	0.2381	0.2254	0.2133	0.2017	0.1906	0.1800	0.1700	0.1604
4	0.4405	0.4231	0.4061	0.3895	0.3733	0.3575	0.3422	0.3272	0.3127	0.2987
5	0.6160	0.5984	0.5809	0.5635	0.5461	0.5289	0.5119	0.4950	0.4783	0.4619
6	0.7622	0.7474	0.7324	0.7171	0.7017	0.6860	0.6703	0.6544	0.6384	0.6224
7	0.8666	0.8560	0.8449	0.8335	0.8217	0.8095	0.7970	0.7841	0.7710	0.7576
8	0.9319	0.9252	0.9181	0.9106	0.9027	0.8944	0.8857	0.8766	0.8672	0.8574
9	0.9682	0.9644	0.9603	0.9559	0.9512	0.9462	0.9409	0.9352	0.9292	0.9228
10	0.9863	0.9844	0.9823	0.9800	0.9775	0.9747	0.9718	0.9686	0.9651	0.9614
11	0.9945	0.9937	0.9927	0.9916	0.9904	0.9890	0.9875	0.9859	0.9841	0.9821
12	0.9980	0.9976	0.9972	0.9967	0.9962	0.9955	0.9949	0.9941	0.9932	0.9922
13	0.9993	0.9992	0.9990	0.9988	0.9986	0.9983	0.9980	0.9977	0.9973	0.9969
14	0.9998	0.9997	0.9997	0.9996	0.9995	0.9994	0.9993	0.9991	0.9990	0.9988
15	0.9999	0.9999	0.9999	0.9999	0.9998	0.9998	0.9998	0.9997	0.9996	0.9996
16	1	1	1	1	0.9999	0.9999	0.9999	0.9999	0.9999	0.9999

Table 3. Normal probability Distribution

Area in upper tail of the normal probability distribution

Z	0.00	0.01	0.02	0.03	0.04	0.05	0.06	0.07	0.08	0.09
0.0	0.5000	0.4960	0.4920	0.4880	0.4840	0.4801	0.4761	0.4721	0.4681	0.4641
0.1	0.4602	0.4562	0.4522	0.4483	0.4443	0.4404	0.4364	0.4325	0.4286	0.4247
0.2	0.4207	0.4168	0.4129	0.4090	0.4052	0.4013	0.3974	0.3936	0.3897	0.3859
0.3	0.3821	0.3783	0.3745	0.3707	0.3669	0.3632	0.3594	0.3557	0.3520	0.3483
0.4	0.3446	0.3409	0.3372	0.3336	0.3300	0.3264	0.3228	0.3192	0.3156	0.3121
0.5	0.3085	0.3050	0.3015	0.2981	0.2946	0.2912	0.2877	0.2843	0.2810	0.2776
0.6	0.2743	0.2709	0.2676	0.2643	0.2611	0.2578	0.2546	0.2514	0.2483	0.2451
0.7	0.2420	0.2389	0.2358	0.2327	0.2296	0.2266	0.2236	0.2206	0.2177	0.2148
0.8	0.2119	0.2090	0.2061	0.2033	0.2005	0.1977	0.1949	0.1922	0.1894	0.1867
0.9	0.1841	0.1814	0.1788	0.1762	0.1736	0.1711	0.1685	0.1660	0.1635	0.1611
1.0	0.1587	0.1562	0.1539	0.1515	0.1492	0.1469	0.1446	0.1423	0.1401	0.1379
1.1	0.1357	0.1335	0.1314	0.1292	0.1271	0.1251	0.1230	0.1210	0.1190	0.1170
1.2	0.1151	0.1131	0.1112	0.1093	0.1075	0.1056	0.1038	0.1020	0.1003	0.0985
1.3	0.0968	0.0951	0.0934	0.0918	0.0901	0.0885	0.0869	0.0853	0.0838	0.0823
1.4	0.0808	0.0793	0.0778	0.0764	0.0749	0.0735	0.0721	0.0708	0.0694	0.0681
1.5	0.0668	0.0655	0.0643	0.0630	0.0618	0.0606	0.0594	0.0582	0.0571	0.0559
1.6	0.0548	0.0537	0.0526	0.0516	0.0505	0.0495	0.0485	0.0475	0.0465	0.0455
1.7	0.0446	0.0436	0.0427	0.0418	0.0409	0.0401	0.0392	0.0384	0.0375	0.0367
1.8	0.0359	0.0351	0.0344	0.0336	0.0329	0.0322	0.0314	0.0307	0.0301	0.0294
1.9	0.0287	0.0281	0.0274	0.0268	0.0262	0.0256	0.0250	0.0244	0.0239	0.0233
2.0	0.0228	0.0222	0.0217	0.0212	0.0207	0.0202	0.0197	0.0192	0.0188	0.0183
2.1	0.0179	0.0174	0.0170	0.0166	0.0162	0.0158	0.0154	0.0150	0.0146	0.0143
2.2	0.0139	0.0136	0.0132	0.0129	0.0125	0.0122	0.0119	0.0116	0.0113	0.0110
2.3	0.0107	0.0104	0.0102	0.0099	0.0096	0.0094	0.0091	0.0089	0.0087	0.0084
2.4	0.0082	0.0080	0.0078	0.0075	0.0073	0.0071	0.0069	0.0068	0.0066	0.0064
2.5	0.0062	0.0060	0.0059	0.0057	0.0055	0.0054	0.0052	0.0051	0.0049	0.0048
2.6	0.0047	0.0045	0.0044	0.0043	0.0041	0.0040	0.0039	0.0038	0.0037	0.0036
2.7	0.0035	0.0034	0.0033	0.0032	0.0031	0.0030	0.0029	0.0028	0.0027	0.0026
2.8	0.0026	0.0025	0.0024	0.0023	0.0023	0.0022	0.0021	0.0021	0.0020	0.0019
2.9	0.0019	0.0018	0.0018	0.0017	0.0016	0.0016	0.0015	0.0015	0.0014	0.0014
3.0	0.0013	0.0013	0.0013	0.0012	0.0012	0.0011	0.0011	0.0011	0.0010	0.0010
3.1	0.0010	0.0009	0.0009	0.0009	0.0008	0.0008	0.0008	0.0008	0.0007	0.0007
3.2	0.0007	0.0007	0.0006	0.0006	0.0006	0.0006	0.0006	0.0005	0.0005	0.0005
3.3	0.0005	0.0005	0.0005	0.0004	0.0004	0.0004	0.0004	0.0004	0.0004	0.0003
3.4	0.0003	0.0003	0.0003	0.0003	0.0003	0.0003	0.0003	0.0003	0.0003	0.0002
3.5	0.0002	0.0002	0.0002	0.0002	0.0002	0.0002	0.0002	0.0002	0.0002	0.0002
3.6	0.0002	0.0002	0.0001	0.0001	0.0001	0.0001	0.0001	0.0001	0.0001	0.0001
3.8	0.0001	0.0001	0.0001	0.0001	0.0001	0.0001	0.0001	0.0001	0.0001	0.0001
4.0	0.0000	0.0000	0.0000	0.0000	0.0000	0.0000	0.0000	0.0000	0.0000	0.0000

Table 4. Percentage point for the Normal probability distribution

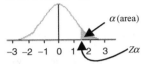

α (area)

-3 -2 -1 0 1 2 3 Zα

The Z-value for selected areas in the upper tail of the Normal distribution

α	Z_α	α	Z_α	α	Z_α	α	Z_α	α	Z_α
0.50	0.0000	0.050	1.6449	0.030	1.8808	0.020	2.0537	0.010	2.3263
0.45	0.1257	0.048	1.6646	0.029	1.8957	0.019	2.0749	0.009	2.3656
0.40	0.2533	0.046	1.6849	0.028	1.9110	0.018	2.0969	0.008	2.4089
0.35	0.3853	0.044	1.7060	0.027	1.9268	0.017	2.1201	0.007	2.4573
0.30	0.5244	0.042	1.7279	0.026	1.9431	0.016	2.1444	0.006	2.5121
0.25	0.6745	0.040	1.7507	0.025	1.9600	0.015	2.1701	0.005	2.5758
0.20	0.8416	0.038	1.7744	0.024	1.9774	0.014	2.1973	0.004	2.6521
0.15	1.0364	0.036	1.7991	0.023	1.9954	0.013	2.2262	0.003	2.7478
0.10	1.2816	0.034	1.8250	0.022	2.0141	0.012	2.2571	0.002	2.8782
0.05	1.6449	0.032	1.8522	0.021	2.0335	0.011	2.2904	0.001	3.0903

Table 5. Percentage points of the Student's *t*-distribution

 α (area)

α = Area in the upper tail of the *t* distribution

υ	0.1	0.05	0.025	0.01	0.005	0.001	0.0005
1	3.0777	6.3137	12.7062	31.8210	63.6559	318.2888	636.5776
2	1.8856	2.9200	4.3027	6.9645	9.9250	22.3285	31.5998
3	1.6377	2.3534	3.1824	4.5407	5.8408	10.2143	12.9244
4	1.5332	2.1318	2.7765	3.7469	4.6041	7.1729	8.6101
5	1.4759	2.0150	2.5706	3.3649	4.0321	5.8935	6.8685
6	1.4398	1.9432	2.4469	3.1427	3.7074	5.2075	5.9587
7	1.4149	1.8946	2.3646	2.9979	3.4995	4.7853	5.4081
8	1.3968	1.8595	2.3060	2.8965	3.3554	4.5008	5.0414
9	1.3830	1.8331	2.2622	2.8214	3.2498	4.2969	4.7809
10	1.3722	1.8125	2.2281	2.7638	3.1693	4.1437	4.5868
11	1.3634	1.7959	2.2010	2.7181	3.1058	4.0248	4.4369
12	1.3562	1.7823	2.1788	2.6810	3.0545	3.9296	4.3178
13	1.3502	1.7709	2.1604	2.6503	3.0123	3.8520	4.2209
14	1.3450	1.7613	2.1448	2.6245	2.9768	3.7874	4.1403
15	1.3406	1.7531	2.1315	2.6025	2.9467	3.7329	4.0728
16	1.3368	1.7459	2.1199	2.5835	2.9208	3.6861	4.0149
17	1.3334	1.7396	2.1098	2.5669	2.8982	3.6458	3.9651
18	1.3304	1.7341	2.1009	2.5524	2.8784	3.6105	3.9217
19	1.3277	1.7291	2.0930	2.5395	2.8609	3.5793	3.8833
20	1.3253	1.7247	2.0860	2.5280	2.8453	3.5518	3.8496
21	1.3232	1.7207	2.0796	2.5176	2.8314	3.5271	3.8193
22	1.3212	1.7171	2.0739	2.5083	2.8188	3.5050	3.7922
23	1.3195	1.7139	2.0687	2.4999	2.8073	3.4850	3.7676
24	1.3178	1.7109	2.0639	2.4922	2.7970	3.4668	3.7454
25	1.3163	1.7081	2.0595	2.4851	2.7874	3.4502	3.7251
26	1.3150	1.7056	2.0555	2.4786	2.7787	3.4350	3.7067
27	1.3137	1.7033	2.0518	2.4727	2.7707	3.4210	3.6895
28	1.3125	1.7011	2.0484	2.4671	2.7633	3.4082	3.6739
29	1.3114	1.6991	2.0452	2.4620	2.7564	3.3963	3.6595
30	1.3104	1.6973	2.0423	2.4573	2.7500	3.3852	3.6460
35	1.3062	1.6896	2.0301	2.4377	2.7238	3.3400	3.5911
40	1.3031	1.6839	2.0211	2.4233	2.7045	3.3069	3.5510
600	1.2830	1.6474	1.9639	2.3326	2.5841	3.1039	3.3068

Table 6. 0.5 % points of the F-distribution

Area (α) in the upper tail of the F distribution for v_1, v_2 degrees pf freedom

v_2								v_1							
	1	2	3	4	5	6	7	8	9	10	12	14	16	17
1	16212.46	19997.36	21614.13	22500.75	23055.82	23439.53	23715.20	23923.81	24091.45	24221.84	24426.73	24572.01	24683.77	24728.48
2	198.5027	199.0120	199.1575	199.2448	199.3030	199.3321	199.3612	199.3758	199.3903	199.3903	199.4194	199.4194	199.4486	199.4486
3	55.5519	49.8003	47.4683	46.1951	45.3911	44.8381	44.4343	44.1250	43.8813	43.6848	43.3865	43.1719	43.0082	42.9391
4	31.3321	26.2844	24.2599	23.1539	22.4563	21.9752	21.6223	21.3522	21.1385	20.9666	20.7046	20.5146	20.3709	20.3108
5	22.7847	18.3136	16.5301	15.5560	14.9394	14.5133	14.2004	13.9607	13.7716	13.6179	13.3846	13.2150	13.0858	13.0326
6	18.6346	14.5442	12.9166	12.0276	11.4637	11.0731	10.7857	10.5656	10.3914	10.2500	10.0345	9.8776	9.7580	9.7086
7	16.2354	12.4037	10.8826	10.0504	9.5220	9.1554	8.8853	8.6779	8.5138	8.3803	8.1764	8.0279	7.9149	7.8678
8	14.6883	11.0426	9.5965	8.8053	8.3019	7.9519	7.6941	7.4958	7.3387	7.2107	7.0149	6.8721	6.7632	6.7180
9	13.6138	10.1068	8.7171	7.9558	7.4710	7.1338	6.8849	6.6932	6.5411	6.4172	6.2273	6.0886	5.9829	5.9388
10	12.8266	9.4269	8.0809	7.3428	6.8724	6.5447	6.3026	6.1159	5.9676	5.8467	5.6614	5.5256	5.4221	5.3790
11	12.2263	8.9121	7.6004	6.8808	6.4217	6.1016	5.8648	5.6821	5.5368	5.4183	5.2363	5.1031	5.0011	4.9586
12	11.7543	8.5097	7.2257	6.5211	6.0711	5.7571	5.5245	5.3451	5.2021	5.0854	4.9063	4.7748	4.6741	4.6322
14	11.0604	7.9217	6.6804	5.9983	5.5622	5.2573	5.0313	4.8566	4.7173	4.6034	4.4281	4.2993	4.2005	4.1591
15	10.7980	7.7007	6.4761	5.8029	5.3722	5.0708	4.8473	4.6743	4.5363	4.4236	4.2497	4.1219	4.0237	3.9827
16	10.5756	7.5138	6.3034	5.6378	5.2116	4.9134	4.6920	4.5206	4.3839	4.2719	4.0993	3.9723	3.8747	3.8338
27	9.3423	6.4886	5.3611	4.7396	4.3402	4.0594	3.8501	3.6875	3.5570	3.4499	3.2840	3.1608	3.0656	3.0256
20	9.9440	6.9865	5.8177	5.1743	4.7615	4.4721	4.2569	4.0900	3.9564	3.8470	3.6779	3.5530	3.4568	3.4164
29	9.2298	6.3958	5.2764	4.6591	4.2621	3.9831	3.7749	3.6131	3.4832	3.3765	3.2110	3.0882	2.9932	2.9532
30	9.1798	6.3546	5.2388	4.6234	4.2276	3.9493	3.7415	3.5801	3.4505	3.3440	3.1787	3.0561	2.9610	2.9211

Table 6. The 1 % points of the F-distribution (continued)

Area (α) in the upper tail of the F distribution for v_1, v_2 degrees pf freedom

v_2	v_1 1	2	3	4	5	6	7	8	9	10	12	14	16
1	4052.18	4999.34	5403.53	5624.26	5763.96	5858.95	5928.33	5980.95	6022.40	6055.93	6106.68	6143.00	6170.01
2	98.5019	99.0003	99.1640	99.2513	99.3023	99.3314	99.3568	99.3750	99.3896	99.3969	99.4187	99.4260	99.4369
3	34.1161	30.8164	29.4567	28.7100	28.2371	27.9106	27.6714	27.4895	27.3449	27.2285	27.0520	26.9238	26.8265
4	21.1976	17.9998	16.6942	15.9771	15.5219	15.2068	14.9757	14.7988	14.6592	14.5460	14.3737	14.2486	14.1540
5	16.2581	13.2741	12.0599	11.3919	10.9671	10.6722	10.4556	10.2893	10.1577	10.0511	9.8883	9.7700	9.6802
6	13.7452	10.9249	9.7796	9.1484	8.7459	8.4660	8.2600	8.1017	7.9760	7.8742	7.7183	7.6050	7.5186
7	12.2463	9.5465	8.4513	7.8467	7.4604	7.1914	6.9929	6.8401	6.7188	6.6201	6.4691	6.3590	6.2751
8	11.2586	8.6491	7.5910	7.0061	6.6318	6.3707	6.1776	6.0288	5.9106	5.8143	5.6667	5.5588	5.4765
9	10.5615	8.0215	6.9920	6.4221	6.0569	5.8018	5.6128	5.4671	5.3511	5.2565	5.1115	5.0052	4.9240
10	10.0442	7.5595	6.5523	5.9944	5.6364	5.3858	5.2001	5.0567	4.9424	4.8491	4.7058	4.6008	4.5204
11	9.6461	7.2057	6.2167	5.6683	5.3160	5.0692	4.8860	4.7445	4.6315	4.5393	4.3974	4.2933	4.2135
12	9.3303	6.9266	5.9525	5.4119	5.0644	4.8205	4.6395	4.4994	4.3875	4.2961	4.1553	4.0517	3.9724
14	8.8617	6.5149	5.5639	5.0354	4.6950	4.4558	4.2779	4.1400	4.0297	3.9394	3.8002	3.6976	3.6187
15	8.6832	6.3588	5.4170	4.8932	4.5556	4.3183	4.1416	4.0044	3.8948	3.8049	3.6662	3.5639	3.4852
16	8.5309	6.2263	5.2922	4.7726	4.4374	4.2016	4.0259	3.8896	3.7804	3.6909	3.5527	3.4506	3.3721
27	7.6767	5.4881	4.6009	4.1056	3.7847	3.5580	3.3882	3.2558	3.1494	3.0618	2.9256	2.8243	2.7458
20	8.0960	5.8490	4.9382	4.4307	4.1027	3.8714	3.6987	3.5644	3.4567	3.3682	3.2311	3.1296	3.0512
29	7.5977	5.4205	4.5378	4.0449	3.7254	3.4995	3.3303	3.1982	3.0920	3.0045	2.8685	2.7672	2.6886
30	7.5624	5.3903	4.5097	4.0179	3.6990	3.4735	3.3045	3.1726	3.0665	2.9791	2.8431	2.7418	2.6632

Table 6. 2.5 % points of the F-distribution (continued)

Area (α) in the upper tail of the F distribution for v_1, v_2 degrees pf freedom

v_2	1	2	3	4	5	6	7	8	9	10	12	14	16
1	647.79	799.48	864.15	899.60	921.83	937.11	948.20	956.64	963.28	968.63	976.72	982.55	986.91
2	38.5062	39.0000	39.1656	39.2483	39.2984	39.3311	39.3557	39.3729	39.3866	39.3984	39.4148	39.4266	39.4357
3	17.4434	16.0442	15.4391	15.1010	14.8848	14.7347	14.6244	14.5399	14.4730	14.4189	14.3366	14.2768	14.2315
4	12.2179	10.6490	9.9792	9.6045	9.3645	9.1973	9.0741	8.9796	8.9046	8.8439	8.7512	8.6837	8.6326
5	10.0069	8.4336	7.7636	7.3879	7.1464	6.9777	6.8530	6.7572	6.6810	6.6192	6.5245	6.4556	6.4032
6	8.8131	7.2599	6.5988	6.2271	5.9875	5.8197	5.6955	5.5996	5.5234	5.4613	5.3662	5.2968	5.2439
7	8.0727	6.5415	5.8898	5.5226	5.2852	5.1186	4.9949	4.8993	4.8232	4.7611	4.6658	4.5961	4.5428
8	7.5709	6.0595	5.4160	5.0526	4.8173	4.6517	4.5285	4.4333	4.3572	4.2951	4.1997	4.1297	4.0761
9	7.2093	5.7147	5.0781	4.7181	4.4844	4.3197	4.1970	4.1020	4.0260	3.9639	3.8682	3.7980	3.7441
11	6.7241	5.2559	4.6300	4.2751	4.0440	3.8806	3.7586	3.6638	3.5879	3.5257	3.4296	3.3588	3.3044
10	6.9367	5.4564	4.8256	4.4683	4.2361	4.0721	3.9498	3.8549	3.7790	3.7168	3.6210	3.5504	3.4963
12	6.5538	5.0959	4.4742	4.1212	3.8911	3.7283	3.6065	3.5118	3.4358	3.3735	3.2773	3.2062	3.1515
14	6.2979	4.8567	4.2417	3.8919	3.6634	3.5014	3.3799	3.2853	3.2093	3.1469	3.0502	2.9786	2.9234
15	6.1995	4.7650	4.1528	3.8043	3.5764	3.4147	3.2934	3.1987	3.1227	3.0602	2.9633	2.8915	2.8360
16	6.1151	4.6867	4.0768	3.7294	3.5021	3.3406	3.2194	3.1248	3.0488	2.9862	2.8891	2.8170	2.7614
20	5.8715	4.4612	3.8587	3.5147	3.2891	3.1283	3.0074	2.9128	2.8365	2.7737	2.6758	2.6030	2.5465
27	5.6331	4.2421	3.6472	3.3067	3.0828	2.9228	2.8021	2.7074	2.6309	2.5676	2.4688	2.3949	2.3373
29	5.5878	4.2006	3.6072	3.2674	3.0438	2.8840	2.7633	2.6686	2.5919	2.5286	2.4295	2.3554	2.2976
30	5.5675	4.1821	3.5893	3.2499	3.0265	2.8667	2.7460	2.6513	2.5746	2.5112	2.4120	2.3378	2.2799

Table 6. 5 % points of the F-distribution (continued)

Area (α) in the upper tail of the F distribution for v_1, v_2 degrees pf freedom

v2 \ v1	1	2	3	4	5	6	7	8	9	10	12	14	16
1	161.45	199.50	215.71	224.58	230.16	233.99	236.77	238.88	240.54	241.88	243.90	245.36	246.47
2	18.5128	19.0000	19.1642	19.2467	19.2963	19.3295	19.3551	19.3709	19.3847	19.3959	19.4125	19.4243	19.4332
3	10.1280	9.5521	9.2766	9.1172	9.0134	8.9407	8.8867	8.8452	8.8123	8.7855	8.7447	8.7149	8.6923
4	7.7086	6.9443	6.5914	6.3882	6.2561	6.1631	6.0942	6.0410	5.9988	5.9644	5.9117	5.8733	5.8441
5	6.6079	5.7861	5.4094	5.1922	5.0503	4.9503	4.8759	4.8183	4.7725	4.7351	4.6777	4.6358	4.6038
6	5.9874	5.1432	4.7571	4.5337	4.3874	4.2839	4.2067	4.1468	4.0990	4.0600	3.9999	3.9559	3.9223
7	5.5915	4.7374	4.3468	4.1203	3.9715	3.8660	3.7871	3.7257	3.6767	3.6365	3.5747	3.5292	3.4944
8	5.3176	4.4590	4.0662	3.8379	3.6875	3.5806	3.5005	3.4381	3.3881	3.3472	3.2839	3.2374	3.2016
9	5.1174	4.2565	3.8625	3.6331	3.4817	3.3738	3.2927	3.2296	3.1789	3.1373	3.0729	3.0255	2.9890
11	4.8443	3.9823	3.5874	3.3567	3.2039	3.0946	3.0123	2.9480	2.8962	2.8536	2.7876	2.7386	2.7009
10	4.9646	4.1028	3.7083	3.4780	3.3258	3.2172	3.1355	3.0717	3.0204	2.9782	2.9130	2.8647	2.8276
12	4.7472	3.8853	3.4903	3.2592	3.1059	2.9961	2.9134	2.8486	2.7964	2.7534	2.6866	2.6371	2.5989
14	4.6001	3.7389	3.3439	3.1122	2.9582	2.8477	2.7642	2.6987	2.6458	2.6022	2.5342	2.4837	2.4446
15	4.5431	3.6823	3.2874	3.0556	2.9013	2.7905	2.7066	2.6408	2.5876	2.5437	2.4753	2.4244	2.3849
16	4.4940	3.6337	3.2389	3.0069	2.8524	2.7413	2.6572	2.5911	2.5377	2.4935	2.4247	2.3733	2.3335
20	4.3513	3.4928	3.0984	2.8661	2.7109	2.5990	2.5140	2.4471	2.3928	2.3479	2.2776	2.2250	2.1840
27	4.2100	3.3541	2.9603	2.7278	2.5719	2.4591	2.3732	2.3053	2.2501	2.2043	2.1323	2.0781	2.0358
29	4.1830	3.3277	2.9340	2.7014	2.5454	2.4324	2.3463	2.2782	2.2229	2.1768	2.1045	2.0500	2.0073
30	4.1709	3.3158	2.9223	2.6896	2.5336	2.4205	2.3343	2.2662	2.2107	2.1646	2.0921	2.0374	1.9946

Table 7. Percentage points for the Chi-squared distribution

Areas (α) in the upper tail of the χ^2 distribution
for v degrees of freedom

v	0.3	0.25	0.2	0.15	0.1	0.05	0.025	0.01	0.005	0.001
1	1.0742	1.3233	1.6424	2.0722	2.7055	3.8415	5.0239	6.6349	7.8794	10.8274
2	2.4079	2.7726	3.2189	3.7942	4.6052	5.9915	7.3778	9.2104	10.5965	13.8150
3	3.6649	4.1083	4.6416	5.3170	6.2514	7.8147	9.3484	11.3449	12.8381	16.2660
4	4.8784	5.3853	5.9886	6.7449	7.7794	9.4877	11.1433	13.2767	14.8602	18.4662
5	6.0644	6.6257	7.2893	8.1152	9.2363	11.0705	12.8325	15.0863	16.7496	20.5147
6	7.2311	7.8408	8.5581	9.4461	10.6446	12.5916	14.4494	16.8119	18.5475	22.4575
7	8.3834	9.0371	9.8032	10.7479	12.0170	14.0671	16.0128	18.4753	20.2777	24.3213
8	9.5245	10.2189	11.0301	12.0271	13.3616	15.5073	17.5345	20.0902	21.9549	26.1239
9	10.6564	11.3887	12.2421	13.2880	14.6837	16.9190	19.0228	21.6660	23.5893	27.8767
10	11.7807	12.5489	13.4420	14.5339	15.9872	18.3070	20.4832	23.2093	25.1881	29.5879
11	12.8987	13.7007	14.6314	15.7671	17.2750	19.6752	21.9200	24.7250	26.7569	31.2635
12	14.0111	14.8454	15.8120	16.9893	18.5493	21.0261	23.3367	26.2170	28.2997	32.9092
13	15.1187	15.9839	16.9848	18.2020	19.8119	22.3620	24.7356	27.6882	29.8193	34.5274
14	16.2221	17.1169	18.1508	19.4062	21.0641	23.6848	26.1189	29.1412	31.3194	36.1239
15	17.3217	18.2451	19.3107	20.6030	22.3071	24.9958	27.4884	30.5780	32.8015	37.6978
16	18.4179	19.3689	20.4651	21.7931	23.5418	26.2962	28.8453	31.9999	34.2671	39.2518
17	19.5110	20.4887	21.6146	22.9770	24.7690	27.5871	30.1910	33.4087	35.7184	40.7911
18	20.6014	21.6049	22.7595	24.1555	25.9894	28.8693	31.5264	34.8052	37.1564	42.3119
19	21.6891	22.7178	23.9004	25.3289	27.2036	30.1435	32.8523	36.1908	38.5821	43.8194
20	22.7745	23.8277	25.0375	26.4976	28.4120	31.4104	34.1696	37.5663	39.9969	45.3142
21	23.8578	24.9348	26.1711	27.6620	29.6151	32.6706	35.4789	38.9322	41.4009	46.7963
22	24.9390	26.0393	27.3015	28.8224	30.8133	33.9245	36.7807	40.2894	42.7957	48.2676
23	26.0184	27.1413	28.4288	29.9792	32.0069	35.1725	38.0756	41.6383	44.1814	49.7276
24	27.0960	28.2412	29.5533	31.1325	33.1962	36.4150	39.3641	42.9798	45.5584	51.1790
25	28.1719	29.3388	30.6752	32.2825	34.3816	37.6525	40.6465	44.3140	46.9280	52.6187
26	29.2463	30.4346	31.7946	33.4295	35.5632	38.8851	41.9231	45.6416	48.2898	54.0511
27	30.3193	31.5284	32.9117	34.5736	36.7412	40.1133	43.1945	46.9628	49.6450	55.4751
28	31.3909	32.6205	34.0266	35.7150	37.9159	41.3372	44.4608	48.2782	50.9936	56.8918
29	32.4612	33.7109	35.1394	36.8538	39.0875	42.5569	45.7223	49.5878	52.3355	58.3006
30	33.5302	34.7997	36.2502	37.9902	40.2560	43.7730	46.9792	50.8922	53.6719	59.7022
31	34.5981	35.8871	37.3591	39.1244	41.4217	44.9853	48.2319	52.1914	55.0025	61.0980
32	35.6649	36.9730	38.4663	40.2563	42.5847	46.1942	49.4804	53.4857	56.3280	62.4873
33	36.7307	38.0575	39.5718	41.3861	43.7452	47.3999	50.7251	54.7754	57.6483	63.8694
34	37.7954	39.1408	40.6756	42.5140	44.9032	48.6024	51.9660	56.0609	58.9637	65.2471
35	38.8591	40.2228	41.7780	43.6399	46.0588	49.8018	53.2033	57.3420	60.2746	66.6192
36	39.9220	41.3036	42.8788	44.7641	47.2122	50.9985	54.4373	58.6192	61.5811	67.9850

Table 8. Table of random numbers

634278	594502	819590	407109	427167	417538	850200	279250	459812	020568
249144	209611	804028	572427	15949	845373	620124	445166	187768	285186
59294	026391	400671	204597	557016	662624	838785	916034	583493	691448
533333	920195	446213	468314	519402	258457	516259	388916	752121	794679
074330	755009	156302	256682	87440	957585	657834	801437	126767	186189
207386	771462	739813	918081	685685	325269	680137	838933	377100	707891
951188	410173	525254	726440	581412	260092	576364	680588	269816	071553
831967	937780	148159	276626	955088	130838	171359	129288	952398	714239
954636	750072	664082	157966	598230	425315	227905	466509	793109	336846
691992	378425	721898	976023	735319	783006	109407	698362	820831	591563
269114	946931	018712	982877	045100	136247	290140	680240	848803	305290
180293	460024	110901	371020	000973	763012	756551	295300	352394	073816
872878	154432	889033	401962	877905	119068	580006	450776	283199	512894
746633	345795	260766	192356	342423	148636	967345	502635	590989	311364
892523	120480	446465	939973	690189	559320	290072	013180	654602	059804
679676	547596	152265	610086	864043	976689	269219	214776	531127	404983
306573	550793	547282	065220	739713	930934	912080	744820	091307	932530
068747	588258	492577	811019	225617	996212	006893	909687	250924	536990
380473	986281	481829	253039	305290	467505	577286	740562	266997	390289
837921	235707	086113	927883	951129	248701	707221	828978	600828	945528
438036	244825	633479	606651	127954	852288	530219	496596	279461	797802
165291	427802	649948	354587	610550	422913	637704	100963	770809	329982
537633	962744	321023	974719	931511	585182	288471	287257	358801	992498
306484	532411	015640	814437	798150	839945	310896	521871	503620	263452
190855	568822	614199	262285	115081	424860	757843	986806	635552	965825
600322	579094	493813	948402	470911	28608	173544	583784	550647	116252
977231	920877	144359	956766	608116	522266	382915	818998	589279	524408
595789	421054	380106	227625	718196	618170	262259	859774	049281	195678
451458	967278	847706	528746	021633	668401	581758	655390	797990	272455
977834	987892	298658	332102	784124	089727	424344	690459	209051	303006
520502	036000	763622	738327	316985	319443	458543	559397	050551	670466
058998	858383	390846	710207	152699	871261	870294	369675	788323	335917

GLOSSARY

Alternative hypothesis
'AND' rules
ANOVA Table
Bayes' Rule
Bias
Blind experiment
Calculator
Census
Central limit theorem
Chi-squared distribution
Clusters
Conditional probability
Conditional probability distribution
Confounding variables
Continuous data
Continuous random variable
Control group
Critical regiom
Critical Z value or percentage point
Data
Datum
Degrees of freedom
Dependent events
Discrete data
Discrete random variable
Discrete uniform probability distribution
Double-blind experiment
Element
Empirical probability distribution
Equal variance
Ethics
Event
Event space
Expected value
Focus group
Independent events

Independent observations
Interviewees
Joint event
Joint probability
'Joint probability' distribution
Judgement samples
Line graph
Lower confidence limit
Marginal probabilities
Marginal probability distribution
Member
Mutually exclusive events
Nominal data
Non-parametric tests
Normality
Null hypothesis
Ogive
One-stage cluster sampling
Ordinal data
'OR' rules
Outcomes
Outlier
Parameter
Parametric tests
Point estimator
Qualitative or categorical data
Quantitative or numeric data
Quota sampling
Random number tables
Random sample
Random variable
Ratio scale
Reduced sample spaces
Researcher
Residual
Respondents

Response bias
'Revised' probabilities
Sample
Sample space
Sampling distribution of proportions
Sampling distribution of the mean
Sampling error
Sampling frame
Sampling methods
Scatter plot or scatter diagram
Self-selected sample
Simple random sample
Skewed
Snowball or opportunity sample

Standard error of estimate
Standard error of proportions
Standard error of the mean
Statistical inference
Statistics
Summary statistic
Survey
Test statistic
Treatment group
Two-stage cluster sampling
Unbiased estimator
Upper confidence limit
Variance
Weighted average

INDEX

CPSIA information can be obtained
at www.ICGtesting.com
Printed in the USA
BVHW010632170819
556100BV00001B/1/P